AF293821

Oberwolfach Seminars

Volume 57

The workshops organized by the *Mathematisches Forschungsinstitut Oberwolfach* are intended to introduce students and young mathematicians to current fields of research. By means of these well-organized seminars, also scientists from other fields will be introduced to new mathematical ideas. The publication of these workshops in the series *Oberwolfach Seminars* (formerly *DMV seminar*) makes the material available to an even larger audience.

Michael Bate • Benjamin Martin •
Gerhard Röhrle

G-Complete Reducibility, Geometric Invariant Theory and Spherical Buildings

Michael Bate
Department of Mathematics
University of York
York, UK

Gerhard Röhrle
Fakultät für Mathematik
Ruhr-Universität Bochum
Bochum, Germany

Benjamin Martin
Institute of Mathematics, School of Natural
and Computing Sciences
University of Aberdeen
Aberdeen, UK

ISSN 1661-237X   ISSN 2296-5041   (electronic)
Oberwolfach Seminars
ISBN 978-3-032-08865-9   ISBN 978-3-032-08866-6   (eBook)
https://doi.org/10.1007/978-3-032-08866-6

Mathematics Subject Classification: 20Gxx, 20E42, 14L24, 14L30

© The Editor(s) (if applicable) and The Author(s), under exclusive license to Springer Nature Switzerland AG
2026

This work is subject to copyright. All rights are solely and exclusively licensed by the Publisher, whether the whole or part of the material is concerned, specifically the rights of translation, reprinting, reuse of illustrations, recitation, broadcasting, reproduction on microfilms or in any other physical way, and transmission or information storage and retrieval, electronic adaptation, computer software, or by similar or dissimilar methodology now known or hereafter developed.
The use of general descriptive names, registered names, trademarks, service marks, etc. in this publication does not imply, even in the absence of a specific statement, that such names are exempt from the relevant protective laws and regulations and therefore free for general use.
The publisher, the authors and the editors are safe to assume that the advice and information in this book are believed to be true and accurate at the date of publication. Neither the publisher nor the authors or the editors give a warranty, expressed or implied, with respect to the material contained herein or for any errors or omissions that may have been made. The publisher remains neutral with regard to jurisdictional claims in published maps and institutional affiliations.

This book is published under the imprint Birkhäuser, www.birkhauser-science.com by the registered company Springer Nature Switzerland AG
The registered company address is: Gewerbestrasse 11, 6330 Cham, Switzerland

If disposing of this product, please recycle the paper.

To the memory of Roger W. Richardson

Preface

The theory of linear algebraic groups in its modern form dates back to work of Chevalley in the 1950s and was developed further in the 1960s by Borel, Bruhat, Tits and others. An important strand of the theory is subgroup structure. The notion of a G-completely reducible subgroup of a connected reductive group G was introduced by Serre in 1997 in the setting of the spherical building associated to G, and developed by him in two further papers. The Classification Theorem for connected reductive groups in terms of root data makes it feasible to prove results by studying the structure of G as we range over all the possible Dynkin types. There is a wealth of work putting this approach into practice and it has led to considerable success in describing the connected reductive subgroups of a simple algebraic group G.

Around two decades ago the authors introduced geometric methods for studying G-complete reducibility using geometric invariant theory. This complements approaches using case-by-case techniques and often leads to conceptual and uniform proofs. Our motivation for writing this book is to explain the geometric approach. The kernel of the book is the set of lecture notes and exercises we used for an Oberwolfach seminar "G-complete reducibility, geometric invariant theory and spherical buildings" we organised in June 2022, but it grew considerably during the writing process. We have tried to keep in the spirit of the seminar notes: to give as complete and self-contained a presentation as possible of some core topics in G-complete reducibility and the geometric methods needed to develop them, along with a sketch of some other topics to give a sense of the breadth and direction of the subject.

The geometric approach is based on work of Roger W. Richardson (1930–1993), who studied conjugacy classes of subgroups of an algebraic group or Lie group G by considering the G-action by simultaneous conjugation on the Cartesian product G^n for $n > 1$. We believe Richardson's work has not received the attention it deserves, so it is fitting that this book is dedicated to his memory.

Our central theme is the application of geometric methods to study subgroups of reductive groups. Many—although not all—of these applications involve G-complete reducibility. Our main tools are geometric invariant theory and spherical buildings. The book is mainly an exposition of material taken from papers written by us alone and with various collaborators. Most of the results are not new, but we have streamlined many of the proofs. The book is aimed mainly at graduate students and early career researchers in algebraic groups, but we hope it will serve as a useful resource for anyone who wants to learn about this geometric approach. To this end, we have provided some exercises at the end of each chapter, along with some hints and partial solutions, and at various points in the text, we have indicated some open problems.

We are grateful to many people working in algebraic groups and related areas for their advice and input during the period we have worked on G-complete reducibility. Particular thanks go to Jean-Pierre Serre for his encouragement over the past 20 years. We thank our current and former PhD students Chris Attenborough, Falk Bannuscher, Sören Böhm, Oliver Goodbourn, Maike Gruchot, Daniel Lond, Sam Martin and Tomohiro Uchiyama; our collaborators Grace Garden, Harry Geranios, Sebastian Herpel, Martin Liebeck, Alastair Litterick, Amnon Neeman, Damian Sercombe, Aner Shalev, David Stewart, Rudolf Tange, Stephan Tillmann, Lewis Topley and Laura Voggesberger; and others including Sean Cotner, Steve Donkin, Robert Guralnick, Gus Lehrer, George McNinch, Bernhard Mühlherr, Donna Testerman and Adam Thomas. Many thanks to the participants at the Oberwolfach seminar in 2022 whose wholehearted and enthusiastic participation convinced us that this book was a good idea, and to the Mathematisches Forschungsinstitut Oberwolfach (MFO) and its staff who have hosted us on this and many other productive occasions. Some material in the book is based on notes written by the second author for a summer school in Novosibirsk in July 2016, and used again in a spring workshop in Bochum in April 2018. He is grateful to the participants and organisers of both these events. Sean Cotner, Richard Hepworth, Wim Hesselink, Alastair Litterick, Jay Taylor and Adam Thomas have given us direct guidance for specific sections of this book. We thank the editorial staff of Birkhäuser for their patience during the preparation of the book. We are grateful to our funders for their support of our work: the Marsden Fund (grants UOC0501, UOC1009), the Engineering and Physical Sciences Research Council (grants EP/C542150/1 and EP/L005328/1), the Deutsche Forschungsgemeinschaft (grants RO 1072/6-1 and RO 1072/22-1), the University of Bochum (Visiting International Professor fellowship), the Royal Society and the London Mathematical Society (for travel grants and other support). Much of our earlier work was carried out at numerous research stays at the MFO as part of their "Research in Pairs" Programme. We also acknowledge funding from the Isaac Newton Institute during their "Algebraic Lie Theory" Programme in Cambridge in 2009 and during their "Geometric and Categorical Lie Theory" Satellite Programme in York in 2025. We are grateful to the Max Planck Institute for Mathematics in Bonn who funded a stay for the first and third authors in 2010. Finally, thanks to our families for their love and support during the writing of this book and the papers that preceded it: Rachel, Daniel and Elizabeth (BM), the many and varied members of the Bate household (MB), and Yvonne (GR).

York, UK Michael Bate
Aberdeen, UK Benjamin Martin
Bochum, Germany Gerhard Röhrle
July 2025

Contents

About the Authors

The authors Michael Bate, Benjamin Martin and Gerhard Röhrle have a longstanding collaboration and friendship (20 years and counting). Together they have written 20 papers in and around this subject area, with a lasting impact on the field of algebraic groups (including subgroup structure, representation theory, geometric invariant theory, spherical buildings) and applications to other areas such as metric geometry and number theory.

Photo credit: Archives of the Mathematisches Forschungsinstitut Oberwolfach

Introduction 1

The theory of linear algebraic groups has developed significantly since its beginnings in the 1950s and 1960s. Algebraic groups have many applications, including to Lie algebras, algebraic geometry, arithmetic geometry and number theory. The description of the irreducible representations of simple algebraic groups in terms of highest weights has become a paradigm for the representation theory of many other algebraic objects.

An early major achievement was the classification of connected reductive algebraic groups over an algebraically closed field k via root data. With this in hand, a natural next step is to study the subgroup structure of these groups. Since the 1980s many researchers have contributed to this research program. One highlight is the description of the maximal positive-dimensional subgroups of a simple algebraic group due to Liebeck and Seitz; it turns out that there are only finitely many conjugacy classes.

A useful tool is the notion of a G-completely reducible subgroup of a connected reductive group G, introduced by J-P. Serre in the 1990s. The prototypical example of a connected reductive group is the general linear group GL_m; one says that a subgroup H of G is completely reducible if H acts completely reducibly on the natural module k^m — that is, if the inclusion of H in GL_m is a semisimple representation of H in the usual sense of representation theory. One can give an intrinsic characterisation of completely reducible subgroups in terms of parabolic subgroups and Levi subgroups of GL_m, without reference to the natural module. This characterisation makes sense for any connected reductive group G, so we can use it as the definition of G-complete reducibility. Serre proved a number of results about G-completely reducible subgroups. He established an important link with Jacques Tits's theory of spherical buildings. The separation of subgroups into G-completely reducible ones and non-G-completely reducible ones turns out to be a helpful organising tool; one can prove results about both types of subgroup. This is one of the themes of this book, and we give many examples.

© The Author(s), under exclusive license to Springer Nature Switzerland AG 2026

M. Bate et al., *G-Complete Reducibility, Geometric Invariant Theory and Spherical Buildings*, Oberwolfach Seminars 57, https://doi.org/10.1007/978-3-032-08866-6_1

In the 1980s R.W. Richardson brought geometric methods to bear on the problem of describing subgroups of G. It is difficult to give the space of subgroups of G a reasonable geometric structure, so Richardson instead studied subgroups indirectly by considering generating tuples.[1] To be more precise, he investigated the geometry of the Cartesian product G^n—the link with subgroups is that given $\mathbf{g} = (g_1, \ldots, g_n) \in G^n$, we obtain a subgroup H of G, namely the subgroup topologically generated by the elements $g_1, \ldots, g_n$. The group G acts on G^n by simultaneous conjugation: we define

$$g \cdot (g_1, \ldots, g_n) = (g g_1 g^{-1}, \ldots, g g_n g^{-1}).$$

If the components of $\mathbf{g}$ generate a subgroup H then for $g \in G$ the components of $g \cdot \mathbf{g}$ generate $g H g^{-1}$, so orbits $G \cdot \mathbf{g}$ give rise to conjugacy classes of subgroups H. Hence one can apply tools from geometric invariant theory to G^n in order to gain information about subgroups, and about conjugacy classes of subgroups. The correspondence between subgroups and generating tuples is not a bijection—a subgroup may admit many generating tuples, or none at all—but these geometric invariant theory methods can yield a surprising amount of information. For instance, Richardson proved that if $\mathrm{char}(k) = 0$ and $\mathbf{g} \in G^n$ generates a subgroup H of G then H is reductive if and only if the orbit $G \cdot \mathbf{g}$ is closed. Much of his work was done in characteristic 0; in positive characteristic, however, he introduced the key notion of a strongly reductive subgroup of G, and proved that if $\mathbf{g} \in G^n$ generates a subgroup H of G then $G \cdot \mathbf{g}$ is closed if and only if H is strongly reductive. With some work one can show that these ideas make sense for non-connected reductive groups as well.

In 2005 the authors proved that G-complete reducibility and strong reductivity are equivalent for subgroups of G. This has made it possible to apply geometric invariant theory methods to the study of G-complete reducibility, using Richardson's original ideas and other techniques: above all, the machinery of optimal destabilising parabolic subgroups developed by Kempf, Rousseau and Hesselink. This complements the methods from earlier work, which have a more algebraic and representation-theoretic flavour. In a nutshell, then, this is the focus of this book: to introduce readers to our geometric approach, developing for the first time in one place all the machinery that is necessary to understand and apply it, and to put these tools into context by linking them with the wider theory of algebraic groups. The book is aimed at early career researchers in algebraic groups, including postgraduate students. However, we hope that collecting all of this material together means that it will also be useful to experts.

---

[1] There is an analogy with Teichmüller theory: rather than studying the moduli space of compact Riemann surfaces Σ of a fixed genus g, one studies the Teichmüller space of **marked** Riemann surfaces. One can identify the latter with a subspace of the character variety $C(\Gamma,\ \mathrm{PSL}_2(\mathbb{R}))$, where $\Gamma = \pi_1(\Sigma)$.

Much of the literature on applying geometric invariant theory methods to algebraic groups is done only in characteristic 0. In our view, this is often a needless restriction, and that is borne out by the material in this book: when the framework is set up correctly, proofs can be derived which work equally well in all settings. On top of this, the notion of G-complete reducibility itself is extremely useful in working out how to extend notions from characteristic 0 to positive characteristic. The authors themselves have been able to prove in arbitrary characteristic several results which were known in characteristic 0, by replacing the hypothesis that a certain subgroup is reductive with the hypothesis that the subgroup is G-completely reducible. In a similar vein, much of the development of the theory of reductive groups has concentrated on *connected* reductive groups. Whilst this is understandable, since these groups have such a nice classification, it is not always necessary (especially in the theory of complete reducibility) and sometimes it actually causes problems: even if one starts by looking at connected subgroups of a connected group, disconnected groups are unavoidable. They crop up as centralisers and normalisers, for example. In this book, as far as possible, we have proved our results in the setting of reductive groups which are not necessarily connected.

Much of the material we present comes from our papers. Very few of the results in the book are new, although we have made some improvements here and there and also streamlined some of the proofs or added more details. Here is a list of the most significant of the results that appear for the first time in this book. We have corrected a serious mistake in the proof of Theorem 6.2.1 from [117], using ideas of Cotner (see Remark 6.8.10). Theorems 8.1.1 and 8.6.7 are strengthenings of results in the literature, as are Theorems 11.6.2 and 11.6.4. Our proof of Theorem 12.10.1 shortens the proof given in [119]. Proposition 12.10.3 corrects a mistake from [119] (see Remark 12.10.5).

We have also highlighted some open problems at various places in the book. For the convenience of the reader here is a list of these: Open Problems 6.9.4, 6.9.5, 6.10.1, 8.9.1, 10.5.6, 11.3.9, 11.4.12, 11.6.1, 11.7.6, 11.7.8 and 12.1.1 (note that the last is of a rather more general nature than the others).

1.1 Expected Background

We expect the reader to be reasonably familiar with the theory of algebraic groups over an algebraically closed field k: in particular, with the main features of the structure theory of connected reductive groups including the notions of maximal torus, Borel subgroup, parabolic subgroup and Levi subgroup. For instance, [77] and the first ten chapters of [168] are a good preparation. An acquaintance with the basic notions of representation theory will be helpful; [77, §31] summarises the key concepts (although only for semisimple groups), and a much more thorough account is given in [82]. We do not assume any familiarity with algebraic groups over non-algebraically closed fields or with spherical buildings. We give an introduction to these topics in Chaps. 11 and 7, respectively. We also do not require any knowledge of geometric invariant theory; the necessary material can be found in Sect. 2.7 and Chap. 3.

1.2   Plan of the Book

Our aim has been to provide complete arguments for almost all of the material pertaining to the geometric approach to complete reducibility over an algebraically closed field, including a complete development of the material on so-called optimal destabilising parabolic subgroups. We have also provided quite detailed coverage of relevant material on spherical buildings and complete reducibility over non-algebraically closed fields, but without giving full proofs in all cases. For Chaps. 5 and 8 in particular, we have built on the presentation given in [120].

Chapter 2 contains preliminary material including some standard notation and the basics of representation theory, but there is also some less standard material such as the section on non-abelian 1-cohomology. Chapter 3 gives an overview of geometric invariant theory, while Chap. 4 is an introduction to G-complete reducibility. The geometric techniques are introduced in Chap. 5, where we give our geometric characterisation of G-complete reducibility (Theorem 5.3.3). These techniques are developed further in Chap. 6, which also contains some related material that, whilst not crucial to the main thread of the book, is closely cognate. Section 6.8 is especially technical, so the reader may wish to gloss over the proofs. Chapter 7 gives an introduction to spherical buildings and discusses the Tits Centre Theorem (Theorem 7.4.1), which is one of the main tools in applications of buildings to geometric invariant theory and G-complete reducibility.

In Chap. 8 we discuss the Kempf-Rousseau-Hesselink optimality theory, a powerful tool in geometric invariant theory. The main result is Theorem 8.1.1, which yields information about subgroups that are **not** G-completely reducible. We give applications of optimality to G-complete reducibility in Chap. 9. In Chap. 10 we discuss aspects of G-complete reducibility that depend on whether the characteristic of k is small or large. In the latter case there are extra tools available, such as the notion of separable subgroups and reductive pairs (Sect. 10.5). Chapter 11 gives a sketch of the theory of G-complete reducibility over non-algebraically closed fields.

Chapter 12 covers a selection of topics related to G-complete reducibility, geometric invariant theory and the interactions between them. These topics include some recent applications and pointers towards some areas for future research.

1.3   How to Avoid Reading this Book

We have tried to give a self-contained presentation of the theory underlying the geometric approach to G-complete reducibility. In this section we present a brief guide through the book for the reader who wants to apply this geometric machinery to the study of subgroups of G, without reading all the details of the proofs.

The core material on applications of our geometric methods to G-complete reducibility is contained in Chap. 4, Sect. 5.4, and Chaps. 9 and 10; the latter concerns phenomena that occur only in large characteristic (or only in small characteristic). Chapter 11 deals with G-complete reducibility over a non-algebraically closed field; the most relevant parts are Sects. 11.3, 11.4 and 11.7. The reader who has some familiarity with geometric invariant theory can proceed directly to Chap. 4, after reading as much of Chap. 2 as they feel they need. Most of the results on R-parabolic and R-Levi subgroups in Sect. 2.9 either are vacuous or are standard facts when G is connected, so the reader who is interested only in the connected case can skip them; they should know, however, the characterisation of parabolic and Levi subgroups in terms of cocharacters (Theorem/Definition 2.9.1) and the notation c_L and c_λ in Remark 2.9.7. They should also note the comment above that non-connected subgroups do turn up often, even when the primary objects of interest are connected reductive groups. The most important results from Chaps. 5 to 8 are Theorems 5.3.3, 5.4.1, 5.4.4, Proposition 5.5.8, Theorems 6.1.1, 6.1.2, 6.2.1, 6.2.2, 7.4.1 and 8.1.1. The reader should at least familiarise themself with the statements of these results.

Section 12.1 contains a brief overview of the subgroup structure theory of connected reductive groups and how it interacts with the theory developed in the rest of the book. Sections 12.2, 12.4 and 12.5 also deal with various aspects of G-complete reducibility over algebraically closed fields.

Preliminaries

2

The purpose of this chapter is to collect a range of preliminary results which are used in the remainder of the book. We make no attempt to be encyclopaedic, so the material is necessarily presented in rather a piecemeal fashion. As explained in the introduction, our assumption is that the reader is familiar with the standard theory of linear algebraic groups over algebraically closed fields, as presented in the books of Borel [32], Humphreys [77] and Springer [168], and so for the most part we simply recall the basic ideas we need as a way of fixing notation; see Sects. 2.3 and 2.4. However, some aspects of the theory are less familiar, so we do give proofs and more specific references when we believe that will help the reader: see for example the material on length functions and cocharacters (Sect. 2.5), R-parabolic subgroups (Sect. 2.9), and non-abelian cohomology (Sect. 2.10). In some cases, we provide details of quite straightforward examples or arguments where they illustrate important ideas which crop up later in the book; examples of this include Example 2.4.2—describing the parabolic subgroups in classical groups—and Sect. 2.8—describing the process of taking a limit along a cocharacter in some easy cases.

This chapter also includes the following: some basic material on convex cones (Sect. 2.2), which is needed in Chaps. 7, 8 and 11; some material on base change and fields of definition for algebraic groups and their actions (Sect. 2.6), which is used throughout the book and generalised to non-algebraically closed fields in Chap. 11; and some preliminary material on actions of algebraic groups (Sect. 2.7), which is picked up in Chap. 3.

2.1   Notation, Conventions and Algebraic Geometry Background

We use the usual notation for the natural numbers $\mathbb{N}$, integers $\mathbb{Z}$, rationals $\mathbb{Q}$, reals $\mathbb{R}$ and complex numbers $\mathbb{C}$. The smallest natural number is 1; we let $\mathbb{N}_0 := \mathbb{N} \cup \{0\}$. If $A \subseteq \mathbb{R}$ then we write $\sup(A)$ (resp., $\inf(A)$) for the supremum (resp., infimum) of A; we take $\sup(A)$

© The Author(s), under exclusive license to Springer Nature Switzerland AG 2026
M. Bate et al., *G-Complete Reducibility, Geometric Invariant Theory and Spherical Buildings*, Oberwolfach Seminars 57, https://doi.org/10.1007/978-3-032-08866-6_2

to be ∞ if A is unbounded above and $-\infty$ if A is empty. Likewise, $\inf(A)$ denotes the infimum of A, and we take $\inf(A)$ to be $-\infty$ if A is unbounded below and ∞ if A is empty. If Γ is a group and $S \subseteq \Gamma$ then we denote by $\langle S \rangle$ the subgroup of Γ generated by S. If V is a vector space then V^* denotes its dual.

All group actions are left actions unless otherwise specified. If a group H acts on a set X and $Y \subseteq X$ then we denote the setwise stabiliser $\{h \in H \mid h \cdot Y \subseteq Y\}$ by $N_H(Y)$ and the pointwise stabiliser $\{h \in H \mid h \cdot y = y \text{ for all } y \in Y\}$ by $C_H(Y)$. In particular, if M and K are subgroups of a group H then we define the centraliser $C_M(K)$ and the normaliser $N_M(K)$ even when K is not a subgroup of M.

For a prime integer p and a positive power $q = p^e$ we let $\mathbb{F}_q$ denote the finite field with q elements. Given a field E we denote by $\overline{E}$ its algebraic closure. Throughout the book,

$$\boxed{k \text{ denotes a field and, unless otherwise specified, we assume } k = \overline{k} \text{ is algebraically closed.}}$$

We write p for the characteristic $\mathrm{char}(k)$: so p is either zero or a prime. All algebraic groups and varieties are quasi-projective as varieties over k. Most varieties in this book are affine, so the word "variety" on its own should be understood to mean "affine variety"; on the occasions when we need to consider non-affine varieties we make it clear by saying "quasi-projective variety", "projective variety", etc.

A variety X is determined by its coordinate algebra $k[X]$; if k' is a k-algebra then the set $X(k')$ of k'-points of X is in one-one correspondence with the set of k-algebra homomorphisms $k[X] \rightarrow k'$. We follow the common convention of identifying X with $X(k)$: so we write, for example, $x \in X$ rather than $x \in X(k)$. (We use the notation $X(k)$ in Sect. 6.7 and Chap. 11 when we are working with non-algebraically closed fields.) For $f \in k[X]$ we denote by X_f the principal affine open subset $\{x \in X \mid f(x) \neq 0\}$. Given a variety X and a point $x \in X$ we can form the tangent space $T_x(X)$ to X at x. A morphism $\phi : X \rightarrow Y$ of varieties gives rise to the comorphism $\phi^* : k[Y] \rightarrow k[X]$, which is a homomorphism of k-algebras, and for each $x \in X$ we also obtain a linear map $d\phi_x : T_x(X) \rightarrow T_{\phi(x)}(Y)$ of tangent spaces, called the differential or derivative of ϕ at x. In the other direction, a homomorphism $\phi^* : k[Y] \rightarrow k[X]$ of coordinate algebras gives rise to a morphism $\phi : X \rightarrow Y$ of varieties.

If $f : X \rightarrow Y$ is a morphism of varieties—e.g., a homomorphism of algebraic groups—then we say that f is injective if it is injective on k-points; we do not insist that the scheme-theoretic fibres are reduced points.

We recall a well-known result (see, e.g., [32, AG.10.1 Thm.]).

Lemma 2.1.1 *Let X and Y be irreducible varieties of dimensions m and n, respectively, and let $f : X \rightarrow Y$ be a dominant map.*

(a) For every $y \in Y$ and every irreducible component Z of $f^{-1}(y)$, $\dim(Z) \geq m - n$.

(b) There is a non-empty open subset U of Y such that for every $y \in Y$ and every irreducible component Z of $f^{-1}(y)$, $\dim(Z) = m - n$.

The morphism $\phi : X \to Y$ is *finite* if $k[X]$ is integral over the image of ϕ^*. The composition of finite morphisms is finite and if $f_1 : X_1 \to Y_1$ and $f_2 : X_2 \to Y_2$ are finite morphisms then $f_1 \times f_2 : X_1 \times X_2 \to Y_1 \times Y_2$ is finite. Finite morphisms are closed, and so a dominant finite morphism is surjective. The following is now clear:

Lemma 2.1.2 *Suppose $\phi : X \to Y$ and $\psi : Y \to Z$ are morphisms such that $\psi \circ \phi$ is finite. Then:*

(i) ϕ is finite;
(ii) if ϕ is dominant then ψ is finite.

Lemma 2.1.3 *Let $\phi : X \to Y$ be a morphism of varieties. Suppose X_1, X_2 are closed subsets of X such that $X = X_1 \cup X_2$. Set $Y_i = \overline{\phi(X_i)}$, and let $j_i : X_i \to Y_i$ be the restriction of ϕ for $i = 1, 2$. Then ϕ is finite if and only if j_1 and j_2 are finite.*

Proof Set $\phi_i = \phi|_{X_i}$ for $i = 1, 2$. Suppose ϕ is finite. Then ϕ_i is finite for $i = 1, 2$, since it is the composition of finite maps $X_i \hookrightarrow X \to Y$. We can factorise ϕ_i as $X_i \xrightarrow{j_i} Y_i \hookrightarrow Y$ for $i = 1, 2$, and Lemma 2.1.2(i) implies that j_i is finite. Conversely, suppose that j_1 and j_2 are finite. Now let $h \in k[X]$. Since $h|_{X_i}$ is integral over $(\phi_i)^*(k[Y_i])$, there exists a monic polynomial $P_i(T)$ with coefficients in $\phi^*(k[Y])$ such that $P_i(h)|_{X_i} = 0$ for $i = 1, 2$. Then $P_1(h)P_2(h) = 0$, so h is integral over $\phi^*(k[Y])$. Hence ϕ is finite. $\square$

The following corollary is immediate.

Corollary 2.1.4 *Let $\phi : X \to Y$ be a morphism of varieties and let $X_1, \ldots, X_t$ be the irreducible components of X. Then ϕ is finite if and only if $\phi|_{X_i}$ is finite for $1 \leq i \leq t$.*

If X is an irreducible variety then we denote the function field of X by $k(X)$. A dominant morphism $\phi : X \to Y$ of irreducible varieties is *separable* if $k(X)/\phi^*(k(Y))$ is a separable field extension. This is automatic when $p = 0$, as then any field extension is separable. The composition of separable maps is separable. We need the following result, which is known as the *derivative criterion for separability*; see [168, Thm. 4.3.6].

Lemma 2.1.5 *Let $\phi : X \to Y$ be a morphism of irreducible varieties. The following are equivalent.*

(a) ϕ is dominant and separable.

(b) There exists $x \in X$ such that x is a smooth point of X, $\phi(x)$ is a smooth point of Y and $d\phi_x \colon T_x(X) \to T_{\phi(x)}(Y)$ is surjective.

(c) There is a non-empty open subset U of X such that for all $x \in U$, x is a smooth point of X, $\phi(x)$ is a smooth point of Y and $d\phi_x \colon T_x(X) \to T_{\phi(x)}(Y)$ is surjective.

We denote by Mat_m the variety of $m \times m$ matrices with entries in k, with coordinate algebra $k[\mathrm{Mat}_m] = k[X_{11}, X_{12}, \ldots, X_{mm}]$. We write $\mathrm{diag}(a_1, \ldots, a_m)$ for the diagonal matrix with entries $a_1, \ldots, a_m$ down the diagonal. The determinant function det is an element of $k[\mathrm{Mat}_m]$ and the group GL_m of invertible $m \times m$ matrices can be identified as the principal open set defined by the non-vanishing of det, so that $k[\mathrm{GL}_m]$ is the localisation of $k[\mathrm{Mat}_m]$ at the determinant.

2.2 Cones

Let $\mathbb{K}$ be a subfield of $\mathbb{R}$. We denote $\mathbb{K} \cap (0, \infty)$ by $\mathbb{K}_+$ and $\mathbb{K} \cap [0, \infty)$ by $\mathbb{K}_{\geq 0}$. Let E be a finite-dimensional vector space over $\mathbb{K}$. Given $v, w \in E$, we define

$$[v, w] = \{tv + (1 - t)w \mid t \in [0, 1] \cap \mathbb{K}\}. \tag{2.2.1}$$

We call a subset E' of E *convex* if for all $v, w \in E'$, $[v, w]$ is contained in E'. We define $\mathbb{K}_+ \cdot v$ to be the subset $\{av \mid a \in \mathbb{K}_+\}$ of E and we call this the *ray through v* in E.

Definition 2.2.1 A *cone* in E is a subset C of E which is closed under scalar multiplication by $\mathbb{K}_+$ (note that we don't insist that 0 belongs to C). It is easy to see that if $C \subseteq E \backslash \{0\}$ is a cone then so is $C \cup \{0\}$.

A cone C is *finitely generated* if there exist $v_1, \ldots, v_r \in C$ for some $r \in \mathbb{N}$ such that

$$C = \left\{ \sum_{i=1}^{r} a_i v_i \ \middle| \ a_i \in \mathbb{K}_{\geq 0} \right\}. \tag{2.2.2}$$

A cone C is *polyhedral* if there exist $\xi_1, \ldots, \xi_t \in E^*$ such that

$$C = \{v \in E \mid \xi_j(v) \geq 0 \text{ for } 1 \leq j \leq t\}. \tag{2.2.3}$$

We say that C is the polyhedral cone *defined by* $\xi_1, \ldots, \xi_t$. (Conversely, a subset C that satisfies (2.2.2) or (2.2.3) is a cone.)

It is clear that finitely generated cones and polyhedral cones are convex, that the intersection of convex cones is a convex cone and that the intersection of finitely many polyhedral cones is a polyhedral cone. If we fix any positive-definite $\mathbb{K}$-valued bilinear form $(\cdot\,,\cdot)$ on E, allowing us to metrise the space E, then we claim that a polyhedral cone is closed. To see this, it is enough to justify the assertion for C defined by a single element $\xi \in E^*$. Given such a ξ, the choice of bilinear form gives an isomorphism $E \cong E^*$ and hence an element $x \in E$ such that $\xi(v) = (x, v)$ for all $v \in E$. Hence C is the closed half-space $\{v \in E \mid (x, v) \geq 0\}$.

A key ingredient in our discussions of the so-called vector building later in the book is the Minkowski-Weyl Theorem. We give a version from [46] cast in the language of this book:

Lemma 2.2.2 *Let E be a finite-dimensional vector space over a subfield $\mathbb{K}$ of $\mathbb{R}$. Then a convex cone in E is polyhedral if and only if it is finitely generated. Therefore, if C is a convex cone in E and we can write $C = C_1 \cup \cdots \cup C_r$ for some polyhedral convex cones C_i, then C is also polyhedral.*

Proof The first statement is precisely the statement of the "Weak Minkowski-Farkas-Weyl Theorem" from [46] (for subfields of $\mathbb{R}$; the theorem in *loc. cit.* is in fact stated over any ordered field). The second statement follows since each C_i, being polyhedral, has a finite generating set, so C is finitely generated by the union of these: if $v \in C$ then $v \in C_i$ for some i, so v can be written as a combination of the generators for C_i. Hence C is polyhedral. $\qquad\square$

2.3 Linear Algebraic Groups

As mentioned earlier, we assume the reader is familiar with the basic theory of linear algebraic groups over an algebraically closed field as set out in [32, 77, 168] (we only consider algebraic groups defined over non-algebraically closed fields towards the end of the book, see Chap. 11). There is a theory of group schemes over k: for instance, see [127] or [82]. For the most part we do not need this, but occasionally we use some ideas from scheme theory: for instance, we do need to consider scheme-theoretic centralisers and normalisers below, but even then we can get away with considering things at the level of Lie algebras. In this book, an *algebraic group* is understood to mean a smooth affine group scheme of finite type over k, which means the same thing as a linear algebraic group in the sense of the standard references. Key examples are the additive group $\mathbb{G}_a$, which we identify with the field k as a group under addition; the multiplicative group $\mathbb{G}_m$, which we identify with the nonzero elements of k as a group under multiplication, denoted k^*; and the group of invertible $m \times m$ matrices GL_m (note that $\mathbb{G}_m \cong \mathrm{GL}_1$). All algebraic groups are linearisable; that is, they can be embedded as closed subgroups of some GL_m (see Remark 3.2.2 below). An *isogeny* is an epimorphism of algebraic groups with finite kernel. If $\phi\colon G_1 \to G_2$ is an isogeny of algebraic groups then there is a bijective correspondence between maximal tori of G_1 and maximal tori of G_2 given by $T_1 \mapsto \phi(T_1)$; this follows from [32, IV.11.14 Prop.(i)].

Let H be an algebraic group. By a subgroup of H we mean a closed subgroup. If we want to consider a possibly non-closed subgroup, we refer to it as an abstract subgroup. We denote the identity component of H by H^0. Recall that the Borel subgroups in a linear algebraic group are all conjugate to each other, and the same is true for the maximal tori. A parabolic subgroup of H is a subgroup P such that H/P is a complete variety, and a subgroup of H is parabolic if and only if it contains a Borel subgroup.

In a connected group the parabolic subgroups are self-normalising; that is, if $H = H^0$ and P is a parabolic subgroup of H, then $P = N_H(P)$. Recall that a torus T is an algebraic group which is isomorphic to a product of copies of the multiplicative group $\mathbb{G}_m$; the number of copies of $\mathbb{G}_m$ is called the rank of T. If T is a subtorus of H, then $C_H(T)$ has finite index in $N_H(T)$; this is known as *rigidity of tori*.

By a representation of an algebraic group H we mean a rational representation: that is, a homomorphism of algebraic groups $\rho : H \to \mathrm{GL}(V)$ for some finite-dimensional vector space V (we restrict attention to finite-dimensional representations in the most part, although below we do see some actions of algebraic groups on coordinate rings, which are generally infinite-dimensional as vector spaces). In this situation we say that V is an H-module, or that V is a representation of H if ρ is understood. A representation V of a torus T is diagonalisable and we obtain the corresponding decomposition of V into weight spaces for the torus; we denote the set of weights by $\Phi_T(V)$.

The Lie algebra of H is denoted by $\mathrm{Lie}(H)$ or (by use of the fraktur font) $\mathfrak{h}$; it can be identified with the tangent space $T_1(H)$. We write $\mathfrak{gl}_m$ for $\mathrm{Lie}(\mathrm{GL}_m)$ and $\mathfrak{sl}_m$ for $\mathrm{Lie}(\mathrm{SL}_m)$. We denote the centre of a Lie algebra $\mathfrak{h}$ by $\mathfrak{z}(\mathfrak{h})$. For every homomorphism $\phi : H_1 \to H_2$ of algebraic groups the differential at the identity induces a homomorphism $d\phi_1 : \mathrm{Lie}(H_1) \to \mathrm{Lie}(H_2)$ of their Lie algebras. A particularly important example of this is given by the inner automorphism $\mathrm{Inn}(h) : H \to H, x \mapsto hxh^{-1}$ for $h \in H$ whose differential is the *adjoint map* $\mathrm{Ad}(h) : \mathfrak{h} \to \mathfrak{h}$. This gives rise to the adjoint representation $\mathrm{Ad} : H \to \mathrm{GL}(\mathfrak{h})$.

If $\mu : H \times H \to H$ is the multiplication map then $(d\mu)_{(1,1)} : \mathfrak{h} \times \mathfrak{h} \to \mathfrak{h}$ is addition; if $\iota : H \to H$ is inversion then $(d\iota)_1 : \mathfrak{h} \to \mathfrak{h}$ is negation. Given $h \in H$ we define $R_h : H \to H; g \mapsto gh$ to be the morphism given by right multiplication by h; similarly, we let $L_h : H \to H; g \mapsto hg$ be left multiplication. Then $\mathrm{Inn}(h) = L_h \circ R_{h^{-1}} = R_{h^{-1}} \circ L_h$. See Exercise 2.1 for an exercise relating the differentials of all of these maps.

We denote by $R_u(H)$ the unipotent radical of H; recall that this is the largest connected normal unipotent subgroup of H. If $h \in H$ and U is a connected normal unipotent subgroup of H^0, then hUh^{-1} is also a connected normal unipotent subgroup of H^0, so we see that $R_u(H) = R_u(H^0)$. We need a couple of results on maximal unipotent subgroups in possibly non-connected algebraic groups. The first is [110, Prop. 3.2].

Proposition 2.3.1 *(i) Any unipotent subgroup of H is contained in a maximal unipotent subgroup of H. In particular, maximal unipotent subgroups of H exist.*
(ii) Any two maximal unipotent subgroups of H are conjugate.

Our final result gives a non-connected analogue of the unipotent radical.

Proposition 2.3.2 *There is a largest normal unipotent subgroup N of H.*

Proof Let N be the intersection of all the maximal unipotent subgroups of H. Then N is a normal unipotent subgroup of H. Let K be any normal unipotent subgroup of H. Then K is contained in some maximal unipotent subgroup U of H by Proposition 2.3.1(i). Since K is normal we have $K \subseteq \bigcap_{h \in H} hUh^{-1}$. But the latter subgroup is N by Proposition 2.3.1(ii). Hence $K \subseteq N$, so we are done. $\qquad\square$

2.4 Reductive Groups

An algebraic group H is *reductive* if $R_u(H) = 1$. The reader is warned that unlike some other sources, we allow reductive groups which are not connected; from time to time we also use the terminology *non-connected reductive group* to indicate this possibility. As observed above, we have $R_u(H) = R_u(H^0)$, and hence H is reductive if and only if H^0 is reductive. Throughout the book,

$$\boxed{G \text{ denotes a reductive algebraic group (henceforth ``a reductive group'').}}$$

We assume the reader is familiar with the classification of connected reductive groups by root data, but recall some of the ingredients below in order to set up notation. We denote by $\kappa(G)$ the number of connected components of G, i.e., $\kappa(G) = [G : G^0]$, the index of G^0 in G.

Example 2.4.1 (The Classical Groups) Key examples of reductive groups are the so-called *classical groups* of matrices: the general linear groups GL_m, the special linear groups SL_m, the orthogonal groups O_m (and special orthogonal groups SO_m), and the symplectic groups Sp_m (for m even). When G is one of the classical groups, we call the space $V = k^m$ of column vectors equipped with the left action of G by matrix multiplication the *natural module* for G.

An algebraic group S, not necessarily connected, is said to be *linearly reductive* if every rational representation of S is semisimple. In characteristic zero, S is linearly reductive if and only if S is reductive in the sense just defined. In characteristic $p > 0$, S is linearly reductive if and only if every element of S is semisimple if and only if S^0 is a torus and $|S/S^0|$ is coprime to p, see [131, §4, Thm. 2]. In all characteristics, therefore, a linearly reductive group is reductive (but not conversely when $p > 0$).

For a connected reductive group G and a maximal torus T of G we let $\Phi = \Phi(G, T)$ denote the root system of G with respect to T (that is, the set of nonzero weights for T in the adjoint module $\mathfrak{g}$). We denote the root datum of G with respect to T by $\Psi(G, T)$. For each root α we have a corresponding root group U_α and a root group homomorphism

$\kappa_\alpha : \mathbb{G}_a \to U_\alpha$. Fixing a Borel subgroup B of G containing T determines a set of simple roots Π and a corresponding set of positive roots Φ^+. For each subset $I \subseteq \Pi$ we can form the subsystem $\Phi(I) \subseteq \Phi$ spanned by I. Then we may define

$$P_I := \langle T, U_\alpha \mid \alpha \in \Phi^+ \cup \Phi(I) \rangle \quad \text{and} \quad L_I := \langle T, U_\alpha \mid \alpha \in \Phi(I) \rangle. \qquad (2.4.1)$$

Then P_I is a parabolic subgroup of G containing B, and L_I is a so-called *Levi subgroup* of P_I: that is, L_I is reductive and $P_I = L_I \ltimes R_u(P_I)$. Further, we have $R_u(P_I) = \langle U_\alpha \mid \alpha \in \Phi^+ \setminus \Phi(I) \rangle$, and every parabolic subgroup containing B is equal to P_I for some $I \subseteq \Pi$. One deduces in particular that all parabolic subgroups in G have such a Levi decomposition. If $I = \{\alpha\}$ then we write P_α instead of $P_{\{\alpha\}}$, and likewise for L_α. Two parabolic subgroups of G are called *opposite* if their intersection is a common Levi subgroup of each one.

The following fact is used a few times in the book: if P and Q are two parabolic subgroups of a connected reductive group G and $R_u(P) \supseteq R_u(Q)$, then $P \subseteq Q$. This follows fairly easily from the results in the previous paragraph.

Example 2.4.2 (Parabolic Subgroups of Classical Groups) Let V be an m-dimensional k-vector space, and suppose $G = \mathrm{GL}(V)$ or $\mathrm{SL}(V)$. Recall that a *flag* in V is a sequence $(U_i)_{1 \leq i \leq r}$ of subspaces U_i with $U_i \subseteq U_{i+1}$ for each i. By convention, we do not include the trivial subspace $\{0\}$ in any flag, or the full space V, and we do not allow repeated entries in a flag; this means the maximal possible length of a flag is $m - 1$. We identify the proper non-trivial subspaces of V with the flags of length 1.

The parabolic subgroups of G are the stabilisers of flags of subspaces. Choosing a basis compatible with a given flag allows one to realise a parabolic subgroup as a "staircase" subgroup of block upper triangular matrices; in particular, the Borel subgroups of G are all conjugate to the subgroup of upper triangular matrices. Two parabolic subgroups are opposite if the corresponding flags are complementary in the following sense: if P is the stabiliser of the flag $(U_i)_{1 \leq i \leq r}$ and Q is the stabiliser of the flag $(W_j)_{1 \leq j \leq s}$ then P and Q are opposite if and only if $r = s$ and $V = U_i \oplus W_{r+1-i}$ for each $1 \leq i \leq r$.

Now suppose $f : V \times V \to k$ is a non-degenerate symmetric or alternating bilinear form on V. A subspace U of V is called *isotropic* if f restricted to U is identically 0; equivalently, if $U \subseteq U^\perp$, where $U^\perp$ is the subspace orthogonal to U under f. Let $G = \mathrm{SO}(V)$ or $\mathrm{Sp}(V)$ denote the corresponding special orthogonal or symplectic group. Then the parabolic subgroups of G are the stabilisers of flags of isotropic subspaces in V. Two parabolic subgroups of G are opposite if the corresponding flags are complementary in the following sense: if P is the stabiliser of the flag $(U_i)_{1 \leq i \leq r}$ and Q is the stabiliser of the flag $(W_j)_{1 \leq j \leq s}$ then P and Q are opposite if and only if $r = s$ and $V = U_i \oplus W_i^\perp$ for each $1 \leq i \leq r$. See Exercise 4.6 for more on this topic.

Suppose $\mathrm{char}(k) = p > 0$, and let G be a connected reductive group with root system Φ and a choice of simple roots Π. Recall that the prime p is said to be *good* for G if it does not divide any of the integers $c_{\alpha\beta}$ appearing in expressions $\beta = \sum_{\alpha\in\Pi} c_{\alpha\beta}\alpha$ where $\beta \in \Phi^+$; if p does divide at least one of these coefficients, p is *bad* for G. A prime is good for G if and only if it is good for each of the simple factors of G, and this notion does not depend on the isogeny type of a simple group, so we just need to list the bad primes for each Dynkin type. There are no bad primes in type A; for types B, C and D the only bad prime is 2; for types E_6, E_7, F_4 and G_2 the bad primes are 2 and 3; for type E_8 the bad primes are 2, 3 and 5. A prime p is *very good* for G if it is good and we do not have $p | (n + 1)$ for any factor of G of type A_n.

For simple G, the *Coxeter number* $h(G)$ is an invariant of G which can be defined in many different ways. It is the order of a Coxeter element (product of the Coxeter generators) in the Weyl group of G. If $\beta = \sum_{\alpha\in\Pi} c_{\alpha\beta}\alpha$ denotes the highest root in the root system of G (with respect to some base Π), then $h(G) = \sum_{\alpha\in\Pi} c_{\alpha\beta} + 1$. Equivalently, we have $h(G) + 1 = \dim(G)/\mathrm{rk}(G)$, where $\mathrm{rk}(G)$ denotes the *rank* of G, i.e., the dimension of a maximal torus of G. For general connected reductive G, we let $h(G)$ denote the maximum of the Coxeter numbers amongst the simple factors of G.

We need to generalise some results that are standard in the connected case to non-connected reductive groups. Note that any normal subgroup of G is reductive (see Exercise 2.3); in particular, $[G, G]$ is reductive.

Lemma 2.4.3 *Let $f : G_1 \to G_2$ be an isogeny of reductive groups. Then $\dim(Z(G_1)) = \dim(Z(G_2))$.*

Proof Let $Z = f^{-1}(Z(G_2)^0)^0$. Since f is surjective, $f(Z(G_1)^0)$ is contained in $Z(G_2)^0$ and $f(Z) = Z(G_2)^0$. We have $\dim(Z) = \dim(Z(G_2))$ since f is an isogeny, so to finish the proof it is enough to show that $Z \subseteq Z(G_1)$. So let $g \in G_1$. Then $f([g, z]) = 1$ for all $z \in Z$, so $\{[g, z] \mid z \in Z\}$ is an irreducible subset of $\ker(f)$. But $\ker(f)$ is finite, so $\{[g, z] \mid z \in Z\} = \{1\}$. Hence $Z \subseteq Z(G_1)$, as required. $\qquad\square$

The following proposition was first proved by Borel and Serre [33, Lem. 5.11 and footnote]; we reproduce some of the details of the proof here. See also Vinberg [192, Prop. 7] and (for a result in a more general context) Brion [40, Thm. 1.1]. We need a preliminary lemma, which we will also use in the proof of Theorem 2.6.2.

Lemma 2.4.4 *Suppose G^0 is a torus. Let H be an abstract subgroup of G, let $H_0 = H \cap G^0$ and let $n = |H/H_0| < \infty$. Suppose that the map $H_0 \to H_0$, $h \mapsto h^n$ is surjective. Then there is a finite subgroup F of H such that $F H_0 = H$.*

Proof Let $\Gamma = H/H_0$. The isomorphism classes of extensions of H_0 by Γ are parametrised by the second cohomology $H^2(\Gamma, H_0)$ (see [41, Ch. IV, Thm. 3.12]); let α denote the element representing G. The map $f : H_0 \to H_0$ given by $x \mapsto x^n$ is surjective by assumption and has finite kernel N, say. Viewing the abelian groups N and H_0 as $\mathbb{Z}\Gamma$-modules, we therefore have a short exact sequence

$$0 \to N \to H_0 \xrightarrow{f} H_0 \to 0$$

which induces an exact sequence in cohomology

$$H^2(\Gamma, N) \to H^2(\Gamma, H_0) \xrightarrow{n} H^2(\Gamma, H_0)$$

(see [41, Ch. III, Prop. 6.1]). Since Γ has order n, the second map on cohomology is 0 by [41, Ch. III, Prop. 10.1], and so we see that α is the image of some $\bar{\alpha} \in H^2(\Gamma, N)$. But then $\bar{\alpha}$ represents an extension F of the finite group N by the finite group Γ, and we can put F inside H via a homomorphism compatible with the projection onto Γ. This completes the proof. $\qquad\qquad\square$

Proposition 2.4.5 *Let G be a reductive group. Then there exists a finite subgroup F of G with $G = FG^0$.*

Proof Let T be a maximal torus of G. Then we can find coset representatives for G/G^0 inside $N_G(T)$, since any element of G can be adjusted by an element of G^0 so that it normalises T. So it is enough to prove that F exists under the assumption that G^0 is a torus. But in this case the result follows from Lemma 2.4.4, taking $H = G$; note that the map $G^0 \to G^0$, $h \mapsto h^n$ is surjective for every $n \in \mathbb{N}$ as k is algebraically closed. $\qquad\square$

We need another useful preliminary result, which is [19, Lem. 2.6].

Lemma 2.4.6 *Let N be a normal subgroup of G. Then there exists a subgroup M of G such that $MN = G$, $M \cap N$ is a finite normal subgroup of M, $M^0 \cap N^0$ is central in both M^0 and N^0, M^0 commutes with N and M^0 is normal in G.*

Proof We start by proving the result in the case that $N = N^0$ is connected. Then the derived group $[N, N]$ is the product of some of the simple factors of G^0. Let M_1 be the product of the remaining simple factors of G^0. Note that since N is normal in G, the simple factors of $[N, N]$ are permuted by G, and hence so are the simple factors of M_1, so M_1 is normal in G. Also, M_1 and N commute.

Let $Z = Z(G^0)^0$, the central torus of G^0. The subtorus $S = (Z \cap N)^0$ is normal in G, and so we can find a complementary torus T which is also normal in G and such that $Z = ST$ (that S is a direct factor of Z is [32, Cor. 8.5]; that T can be chosen normal in G follows by an averaging argument, see [117, Lem. 2.1]). Now let F be as in Proposition 2.4.5, so that F is finite and $G = FG^0$, and set $M = FTM_1$. Then the desired properties follow easily; note that M^0 is normal in G, since $M^0 = TM_1$ and M_1 and T are both normal.

Finally, to deal with the case that N is not connected, note that since N^0 is characteristic in N, N^0 is normal in G, so we can find a subgroup M satisfying the conclusions of the lemma for N^0 by the previous two paragraphs. But then $MN = G$ since $MN^0 = G$; $M \cap N$ is a finite extension of $M \cap N^0$, so $M \cap N$ is finite; and $M^0 \cap N^0$ is central in M^0 and N^0 by construction. To show that N commutes with M^0, it is enough to show that for any $x \in N$ the set $\{[x, y] \mid y \in M^0\}$ is trivial. This set is certainly connected, and it is also contained in the finite set $M^0 \cap N \subseteq M \cap N$ since M^0 and N are normal in G, so we are done. $\qquad\square$

Corollary 2.4.7 *Let G be a reductive group. Then $[G, G] \cap Z(G)$ is finite, G is a finite extension of $[G, G]Z(G)$, and $\dim(G) = \dim([G, G]) + \dim(Z(G))$.*

Proof Let $N = Z(G)$, and pick M as in Lemma 2.4.6. Since $MN = G$ it follows that $[G, G] = [M, M]$. Since $M \cap N$ is finite, $[G, G] \cap Z(G)$ is finite.

Now let $N = [G, G]$, and pick M as in Lemma 2.4.6. Then M^0 is normal in G, so for any $g \in G$, $\{[g, m] \mid m \in M^0\}$ is an irreducible subset of $M^0 \cap [G, G]$. But $M^0 \cap [G, G]$ is finite, so $[g, m] = 1$ for all $g \in G$ and all $m \in M^0$. Hence $M^0 \subseteq Z(G)$. It follows that $Z(G)N \supseteq Z(G)^0 N \supseteq M^0 N \supseteq G^0$ since $MN = G$. So G is a finite extension of $[G, G]Z(G)$ and $\dim(Z(G)^0) + \dim(N) = \dim(G)$, as required. $\qquad\square$

The following is [117, Lem. 6.8]. The proof there is rather involved, so we give an alternative shorter argument.

Proposition 2.4.8 *If H is a reductive subgroup of a reductive group G then $N_G(H)$ is a finite extension of $HC_G(H)$. In particular, $N_G(H)^0 = H^0 C_G(H)^0$.*

Proof Let $U = R_u(N_G(H))$, let $\pi : N_G(H) \to N_G(H)/U$ denote the quotient map, and let $N = \pi(H)$. Since N is a normal subgroup of the reductive group $N_G(H)/U$, we can find a subgroup M of $N_G(H)/U$ as in Lemma 2.4.6 with $N_G(H)/U = MN$. Let $C = \pi^{-1}(M)$ denote the preimage of M in $N_G(H)$. Then $N_G(H) = HC$.

Now for $x \in C^0$ and $h \in H$ we have $\pi(xhx^{-1}h^{-1}) = 1$ since M^0 commutes with N, and hence $xhx^{-1}h^{-1} \in U \cap H$. In fact, for each $h \in H$, since C^0 is connected, the image of the map $C^0 \to U \cap H, x \mapsto xhx^{-1}h^{-1}$ is connected and contains 1, so lies in $(U \cap H)^0$. But H is reductive, so $(U \cap H)^0 = \{1\}$. Thus we see that $C^0 \subseteq C_G(H)$, and we're done. $\qquad\square$

The following result is proved in [146, Prop. 10.1.5]. Later in the book we provide an alternative proof using the machinery of G-complete reducibility: see Theorem 5.4.1 and Remark 5.4.2.

Proposition 2.4.9 *Let S be a linearly reductive subgroup of G. Then $C_G(S)$ and $N_G(S)$ are reductive.*

We now briefly discuss the notion of a maximal subgroup. Recall our convention that subgroup means closed subgroup. A subgroup H of G is called *maximal* if H is proper and H is not properly contained in any other proper subgroup of G (that is, H is maximal among the proper subgroups of G). We do not assume that maximal subgroups are connected. A maximal connected subgroup is a proper connected subgroup of G which is not properly contained in any other proper connected subgroup of G. (This is not *a priori* the same thing as a connected maximal subgroup, which is a maximal subgroup that happens to be connected.) Clearly a conjugate of a maximal subgroup is also maximal, and likewise for maximal connected subgroups. Not every proper subgroup is contained in a maximal subgroup: see Proposition 6.1.4.

It follows from an easy dimension argument that any proper connected subgroup of G is contained in a maximal connected subgroup. The theory of maximal subgroups is complicated; we give a further short discussion in Sect. 12.1. Here is a relatively straightforward result we will need later.

Lemma 2.4.10 *Let T be a maximal torus of G. Then there are only finitely many subgroups H of G such that $T \subseteq H$. In particular, every such subgroup is contained in a maximal subgroup of G.*

Proof The group H^0 is T-stable, so it is generated by T along with the root subgroups U_α for $\alpha \in \Phi(G, T)$ such that $U_\alpha \subseteq H^0$. Since $\Phi(G, T)$ is finite, there are only finitely many possibilities for H^0. Every connected component of H meets $N_H(T)$ by conjugacy of maximal tori in H, so H is generated by H^0 along with $N_H(T)$. But there are only finitely many subgroups of $N_G(T)$ that contain T since $N_G(T)/T$ is finite. This proves the first assertion, and the second follows immediately. $\qquad\square$

2.5   Cocharacters and Length Functions

Let H be an algebraic group. We denote the set of characters of H by X_H and the set of cocharacters by Y_H; characters are homomorphisms $H \to \mathbb{G}_m$ and cocharacters are homomorphisms $\mathbb{G}_m \to H$. If M is a subgroup of H then there is an obvious inclusion $Y_M \subseteq Y_H$ via the inclusion $M \subseteq H$. In particular, since the image of a cocharacter is a connected subgroup of H, we have $Y_H = Y_{H^0}$. If $\lambda \in Y_H$ and M is a subgroup of H such that $\mathrm{Im}(\lambda) \subseteq M$ then we say that λ *evaluates in M*.

If T is a torus then X_T and Y_T both have the structure of abelian group which we write additively in both cases, so that $X_T \cong \mathbb{Z}^n \cong Y_T$, where n is the rank of T. A cocharacter λ is *indivisible* if it cannot be written as $n\mu$ for some cocharacter μ and some $n \in \mathbb{N}$ with $n > 1$. We denote the pairing between X_T and Y_T with angle brackets: that is, for each $\lambda \in Y_T, \chi \in X_T$ we define $\langle \lambda, \chi \rangle \in \mathbb{Z}$ by $(\chi \circ \lambda)(a) = a^{\langle \lambda, \chi \rangle}$ for each $a \in k^*$. Note that if $T \subseteq H$ is a subtorus of an algebraic group H, then the conjugation action of H induces maps $Y_T \to Y_{hTh^{-1}}, \lambda \mapsto h \cdot \lambda$ and $X_T \to X_{hTh^{-1}}, \chi \mapsto h \cdot \chi$, where we define

$(h \cdot \lambda)(a) = h\lambda(a)h^{-1}$ for each $a \in k^*$ and $(h \cdot \chi)(t') = \chi(h^{-1}t'h)$ for each $t' \in hTh^{-1}$. With these definitions, we have $((h \cdot \chi) \circ (h \cdot \lambda))(a) = (\chi \circ \lambda)(a)$ for each $a \in k^*$, and hence the pairing between X_T and Y_T is H-invariant in the following sense:

$$\langle h \cdot \lambda, h \cdot \chi \rangle = \langle \lambda, \chi \rangle. \tag{2.5.1}$$

It is clear that if $\phi : H_1 \to H_2$ is an isogeny, then the map $\lambda \mapsto \phi \circ \lambda$ is an injective map from Y_{H_1} to Y_{H_2}. Since $(H_1)^0$ and $(H_2)^0$ have the same rank, the image of this map has finite index. Therefore,

$$\text{for all } \lambda \in Y_{H_2} \text{ there exist } n \in \mathbb{N}, \mu \in Y_{H_1} \text{ such that } \phi \circ \mu = n\lambda. \tag{2.5.2}$$

Definition 2.5.1 Let G be a reductive group. A *length function* on Y_G is a G-invariant function $\| \cdot \| : Y_G \to \mathbb{Z}, \lambda \mapsto \|\lambda\|$, such that for each maximal torus T of G the restriction of $\| \cdot \|$ to Y_T is the norm arising from a positive-definite[1] W-invariant $\mathbb{Z}$-valued bilinear form on Y_T, where $W = N_G(T)/T$ is the Weyl group of G.

A length function always exists. We can construct one using a standard averaging argument as follows. Fix a maximal torus T_0 of G and pick any positive-definite $\mathbb{Z}$-valued bilinear form $(\cdot, \cdot)$ on Y_{T_0}. To make this W-invariant, we can average: let $W = N_G(T_0)/T_0$ be the Weyl group and define a new positive-definite $\mathbb{Z}$-valued bilinear form $(\cdot, \cdot)_0$ on Y_{T_0} by

$$(\lambda, \mu)_0 = \sum_{w \in W} (w \cdot \lambda, w \cdot \mu).$$

It is clear that $(\cdot, \cdot)_0$ is W-invariant in the sense that $(w \cdot \lambda, w \cdot \mu)_0 = (\lambda, \mu)_0$ for any $\lambda, \mu \in Y_{T_0}$ and any $w \in W$.

Let $\| \cdot \|_0$ be the norm on Y_{T_0} associated to $(\cdot, \cdot)_0$. Now we can define our length function $\| \cdot \|$ as follows: if T is any maximal torus of G and $\lambda \in Y_T$, then we define $\|\lambda\| = \|g \cdot \lambda\|_0$, where $g \in G$ is such that $gTg^{-1} = T_0$. This is well-defined by the W-invariance of $\| \cdot \|_0$, since if $g_1, g_2 \in G$ and $g_1 T g_1^{-1} = g_2 T g_2^{-1} = T_0$ then $g_2 = ng_1$ for some $n \in N_G(T_0)$.

It turns out that if G is semisimple then there is a canonical choice of length function (see [37, Ch. VI, §1, no. 12]).

It is clear that for $H \subseteq G$ a reductive subgroup of G, the restriction of a length function on Y_G to $Y_H \subseteq Y_G$ gives a length function on Y_H.

Example 2.5.2 Let $G = \mathrm{GL}_m$ and let T be the maximal torus of diagonal elements. We show how to construct a length function on Y_G. Any $\lambda \in Y_T$ has the form $\lambda(a) = $

---

[1] We adopt the usual convention that a positive-definite bilinear form is symmetric.

$\mathrm{diag}(a^{n_1}, \ldots, a^{n_m})$ for some $n_1, \ldots, n_m \in \mathbb{Z}$. So we have an isomorphism of $\mathbb{Z}$-modules from Y_T to $\mathbb{Z}^m$ which maps λ as above to $(n_1, \ldots, n_m)$. Define a positive-definite $\mathbb{Z}$-valued bilinear form $(\cdot, \cdot)$ on $\mathbb{Z}^m$ by

$$\big((n_1, \ldots, n_m), (n'_1, \ldots, n'_m)\big) = \sum_{i=1}^{m} n_i n'_i.$$

(This is just the usual dot product between vectors of length m.) Since the Weyl group $W = S_m$ acts on T by permuting the elements along the diagonal, $(\cdot, \cdot)$ is W-invariant. Hence $(\cdot, \cdot)$ yields a length function $\| \cdot \|$ for G.

A similar construction works for SL_m: we identify $Y_{T \cap \mathrm{SL}_m}$ with the subspace of $\mathbb{Z}^m$ given by $n_1 + \cdots + n_m = 0$. More generally, if H is any subgroup of GL_m then we obtain a length function on Y_H by restricting $\| \cdot \|$, by the comment just before the example.

Lemma 2.5.3 *Let H be a simple factor of G and let $\| \cdot \|_1$ and $\| \cdot \|_2$ be length functions on Y_G. Then there exists $a > 0$ such that $(\| \cdot \|_2)|_H = a(\| \cdot \|_1)|_H$. In particular, if G is simple then length functions are unique up to a positive scalar multiple.*

Proof Since we are only interested in the value of the length function on Y_H, we may assume $H = G$. By G-invariance, a length function on Y_G is determined by its restriction to Y_T for any maximal torus T, so we fix such a T. Let $(\cdot, \cdot)_1$ and $(\cdot, \cdot)_2$ be the $\mathbb{Z}$-valued bilinear forms on Y_T corresponding to the two length functions, giving rise to two $\mathbb{R}$-valued bilinear forms on the real space $V := Y_T \otimes_{\mathbb{Z}} \mathbb{R}$. Since G is simple, the Weyl group $W = N_G(T)/T$ is irreducible, and its faithful action on V is the reflection representation (see Exercise 7.1(ii) below). This representation V is absolutely irreducible, so in particular the only endomorphisms of V which commute with the action of W are scalars; see [1, Ex. 1.101]. Since a length function is W-invariant, the corresponding $\mathbb{R}$-valued form on V is W-invariant, and hence our two forms give rise to two $\mathbb{R}W$-module isomorphisms $V \to V^*$ given by $v \mapsto (v, -)_i$ for $i = 1, 2$. Any two such isomorphisms differ by a scalar (by absolute irreducibility). Pulling this conclusion back through, we see that the two length functions must also differ by a scalar. $\qquad\square$

Note that since a length function comes from an integer-valued form on Y_T, only scalar multiples of a given form which preserve that integrality give rise to new length functions. In particular: multiplying a length function by a positive integer does give another length function; multiplying by a positive rational number may or may not give another length function; multiplying by an irrational number definitely does not give another length function!

Remark 2.5.4 Suppose $\Gamma \subseteq \mathrm{Aut}(G)$ is a group of automorphisms of the reductive group G such that the image of Γ in $\mathrm{Out}(G) = \mathrm{Aut}(G)/\mathrm{Inn}(G)$ is finite. Then, by the same sort of averaging argument as in the proof of existence of a length function, we may find a length function on G which is Γ-invariant. In particular, if G is semisimple then we can choose $\|\cdot\|$ to be $\mathrm{Aut}(G)$-invariant, since $\mathrm{Aut}(G)$ is a finite extension of $\mathrm{Inn}(G)$ in this case. On the other hand, if $\dim(Z(G)) \geq 2$ then it can happen that $\mathrm{Aut}(G)$ is infinite: for example, the group $\mathrm{SL}_m(\mathbb{Z})$ acts on a rank m torus T by automorphisms, and T does not admit an $\mathrm{Aut}(T)$-invariant length function (see Exercise 2.4).

2.6 Algebraically Closed Extension Fields

Often in the theory of algebraic groups, one fixes an algebraically closed ground field k once and for all. Some of the phenomena we will study in Chap. 6, however, can change when one passes from k to an algebraically closed extension field E/k. We record some material on base change and fields of definition for varieties and algebraic groups. We refer the reader to Chap. 11, where we consider extensions of arbitrary fields, for more details. At this stage we restrict attention to extensions of algebraically closed fields only.

First suppose that X is a variety over k, and E/k is an algebraically closed field extension of k. Then we can form a variety denoted X_E over E by *base change*.[2] As a variety over E, X_E is the variety with coordinate algebra $E[X_E] := k[X] \otimes_k E$. We stick with our convention that we identify a variety over an algebraically closed field with its set of points; hence we write X_E rather than $X_E(E)$ or $X(E)$ for the set of E-points. Since k is a subfield of E, we may regard X as a subset of X_E; note that X is dense in X_E. If Z is another variety then $(X \times Z)_E \cong X_E \times Z_E$. If Y is a closed subvariety of X then we may regard Y_E as a closed subvariety of X_E. We say that a closed subvariety Z of X_E is *k-defined* if $Z = Y_E$ for some closed subvariety Y of X. It can be shown that Z is k-defined if and only if $X \cap Z$ is dense in Z. Here is a concrete criterion: a closed subvariety Z of X_E is k-defined if and only if it is the set of common zeroes of some elements of $k[X]$.

If $f : Y \to X$ is a map of k-varieties then we get a map $f_E : Y_E \to X_E$ by base change. Given a map $h : X_E \to Y_E$, h is of the form f_E for some map $f : X \to Y$ if and only if $h(X) \subseteq Y$. In this case we say that h is *k-defined*.

Now suppose H is an algebraic k-group. The group structure on H corresponds to a coalgebra structure on $k[H]$, giving a coalgebra structure on $E[H_E]$; hence H_E acquires the structure of an algebraic group over E. Note that H is a subgroup of H_E and H_E is reductive if and only if H is. If T is a maximal torus of H, then T_E is a maximal torus of H_E; similarly for parabolic subgroups and their Levi subgroups, etc. In the case of a reductive group G, the roots systems $\Phi(G, T)$ and $\Phi(G_E, T_E)$ are canonically isomorphic, as are the root data $\Psi(G, T)$ and $\Psi(G_E, T_E)$. We record a useful fact: if M and K are subgroups of H and M_E and K_E are H_E-conjugate then M and K are H-conjugate. (For

---

[2] In this context base change is also referred to as *extension of scalars*.

the closed set $\{h \in H_E \mid h M_E h^{-1} = K_E\}$ of H_E is k-defined and non-empty, so it has a k-point.)

We need to consider subfields of k as well as extension fields. Suppose X is a variety over k and k_0 is an algebraically closed subfield of k. Then we say that X *admits a k_0-structure* if there is a variety X_0 over k_0 such that $X \cong (X_0)_k$; we call X_0 a k_0-*descent of* X. We can identify X_0 with a subset of X. Similarly, if H is an algebraic k-group and k_0 is an algebraically closed subfield of k, we say that H *admits a k_0-structure* if there is an algebraic k_0-group H_0 such that $H \cong (H_0)_k$ as algebraic groups; we call H_0 a k_0-*descent of H* and we can identify H_0 with an abstract subgroup of H. If H is an algebraic k-group then we have two different notions of k_0-descent—a k_0-descent as an algebraic group and a k_0-descent as a variety—but the meaning will always be clear from the context. Note that if H_0 is a k_0-descent of H as a variety and H_0 is an abstract subgroup of H then we can give H_0 the unique structure of an algebraic k_0-group in such a way that it becomes a k_0-descent of H_0 as an algebraic group.

Suppose $f\colon X \to Y$ is an isomorphism of varieties and Y has a k_0-descent Y_0. We can port this structure across to get a k_0-descent X_0 of X using f: we take X_0 to be the variety whose coordinate ring $k_0[X_0]$ is $f^*(k_0[Y_0])$. Since $k[Y] \cong k_0[Y_0] \otimes_{k_0} k$ and f is an isomorphism, we have $k[X] \cong k_0[X_0] \otimes_{k_0} k$, so this definition makes sense. It follows from the construction that $f(X_0) = Y_0$.

Example 2.6.1 Let H be a torus over k and let k_0 be an algebraically closed subfield of k. Then $H \cong (k^*)^r$ for some r. It is not hard to show that $(k_0^*)^r$ is a k_0-descent of H, and it is in fact the only k_0-descent of H; moreover, any automorphism of H is k_0-defined. If $t \in H$ and t is of finite order then, writing $t = (x_1, \ldots, x_r)$ for some $x_1, \ldots, x_r \in k^*$, we see that the x_i all have finite order. Since k_0 is algebraically closed, this implies that $t \in (k_0^*)^r$. This shows that H_0 contains every torsion element of H. Moreover, any $\lambda \in Y_T$ is automatically k_0-defined.

The notion of descent is of particular importance when k_0 is the algebraic closure $\overline{\mathbb{Q}}$ or $\overline{\mathbb{F}_p}$ of the prime field of k, and H is a reductive group. We have the following:

Theorem 2.6.2 *Let k_0 be an algebraically closed subfield of k. Then G admits a k_0-structure.*

Proof It's enough to consider the case that k_0 is the algebraic closure of the prime field. For connected G, the result follows quickly from the classification of reductive groups by root data (see [168, Thm. 9.6.2, Thm. 10.1.1], for example): any connected reductive group is isomorphic to one which can clearly be constructed over k_0.

For non-connected G, fix a k_0-descent M of G^0 and a maximal torus S of M; recall that we may regard M as a subgroup of G^0. We prove there exists a finite subgroup F of $N_G(M)$ such that F meets every connected component of G and $F \cap G^0 \subseteq M$. If G^0 is a torus then we can take F as in Proposition 2.4.5, and it follows from Example 2.6.1 that F has the desired properties, but for arbitrary G we need a more complicated argument. Given such an F, we complete the proof as follows. Let C be any connected component of G. Pick $g \in F \cap C$. Left multiplication by g^{-1} gives an isomorphism from C to G^0, and we can port the k_0-descent M of G^0 over to obtain a k_0-descent gM of C. Since $F \cap G^0 \subseteq M$, gM does not depend on the choice of g. It is now straightforward to check that these k_0-descents glue together to give a k_0-descent $FM = \bigcup_{h \in F} hM$ of G as a variety. It follows from the construction that FM is an abstract subgroup of G, so it is a k_0-descent of G as an algebraic group, as required.

We now prove the existence of such an F. Set $T = S_k$, a maximal torus of G, and set $H = N_G(M) \cap N_G(T)$. We claim that H meets every connected component of G. To see this, let $g \in N_G(T)$. Then g gives rise to an automorphism ψ of the root datum $\Psi(G^0, T)$ of G^0 with respect to T under base change. We may regard ψ as an automorphism of $\Psi(M, S)$, as this root datum is canonically isomorphic to $\Psi(G^0, T)$. By the Isomorphism Theorem [168, Thm. 9.6.2], there exists $\phi \in \mathrm{Aut}(M)$ such that ϕ stabilises S and induces ψ on $\Psi(M, S)$. So $\phi_k \in \mathrm{Aut}(G)$ stabilises T and induces ψ on $\Psi(G^0, T)$. Applying the Isomorphism Theorem again, there exists $t \in T$ such that $gt = \phi_k$. Then gt belongs to $N_G(M)$, and the claim now follows because $N_G(T)$ meets every connected component of G.

Set $H_0 = H \cap T$ and set $n = |H/H_0|$. We claim there exists a finite subgroup F of H such that $FH_0 = H$. By Lemma 2.4.4, it is enough to prove that the map $H_0 \to H_0$, $h \mapsto h^n$ is surjective. So let $l \in H_0$. Conjugation by l gives an element β of $\mathrm{Aut}(G^0)$. Since l maps M into M, β is of the form α_E for some $\alpha \in \mathrm{Aut}(M)$. Since β stabilises T, α stabilises S. But α gives the trivial automorphism of $\Psi(M, S)$, so α is inner, so by the Isomorphism Theorem there exists $s \in S$ such that α is conjugation by s. Thus $l = sz$ for some $z \in Z(G^0)$. By Proposition 2.4.5 there is a finite subgroup F' of $Z(G^0)$ such that $Z(G^0) = F'Z(G^0)^0$, so we can write $z = z_0 z_1$ for some $z_0 \in F'$ and some $z_1 \in Z(G^0)^0$. Now z_0 is a torsion element of T, so $z_0 \in S$ by Example 2.6.1. Hence, replacing s with sz_0 and z with z_1, we can assume without loss that $z \in Z(G^0)^0$. We can choose $\tilde{s} \in S$ and $\tilde{z} \in Z(G^0)$ such that $\tilde{s}^n = s$ and $\tilde{z}^n = z$. Then $\tilde{s}\tilde{z} \in H$ and $(\tilde{s}\tilde{z})^n = \tilde{s}^n \tilde{z}^n = sz = l$. This proves the claim.

To finish the proof, it is enough to show that $F \cap G^0 \subseteq M$. So let $h \in F \cap G^0$. Let $r = |h|$. Since the Weyl groups of M and G^0 are canonically isomorphic, there exist $m \in N_M(S)$ and $t \in T$ such that $th = m$. We have

$$t^r = (mh^{-1})^r = m(h^{-1}mh)(h^{-2}mh^2) \cdots (h^{1-r}mh^{r-1})h^{-r}$$

$$= m(h^{-1}mh)(h^{-2}mh^2) \cdots (h^{1-r}mh^{r-1}) \in M,$$

so $t^r \in M \cap T = S$. Since $t \in T$ and k_0 is algebraically closed, t belongs to S. Hence $h = t^{-1}m \in M$, as required. $\qquad\square$

The following result is very useful later on; it allows us to approximate a reductive group by finite subgroups in positive characteristic.

Proposition 2.6.3 *Suppose* $\mathrm{char}(k) = p > 0$. *Fix an* $\overline{\mathbb{F}_p}$*-descent* G_0 *of* G *(there is at least one by Theorem 2.6.2). Then there is an ascending chain* $M_1 \subseteq M_2 \subseteq \cdots$ *of finite subgroups of* G_0 *such that* $\bigcup_{n=1}^{\infty} M_n = G_0$. *In particular,* $\bigcup_{n=1}^{\infty} M_n$ *is dense in* G.

Proof Choose an embedding of G_0 in GL_m over $\overline{\mathbb{F}_p}$. Base change gives an embedding of G in GL_m over k. Then set $M_n = G_0 \cap \mathrm{GL}_m(\mathbb{F}_{p^{n!}})$ for each n. It is clear that $M_1 \subseteq M_2 \subseteq \cdots$ and that $\bigcup_{n=1}^{\infty} M_n = G_0$. But G_0 is dense in G, so we are done. $\square$

2.7   Actions and Orbits

Let H be an algebraic group and let X be a variety. An *action* of H on X is a function $H \times X \to X$ which is an action of the abstract group H on the set X and is also a morphism of varieties (recall our convention that actions are left actions unless otherwise specified). We call X an *H-variety*. If H' is a subgroup of H then the restriction of the action to $H' \times X$ gives an action of H' on X. The *kernel* of the action is the normal subgroup $\{h \in H \mid h \cdot x = x \text{ for all } x \in X\}$.

Given $x \in X$, we define $H_x = \{h \in H \mid h \cdot x = x\}$; we call this the *stabiliser of* x. For $x \in X$ and $h \in H$ we have

$$H_{h \cdot x} = h H_x h^{-1}. \tag{2.7.1}$$

For $x \in X$ we define the *orbit of* x by $H \cdot x = \{h \cdot x \mid h \in H\}$. We say that x, x' are *conjugate* if they belong to the same orbit. For each $x \in X$ we define the *orbit map* $\kappa_x \colon H \to H \cdot x$ by $\kappa_x(h) = h \cdot x$. We define $\mathfrak{h}_x = \ker((d\kappa_x)_1)$, where we recall that $(d\kappa_x)_1$ denotes the differential of κ_x at the identity $1 \in H$. We call $\mathfrak{h}_x$ the *infinitesimal stabiliser of* x. Note that $\mathrm{Lie}(H_x) \subseteq \mathfrak{h}_x$ because the restriction of κ_x to H_x is constant. If $p > 0$ then it can happen that this inclusion is proper; we explore this phenomenon further in Sect. 3.5.

For different points in the same orbit, there is a straightforward link between the orbit maps, as follows. For any $x \in X$ and $h, h' \in H$, we have $\kappa_x(h'h) = (h'h) \cdot x = h' \cdot (h \cdot x) = \kappa_{h \cdot x}(h')$, so

$$\kappa_x \circ R_h = \kappa_{h \cdot x}, \tag{2.7.2}$$

where we recall that $R_h : H \to H$ is right multiplication by h.

Example 2.7.1   (i). Let $X = H$. Then H acts on X by conjugation: $h \cdot x = hxh^{-1}$.

(ii). Let $X = \mathfrak{h} = \mathrm{Lie}(H)$. Then H acts on X via the adjoint action: $h \cdot x = \mathrm{Ad}(h)(x)$.

(iii). Let V be an H-module. We obtain an action of H on V given by $h \cdot v = \rho(h)v$. We also get an action of H on V^* given by $(h \cdot f)(v) = f(h^{-1} \cdot v)$ for $f \in V^*$ and $v \in V$.

(iv). Let $m \in \mathbb{N}$, let $H = \mathrm{GL}_m$ and let $X = \mathrm{Mat}_m$. Then H acts on X by conjugation: $h \cdot A = hAh^{-1}$. Under the usual identification of Mat_m with $\mathrm{Lie}(\mathrm{GL}_m)$, this conjugation action coincides with the adjoint action, so we have a special case of (ii).

Example 2.7.2 We give a generalisation of Example 2.7.1(i) which plays a key role in our geometric approach. Let $n \in \mathbb{N}$. We define an action of H on H^n by

$$h \cdot (h_1, \ldots, h_n) = (hh_1h^{-1}, \ldots, hh_nh^{-1}).$$

We call this the action by *simultaneous conjugation*. Below we often use boldmath notation **h** to denote a tuple $(h_1, \ldots, h_n) \in H^n$. For future reference, note that the irreducible components of H^n are the subsets of the form $C_1 \times \cdots \times C_n$, where each C_i is a connected component of H. Hence H^n is a smooth pure-dimensional variety of dimension $n \dim(H)$.

Remark 2.7.3 In order to avoid repeating ourselves, we set the following convention for the rest of the book: given any $n \in \mathbb{N}$ (including the case $n = 1$) and an algebraic group H, when H acts on H^n it should be assumed that the action is the simultaneous conjugation action unless it is otherwise indicated.

If X_1 is an H_1-variety and X_2 is an H_2-variety then there is an action of $H_1 \times H_2$ on $X_1 \times X_2$ given by

$$(h_1, h_2) \cdot (x_1, x_2) = (h_1 \cdot x_1, h_2 \cdot x_2).$$

We call this the *product action* of $H_1 \times H_2$. Now suppose that X_1 and X_2 are H-varieties. We have an action of H on $X_1 \times X_2$ given by

$$h \cdot (x_1, x_2) = (h \cdot x_1, h \cdot x_2).$$

We refer to this as the *diagonal action* of H; it is the restriction of the product action of $H \times H$ to the diagonal subgroup. For instance, the simultaneous conjugation action from Example 2.7.2 arises by iterating this construction repeatedly.

We require the following standard result [132, Lem. 3.7].

Lemma 2.7.4 *Let X be an H-variety.*

(i) For any $x \in X$, H_x is a subgroup of H. The function $x \mapsto \dim(H_x)$ is upper semi-continuous: that is, for any $m \in \mathbb{N}_0$ the set $\{x \in X \mid \dim(H_x) \geq m\}$ is closed.
(ii) Let $x \in X$. Then $\overline{H \cdot x}$ is a union of H-orbits and $H \cdot x$ is an open subset of $\overline{H \cdot x}$.

Remark 2.7.5 It follows from Lemma 2.7.4(ii) that if $x \in X$ then the orbit $H \cdot x$ is a quasi-affine variety, so it makes sense to talk about $\dim(H \cdot x)$ (note that $H \cdot x$ is smooth and every irreducible component of $H \cdot x$ has the same dimension, since H acts transitively on $H \cdot x$). It also follows that $\overline{H \cdot x} \backslash H \cdot x$ is a union of H-orbits C such that $\dim(C) < \dim(H \cdot x)$. We deduce that any H-orbit of minimal dimension in X is closed. In particular, closed orbits exist and if every orbit has the same dimension then all orbits are closed.

Lemma 2.7.6 *Let X be an H-variety and let $x \in X$.*

(i) $\dim(H \cdot x) = \dim(H) - \dim(H_x)$.
(ii) If $y \in \overline{H \cdot x} \setminus H \cdot x$, then $\dim(H_y) > \dim(H_x)$.

Proof For (i), the orbit map κ_x is surjective and each fibre has dimension $\dim(H_x)$ since it is conjugate to H_x, by (2.7.1). For (ii), if $y \in \overline{H \cdot x} \setminus H \cdot x$, it follows from Remark 2.7.5 that $\dim(H \cdot y) < \dim(H \cdot x)$, and then $\dim(H_y) > \dim(H_x)$ by (i). $\square$

Corollary 2.7.7 *Let X be an H-variety. Then the function $x \mapsto \dim(H \cdot x)$ is lower semi-continuous: that is, for any $m \in \mathbb{N}_0$ the set $\{x \in X \mid \dim(H \cdot x) \leq m\}$ is closed.*

Proof This follows from Lemmas 2.7.6 and 2.7.4(i). $\square$

A fundamental fact about orbit maps is that they are open: see for example [168, Thm. 5.3.2(i)].

Lemma 2.7.8 *Suppose X is an H-variety, let $x \in X$, and let $\kappa_x : H \to H \cdot x$ be the orbit map. Then κ_x is an open map.*

Let X_1 and X_2 be H-varieties and let $f : X_1 \to X_2$ be a map. We say that f is H-*equivariant* or just *equivariant* if $f(h \cdot x) = h \cdot f(x)$ for all $x \in X_1$ and all $h \in H$. If Y is a variety and $f' : X_1 \to Y$ is a map then f' is H-*invariant* or just *invariant* if $f'(h \cdot x) = f'(x)$ for all $x \in X_1$ and all $h \in H$.

Remark 2.7.9 We consider the behaviour of orbits and stabilisers under algebraically closed field extensions in Sect. 3.9.

2.8   Limits and Cocharacters

We now introduce the notion of limits.

Definition 2.8.1 Let $f: k^* \to X$ be a morphism of varieties. Viewing k^* as an open subvariety of k, we say that $\lim_{a \to 0} f(a)$ *exists* if f extends to a morphism $\widehat{f}: k \to X$. If the limit exists, we set $\lim_{a \to 0} f(a) = \widehat{f}(0)$. Note that $\widehat{f}$, if it exists, is unique, because k^* is dense in k.

Lemma 2.8.2 *The following are some basic properties of limits.*

(i) *If $f: k^* \to X$ and $h: X \to Y$ are morphisms of varieties and $x := \lim_{a \to 0} f(a)$ exists then $\lim_{a \to 0}(h \circ f)(a)$ exists, and $\lim_{a \to 0}(h \circ f)(a) = h(x)$.*

(ii) *Let $f: k^* \to X$ be a morphism of varieties and let $h: X \to Y$ be a closed embedding of varieties. If $y := \lim_{a \to 0}(h \circ f)(a)$ exists then $x := \lim_{a \to 0} f(a)$ exists, and $h(x) = y$.*

(iii) *If $f_1: k^* \to X_1$ and $f_2: k^* \to X_2$ are morphisms then $\lim_{a \to 0}(f_1 \times f_2)(a)$ exists if and only if $x_1 := \lim_{a \to 0} f_1(a)$ and $x_2 := \lim_{a \to 0} f_2(a)$ exist, and in this case $\lim_{a \to 0}(f_1 \times f_2)(a) = (x_1, x_2)$.*

Proof   (i). If $\widehat{f}: k \to X$ is an extension of f then $h \circ \widehat{f}: k \to Y$ is an extension of $h \circ f$, so $\lim_{a \to 0}(h \circ f)(a)$ exists and $\lim_{a \to 0}(h \circ f)(a) = (h \circ \widehat{f})(0) = h(\widehat{f}(0))$.

(ii). By hypothesis, $h \circ f$ extends to a map $F: k \to Y$. Now $F(k^*) = (h \circ f)(k^*) \subseteq h(X)$, so $F(k) \subseteq h(X)$ since $h(X)$ is closed in Y. Hence we may regard F as a map from k to $h(X)$. As h gives an isomorphism from X onto $h(X)$, we obtain a map $F': k \to X$ such that $h \circ F' = F$. For any $a \in k^*$, $h(F'(a)) = F(a) = h(f(a))$, so $F'(a) = f(a)$. Hence $\lim_{a \to 0} f(a)$ exists and equals $F'(0)$. We have $h(F'(0)) = F(0) = y$, as required.

(iii). We leave the proof of this as Exercise 2.5. $\square$

Example 2.8.3 Let $n \in \mathbb{Z}$. Define $f: k^* \to k$ by $f(a) = a^n$. If $n > 0$ then the morphism $\widehat{f}: k \to k$ given by $f(a) = a^n$ is an extension of f, so $\lim_{a \to 0} f(a)$ exists and equals $0^n = 0$. Likewise, if $n = 0$ then $\lim_{a \to 0} f(a)$ exists and equals 1 (as usual, we interpret a^0 as 1 for any a). On the other hand, if $n < 0$ then $\lim_{a \to 0} f(a)$ does **not** exist. For suppose otherwise. Define $h: k^* \to k$ by $h(a) = a^{-n}$. Then fh is the constant function 1, so $\lim_{a \to 0}(fh)(a) = 1$. By Lemma 2.8.2(ii), $\lim_{a \to 0}(f \times h)(a) = (\lim_{a \to 0} f(a), \lim_{a \to 0} h(a))$. Since multiplication is a morphism from k^2 to k, Lemma 2.8.2(i) implies that

$$1 = \lim_{a \to 0}(fh)(a) = \left(\lim_{a \to 0} f(a) \right) \left(\lim_{a \to 0} h(a) \right).$$

But $\lim_{a \to 0} h(a) = 0$ as $-n > 0$, so we get a contradiction.

Let $f \colon k^* \to X$ be a morphism of varieties and let E/k be an extension of algebraically closed fields. Base change gives a map $f_E \colon E^* \to X_E$. If f extends to a map $\widehat{f} \colon k \to X$ then $\left(\widehat{f}\right)_E \colon E \to X_E$ is an extension of f_E. Conversely, one can show that if f_E extends to a map $F \colon E \to X_E$ then $F = h_E$ for some map $h \colon k \to X$. Then f_E and h_E agree on the dense subset k^* of E^*, so $f = h|_{k^*}$, so h is an extension of f. This shows that $\lim_{a \to 0} f(a)$ exists if and only if $\lim_{a \to 0} f_E(a)$ does.

Here is our main application. Let X be an H-variety for an algebraic group H, let $x \in X$ and let $\lambda \in Y_H$. Define $f \colon k^* \to X$ by $f(a) = \lambda(a) \cdot x$. We write $\lim_{a \to 0} \lambda(a) \cdot x$ rather than $\lim_{a \to 0} f(a)$ for this limit, and we will sometimes speak of "taking the limit along λ". It is important to know in many contexts when $\lim_{a \to 0} \lambda(a) \cdot x$ exists. Note that $\lim_{a \to 0} \lambda(a) \cdot x$, if it exists, belongs to $\overline{H \cdot x}$; see Exercise 2.6. If E/k is an algebraically closed field extension, $x \in X$ and $\lambda \in Y_H$ then λ_E belongs to Y_{H_E} and it follows from the previous paragraph that $\lim_{a \to 0} \lambda(a) \cdot x$ exists if and only if $\lim_{a \to 0} \lambda_E(a) \cdot x$ does.

Example 2.8.4 A fundamental example of taking limits along cocharacters is given by the case of an H-module V for an algebraic group H. In this case, given any maximal torus T of V, we get a finite set of weights $\Phi_T(V) \subseteq X_T$, and we have

$$V = \bigoplus_{\chi \in \Phi_T(V)} V_\chi, \quad \text{where } V_\chi = \{v \in V \mid t \cdot v = \chi(t)v \text{ for each } t \in T\}. \tag{2.8.1}$$

For any $v \in V$ we can therefore write uniquely $v = \sum_{\chi \in \Phi_T(V)} v_\chi$ with $v_\chi \in V_\chi$, and we call the set of χ such that $v_\chi \neq 0$ the *support* of v, denoted $\mathrm{supp}_T(v)$. In particular, $\mathrm{supp}_T(0) = \varnothing$. Now for $\lambda \in Y_T$ and $\chi \in \Phi_T(V)$, we get $\lambda(a) \cdot v = a^{\langle \lambda, \chi \rangle} v$ for each $v \in V_\chi$. Hence for $v \in V$, $\lim_{a \to 0} \lambda(a) \cdot v$ exists if and only if $\langle \lambda, \chi \rangle \geq 0$ for all $\chi \in \mathrm{supp}_T(v)$, and $\lim_{a \to 0} \lambda(a) \cdot v = 0$ if and only if $\langle \lambda, \chi \rangle > 0$ for all $\chi \in \mathrm{supp}_T(v)$; see Example 2.8.3.

With this in mind, we set

$$V_{\lambda,0} = \sum_{\langle \lambda, \chi \rangle = 0} V_\chi, \qquad V_{\lambda,>0} = \sum_{\langle \lambda, \chi \rangle > 0} V_\chi, \qquad V_{\lambda,\geq 0} = \sum_{\langle \lambda, \chi \rangle \geq 0} V_\chi = V_{\lambda,0} \oplus V_{\lambda,>0},$$

and similarly we define $V_{\lambda,\leq 0}$ and $V_{\lambda,<0}$. Then $\lim_{a \to 0} \lambda(a) \cdot v$ exists if and only if $v \in V_{\lambda,\geq 0}$, and the map taking points in $V_{\lambda,\geq 0}$ to their limits is simply the linear projection from the subspace $V_{\lambda,\geq 0}$ to the subspace $V_{\lambda,0}$ factoring out the subspace $V_{\lambda,>0}$.

Now suppose $\mu \in Y_T$ is another cocharacter. Then because λ and μ commute we can in particular form the sum $n\lambda + \mu$ for $n \in \mathbb{N}$. Note that if we choose n sufficiently large, then $\langle \lambda, \chi \rangle > 0$ implies $\langle n\lambda + \mu, \chi \rangle > 0$ and $\langle \lambda, \chi \rangle < 0$ implies $\langle n\lambda + \mu, \chi \rangle < 0$. Thus we see that for sufficiently large n we get $V_{n\lambda+\mu,\geq 0} \subseteq V_{\lambda,\geq 0}$ and $V_{n\lambda+\mu,0} = V_{\lambda,0} \cap V_{\mu,0}$. We use this idea a fair amount in what follows.

Example 2.8.5 A key case of the previous example is where $H = \mathrm{GL}_m$ and $V = \mathrm{Mat}_m = \mathrm{Lie}(H)$ is the space of all $m \times m$ matrices, with H acting by conjugation (i.e., the adjoint action). Then for the diagonal maximal torus T, the 0-weight space is $\mathrm{Lie}(T)$, the subspace of diagonal matrices. Each of the other T-weight spaces is a root space and is spanned by an elementary matrix E_{ij} for some $1 \leq i, j \leq m$ with $i \neq j$; denote the corresponding root by α_{ij}. A cocharacter $\lambda \in Y_T$ has the form $\lambda(a) = \mathrm{diag}(a^{r_1}, \ldots, a^{r_m})$ with $r_i \in \mathbb{Z}$, and an easy calculation reveals that $\langle \lambda, \alpha_{ij} \rangle = r_i - r_j$. As in the previous example, this makes it straightforward to work out for which matrices X the limit $\lim_{a \to 0} \lambda(a) \cdot X$ exists: this limit exists if and only if the (i, j)-entry $X_{ij} = 0$ whenever $r_i < r_j$.

Note that if P is any standard block upper-triangular parabolic subgroup of GL_m, then we can find a suitable diagonal cocharacter λ such that

$$P = V_{\lambda, \geq 0} \cap H = \left\{ h \in H \ \middle| \ \lim_{a \to 0} \lambda(a) h \lambda(a)^{-1} \text{ exists} \right\}.$$

Since every parabolic subgroup of H is conjugate to such a P, we see that every parabolic subgroup of H arises in a similar way by choosing a suitable cocharacter $\lambda \in Y_H$. Conversely, for every cocharacter $\lambda \in Y_H$, the intersection $P = V_{\lambda, \geq 0} \cap H$ is a parabolic subgroup of H, and the subspace $V_{\lambda, \geq 0}$ is a subalgebra of Mat_m (viewed both as an associative algebra and a Lie algebra); it is nothing other than $\mathrm{Lie}(P)$.

We write $\mathcal{P}_\lambda = V_{\lambda, \geq 0}$ and $\mathcal{L}_\lambda = V_{\lambda, 0}$ for the subalgebras of Mat_m arising in this way, and $c_\lambda : \mathcal{P}_\lambda \to \mathcal{L}_\lambda$ for the surjective homomorphism of algebras given by taking the limit along λ. We write $\mathcal{N}_\lambda = V_{\lambda, >0}$; clearly $\mathcal{N}_\lambda$ is the kernel of c_λ. Similarly, we write $P_\lambda = \mathcal{P}_\lambda \cap H$ and $L_\lambda = \mathcal{L}_\lambda \cap H$ for the corresponding subgroups of H, and we keep the notation $c_\lambda : P_\lambda \to L_\lambda$ for the corresponding surjective algebraic group homomorphism; note that L_λ is a Levi subgroup of the parabolic subgroup P_λ, and it is not hard to check that the kernel of the map c_λ (as a group homomorphism, not an algebra homomorphism) is the unipotent radical $R_u(P_\lambda)$.

Finally, note that the observation at the end of the previous example shows that if λ and μ are two commuting cocharacters of H, then for sufficiently large n we have $P_{n\lambda + \mu} \subseteq P_\lambda$.

2.9   R-Parabolic Subgroups and R-Levi Subgroups

There is a well-known characterisation of parabolic subgroups of a connected reductive group G in terms of cocharacters of G, see [168, Prop. 8.4.5] for example; see also Example 2.8.5 for how this works when $G = \mathrm{GL}_m$. For non-connected G we take this as a *definition*, as in [147, 2.1–2.3].

Theorem/Definition 2.9.1 *Let G be a reductive group and suppose $\lambda \in Y_G$. Define*

$$P_\lambda := \left\{ g \in G \ \middle| \ \lim_{a \to 0} \lambda(a) g \lambda(a)^{-1} \ exists \right\},$$

$$L_\lambda := \left\{ g \in G \ \middle| \ \lim_{a \to 0} \lambda(a) g \lambda(a)^{-1} = g \right\},$$

$$U_\lambda := \left\{ g \in G \ \middle| \ \lim_{a \to 0} \lambda(a) g \lambda(a)^{-1} = 1 \right\}.$$

Then:

(i) *P_λ is a parabolic subgroup of G, $U_\lambda = R_u(P_\lambda)$, and $P_\lambda = L_\lambda \ltimes R_u(P_\lambda)$.*

(ii) *P_λ^0 is a parabolic subgroup of G^0 and L_λ^0 is a Levi subgroup of P_λ^0.*

(iii) *The map $c_\lambda : P_\lambda \to L_\lambda$ defined by*

$$c_\lambda(g) := \lim_{t \to 0} \lambda(t) g \lambda(t)^{-1}$$

is a surjective homomorphism of algebraic groups. Moreover, $L_\lambda = C_G(\mathrm{Im}(\lambda))$ is the set of fixed points of c_λ and $R_u(P_\lambda)$ is the kernel of c_λ.

(iv) *P_λ is a proper subgroup of G if and only if $\lambda(k^*) \not\subseteq Z(G)$.*

(v) *If $\mu \in Y_G$ commutes with λ, then for sufficiently large n we have $P_{n\lambda + \mu} \subseteq P_\lambda$.*

(vi) *For any $g \in G$, $P_{g \cdot \lambda} = g P_\lambda g^{-1}$ and $L_{g \cdot \lambda} = g L_\lambda g^{-1}$.*

We call the parabolic subgroups of G arising in this way the Richardson parabolic subgroups *or* R-parabolic subgroups *of G. Given an R-parabolic subgroup P, the subgroups of the form $L = L_\lambda$ for $\lambda \in Y_G$ with $P = P_\lambda$ are called the* Richardson Levi subgroups *or* R-Levi subgroups *of P. However, when we speak of "an R-Levi subgroup of G", we mean "an R-Levi subgroup of some R-parabolic subgroup of G".*

Proof We may embed G inside some GL_m. Now the facts that P_λ, L_λ and $U_\lambda = R_u(P_\lambda)$ are subgroups of G, and the decomposition $P_\lambda = L_\lambda \ltimes R_u(P_\lambda)$ follow from the corresponding results in GL_m, and for GL_m the results are explained by Example 2.8.5. The fact that P_λ is parabolic in G follows from the fact that P_λ^0 is parabolic in G^0, which is part of [168, Prop. 8.4.5]. Concretely, choosing some maximal torus T of G with $\lambda \in Y_T$, we can realise P_λ^0 (resp., L_λ^0) as the subgroup of G^0 generated by T and the root groups U_α with $\alpha \in \Phi(G, T)$ satisfying $\langle \lambda, \alpha \rangle \geq 0$ (resp., $\langle \lambda, \alpha \rangle = 0$). Similarly, $R_u(P_\lambda)$ is generated by the root groups U_α with $\langle \lambda, \alpha \rangle > 0$. This establishes (i)–(iii), and (iv) follows from (iii). Part (v) was also established for GL_m in Example 2.8.5, and this carries over to G. For (vi), Lemma 2.8.2(i) implies that if $h' := \lim_{a \to 0} \lambda(a) \cdot h$ exists then $\lim_{a \to 0}(g \cdot \lambda)(a) \cdot (ghg^{-1}) = \lim_{a \to 0} g \lambda(a) h \lambda(a)^{-1} g^{-1}$ exists and equals $gh'g^{-1}$. Hence $g P_\lambda g^{-1} \subseteq P_{g \cdot \lambda}$. The reverse implication follows similarly, so $P_{g \cdot \lambda} = g P_\lambda g^{-1}$. It is immediate from part (iii) that $L_{g \cdot \lambda} = g L_\lambda g^{-1}$. $\qquad \square$

Remark 2.9.2 There are only finitely many conjugacy classes of R-parabolic subgroups. This follows from Theorem 2.9.1(ii) and because G^0 has only finitely many parabolic subgroups up to G^0-conjugacy. By Lemma 2.4.10 there are only finitely many R-parabolic subgroups containing a fixed maximal torus of G. Moreover, if P is an R-parabolic subgroup of G then $N_{G^0}(P^0) = P^0$, so $N_G(P^0)$ is a finite extension of P^0; hence P^0 has only finitely many overgroups.

Remark 2.9.3 Let E/k be an algebraically closed field extension and let $\lambda \in Y_G$. Then $\mathrm{Im}(\lambda_E) = \mathrm{Im}(\lambda)_E$ is k-defined, so $L_{\lambda_E} = C_{G_E}(\lambda_E)$ is k-defined; in fact, we have $L_{\lambda_E} = (L_\lambda)_E$. We have an analogous result for the R-parabolic subgroup as well. To see this, we can choose an embedding of G in some GL_m such that λ evaluates in the subgroup of diagonal matrices. It follows from the description of parabolic subgroups in terms of block matrices in Example 2.8.5 that $P_{\lambda_E}(\mathrm{GL}_m)$ is k-defined, which implies that $P_{\lambda_E} = P_{\lambda_E}(\mathrm{GL}_m) \cap G_E$ is k-defined. Hence $P_{\lambda_E} \cap G$ is dense in P_{λ_E}. But if $g \in G$ then $\lim_{a \to 0} \lambda_E(a) g \lambda_E(a)^{-1}$ exists if and only if $\lim_{a \to 0} \lambda(a) g \lambda(a)^{-1}$ exists, so $P_{\lambda_E} \cap G = P_\lambda$. We deduce that $P_{\lambda_E} = (P_\lambda)_E$. It is straightforward to show that $c_{\lambda_E} = (c_\lambda)_E$. We leave it to the reader to prove using Example 2.6.1 and Lemma 2.9.10 below that the map $P \mapsto P_E$ gives a bijection from the set of G-conjugacy classes of R-parabolic subgroups of G to the set of G_E-conjugacy classes of R-parabolic subgroups of G_E.

Definition 2.9.4 We say that R-parabolic subgroups P and Q are *opposite* if their intersection is a common R-Levi subgroup of each one. In this case we say that P and Q are *opposite with respect to L*, where $L = P \cap Q$.

Remark 2.9.5 Clearly P_λ and $P_{-\lambda}$ are opposite for any $\lambda \in Y_G$, so for every R-parabolic subgroup P and every R-Levi subgroup L of P, there is an R-parabolic subgroup opposite to P with respect to L. If P and Q are opposite with respect to L then $P^0 \cap Q^0 = L^0$, so Q^0 is opposite to P^0 with respect to L^0; recall that L^0 is a Levi subgroup of P^0 and Q^0 by Theorem/Definition 2.9.1(ii). Note that there is a unique opposite to P with respect to L (Exercise 2.8).

Example 2.9.6 Whilst it is true in the connected case that all parabolic subgroups are R-parabolic subgroups, this is not the case when G is disconnected. For a trivial example let G be any finite group viewed as a reductive algebraic group. Then the only cocharacter is the trivial cocharacter, so the only R-parabolic subgroup is G itself. However, all of the subgroups of G are parabolic in the sense that the quotient varieties are complete (being finite). Any such example can obviously be souped up to a positive-dimensional example by taking the direct product with a connected reductive group.

Remark 2.9.7 The maps $c_\lambda : P_\lambda \to L_\lambda$ are very important in what follows. Given a pair (P, L) consisting of an R-parabolic P and an R-Levi subgroup L of P, there may be several choices of $\lambda \in Y_G$ with $P = P_\lambda$ and $L = L_\lambda$, but for all such choices the map c_λ is the same: it is just the projection $P \to L$ corresponding to the semidirect product decomposition $P = L \ltimes R_u(P)$. Therefore, when the choice of λ is not important, we allow ourselves to write $c_L : P \to L$ for this map. If $n \in \mathbb{N}$ then we also write c_λ for the map $P_\lambda^n \to L_\lambda^n$.

Note that if $H \subseteq P$ is any subgroup of P then $\dim(c_L(H)) \leq \dim(H)$.

Let H be a reductive subgroup of G. There is a natural inclusion $Y_H \subseteq Y_G$ coming from the inclusion $H \subseteq G$. If necessary, we distinguish between the ambient groups we are working in by writing $P_\lambda(H)$, $P_\lambda(G)$, etc., for $\lambda \in Y_H$, but usually we just write P_λ rather than $P_\lambda(G)$. It is obvious from the definition that if $\lambda \in Y_H$, then $P_\lambda(H) = P_\lambda \cap H$ and likewise for $L_\lambda(H)$ and $R_u(P_\lambda(H))$. We use this in our next result.

Lemma 2.9.8 *Suppose H is a reductive subgroup of G. Then for every pair (Q, M) consisting of an R-parabolic subgroup Q of H and an R-Levi subgroup M of Q, there exists a pair (P, L) consisting of an R-parabolic subgroup P of G and an R-Levi subgroup L of P such that $Q = P \cap H$ and $M = L \cap H$. Further, in this situation we have $R_u(Q) = R_u(P) \cap H$.*

Proof Just choose $\lambda \in Y_H$ such that $Q = P_\lambda(H)$ and $M = L_\lambda(H)$. Then the result follows by setting $P = P_\lambda$ and $L = L_\lambda$. $\square$

We also have the following:

Lemma 2.9.9 *Suppose H is a reductive subgroup of G and $\lambda \in Y_H$ is such that $H \subseteq P_\lambda$. Then $H \subseteq L_\lambda$.*

Proof Note that $H \subseteq P_\lambda$ implies $P_\lambda(H) = H$. But H is reductive, so $R_u(P_\lambda(H)) = R_u(H) = \{1\}$ and we see from Theorem 2.9.1(i) that $H = L_\lambda(H) \subseteq L_\lambda$. $\square$

We record the analogues for R-parabolic subgroups of several results which are standard for parabolic subgroups in connected reductive groups.

Lemma 2.9.10 *Let P be an R-parabolic subgroup of G and let T be a maximal torus of P. Then:*

(i) *There is $\lambda \in Y_T$ such that $P = P_\lambda$.*
(ii) *L_λ is the unique R-Levi subgroup of P containing T.*
(iii) *$R_u(P)$ acts simply transitively by conjugation on the set of R-Levi subgroups of P.*
(iv) *Let L be an R-Levi subgroup of P. The R-parabolic subgroups of G contained in P are all of the form $Q R_u(P)$, where Q is an R-parabolic subgroup of L. Conversely, if Q is an R-parabolic subgroup of L, then $Q R_u(P)$ is an R-parabolic subgroup of G.*

Further, if $L = L_\lambda$ as above, then the sub-R-parabolic subgroups of P all have the form $P_{n\lambda+\mu}$ where $\mu \in Y_L$ and $n \in \mathbb{N}$ is chosen to be sufficiently large.

(v) If $P' \subseteq P$ are two R-parabolic subgroups of G, then every R-Levi subgroup of P' is contained in a unique R-Levi subgroup of P. Hence, for any subgroup $H \subseteq P'$, if H is contained in an R-Levi subgroup of P' then H is contained in an R-Levi subgroup of P.

Proof The results claimed in (i) and (ii) for connected G are [168, Cor. 8.4.4, Prop. 8.4.5]. For (i) in the general case, choose $\mu \in Y_G$ such that $P = P_\mu$. Then $\mu \in Y_{T'}$ for some maximal torus T' of P. Choosing any $g \in P$ such that $gT'g^{-1} = T$, we have $\lambda = g \cdot \mu \in Y_T$ and $P_\lambda = P_{g \cdot \mu} = gP_\mu g^{-1} = P$ by Theorem 2.9.1(vi), proving (i). Now $\lambda \in Y_T$ implies $T \subseteq L_\lambda$, which is an R-Levi subgroup of P, so there is at least one R-Levi subgroup containing T. For the uniqueness statement in (ii), if M is another R-Levi subgroup of P containing T then M is stable under conjugation by λ and hence $c_\lambda(M) \subseteq M \cap L_\lambda$. Since $P = M \ltimes R_u(P)$ it follows that $M = L_\lambda$, as required.

For (iii) note that since P acts by conjugation on its R-Levi subgroups and also acts transitively by conjugation on its maximal tori, the fact that $R_u(P)$ acts transitively on the set of R-Levi subgroups follows from (ii) and the fact that $P = L \ltimes R_u(P)$. Now let $P = P_\lambda$ and suppose $u \in R_u(P)$ normalises $L = L_\lambda$. Then the elements $\lambda(a)u\lambda(a)^{-1}$ also normalise L, as does the limit point 1, which shows that u lies in $N_G(L)^0$. Now $L = C_G(S)$ for some torus S, so $N_G(L)^0$ normalises S; it follows that $N_G(L)^0$ is contained in $C_G(S) = L$, by rigidity of tori, and we conclude that $u \in R_u(P) \cap L = \{1\}$, which completes the proof of (iii).

For the results in (iv), since all the R-Levi subgroups of P are conjugate by $R_u(P)$, we may as well assume straight away that $L = L_\lambda$. If P' is an R-parabolic subgroup of G contained in P, then since $(P')^0 \subseteq P^0$ we get $R_u(P') \supseteq R_u(P)$. If we let S be a maximal torus of P', then S is a maximal torus of P also, so contained in some R-Levi subgroup of P, so some $R_u(P)$-conjugate of S is contained in L, by (i) and (ii). Therefore, we may find $\mu \in Y_L$ with $P' = P_\mu$. Let $Q = P_\mu(L) = P' \cap L$. It is clear that $QR_u(P) \subseteq P'$. On the other hand, since $R_u(P) \subseteq P'$ and $P = L \ltimes R_u(P)$, we have $P' \subseteq (P' \cap L)R_u(P) = QR_u(P)$, so we see that $P' = QR_u(P)$.

Now if Q is an R-parabolic subgroup of L, we can write $Q = P_\mu(L)$ for some $\mu \in Y_L$. By Theorem/Definition 2.9.1(v), we can choose n large enough so that $P_{n\lambda+\mu} \subseteq P_\lambda$. It is clear that $Q \subseteq P_{n\lambda+\mu}$ and also $R_u(P) \subseteq R_u(P_{n\lambda+\mu}) \subseteq P_{n\lambda+\mu}$, since $P^0_{n\lambda+\mu} \subseteq P^0$ is a containment of parabolic subgroups of G^0. So $QR_u(P) \subseteq P_{n\lambda+\mu}$. But conversely, since λ is central in L, we have that λ evaluates in Q and hence $c_\lambda(QR_u(P)) = Q = c_\lambda(P_{n\lambda+\mu})$. This is enough to conclude that $P_{n\lambda+\mu} = QR_u(P)$; hence, in particular, $QR_u(P)$ is an R-parabolic subgroup of G. This completes the proof of (iv).

For (v), using part (iv) we can write $P' = P_{n\lambda+\mu}$ for suitable $n \in \mathbb{N}$ and $\mu \in Y_{L_\lambda}$. But then $L_{n\lambda+\mu}$ is an R-Levi subgroup of $P' = P_{n\lambda+\mu}$, and it is contained in L_λ. Since every other R-Levi subgroup of P' is P'-conjugate to $L_{n\lambda+\mu}$ by (iii), and $P' \subseteq P$, it follows that every R-Levi subgroup of P' is contained in an R-Levi subgroup of P. Uniqueness follows from (ii), since a maximal torus of P' is a maximal torus of P. $\qquad\square$

Remark 2.9.11 The converse to the final conclusion of part (v) above is not true in general. For a trivial example, take $P = G$ and let P' be any R-parabolic subgroup of G such that $(P')^0$ is proper in G^0. Then $H = P'$ is in an R-Levi subgroup of P but not one of P'. See Lemma 2.9.20 for a situation where the converse does hold.

Since it is used several times in what follows, we recast part of Lemma 2.9.10 as a separate lemma:

Lemma 2.9.12 *Let P be an R-parabolic subgroup of G and let L be an R-Levi subgroup of P. Then the map $Q \mapsto Q R_u(P)$ gives a bijection from the set of R-parabolic subgroups of L to the set of R-parabolic subgroups of G contained in P, with inverse $R \mapsto R \cap L$.*

Proof That every sub-R-parabolic subgroup of G contained in P has the form $Q R_u(P)$ is part of Lemma 2.9.10(iv), and the rest of the result follows easily. $\qquad\square$

Corollary 2.9.13 *Suppose P and Q are two R-parabolic subgroups of G. Then:*

(i) $P \cap Q$ contains a maximal torus of G.
(ii) $(P \cap Q)R_u(P) \subseteq P$ is an R-parabolic subgroup of G, and it equals P if and only if Q contains an R-Levi subgroup of P.

Proof Part (i) follows from the connected case, which is [32, Cor. 14.13]. For (ii), let T be a maximal torus of $P \cap Q$ and write $P = P_\lambda$, $Q = P_\mu$ for $\lambda, \mu \in Y_T$, using Lemma 2.9.10(i). Then $\mu \in Y_{L_\lambda}$ and $Q \cap L_\lambda = P_\mu(L_\lambda)$. Now $P_{n\lambda+\mu}$ is an R-parabolic subgroup contained in P for sufficiently large n by Theorem/Definition 2.9.1(v). It is clear that $P \cap Q \subseteq P_{n\lambda+\mu}$, and since $R_u(P) \subseteq P_{n\lambda+\mu}$ also we have $(P \cap Q)R_u(P) \subseteq P_{n\lambda+\mu}$. But $c_\lambda(P_{n\lambda+\mu}) = P_\mu(L_\lambda) \subseteq P \cap Q$, so we can deduce the reverse inclusion (and hence, in particular, see that $(P \cap Q)R_u(P)$ is an R-parabolic subgroup of G). For the final part of (ii), note that $P_{n\lambda+\mu} = P_\lambda$ if and only if $L_{n\lambda+\mu} = L_\lambda$ if and only if μ is central in L_λ if and only if $L_\lambda \subseteq L_\mu \subseteq P_\mu = Q$, using Theorem/Definition 2.9.1(iv) for the second equivalence. $\qquad\square$

Corollary 2.9.14 *Suppose $\lambda \in Y_G$ and T is a maximal torus of G with $T \subseteq P = P_\lambda$. Then there exists $u \in R_u(P)$ with $u \cdot \lambda \in Y_T$.*

Proof From Lemma 2.9.10(ii) we know that there is a unique R-Levi subgroup L of P containing T, and there is a unique $u \in R_u(P_\lambda)$ such that $uL_\lambda u^{-1} = L_{u\cdot\lambda} = L$. But then $u \cdot \lambda$ evaluates in $Z(L)^0 \subseteq T$, as required. $\qquad\square$

Corollary 2.9.15 *Let S be a torus in G. Then $C_G(S)$ is an R-Levi subgroup of some R-parabolic subgroup of G.*

Proof Given a subtorus S of G, we have $C_G(S)^0 = C_{G^0}(S)$ since the latter group is a connected reductive subgroup of G^0 by [32, Cor. 13.17(2)]. Thus $C_G(S)$ is reductive. Now we prove the rest of the result by descending noetherian induction (i.e., using the fact that a descending chain of closed subsets in G must eventually stabilise). The base case is that S is central in G: then $C_G(S) = L_\lambda$ where λ is the trivial cocharacter. Otherwise, there exists some $\lambda \in Y_S$ with L_λ proper in G. Then since $\lambda \in Y_S$ we infer $C_G(S) \subseteq L_\lambda$ and so we can write $C_G(S) = L_\mu(L_\lambda)$ by induction. For sufficiently large n we have $L_\mu(L_\lambda) = L_{n\lambda+\mu}$ by the arguments in the proof of Lemma 2.9.10(iv). $\qquad\square$

Lemma 2.9.16 *Let $\lambda \in Y_G$ and let $g \in G$. Suppose $g \cdot \lambda = u \cdot \lambda$ for some $u \in R_u(P_\lambda)$. Then $g \in P_\lambda$.*

Proof Since $g \cdot \lambda = u \cdot \lambda$, we get $(u^{-1}g) \cdot \lambda = \lambda$. Hence $u^{-1}g \in L_\lambda \subseteq P_\lambda$. Since u belongs to P_λ, g must belong to P_λ as well. $\qquad\square$

We also record the following rather striking result which relates G-conjugacy of cocharacters to H-conjugacy when H is a reductive subgroup G.

Corollary 2.9.17 *Let H be a reductive subgroup of G and let $\lambda, \mu \in Y_H$. Suppose there exists $u \in R_u(P_\lambda)$ such that $u \cdot \lambda = \mu$. Then $u \in R_u(P_\lambda(H))$.*

Proof The key is the fact that u is unique, which follows from Lemma 2.9.10(iii). Since $\mu = u \cdot \lambda$, we obtain $P_\mu = P_\lambda$. Since both λ and μ are cocharacters of H, we can also conclude that $P_\lambda(H) = P_\mu(H)$, and hence $L_\mu(H)$ and $L_\lambda(H)$ are two R-Levi subgroups of $P_\lambda(H)$. This means that $u'L_\lambda(H)(u')^{-1} = L_{u'\cdot\lambda}(H) = L_\mu(H)$ for some $u' \in R_u(P_\lambda(H))$, by another application of Lemma 2.9.10(iii). We can therefore find maximal tori $S \subseteq T$ of $P_\lambda(H) \subseteq P_\lambda$ with $\mu, u' \cdot \lambda \in Y_S \subseteq Y_T$. Then $L_\mu = uL_\lambda u^{-1}$ and $L_{u'\cdot\lambda} = u'L_\lambda(u')^{-1}$ are two R-Levi subgroups of P_λ containing T, so they are equal by Lemma 2.9.10(ii). A final application of Lemma 2.9.10(iii) implies that $u = u'$, and this gives the result claimed. $\qquad\square$

We also require the following in several places.

Lemma 2.9.18 *Suppose P is a parabolic subgroup of G^0. Then $N_G(P)$ is an R-parabolic subgroup of G, and $N_G(P)^0 = P$. Hence $R_u(N_G(P)) = R_u(P)$.*

Proof Let $Q = N_G(P)$. We may write $P = P_\lambda(G^0)$ for some $\lambda \in Y_{G^0} = Y_G$. Note that since $P_\lambda^0 = P$ we have $P_\lambda \subseteq Q$. Choose a maximal torus T of P such that $\lambda \in Y_T$. Note that for any $g \in Q$, gTg^{-1} is a maximal torus of P, and hence the conjugacy of maximal tori in P implies that Q is generated by P and $N_Q(T)$. Rigidity of tori implies that we can choose a finite set $g_1, \ldots, g_r \in Q$ of coset representatives for $N_Q(T)/T$, and we can set $\mu := g_1 \cdot \lambda + \cdots + g_r \cdot \lambda$. Then it is not hard to check that P_μ contains P and $N_Q(T)$, and hence $P_\mu = Q$. The final statements follow because $N_G(P)^0 \subseteq N_{G^0}(P) = P$, since P is parabolic in G^0. $\qquad\qquad\square$

Lemma 2.9.19 *Let P be an R-parabolic subgroup of G. Then the R-Levi subgroups of P are precisely the subgroups of the form $N_P(M)$ for M a Levi subgroup of P^0.*

Proof If L is an R-Levi subgroup of P then L^0 is a Levi subgroup of P^0 by Theorem 2.9.1(ii). Since $N_{P^0}(L^0) = L^0$, we have $N_P(L^0)^0 = L^0$. Now let M be a Levi subgroup of P^0 and let T be a maximal torus of M. By Lemma 2.9.10(ii) there is a unique R-Levi subgroup L of P such that $T \subseteq L$. The Levi subgroups L^0 and M of P^0 both contain T, so by Lemma 2.9.10(ii) (applied to P^0 inside G^0) they are equal; in particular, L normalises M. Now $N_P(T)$ normalises M, again by Lemma 2.9.10, so $N_P(M)$ meets every component of P. Since $P = L \ltimes R_u(P)$, L also meets every component of P. It follows that $L = N_P(M)$. $\qquad\qquad\square$

Lemma 2.9.20 *Let P and Q be R-parabolic subgroups of G such that $P \subseteq Q$ and $P^0 = Q^0$. If L is an R-Levi subgroup of P then $N_Q(L^0)$ is an R-Levi subgroup of Q, and all the R-Levi subgroups of Q are of this form. In particular, a subgroup H of P is contained in an R-Levi subgroup of P if and only if it is contained in an R-Levi subgroup of Q.*

Proof By the previous lemma, the R-Levi subgroups of P (resp., Q) have the form $N_P(M)$ (resp., $N_Q(M)$) for Levi subgroups M of $P^0 = Q^0$. The result follows easily. $\quad\square$

Lemma 2.9.21 *Let $f : G_1 \to G_2$ be an isogeny of reductive groups and let $\lambda \in Y_{G_1}$. Set $\mu = f \circ \lambda \in Y_{G_2}$. Then*

   (i) $f(P_\lambda) = P_\mu$, $f(L_\lambda) = L_\mu$;
   (ii) $f^{-1}(P_\mu) = P_\lambda$, $f^{-1}(L_\mu) = L_\lambda$;
   (iii) L_λ is a proper subgroup of G_1 if and only if L_μ is a proper subgroup of G_2;
   (iv) $f(R_u(P_\lambda)) = R_u(P_\mu)$.

Proof The kernel of f is a finite normal subgroup of G_1, and so is centralised by the connected group G_1^0. This means that $\ker(f)$ commutes with every cocharacter of G_1, and hence is contained in every R-Levi subgroup and every R-parabolic subgroup of G_1. It is clear using Lemma 2.8.2 that: (a) $f(P_\lambda) \subseteq P_\mu$; (b) $f(L_\lambda) \subseteq L_\mu$; (c) $f(R_u(P_\lambda)) \subseteq R_u(P_\mu)$. By (c), since f is an isogeny, we have $\dim(R_u(P_\lambda)) \le \dim(R_u(P_\mu))$. Now $f^{-1}(P_\mu)^0$ is an overgroup of P_λ^0, hence is a parabolic subgroup of G^0, so $R_u(f^{-1}(P_\mu)) \subseteq R_u(P_\lambda)$. But this implies that $\dim(R_u(P_\mu)) \le \dim(R_u(P_\lambda))$, and we can conclude that

$\dim(P_\lambda) = \dim(P_\mu)$. It follows that the inclusions in (a), (b) and (c) are all equalities, and the rest of the result now follows. $\qquad\square$

Remark 2.9.22 It follows that if $f : G_1 \to G_2$ is an isogeny, then the preimage of an R-parabolic (resp., R-Levi) subgroup of G_2 is an R-parabolic (resp., R-Levi) subgroup of G_1. To be precise, if $\mu \in Y_{G_2}$ then by (2.5.2) there exist $\lambda \in Y_{G_1}$ and $n \in \mathbb{N}$ such that $\phi \circ \lambda = n\mu$; we have $f^{-1}(P_\mu) = P_\lambda$ and $f^{-1}(L_\mu) = L_\lambda$ by Lemma 2.9.21(ii).

We finish this section with a generalisation to non-connected G of the famous Borel-Tits Theorem [36, Thm. 2.5] on unipotent subgroups of reductive groups. The original result is a fundamental tool in the analysis of the subgroup structure of (connected) reductive groups, since it implies in particular that a maximal subgroup is either reductive or parabolic.

Theorem 2.9.23 *Suppose U is a unipotent subgroup of G^0. Then there exists an R-parabolic subgroup $P = P_{\mathrm{BT}}(U)$ of G such that $U \subseteq R_u(P)$ and $N_G(U) \subseteq P$.*

Proof We first sketch the original proof of Borel-Tits in the case $G = G^0$ is connected. Form two ascending chains of subgroups:

$$U = U_1 \subseteq U_2 \subseteq \dots,$$

$$N_1 \subseteq N_2 \subseteq \dots,$$

where $N_i = N_G(U_i)$ and $U_{i+1} = R_u(N_i)$ for each $i \geq 1$. Each U_i is a closed connected subgroup of G, so the first sequence must stabilise for dimension reasons; hence the second sequence stabilises too. We therefore find subgroups U' and N' of G such that $U \subseteq U' = R_u(N')$ and $N_G(U) \subseteq N' = N_G(U')$. The meat of the proof is to show that a subgroup of G which is the normaliser of its own unipotent radical is a parabolic subgroup of G.

Now suppose G is not connected. Since the Borel-Tits parabolic subgroup P of G^0 is stable under automorphisms of G^0 which normalise U, we can conclude that $N_G(U) \subseteq N_G(P)$. But $N_G(P)$ is an R-parabolic subgroup of G with $R_u(N_G(P)) = R_u(P)$ by Lemma 2.9.18, so we are done in this case too. $\qquad\square$

2.10 Non-Abelian 1-Cohomology

We need an important cohomological result of Richardson. In order to provide a proof we first recall some basic material (see [146, §6] for more details).

Let H and U be algebraic groups and suppose H acts on U by group automorphisms: that is, U is an H-variety, and the action is defined via a homomorphism $H \to \mathrm{Aut}(U)$. We call a map of varieties $\alpha \colon H \to U$ a 1-*cocycle* if it satisfies the following condition:

$$\alpha(hh') = \alpha(h)(h \cdot \alpha(h')) \text{ for all } h, h' \in H. \tag{2.10.1}$$

We refer to (2.10.1) as the *cocycle equation*. We denote the set of all 1-cocycles by $Z^1(H, U)$. It contains an obvious *trivial 1-cocycle*, given by $\alpha(h) = 1$ for all $h \in H$. For any $\alpha \in Z^1(H, U)$ and any $h \in H$, we have $\alpha(1) = 1$ (easy exercise) and hence $1 = \alpha(h^{-1}h) = \alpha(h^{-1})(h^{-1} \cdot \alpha(h))$ by (2.10.1), from which we deduce that

$$\alpha(h^{-1}) = \left(h^{-1} \cdot \alpha(h)\right)^{-1}.$$

We define an equivalence relation on $Z^1(H, U)$ by $\alpha_1 \sim \alpha_2$ if there exists $u \in U$ such that $\alpha_2(h) = u\alpha_1(h)(h \cdot u^{-1})$ for all $h \in H$; in this case we say that α_1 and α_2 are *cohomologous*. The equivalence classes of this relation are the *1-cohomology classes*, and the set of these classes is denoted $H^1(H, U)$. Given $\alpha \in Z^1(H, U)$, we denote by $\bar{\alpha} \in H^1(H, U)$ the 1-cohomology class of α; we give the 1-cohomology class of the trivial cocycle the label 1, and call it the *trivial cohomology class*. For each $u \in U$ we can define $\chi_u \in Z^1(H, U)$ by $\chi_u(h) = u(h \cdot u)^{-1}$ for all $h \in H$. Then a 1-cocycle lies in the trivial cohomology class if and only if it is of the form χ_u for some $u \in U$. The set of χ_u is denoted $B^1(H, U)$ and is called the set of 1-*coboundaries*. We say that $H^1(H, U)$ is *trivial* if $H^1(H, U) = \{1\}$, which is the same as saying that $Z^1(H, U) = B^1(H, U)$. See Exercise 2.9 for more details.

In the special case when U is abelian it is easy to see that $Z^1(H, U)$ has the structure of an abelian group with respect to pointwise addition of 1-cocycles: that is, we define $(\alpha_1 + \alpha_2)(h) = \alpha_1(h)\alpha_2(h)$ for each $\alpha_1, \alpha_2 \in Z^1(H, U)$ and $h \in H$. Then the set $B^1(H, U)$ of 1-coboundaries is a subgroup of $Z^1(H, U)$ since

$$(\chi_{u_1} + \chi_{u_2})(h) = \chi_{u_1}(h)\chi_{u_2}(h) = u_1(h \cdot u_1)^{-1}u_2(h \cdot u_2)^{-1}$$

$$= (u_1u_2)(h \cdot (u_1u_2))^{-1} = \chi_{u_1u_2}(h)$$

for each $h \in H$, so $\chi_{u_1} + \chi_{u_2} = \chi_{u_1u_2}$ (note we use the fact that multiplication in U is abelian in an essential way here). Therefore we may identify $H^1(H, U)$ with the quotient $Z^1(H, U)/B^1(H, U)$, so $H^1(H, U)$ also has the structure of an abelian group. If, moreover, U is a vector space over k then $Z^1(H, U)$ and $B^1(H, U)$ with the obvious scalar multiplication are vector spaces over k, so $H^1(H, U)$ is also a vector space over k. When U is abelian we refer to *abelian cohomology*. In general we use the terminology "non-abelian cohomology" to signify that U need not be abelian.

Let U_1, U_2 be algebraic groups on which H acts by group automorphisms and let $f: U_1 \to U_2$ be an H-equivariant homomorphism. It is easy to check that we get a map $Z^1(f): Z^1(H, U_1) \to Z^1(H, U_2)$ given by $Z^1(f)(\alpha) = f \circ \alpha$. Further, if $\alpha_1 \sim \alpha_2$, then $Z^1(f)(\alpha_1) \sim Z^1(f)(\alpha_2)$. To see this, suppose that $u \in U_1$ is such that $\alpha_2(h) = u\alpha_1(h)(h \cdot u^{-1})$ for all $h \in H$. Then we see that

$$(f \circ \alpha_2)(h) = f(\alpha_2(h)) = f(u\alpha_1(h)(h \cdot u^{-1})) = f(u)(f \circ \alpha_1)(h)(h \cdot f(u)^{-1}),$$

which gives the result. We deduce that $Z^1(f)$ descends to give a well-defined map on 1-cohomology: $H^1(f)\colon H^1(H, U_1) \to H^1(H, U_2)$. Note that if f is the trivial homomorphism, then $H^1(f)$ is the constant map with value 1. There is also an obvious functoriality: if $g : U_2 \to U_3$ is another H-equivariant homomorphism, then $Z^1(g \circ f) = Z^1(g) \circ Z^1(f)$ and $H^1(g \circ f) = H^1(g) \circ H^1(f)$.

Lemma 2.10.1 *Keeping the set-up above, suppose N is an H-stable normal subgroup of U, so that H acts on N and U/N as well. Let $i\colon N \to U$ be the inclusion and let $\pi\colon U \to U/N$ be the canonical projection. Consider the sequence*

$$H^1(H, N) \xrightarrow{\;H^1(i)\;} H^1(H, U) \xrightarrow{\;H^1(\pi)\;} H^1(H, U/N).$$

Then $H^1(\pi)^{-1}(1) = \mathrm{Im}(H^1(i))$.

Proof The composition $\pi \circ i$ is the trivial map from N to U/N, so $H^1(\pi \circ i) = H^1(\pi) \circ H^1(i)$ is the constant map with value 1 from $H^1(H, N)$ to $H^1(H, U/N)$. Hence the preimage $H^1(\pi)^{-1}(1)$ contains the image of $H^1(i)$.

For the converse, let $\beta \in Z^1(H, U)$ such that $H^1(\pi)(\overline{\beta}) = 1$. Then $Z^1(\pi)(\beta) \sim 1$, so there exists $x \in U/N$ such that $\pi(\beta(h)) = (Z^1(\pi)(\beta))(h) = x(h \cdot x)^{-1}$ for all $h \in H$. Choose $u \in U$ with $\pi(u) = x$ and define $\beta' \in Z^1(H, U)$ by $\beta'(h) = u^{-1}\beta(h)(h \cdot u)$. Note that $\beta' \sim \beta$, so $\overline{\beta'} = \overline{\beta}$. For any $h \in H$, we conclude that

$$\pi(\beta'(h)) = \pi(u^{-1})\pi(\beta(h))\pi(h \cdot u) = x^{-1}x(h \cdot x)^{-1}(h \cdot x) = 1.$$

This implies that $\beta'(h) \in N$ for each $h \in H$, so it makes sense to define $\alpha \in Z^1(H, N)$ by $\alpha(h) = \beta'(h)$ and deduce that $\beta' = Z^1(i)(\alpha)$. It follows that $H^1(i)(\overline{\alpha}) = \overline{\beta'} = \overline{\beta}$, so $\overline{\beta} \in \mathrm{Im}(H^1(i))$, as required. $\qquad\qquad\square$

Lemma 2.10.2 ([146, Lem. 6.2.5]) *Let H be a linearly reductive group acting by group automorphisms on an abelian unipotent group U. Then $H^1(H, U)$ is trivial.*

Proof Suppose first that $\mathrm{char}(k) = 0$. Then U is a vector group and H acts linearly on U [168, Thm. 3.4.7]. It follows that $H^1(H, U)$ is trivial, [127, Prop. 15.15]. So suppose that $\mathrm{char}(k) = p > 0$. Let $\alpha \in Z^1(H, U)$. Let $M_1 \subseteq M_2 \subseteq \cdots$ be as in Proposition 2.6.3 and let $\alpha_n = \alpha|_{M_n}$. Fix $n \in \mathbb{N}$. We claim that $\alpha_n \sim 1$. Let $r_n = |M_n|$; then r_n is coprime to p since M_n is linearly reductive. Set $u = \prod_{\gamma \in M_n} \alpha_n(\gamma)$. Then since we are multiplying over all elements of M_n, for any $\delta \in M_n$ we also have

$$u = \prod_{\gamma \in M_n} \alpha_n(\delta\gamma) = \prod_{\gamma \in M_n} \alpha_n(\delta)(\delta \cdot \alpha_n(\gamma)) = \alpha_n(\delta)^{r_n}(\delta \cdot u),$$

where we use the cocycle equation (2.10.1) and the fact that U is abelian. This implies that $u(\delta \cdot u)^{-1} = \alpha_n(\delta)^{r_n}$ for all $\delta \in M_n$. Since r_n is coprime to p, the map $x \mapsto x^{r_n}$ is a bijection on U. Hence there exists $w \in U$ such that $w^{r_n} = u$, and we have $(w(\delta \cdot w)^{-1})^{r_n} = u(\delta \cdot u)^{-1} = \alpha_n(\delta)^{r_n}$ for all $\delta \in M_n$. But then $\alpha_n(\delta) = w(\delta \cdot w)^{-1}$ for all $\delta \in M_n$, which proves the claim.

Define $C_n = \{u \in U \mid \alpha(\gamma) = u(\gamma \cdot u)^{-1} \text{ for all } \gamma \in M_n\}$. Then each C_n is non-empty by the claim above and $C_1 \supseteq C_2 \supseteq \cdots$. But the C_n are closed subsets, so this chain must eventually become constant. Hence there exists $u \in U$ such that $\alpha(h) = u(h \cdot u)^{-1}$ for all $h \in \bigcup_{n=1}^{\infty} M_n$. Since $\bigcup_{n=1}^{\infty} M_n$ is dense in H, we have $\alpha(h) = u(h \cdot u)^{-1}$ for all $h \in H$. This completes the proof. $\qquad\square$

Theorem 2.10.3 *Let H be a linearly reductive group acting by group automorphisms on a unipotent group U. Then $H^1(H, U)$ is trivial.*

Proof The unipotent group U is solvable, hence the derived series for U has finite length and terminates at 1. We use induction on the length of the derived series. The result is trivially true if $U = \{1\}$. For the general case, the derived group $[U, U]$ has shorter derived series than U, and so by induction $H^1(H, [U, U]) = \{1\}$. Further, $U/[U, U]$ is abelian, so we can apply Lemma 2.10.2 to deduce that $H^1(H, U/[U, U]) = \{1\}$. But then $H^1(H, U) = \{1\}$ by Lemma 2.10.1, so we are done. $\qquad\square$

Here is our main application of this formalism. Let M be an algebraic group and suppose M acts by automorphisms on an algebraic group U. Let $\sigma : H \to M$ be a homomorphism of algebraic groups. We allow H to act on U by $h \cdot u = \sigma(h) \cdot u$. Given a map $\alpha : H \to U$, define $\rho_\alpha : H \to M \ltimes U$ by

$$\rho_\alpha(h) = \alpha(h)\sigma(h).$$

Note that by rearranging the above equation we can also write $\rho_\alpha(h) = \sigma(h)(h^{-1} \cdot \alpha(h))$. We denote the canonical projection from $M \ltimes U$ to M by π.

Lemma 2.10.4 *Let M, U, H, σ, α and ρ_α be as above.*

(a) *ρ_α is a homomorphism if and only if α belongs to $Z^1(H, U)$.*

(b) *Let $u \in U$. Then $u \cdot \sigma = \rho_\alpha$ if and only if $\alpha = \chi_u$, where $u \cdot \sigma : H \to M \ltimes U$ is defined by $(u \cdot \sigma)(h) = u\sigma(h)u^{-1}$.*

(c) *Suppose $H^1(H, U)$ is trivial. Let $\rho : H \to M \ltimes U$ be a homomorphism such that $\pi \circ \rho = \sigma$. Then ρ is U-conjugate to σ.*

(d) *Suppose that for every homomorphism $\rho : H \to M \ltimes U$ such that $\pi \circ \rho = \sigma$, ρ is U-conjugate to σ. Then $H^1(H, U)$ is trivial.*

Proof (a). For $h_1, h_2 \in H$ we have

$$\rho_\alpha(h_1 h_2) = \alpha(h_1 h_2)\sigma(h_1 h_2) = \alpha(h_1 h_2)\sigma(h_1)\sigma(h_2),$$

and

$$\rho_\alpha(h_1)\rho_\alpha(h_2) = \alpha(h_1)\sigma(h_1)\alpha(h_2)\sigma(h_2).$$

These two are equal if and only if

$$\alpha(h_1 h_2) = \alpha(h_1)\sigma(h_1)\alpha(h_2)\sigma(h_1)^{-1} = \alpha(h_1)(h_1 \cdot \alpha(h_2)),$$

that is if and only if $\alpha \in Z^1(H, U)$.

(b). For $h \in H$, $u\sigma(h)u^{-1} = \alpha(h)\sigma(h)$ if and only if $\alpha(h) = u\sigma(h)u^{-1}\sigma(h)^{-1} = \chi_u(h)$.

(c). Let $\rho\colon H \to M \ltimes U$ be a homomorphism such that $\pi \circ \rho = \sigma$. Then for each $h \in H$ we can write $\rho(h) = \sigma(h)\beta(h)$ for some $\beta(h) \in U$. Then define a map $\alpha\colon H \to U$ by $\alpha(h) = h^{-1} \cdot \beta(h)$, so that $\rho(h) = \rho_\alpha(h)$. By part (a), α belongs to $Z^1(H, U)$. But we're assuming that $H^1(H, U) = \{1\}$, so $\alpha = \chi_u$ for some $u \in U$, so $\rho = u \cdot \sigma$ by (b).

(d). Given any $\alpha \in Z^1(H, U)$, the corresponding ρ_α is a homomorphism by (a), and hence is U-conjugate to σ by hypothesis. But then (b) implies that $\alpha = \chi_u$ for some $u \in U$, so $H^1(H, U)$ is trivial.

$\square$

Remark 2.10.5 Lemma 2.10.4 yields an alternative proof of some special cases of Theorem 2.10.3. For instance, suppose H is a torus and U is unipotent. Take $M = H$ and σ to be the identity. Let $\alpha \in Z^1(H, U)$. Then $\mathrm{Im}(\rho_\alpha)$ is a torus, so by conjugacy of maximal tori in $H \ltimes U$, there exist $h \in H$ and $u \in U$ such that $hu\mathrm{Im}(\rho_\alpha)u^{-1}h^{-1} \subseteq H$. Without loss we can take $h = 1$. We see that $u \cdot \rho_\alpha = \sigma$, so $\alpha = \chi_u$ by Lemma 2.10.4(b). Hence $H^1(H, U) = 1$.

Now suppose that $\mathrm{char}(k) = p > 0$, H is linearly reductive and finite and U is unipotent. Take $M = H$ and σ to be the identity. Let $\alpha \in Z^1(H, U)$. Without loss we can replace U with the subgroup generated by $\mathrm{Im}(\alpha)$, so we can assume U is finite. Since $|H|$ is coprime to p, U is the unique Sylow p-subgroup of $H \ltimes U$ and H is a complement to U. Now ρ_α is injective, so $\mathrm{Im}(\rho_\alpha)$ is also a complement to U. The Schur-Zassenhaus Theorem [80, Thm. 3.5] implies that any two complements to U are conjugate, and hence there is some $u \in U$ such that $H = u\mathrm{Im}(\rho_\alpha)u^{-1}$. We see that $u \cdot \rho_\alpha = \sigma$, so $\alpha = \chi_u$ by Lemma 2.10.4(b). Hence $H^1(H, U) = 1$.

Lemma 2.10.4 plays a crucial part in the subgroup structure investigations discussed in Sect. 12.1: in this case we take M to be a Levi subgroup of a parabolic subgroup P of G and we take U to be $R_u(P)$. Here is another application, which we will use in Sect. 6.7.

Proposition 2.10.6 *Let F be a finite group and let A be an abelian group (written additively) on which F acts by group automorphisms. Suppose that the map $A \to A$, $a \mapsto |F|a$ is surjective and has finite kernel. Then $H^1(F, A)$ is finite.*

Proof Let $\alpha \in Z^1(F, A)$ and set $a = \sum_{\gamma \in F} \alpha(\gamma)$. For any $\delta \in F$ we have

$$\delta \cdot a = \sum_{\gamma \in F} \delta \cdot \alpha(\gamma) = \sum_{\gamma \in F} (\alpha(\delta\gamma) - \alpha(\delta)) = a - |F|\alpha(\delta), \qquad (2.10.2)$$

where the second equality comes from the 1-cocycle relation and the third one follows because $\sum_{\gamma \in F} \alpha(\delta\gamma)$ is the same as $\sum_{\gamma \in F} \alpha(\gamma)$—we are just summing up the terms in a different order. By hypothesis, there exists $a' \in A$ such that $|F|a' = a$. We deduce from (2.10.2) that $|F|\alpha(\delta) = |F|(a' - \delta \cdot a')$. Hence $|F|(\alpha - \chi_{a'}) = 0$. This shows that α is cohomologous to a 1-cocycle which takes values in the finite set $\{b \in A \mid |F|b = 0\}$. Since F is finite, there are only finitely many such 1-cocycles. We conclude that $H^1(F, A)$ is finite. $\qquad \square$

Corollary 2.10.7 *Let J be a finite group acting on a torus S, and let F be a finite group. Then there are only finitely many S-conjugacy classes of homomorphisms $\rho \colon F \to J \ltimes S$.*

Proof Let $\pi \colon J \ltimes S \to J$ be the canonical projection. Let $\sigma \colon F \to J$ be a homomorphism. We allow F to act on S via σ. By Lemma 2.10.4, there is a bijection between $H^1(F, S)$ and the set of S-conjugacy classes of homomorphisms $\rho \colon F \to J \ltimes S$ such that $\pi \circ \rho = \sigma$. But S satisfies the hypotheses of Proposition 2.10.6, so $H^1(F, S)$ is finite. Since there are only finitely many possibilities for σ, the result follows. $\qquad \square$

Now we give our other main application of 1-cohomology. The set-up is as follows: X is an H-variety, $x \in X$ and S is a linearly reductive subgroup of H_x. Set $L = C_H(S)$, and suppose U is a unipotent subgroup of H which is normalised by S.

Proposition 2.10.8 *With notation as just introduced, if x' is an S-fixed point of X such that x' is U-conjugate to x, then x' is $(U \cap L)$-conjugate to x.*

Proof Let $u' \in U$ such that $x' = u' \cdot x$. For any $s \in S$,

$$(u')^{-1} s u' s^{-1} \cdot x = (u')^{-1} s u' \cdot x = (u')^{-1} s \cdot x' = (u')^{-1} \cdot x' = x,$$

so $\alpha(s) := (u')^{-1} s u' s^{-1}$ is an element of U which fixes x. This gives a map $\alpha \colon S \to U_x$. For any $s, t \in S$,

$$u'\alpha(st) = stu't^{-1}s^{-1} = su'\alpha(t)s^{-1} = su's^{-1}s\alpha(t)s^{-1} = u'\alpha(s)(s \cdot \alpha(t)),$$

so α belongs to $Z^1(S, U_x)$. Now $H^1(S, U_x)$ is trivial by Theorem 2.10.3, so $\alpha = \chi_u$ for some $u \in U_x$. We have $u'u \cdot x = u' \cdot x = x'$ and

$$su'us^{-1} = su's^{-1}sus^{-1} = u'\alpha(s)(s \cdot u) = u'u(s \cdot u)^{-1}(s \cdot u) = u'u$$

for all $s \in S$, so $u'u \in U \cap L$. Hence x' is $(U \cap L)$-conjugate to x. $\qquad\square$

We can apply the previous result in particular when G is a reductive group and U is the unipotent radical of an R-parabolic subgroup of G.

Proposition 2.10.9 *Suppose X is a G-variety and $x \in X$. Let S be a linearly reductive subgroup of G_x. Let $\lambda \in Y_{C_G(S)}$ such that $x' = \lim_{a\to 0} \lambda(a) \cdot x$ exists. If x' is $R_u(P_\lambda)$-conjugate to x, then x' is $R_u(P_\lambda(C_G(S)))$-conjugate to x.*

Proof Since $\lambda \in Y_{C_G(S)}$ and S is central in $C_G(S)$, we see that $S \subseteq P_\lambda$ and hence S normalises $R_u(P_\lambda)$. Moreover, S fixes all the points $\lambda(a) \cdot x$ for $a \in k^*$, so S also fixes x'. Now the result follows from Proposition 2.10.8. $\qquad\square$

Corollary 2.10.10 (Levi Descent for Orbits) *Suppose X is a G-variety and $x \in X$. Let S be a torus in G_x and set $L = C_G(S)$. Let $\lambda \in Y_L$ such that $x' := \lim_{a\to 0} \lambda(a) \cdot x$ exists. If x' is $R_u(P_\lambda)$-conjugate to x, then x' is $R_u(P_\lambda(L))$-conjugate to x.*

Proof This follows immediately from Proposition 2.10.9. $\qquad\square$

2.11 Historical Remarks and References

We have given precise references for many of the results above, so we do not repeat those here. It is important, however, to acknowledge also the key early paper [35], which contains many of the structural results on (connected) reductive groups we have collected above, including versions for parabolic subgroups of connected reductive groups of almost all of the results we have proved in the non-connected setting in Sect. 2.9. In terms of the development of some of the ideas in the chapter, the descriptions of the R-parabolic and R-Levi subgroups of G in terms of cocharacters in Theorem/Definition 2.9.1 are due to Richardson [147, 2.1–2.3], and these ideas were developed in [117], where R-parabolic subgroups were called generalised parabolic subgroups, and in [18, §6]. Lemma 2.9.18 is [117, Prop. 5.4(a)], while Lemma 2.9.21 is taken from [18, Lem. 2.11]. Our approach to length functions follows that of Kempf [89, §2]; length functions are called *norms* in [75, (1.3)] and [21, §2.2]. The proofs of Lemma 2.4.4, Proposition 2.4.5 and Theorem 2.6.2 are based on the proof of [117, Prop. 3.2]. The material on non-abelian cohomology in Sect. 2.10 is mostly a modified version of that in [146, §6]. In particular, Theorem 2.10.3 is [146, Prop. 6.1].

2.12 Exercises

Exercise 2.1 Let H be an algebraic group and let $h \in H$. Define $\alpha, \beta \colon H \to H$ by $\alpha(m) = mhm^{-1}$ and $\beta(m) = mhm^{-1}h^{-1}$. Show that for any $X \in \mathfrak{h}$,

$$(d\alpha)_1(X) = (dR_h)_1(X - \mathrm{Ad}(h)(X))$$

and

$$(d\beta)_1(X) = X - \mathrm{Ad}(h)(X).$$

[*Hint*: Use the facts, recalled in Sect. 2.3 above, that the differential of the group multiplication at $(1, 1)$ is addition, and the differential of inversion is negation.]

Exercise 2.2 Show that if H is a connected algebraic group then the largest normal unipotent subgroup of H is $R_u(H)$. (The point of this exercise is that this subgroup is **connected**.)

Exercise 2.3 Show that an extension of reductive groups is reductive. Show that if H is reductive then so is any normal subgroup of H.

Exercise 2.4 Let T be a torus of rank m, and write elements of T as $t = (t_1, \ldots, t_m)$ with $t_i \in k^*$. Let $A = (a_{ij}) \in \mathrm{SL}_m(\mathbb{Z})$.

(i) Define a map $\phi_A \colon T \to T$ as follows: given $t = (t_1, \ldots, t_m) \in T$, $\phi_A(t)$ has as its ith entry

$$\phi_A(t)_i = \prod_{j=1}^{m} t_j^{a_{ij}}.$$

  Show that ϕ_A is an automorphism of T.

(ii) Show that if $m \geq 2$ then the only $\mathbb{Z}$-valued $\mathrm{Aut}(T)$-invariant symmetric bilinear form on Y_T is the zero form.

Exercise 2.5 Complete the proof of Lemma 2.8.2 by proving part (iii). That is, suppose $f_1 \colon k^* \to X_1$ and $f_2 \colon k^* \to X_2$ are morphisms, and show that $\lim_{a \to 0}(f_1 \times f_2)(a)$ exists if and only if $x_1 := \lim_{a \to 0} f_1(a)$ and $x_2 := \lim_{a \to 0} f_2(a)$ exist, and in this case $\lim_{a \to 0}(f_1 \times f_2)(a) = (x_1, x_2)$.

Exercise 2.6 Suppose H is an algebraic group and X is an H-variety. Let $x \in X$ and $\lambda \in Y_H$ be such that $x' := \lim_{a \to 0} \lambda(a) \cdot x$ exists. Show that $x' \in \overline{H \cdot x}$.

Exercise 2.7 Let V be an H-module, let T be a maximal torus of H, let $\lambda \in Y_T$ and recall the notation of Example 2.8.4. Let $g \in H$.

(i) Show that $\chi \in \Phi_T(V)$ if and only if $g \cdot \chi \in \Phi_{gTg^{-1}}(V)$, and in this case express the weight space $V_{g \cdot \chi}$ in terms of the weight space V_χ.

(ii) Deduce that $V_{g \cdot \lambda, \geq 0} = g \cdot V_{\lambda, \geq 0}$, and similarly for $V_{\lambda, 0}$, $V_{\lambda, > 0}$, etc.

Exercise 2.8 Let P be an R-parabolic subgroup of G and let L be an R-Levi subgroup of P. Show that there is a unique opposite to P with respect to L.

Exercise 2.9 Let H and U be algebraic groups and suppose H acts on U by group automorphisms.

(i) Let $\alpha \in Z^1(H, U)$ and let $u \in U$. Show that the map $\beta \colon H \to U$ given by $\beta(h) = u\alpha(h)(h \cdot u)^{-1}$ belongs to $Z^1(H, U)$.

(ii) Let $\alpha, \beta \in Z^1(H, U)$. Define $\alpha \sim \beta$ if there exists $u \in U$ such that

$$\beta(h) = u\alpha(h)(h \cdot u)^{-1}$$

for all $h \in H$. Show that this gives an equivalence relation on $Z^1(H, U)$.

Exercise 2.10 Let M, U, H and σ be as in Lemma 2.10.4. Let $\alpha, \beta \in Z^1(H, U)$. Prove that ρ_α and ρ_β are U-conjugate if and only if $\overline{\alpha} = \overline{\beta}$. (This generalises Lemma 2.10.4(b).)

Geometric Invariant Theory

3

Suppose G acts on a variety X. A fundamental problem in geometric invariant theory is to determine the closed orbits of G in X. These orbits correspond to the points in the quotient variety $X /\!\!/ G$, so this is the first step towards understanding the geometry of the quotient. Often it is of particular interest to find an open dense set of points on which the quotient map $\pi \colon X \to X /\!\!/ G$ is especially well-behaved. Moreover, once the closed orbits are known, one can study degeneration phenomena: the way in which a point in a non-closed orbit can be brought inside a closed orbit by taking a limit along a cocharacter.

An important family of examples arises as follows. Take G to be a subgroup of $\mathrm{GL}(V)$ for some finite-dimensional vector space V, choose a subvariety Y of $\mathrm{End}(V)$ that is stable under conjugation by G, and take X to be Y^n for some $n \in \mathbb{N}$, where G acts on Y^n by simultaneous conjugation. Typically, Y carries some algebraic structure: it might be a subgroup of $\mathrm{GL}(V)$, or a Lie subalgebra or associative subalgebra of $\mathrm{End}(V)$. For instance, if $G = \mathrm{GL}(V)$ and $Y = \mathrm{End}(V)$ then, by work of Kraft, [91, Prop. 4.4], for any $x = (y_1, \ldots, y_n) \in Y^n = X$, the orbit $G \cdot x$ is closed if and only if V is semisimple as an A-module, where A is the associative subalgebra of $\mathrm{End}(V)$ generated by $y_1, \ldots, y_n$; moreover, if $G \cdot x$ is not closed, then the degeneration process referred to above is the so-called "semisimplification", in which one replaces the A-module V with the direct sum of its composition factors, [91, Prop. 4.5].

The purpose of this chapter is to collect together some of the material on geometric invariant theory which allows us to make the preceding remarks more concrete, and which forms a key component of the toolkit used throughout the rest of the book. We start in Sect. 3.1 by recalling some material on quotients of varieties by algebraic group actions, including the important fact that every G-variety X has a categorical quotient $X /\!\!/ G$. After that we extend some of our preliminary material on orbits and the process of taking limits along cocharacters so that we can state the Hilbert-Mumford Theorem 3.3.2 in Sect. 3.3. This provides the key link between closed orbits and R-parabolic subgroups which is

© The Author(s), under exclusive license to Springer Nature Switzerland AG 2026

M. Bate et al., *G-Complete Reducibility, Geometric Invariant Theory and Spherical Buildings*, Oberwolfach Seminars 57, https://doi.org/10.1007/978-3-032-08866-6_3

exploited throughout the rest of the book. In the final sections of the chapter we collect some related results, including material on nullcones, separable orbits and so-called visible G-varieties.

3.1   Quotient Varieties

One of the main goals of geometric invariant theory is to construct quotient varieties. There is a vast and elaborate theory, but we recall only the parts that we need. We wish to describe the construction of the categorical quotient $X /\!/ G$ when X is affine. At several points, however, we need to consider non-affine varieties, so we broaden our discussion to the general quasi-projective case. We retain our usual convention that "variety" with no further qualification means "affine variety".

To begin, let X be a quasi-projective variety equipped with an action of an algebraic group H. We would like to give the set of orbits $\{H \cdot x \mid x \in X\}$ the structure of a variety in such a way that the map $x \mapsto H \cdot x$ is a map of varieties and the fibres of this map are precisely the H-orbits. In general, this is impossible. For if $\phi \colon X \to Y$ is a map of varieties and ϕ is constant on each H-orbit then ϕ is constant on the closure of each H-orbit; hence the fibres of ϕ cannot be the H-orbits unless every H-orbit is closed. So in order to construct quotient varieties, some compromises must be made.

Definition 3.1.1 Let X be a quasi-projective variety and H an algebraic group acting on X. Let Y be a quasi-projective variety and let $\phi \colon X \to Y$ be a map of varieties. We call (Y, ϕ) a *categorical quotient* of X if the following condition holds: for every quasi-projective variety Z and every H-invariant map $f \colon X \to Z$, there is a unique map $\widehat{f} \colon Y \to Z$ such that $f = \widehat{f} \circ \phi$, i.e.,

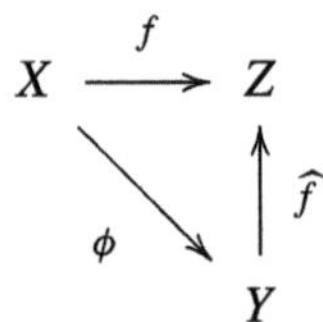

commutes; see also [8, §2.1]. We call (Y, ϕ) an *orbit space* if (Y, ϕ) is a categorical quotient and the fibres of ϕ are the H-orbits.

If a categorical quotient exists, it is unique up to the obvious notion of isomorphism; see Exercise 3.1.

Example 3.1.2   (i). Let V be a finite-dimensional vector space and let k^* act on the quasi-affine variety $V \backslash \{0\}$ by scalar multiplication in the usual way. Let π be the canonical projection from $V \backslash \{0\}$ to the projective space $\mathbb{P}(V)$. Then it can be shown that $(\mathbb{P}(V), \pi)$ is an orbit space (this follows from the discussions in [132, Ch. 3, §4]). Note that $\mathbb{P}(V)$ is not affine.

(ii). Now let k^* act on all of V by scalar multiplication. Suppose $f: V \to Z$ is a k^*-invariant map to a variety Z. Then f is constant on the closure of each k^*-orbit. But 0 belongs to the closure of every k^*-orbit, so f must be constant. It follows from this that (Y, ϕ) is a categorical quotient of V, where Y is a singleton set and ϕ is the constant map from V to Y. In this case, however, ϕ does not separate distinct k^*-orbits of V, so (Y, ϕ) is not an orbit space.

Example 3.1.3 Let H be an algebraic group and let M be a subgroup of H. Then M acts on H by $m \cdot h = hm^{-1}$ and the set H/M of left cosets may be given the structure of a quasi-projective variety in such a way that $(H/M, \pi)$ is an orbit space, where the map $\pi: H \to H/M$ is the canonical projection (see [32, II.6.8 Thm.]). For instance, if H is connected and reductive then H/M is projective if and only if M is a parabolic subgroup of H. If M is normal in H then H/M is affine and inherits the structure of an algebraic group. It follows from the construction in [32, II.6.8 Thm.] that the action of H on H by left multiplication descends to give an action of H on H/M (see Exercise 3.3 for the case when M is reductive).

In the case of a reductive group acting on an affine variety there is a well-known construction of a categorical quotient. The construction rests on Hilbert's "fundamental theorem of invariant theory", for $k = \mathbb{C}$ and $\mathrm{GL}(V)$, generalised to an arbitrary semisimple algebraic group in characteristic 0 by Mumford and others (Cartier, Iwahori, Nagata and Weyl), and Haboush [72] in full generality; see also [132, Thm. 3.4] and [130, Thm. A.1.0]. We first recall that for any algebraic group H and any H-variety X there is an induced action of H on the coordinate algebra $k[X]$: for $h \in H$ and $f \in k[X]$ we define $h \cdot f \in k[X]$ by $(h \cdot f)(x) = f(h^{-1} \cdot x)$ for each $x \in X$. It is straightforward to check that this gives an action of H on $k[X]$ by k-algebra automorphisms. This action is *locally finite* in the following sense: for all $r \in \mathbb{N}$ and all $f_1, \ldots, f_r \in k[X]$, there is a finite-dimensional H-stable subspace V of $k[X]$ such that $f_1, \ldots, f_r \in V$ (see [132, Lem. 3.1]). We let $k[X]^H$ denote the subring of H-invariant functions; i.e.,

$$k[X]^H := \{f \in k[X] \mid h \cdot f = f \text{ for all } h \in H\}.$$

For a reductive group, we have the following.

Theorem 3.1.4 *Let X be a G-variety. Then the ring of invariants $k[X]^G$ is a finitely generated k-algebra.*

We define $X /\!/ G$ to be the variety with coordinate ring $k[X]^G$. The inclusion of $k[X]^G$ in $k[X]$ gives rise to a map $\pi_X: X \to X /\!/ G$; we call this the *canonical projection*. If there is more than one reductive group acting then we write $\pi_{X,G}$ rather than π_X if necessary to avoid confusion.

Proposition 3.1.5 *Let X be a G-variety.*

(i) *$(X/\!/G, \pi_X)$ is a categorical quotient.*
(ii) *π_X is surjective.*
(iii) *Let X_1, X_2 be disjoint closed G-stable subsets of X. Then there exists $f \in k[X]^G$ such that $f|_{X_1} = 1$ and $f|_{X_2} = 0$. In particular, $\pi_X(X_1)$ and $\pi_X(X_2)$ are disjoint.*
(iv) *If Y is a closed G-stable subset of X then $\pi_X(Y)$ is closed.*
(v) *For any $x \in X$, there is a unique closed orbit contained in $\overline{G \cdot x}$.*
(vi) *π_X gives a bijection $O \mapsto \pi_X(O)$ from the set of closed G-orbits in X to the set of points of $X/\!/G$.*

Sketch Proof For part (i), we need to prove that the categorical quotient property holds. Let Y be a variety and let $f : X \to Y$ be a G-invariant map. It is clear that the comorphism f^* maps $k[Y]$ into $k[X]^G$, so we get a map $\widehat{f} : X/\!/G \to Y$ such that $f = \widehat{f} \circ \pi_X$. This deals with the case that Y is affine; for the extension to general quasi-projective Y, see [132, §3]. Parts (ii) and (iii) follow from [132, Thm. 3.5(ii)] and [132, Lem. 3.3], respectively, and surjectivity of π_X also implies the uniqueness of $\widehat{f}$ in (i). Part (iv) is [132, Thm. 3.5(iv)]. Since elements of $k[X]^G$ are constant on orbit closures, part (v) and the injectivity of the map $O \mapsto \pi_X(O)$ both follow from part (iii) (note that if $x \in X$ then $\overline{G \cdot x}$ contains at least one closed orbit by Remark 2.7.5). If $z \in X/\!/G$ then by part (ii) there exists $x \in X$ such that $z = \pi_X(x)$. By part (v) there is a unique closed orbit O contained in $\overline{G \cdot x}$, and we have $\pi_X(O) = \pi_X(x) = z$. Hence the map $O \mapsto \pi_X(O)$ is surjective, so (vi) is proved. □

Example 3.1.6 Here is an important example which we will return to in Chap. 6. Recall from Example 2.7.2 that G acts on G^n by simultaneous conjugation, so we can form the quotient $G^n/\!/G$. We denote the canonical projection from G^n to $G^n/\!/G$ by π_{G^n}.

Example 3.1.7 Let X be a G-variety and suppose H is a reductive subgroup of G. The map $\pi_{X,G} : X \to X/\!/G$ is H-invariant, so there is a unique map $j : X/\!/H \to X/\!/G$ such that $\pi_{X,G} = j \circ \pi_{X,H}$.

In practice it is often extremely difficult to compute the ring of invariants explicitly, but some cases are known. In Sect. 5.6 we look at some examples where $G = \mathrm{GL}_m$ or SL_m acting by simultaneous conjugation on $(\mathrm{Mat}_m)^n$, $(\mathrm{GL}_m)^n$, $(\mathrm{SL}_m)^n$.

Remark 3.1.8 Let X be a G-variety and let $f \in k[X]^G$. The open affine set X_f is G-stable, so we may regard it as a G-variety. By [132, Thm. 3.5] we may identify $X_f/\!/G$ with $\pi_G(X_f)$. Here is an example which we will discuss in Sect. 5.6. Let X be Mat_m with the conjugation action and let f be det: then X_f is GL_m with the conjugation action.

Proposition 3.1.9 *Let X be a G-variety.*

(i) *If X is irreducible and normal, then $X /\!\!/ G$ is normal.*
(ii) *If G is finite, then π_X is finite.*

Proof Part (i) is [8, 2.1.9(a)], and part (ii) follows from the discussion following [13, Rem. 2.4]. $\qquad\square$

Example 3.1.10 Let $H = X = \mathrm{SL}_2$ with H acting by conjugation. Define $\lambda \in Y_H$ by $\lambda(a) = \mathrm{diag}(a, a^{-1})$. Let

$$x_1 = \begin{pmatrix} 1 & 1 \\ 0 & 1 \end{pmatrix}.$$

Then

$$\lambda(a) \cdot x_1 = \lambda(a)x_1\lambda(a)^{-1} = \begin{pmatrix} a & 0 \\ 0 & a^{-1} \end{pmatrix}\begin{pmatrix} 1 & 1 \\ 0 & 1 \end{pmatrix}\begin{pmatrix} a^{-1} & 0 \\ 0 & a \end{pmatrix} = \begin{pmatrix} 1 & a^2 \\ 0 & 1 \end{pmatrix}.$$

We see that $\lim_{a \to 0} \lambda(a) \cdot x_1$ exists and equals I (the identity matrix), so $I \in \overline{H \cdot x_1}$ by Exercise 2.6 and it follows that $\pi_X(x_1) = \pi_X(I)$. This shows that $X /\!\!/ H$ is not an orbit space: the orbit $H \cdot x_1$ coalesces with the closed orbit $H \cdot I$ in the quotient, giving an illustration of the problem discussed at the start of the section.

The result in the next example is due to Richardson [145, Thm. A].

Example 3.1.11 Let H be an algebraic group and let K be a reductive subgroup of H. Then the categorical quotient H/K from Example 3.1.3 coincides with the quotient $H /\!\!/ K$ defined above by the uniqueness of categorical quotients, so H/K is affine. (Note that although the statement of [145, Thm. A] has the hypothesis that H is reductive, Richardson himself observes that the argument above goes through without that assumption.) Conversely, if H is reductive and H/K is affine then it follows from [145, Thm. A] that K is reductive.

Example 3.1.12 Let X be an H-variety and let $x \in X$. Recall that H_x acts on H (Example 3.1.3) and $H \cdot x$ has the structure of a quasi-affine variety (Remark 2.7.5). The map κ_x is H_x-invariant, so it gives rise to a map $\widehat{\kappa_x}$ from H/H_x to $H \cdot x$. Clearly $\widehat{\kappa_x}$ is surjective and H-equivariant (with respect to the H-action on H/H_x described in Example 3.1.3). If $\widehat{\kappa_x}(h_1 H_x) = \widehat{\kappa_x}(h_2 H_x)$ then $\kappa_x(h_1) = \kappa_x(h_2)$, so $h_1 \cdot x = h_2 \cdot x$. Hence $h_2 = h_1 z$ for some $z \in H_x$, so $h_1 H_x = h_2 H_x$. This shows that $\widehat{\kappa_x}$ is a bijection. It follows from the proof of [146, Lem. 10.1.3] that $\widehat{\kappa_x}$ is finite. Applying Lemma 2.1.5, we see that if κ_x is separable then $\widehat{\kappa_x}$ is separable; Zariski's Main Theorem [127, Thm. 5.36] then implies that $\widehat{\kappa_x}$ is an isomorphism of varieties.

Remark 3.1.13 One can define what it means for a group scheme $\mathcal{H}$ to act on a quasi-projective variety X, and there is an obvious notion of a categorical quotient. In particular, if $\mathcal{H}$ is a subgroup scheme of an algebraic group H then $\mathcal{H}$ acts on H by right inverse multiplication and there exists a quasi-projective categorical quotient, which we denote by $H/\mathcal{H}$ (see [127, §5.2]). Now let X be an H-variety and let $x \in X$. The orbit map κ_x gives rise to a map $\widetilde{\kappa_x}$ from $H/\mathcal{H}_x$ to $H \cdot x$. It can be shown that $\widetilde{\kappa_x}$ is an isomorphism: see [82, Ch. I.5].

Lemma 3.1.14 *Suppose the connected group H acts on the variety X. Suppose some Borel subgroup B of H fixes a point $x \in X$. Then H fixes x.*

Proof Because B fixes x, there is an induced map $H/B \to X$ given by $hB \mapsto h \cdot x$. Since H/B is projective and irreducible and X is affine, this map is constant. Hence H fixes x. $\qquad\qquad\square$

We finish this section with a useful criterion for a stabiliser to be reductive.

Proposition 3.1.15 *Let X be an H-variety and let $x \in X$.*

(i) *If H_x is reductive then $H \cdot x$ is affine.*
(ii) *Suppose H is reductive. If $H \cdot x$ is affine then H_x is reductive. In particular, H_x is reductive if $H \cdot x$ is closed.*

Proof Without loss we can reduce to the case when H is connected. The map $\widehat{\kappa_x}$ is finite and bijective by Example 3.1.12. It follows from the proof of [146, Lem. 10.1.3] that H/H_x is affine if $H \cdot x$ is affine, and [32, AG.18.3(2)] implies that $H \cdot x$ is affine if H/H_x is affine. Part (i) and the first assertion of (ii) now follow from Example 3.1.11. The second assertion of (ii) holds since if $H \cdot x$ is closed in the affine variety X then $H \cdot x$ is affine. $\quad\square$

3.2   Linearisation and Basic Limit Calculations

In this section we recall the fact that any H-variety can be linearised. In conjunction with the ideas set up in Example 2.8.4, this allows us to deduce some basic properties of the process of taking limits in such varieties along cocharacters. These results are used many times in the rest of the book.

Lemma 3.2.1 *Let X be an H-variety and let S be a possibly empty closed H-stable subscheme of X. Then there is an H-equivariant map ψ from X to a rational H-module*

V such that the scheme-theoretic preimage $\psi^{-1}(0)$ is equal to S. Moreover, if $S = \varnothing$ or S is a singleton set then we can take ψ to be a closed embedding.

Proof Choose $f_1, \ldots, f_t \in k[X]$ such that the f_i generate the ideal I corresponding to S. Recall from the previous section that H acts on $k[X]$ by k-algebra automorphisms, and the action is locally finite. In particular, the subspace W spanned by the elements $\{h \cdot f_i \mid 1 \leq i \leq t, h \in H\}$ is finite-dimensional. Let $V = W^*$. Now $k[V]$ is the symmetric algebra $S(W^*)$ and the inclusion of W^* in $k[X]$ gives rise to a k-algebra homomorphism from $k[V]$ to $k[X]$. This corresponds to the map of varieties $\psi \colon X \to V$ given by $\psi(v) = \alpha_v$, where $\alpha_v(f) = f(v)$ for $f \in W$. For any $v \in V$ and $h \in H$ and for any $f \in W$,

$$\psi(h \cdot v)(f) = \alpha_{h \cdot v}(f) = f(h \cdot v) = (h^{-1} \cdot f)(v) = \alpha_v(h^{-1} \cdot f) = (h \cdot \alpha_v)(f),$$

so ψ is H-equivariant. The ideal of $\{0\}$ in $k[V]$ is generated by W^*, so the ideal of $\psi^{-1}(0)$ is generated by $\psi^*(W^*)$. But $\psi^*(W^*)$ generates I, so $\psi^{-1}(0) = S$, as required.

If $S = \{x\}$ for some H-fixed $x \in X$ then we can take the f_i to be a set of generators for $k[X]$, chosen so that $f_i(x) = 0$ for each i. If S is empty then we can take $f_1 = 1$ and $f_2, \ldots, f_t$ to be a set of generators for $k[X]$. In both cases the homomorphism from $k[V]$ to $k[X]$ is surjective, so ψ is a closed embedding. $\qquad\square$

Remark 3.2.2 The construction of Lemma 3.2.1 gives a homomorphism from H to $\mathrm{GL}(V)$. We can apply this idea to linearise the group H. We have an action of H on itself by left multiplication, so we get a closed H-equivariant embedding of H in an H-module V (taking $S = \varnothing$), which gives a homomorphism $\rho \colon H \to \mathrm{GL}(V)$. It is not hard to see that ρ is an embedding of H in $\mathrm{GL}(V)$.

Remark 3.2.3 The case of non-empty S in Lemma 3.2.1 is particularly important for the optimality arguments in Chap. 8, and for those it is also important that we can take S to be a subscheme of X (as opposed to just a subvariety); see for example the proof of Lemma 8.6.12. However, for most of the applications in this chapter, we simply need the map ψ to be a closed embedding, and then we can take S to be empty.

We now collect several basic properties of limits and their interactions with R-parabolic subgroups, all of which use the idea of linearising the action to facilitate the proofs. Recall the notation and discussion of Example 2.8.4. We introduce a piece of terminology: given an H-variety X, a point $x \in X$ and a cocharacter $\lambda \in Y_H$, we say λ *fixes* x or x is λ-*fixed* if $\lambda(a) \cdot x = x$ for all $a \in k^*$ (equivalently, if $\lambda \in Y_{H_x}$). Note that if V is an H-module, then the space $V_{\lambda,0}$ is precisely the subspace of λ-fixed points.

Lemma 3.2.4 *Let X be an H-variety. For any $\lambda \in Y_H$ and $x \in X$ such that $x' = \lim_{a \to 0} \lambda(a) \cdot x$ exists, x' is λ-fixed.*

Proof This is obvious for an H-module V, since the limit lies in $V_{\lambda,0}$, the space of λ-fixed points. But we can deduce the result in general from this case, by Lemma 3.2.1 (taking $S = \varnothing$). $\qquad\square$

Lemma 3.2.5 *Suppose X is an H-variety, $x \in X$ and $\lambda, \mu \in Y_H$ are commuting cocharacters such that $x_1 = \lim_{a\to 0} \lambda(a) \cdot x$ and $x_2 = \lim_{a\to 0} \mu(a) \cdot x$ both exist. Then $\lim_{a\to 0} \lambda(a) \cdot x_2$, $\lim_{a\to 0} \mu(a) \cdot x_1$ and $\lim_{a\to 0}(\lambda + \mu)(a) \cdot x$ all exist, and these limits are equal.*

Proof We may assume X is a closed H-stable subset of an H-module V by Lemma 3.2.1. Since x_1 exists, $x \in V_{\lambda,\geq 0}$, and taking the limit along λ is just the projection onto $V_{\lambda,0}$. Since λ and μ commute, $V_{\lambda,\geq 0}$ is $\mu(k^*)$-stable, and hence (because a subspace is closed) $x_2 \in V_{\lambda,\geq 0}$, so $\lim_{a\to 0} \lambda(a) \cdot x_2$ does indeed exist. Swapping the roles of λ and μ we see that $\lim_{a\to 0} \mu(a) \cdot x_1$ exists. Now we can see that the iterated limits are the same, since they arise by doing two projections one after the other, and those projection maps commute. In fact, this common value is the projection onto $V_{\lambda,0} \cap V_{\mu,0}$. Now $x \in V_{\lambda,\geq 0} \cap V_{\mu,\geq 0} \subseteq V_{\lambda+\mu,\geq 0}$, so $\lim_{a\to 0}(\lambda + \mu)(a) \cdot x$ exists and equals the projection of x onto $V_{\lambda+\mu,0}$. But $V_{\lambda+\mu,0} \cap (V_{\lambda,\geq 0} \cap V_{\mu,\geq 0}) = V_{\lambda,0} \cap V_{\mu,0}$, so this limit is equal to the other two. $\qquad\square$

We now collect some results valid when the group acting is reductive, so we recall that G always denotes a reductive group. The first is a basic result on the behaviour of weights.

Lemma 3.2.6 *Let G be reductive, V a rational G-module, T a maximal torus of G, and $\lambda \in Y_T$. Let $\chi \in \Phi_T(V)$, let $v \in V_\chi$ and let $g \in P_\lambda$ (resp., $g \in R_u(P_\lambda)$). Then for all $\chi' \in \mathrm{supp}_T(g \cdot v - v)$, we have $\langle \lambda, \chi' \rangle \geq \langle \lambda, \chi \rangle$ (resp., $\langle \lambda, \chi' \rangle > \langle \lambda, \chi \rangle$).*

Sketch Proof Recall the decomposition $P_\lambda = L_\lambda \ltimes R_u(P_\lambda)$. Acting by elements of L_λ does not change the λ-weight, so the key is to prove the result for $g \in R_u(P_\lambda)$. In particular, since $R_u(P_\lambda) \subseteq G^0$, we can reduce to the case when G is connected. Since $R_u(P_\lambda)$ is generated by the root groups it contains, the key is to consider a single root element, and in this case the result is standard; see [31, Lem. 5.2]. $\qquad\square$

For the next three results, let X be a G-variety, let $x \in X$, and let $\lambda \in Y_G$.

Lemma 3.2.7 *Suppose $x' = \lim_{a\to 0} \lambda(a) \cdot x$ exists. Let $u \in R_u(P_\lambda)$. Then*

$$\lim_{a\to 0} \lambda(a) \cdot (u \cdot x) = x' \quad \text{and} \quad \lim_{a\to 0}(u \cdot \lambda)(a) \cdot x = u \cdot x'.$$

In particular, both of these limits exist.

Proof We use Lemma 2.8.2 repeatedly. For any $a \in k^*$ we can write

$$\lambda(a) \cdot (u \cdot x) = (\lambda(a)u\lambda(a)^{-1}) \cdot (\lambda(a) \cdot x).$$

Since $\lim_{a\to 0} \lambda(a)u\lambda(a)^{-1} = 1$ and $x' = \lim_{a\to 0} \lambda(a) \cdot x$, the limit of the RHS exists as $a \to 0$ and that limit equals x'; hence the limit as $a \to 0$ on the LHS exists and equals x'. Now we may write

$$x' = \lim_{a\to 0} \lambda(a) \cdot x = \lim_{a\to 0} \lambda(a) \cdot (u^{-1} \cdot x) = \lim_{a\to 0} (\lambda(a)u^{-1}) \cdot x.$$

Acting on both sides with u gives the second result claimed. $\qquad\square$

Lemma 3.2.8 *Let $u \in R_u(P_\lambda)$. Then $\lim_{a\to 0} \lambda(a) \cdot x$ exists and equals $u \cdot x$ if and only if $u^{-1} \cdot \lambda$ fixes x.*

Proof By Lemma 3.2.4, if $\lim_{a\to 0} \lambda(a) \cdot x$ exists and equals $u \cdot x$, then λ fixes $u \cdot x$, and hence $u^{-1} \cdot \lambda$ fixes x. Conversely, if $u^{-1} \cdot \lambda$ fixes x, then $\lim_{a\to 0}(u^{-1} \cdot \lambda)(a) \cdot x = x$, and in particular this limit exists. Since for each $a \in k^*$, we have

$$(u^{-1} \cdot \lambda)(a) \cdot x = u^{-1} \cdot (\lambda(a)u\lambda(a)^{-1}) \cdot (\lambda(a) \cdot x),$$

and since $\lim_{a\to 0}(\lambda(a)u\lambda(a)^{-1}) = 1$ because $u \in R_u(P_\lambda)$, we can quickly deduce from Lemma 3.2.7 that $\lim_{a\to 0} \lambda(a) \cdot x$ exists and equals $u \cdot x$. $\qquad\square$

Lemma 3.2.9 *Suppose $x' := \lim_{a\to 0} \lambda(a)\cdot x$ exists, and suppose $z \in P_{-\lambda}$ and $u \in R_u(P_\lambda)$ are such that $zu \cdot x$ is λ-fixed. Then $x' = u \cdot x$.*

Proof We may assume X lies inside a G-module V by Lemma 3.2.1. Write $z = yl$, where $y \in R_u(P_{-\lambda})$ and $l \in L_\lambda$. Lemma 3.2.6 implies that $V_{\lambda,\leq 0}$ is $P_{-\lambda}$-stable, and the hypothesis that $zu \cdot x$ is λ-fixed implies that $ylu \cdot x \in V_{\lambda,0}$. Thus,

$$lu \cdot x = y^{-1}(ylu) \cdot x \in V_{\lambda,\leq 0}.$$

On the other hand, $lu \cdot x \in V_{\lambda,\geq 0}$, since $x \in V_{\lambda,\geq 0}$ and this subspace is P_λ-stable by Lemma 3.2.6 again. So $lu \cdot x \in V_{\lambda,\leq 0} \cap V_{\lambda,\geq 0} = V_{\lambda,0}$, and in particular we have $\lim_{a\to 0} \lambda(a) \cdot (lu \cdot x) = lu \cdot x$. But λ commutes with l and sends u to 1 in the limit, so we have $\lim_{a\to 0} \lambda(a) \cdot (lu \cdot x) = l \cdot x'$. Thus we conclude that $lu \cdot x = l \cdot x'$, and so $u \cdot x = x'$ as required. $\qquad\square$

3.3   Limits and Closed Orbits

We now give a criterion for orbits to be closed in terms of limits of cocharacters.

Definition 3.3.1 Let X be an H-variety and let $x \in X$.

(a) Let $O \subseteq X$ be an H-orbit. We say that O is *accessible from* x if there exists $\lambda \in Y_H$ such that $x' := \lim_{a \to 0} \lambda(a) \cdot x$ belongs to O. (For example, $H \cdot x$ is accessible from x: just take $\lambda = 0$.)
(b) We say that the orbit $H \cdot x$ is *cocharacter-closed* if $H \cdot x$ is the only H-orbit that is accessible from x.

Theorem 3.3.2 (Hilbert-Mumford Theorem) *Let X be a G-variety and let $x \in X$.*

(a) *Let O be the unique closed orbit contained in $\overline{G \cdot x}$. Then O is accessible from x.*
(b) *$G \cdot x$ is closed if and only if $G \cdot x$ is cocharacter-closed.*

Sketch Proof The proof of part (a) is difficult: see [89, Thm. 1.4]. For (b), if a G-orbit O is accessible from x then there exists $\lambda \in Y_G$ such that $x' := \lim_{a \to 0} \lambda(a) \cdot x$ belongs to O. Then $x' \in \overline{G \cdot x}$ by Exercise 2.6, so $O \subseteq \overline{G \cdot x}$. Hence if $G \cdot x$ is closed then $G \cdot x$ is cocharacter-closed. The converse follows from part (a). $\square$

Example 3.3.3 Let $X = G$ with the conjugation action and let $x \in X$. We claim that $1 \in \overline{G \cdot x}$ if and only if $x \in G^0$ and x is unipotent. By the Hilbert-Mumford Theorem, $1 \in \overline{G \cdot x}$ if and only if there exists $\lambda \in Y_G$ such that $\lim_{a \to 0} \lambda(a) \cdot x = 1$. But if $\lambda \in Y_G$ then $\lim_{a \to 0} \lambda(a) \cdot x = 1$ if and only if $x \in R_u(P_\lambda)$. Now $R_u(P_\lambda)$ is contained in G^0 and consists of unipotent elements. Conversely, any unipotent $x \in G^0$ belongs to $R_u(P_\lambda)$ for some $\lambda \in Y_G$: we can choose a Borel subgroup B of G^0 such that $x \in R_u(B)$, and take $\lambda \in Y_G$ such that $P_\lambda = N_G(B)$ (see Lemma 2.9.18). This proves the claim.

More generally, let $X = G^n$ with the simultaneous conjugation action of G. A similar argument shows that for any $(g_1, \dots, g_n) \in G^n$, $(1, \dots, 1)$ belongs to $\overline{G \cdot (g_1, \dots, g_n)}$ if and only if the g_i all belong to a common maximal unipotent subgroup of G^0.

Example 3.3.4 Suppose G is connected and let $g \in G$. Then the conjugacy class $G \cdot g$ is closed if and only if g is semisimple. This is well known (see, e.g., [79, Cor. 1.7]).

Remark 3.3.5 The authors gave an incorrect statement of Theorem 3.3.2 in [18, Thm. 2.3], where we claimed that *every* orbit O in $\overline{G \cdot x}$ is accessible from x. This is *not* always true, as is noted in [17, Ex. 11.1]. Examples like this also occur "in nature": suppose G is a simple group and X is the variety of unipotent elements of G. Then the orbit $G \cdot x$ of a regular unipotent element x is open and dense in X, so $\overline{G \cdot x} = X$. However, if X contains a distinguished unipotent element y which is not regular, then $G \cdot y$ is not accessible from $G \cdot x$. To see this last claim, note that if $y = \lim_{a \to 0} \lambda(a) \cdot x$, then y is centralised by the torus $\mathrm{Im}(\lambda)$ by Lemma 3.2.4. Since a distinguished unipotent element (by definition) is not centralised by any non-central torus of G, we deduce that the only way that y can be distinguished is if λ is central, and hence $y = x$.

There are non-regular distinguished elements in most simple groups away from type A (in type A, the only distinguished unipotent elements are the regular ones). For example, the orbits of unipotent elements in a symplectic group Sp_{2m} can be labelled by the partitions of $2m$ for which each odd part occurs with an even multiplicity, e.g., see [102, Thm. 3.1]. The distinguished orbits are those corresponding to partitions of $2m$ into distinct even parts, e.g., see [102, Prop. 3.5], and the regular orbit is the one corresponding to the partition $(2m)$, so there are non-regular distinguished unipotent elements in Sp_{2m} for any $m \geq 3$.

Theorem 3.3.6 *Let X be a G-variety and let $x \in X$. Let $\lambda \in Y_G$ and suppose $x' :=$ $\lim_{a \to 0} \lambda(a) \cdot x$ exists and is G-conjugate to x. Then x' is $R_u(P_\lambda)$-conjugate to x.*

Proof Note that $G \cdot x$ breaks up into a finite disjoint union of G^0-orbits, each of which is therefore open and closed in $G \cdot x$. Since each $\lambda(a) \cdot x$ in fact lies in $G^0 \cdot x$, the condition that $x' \in G \cdot x$ therefore implies that $x' \in G^0 \cdot x$. Thus we can reduce to the case that G is connected. We can also embed X G-equivariantly inside a G-module V, by Lemma 3.2.1.

We first aim to show that x' is P_λ-conjugate to x. Choose a maximal torus T such that $\lambda \in Y_T$, and let B be a Borel subgroup of G contained in P_λ and containing T. Let B^- be the opposite Borel with $B \cap B^- = T$. The set $P_\lambda R_u(P_{-\lambda})$ contains the so-called "big cell" BB^-, which is an open neighbourhood of 1 in G [32, Cor. 14.14]. The orbit map $\kappa_{x'} \colon G \to G \cdot x'$ is an open map by Lemma 2.7.8, so $P_\lambda R_u(P_{-\lambda}) \cdot x'$ contains an open neighbourhood of x' in $G \cdot x'$. Since $x' = \lim_{a \to 0} \lambda(a) \cdot x$ belongs to the closure of $\mathrm{Im}(\lambda) \cdot x$, there exists $b \in k^*$ such that $\lambda(b) \cdot x \in P_\lambda R_u(P_{-\lambda}) \cdot x'$: say, $\lambda(b) \cdot x = gu \cdot x'$ for some $g \in P_\lambda$ and some $u \in R_u(P_{-\lambda})$. This gives

$$h \cdot x = u \cdot x',$$

where $h := g^{-1} \lambda(b) \in P_\lambda$. Write $h = lu'$ for $l \in L_\lambda$ and $u' \in R_u(P_\lambda)$. We have $\lim_{a \to 0} \lambda(a) \cdot (u' \cdot x) = x'$ by Lemma 3.2.7. Since l commutes with λ, we therefore get

$$\lim_{a \to 0} \lambda(a) \cdot (u \cdot x') = \lim_{a \to 0} \lambda(a) \cdot (h \cdot x) = l \cdot x'.$$

In particular, the first limit exists, even though $\lim_{a \to 0} \lambda(a) u \lambda(a)^{-1}$ may not.

Now λ and $-\lambda$ fix x', by Lemma 3.2.4. Consequently, by Lemma 3.2.6 applied to $-\lambda$, we see that $\mathrm{supp}_T(u \cdot x' - x')$ consists of weights χ satisfying $\langle -\lambda, \chi \rangle > 0$. Hence $\langle \lambda, \chi \rangle = -\langle -\lambda, \chi \rangle < 0$ for all $\chi \in \mathrm{supp}_T(u \cdot x' - x')$. But $\lim_{a \to 0} \lambda(a) \cdot (u \cdot x')$ exists, so $\lim_{a \to 0} \lambda(a) \cdot (x' - u \cdot x')$ exists, so any weight in $\mathrm{supp}_T(u \cdot x' - x')$ must have non-negative pairing with λ. This forces $\mathrm{supp}_T(u \cdot x' - x')$ to be empty, which is to say that $u \cdot x' = x'$. Thus $x' = h \cdot x$ is P_λ-conjugate to x, as required.

To finish, we prove that $x' \in R_u(P_\lambda) \cdot x$. Recall that we have written $h = lu'$ with $l \in L_\lambda$ and $u' \in R_u(P_\lambda)$. Then $h \cdot x = x'$ gives $u' \cdot x = l^{-1} \cdot x'$. The RHS of this is

λ-fixed, since $l^{-1} \in L_\lambda$ and x' is λ-fixed, so if we take limits of both sides along λ we get $x' = l^{-1} \cdot x'$, where the limit for the LHS is computed using Lemma 3.2.7. So actually $x' = u' \cdot x$ and we are done. $\square$

We can now deduce the following important corollaries of Proposition 2.10.9 above.

Corollary 3.3.7 *Suppose X is a G-variety, $x \in X$, and S is a linearly reductive subgroup of G_x. If $G \cdot x$ is closed then $C_G(S) \cdot x$ is closed.*

Proof Note that $C_G(S)$ is reductive by Proposition 2.4.9. Let $\lambda \in Y_{C_G(S)}$ be such that $x' = \lim_{a \to 0} \lambda(a) \cdot x$ exists. Then since $G \cdot x$ is closed by hypothesis, x' is G-conjugate to x. By Theorem 3.3.6, x' is $R_u(P_\lambda)$-conjugate to x, and hence by Proposition 2.10.9, x' is $R_u(P_\lambda(C_G(S)))$-conjugate to x. In particular, then, $x' \in C_G(S) \cdot x$, so $C_G(S) \cdot x$ is closed, thanks to the Hilbert-Mumford Theorem 3.3.2. $\square$

Corollary 3.3.8 (Levi Descent for Closed Orbits) *Suppose X is a G-variety, $x \in X$, and $L = C_G(S)$ with S a torus of G_x. If $G \cdot x$ is closed then $L \cdot x$ is closed.*

Proof This follows immediately from Corollary 3.3.7. $\square$

3.4   Nullcones and Invariants

Definition 3.4.1 Let V be a G-module. We define the *nullcone* $\mathcal{N}(V)$ or $\mathcal{N}_G(V)$ of V to be $\pi_V^{-1}(\pi_V(0))$. (Here $\pi_V : V \to V /\!/ G$ is the canonical projection as in Sect. 3.1.)

It follows from the Hilbert-Mumford Theorem 3.3.2 that for any $v \in V$, v belongs to $\mathcal{N}(V)$ if and only if there exists $\lambda \in Y_G$ such that $\lim_{a \to 0} \lambda(a) \cdot v = 0$.

Example 3.4.2 Let $m, n \in \mathbb{N}$ and let $M_{m,n} := (\mathrm{Mat}_m)^n$. Then GL_m acts on $M_{m,n}$ by simultaneous conjugation. Let $(A_1, \ldots, A_n) \in M_{m,n}$. Then $(A_1, \ldots, A_n) \in \mathcal{N}(M_{m,n})$ if and only if $A_1, \ldots, A_n \in N_\lambda$ for some $\lambda \in Y_{\mathrm{GL}_m}$ (see Example 2.8.5). In this case the subalgebra of Mat_m generated by the A_i consists of nilpotent matrices. Conversely, if the subalgebra $\mathcal{A}$ of Mat_m generated by the A_i consists of nilpotent matrices then, by [78, Cor. 3.3], there exists $g \in \mathrm{GL}_m$ such that $g\mathcal{A}g^{-1}$ is contained in the subalgebra of strictly upper triangular matrices; then $\mathcal{A} \subseteq N_{g^{-1} \cdot \lambda}$, where $\lambda \in Y_{\mathrm{GL}_m}$ is given by $\lambda(a) = \mathrm{diag}(a^m, a^{m-1}, \ldots, a)$.

The coordinate ring $k[V]$ is graded by degree and the G-action preserves the grading. It follows that $k[V]^G$ is graded and a function $f \in k[G]$ is G-invariant if and only if each homogeneous part of f is G-invariant. Let $k[V]_+^G$ be the ideal of $k[V]^G$ generated by the homogeneous elements of positive degree. Then $\mathcal{N}(V)$ is the set of zeroes of $k[V]_+^G$: see Exercise 3.7.

The following result, which we use more than once, is [58, Lem. 2.5.5].

Proposition 3.4.3 *Let V be a G-module. Let $f_1, \ldots, f_t$ be homogeneous elements of $k[V]_+^G$ such that the nullcone $\mathcal{N}(V)$ is the set of zeroes of $f_1, \ldots, f_t$. Let $R = k[f_1, \ldots, f_t]$. Then $k[V]^G$ is a finitely generated R-module.*

3.5 Separable Orbits

Recall from Sect. 2.7 that for an H-variety X and $x \in X$ we let $\kappa_x : H \to H \cdot x$ denote the orbit map, and we define $\mathfrak{h}_x = \ker((d\kappa_x)_1)$.

Definition 3.5.1 Let H be an algebraic group and let X be an H-variety. Let $x \in X$. We say that the orbit $H \cdot x$ is *separable* if the restriction of κ_x to H^0—regarded as a map onto $H^0 \cdot x$—is separable. (This does not depend on the choice of the element x in the orbit by (2.7.2), so the definition makes sense. Recall also that $H^0 \cdot x$ is a quasi-affine variety, and $H^0 \cdot x$ is irreducible since H^0 is connected.)

We give a criterion for separability in terms of the stabilisers H_x and $\mathfrak{h}_x$.

Proposition 3.5.2 *Let X be an H-variety and let $x \in X$. Then the following are equivalent:*

(a) $\mathrm{Lie}(H_x) = \mathfrak{h}_x$.
(b) $\dim(H_x) = \dim(\mathfrak{h}_x)$.
(c) $H \cdot x$ *is separable.*

Proof The equivalence of (a) and (b) is clear since $\mathrm{Lie}(H_x) \subseteq \mathfrak{h}_x$ always (as noted in Sect. 2.7). Since H is smooth, all points of the orbit are smooth, so we can apply the derivative criterion for separability (Lemma 2.1.5) to κ_x to deduce the result, using the fact that $\dim(H \cdot x) = \dim(H) - \dim(H_x)$ by Lemma 2.7.6(i); note that if κ_x is separable then so is $\kappa_{x'}$ for all $x' \in H \cdot x$, by (2.7.2). $\qquad\square$

Remark 3.5.3 There is a scheme-theoretic formulation of orbit separability; we do not need it below but it is often mentioned in the literature. Given $x \in X$, we define $\mathcal{H}_x$ to be the scheme-theoretic stabiliser of x in H. Then $\mathcal{H}_x$ is a group scheme. If $\mathcal{H}_x$ is a variety— that is, if it is reduced as a scheme—then $\mathcal{H}_x$ is a subgroup of H in our usual sense, and $\mathcal{H}_x = H_x$. Otherwise H_x is a proper subgroup subscheme of $\mathcal{H}_x$.

The scheme-theoretic formulation of the separability criterion is as follows: $H \cdot x$ is separable if and only $\mathcal{H}_x$ is smooth. Note that $\mathcal{H}_x$ is smooth if and only if it is reduced

[127, Prop. 1.28], and so this criterion follows from Proposition 3.5.2 if we can show that $\mathrm{Lie}(\mathcal{H}_x) = \mathfrak{h}_x$. This in turn we can prove by an argument similar to that in [127, Prop. 10.31]: we may embed X H-equivariantly in some H-module V, by Lemma 3.2.1, and then we see that for $G = \mathrm{GL}(V)$ and $x \in V$ we have $\mathrm{Lie}(\mathcal{G}_x) = \{A \in \mathrm{End}(V) \mid Ax = 0\} = \mathfrak{g}_x$, where $\mathcal{G}_x$ is the scheme-theoretic stabiliser for x in $\mathrm{GL}(V)$. The equality $\mathrm{Lie}(\mathcal{H}_x) = \mathfrak{h}_x$ follows.

When $p = 0$, every orbit map is separable by the results in Sect. 2.1. One can see this from the scheme-theoretic characterisation above as well, because every group scheme of finite type over k is smooth [57, II, §6, 1.1].

3.6 Equivariant Maps and Inclusions

Let X and Y be G-varieties and let $\phi\colon Y \to X$ be a G-equivariant map. Then the map $Y \to X/\!/G$ given by $Y \xrightarrow{\phi} X \xrightarrow{\pi_X} X/\!/G$ has a unique factorisation $Y \xrightarrow{\pi_Y} Y/\!/G \to X/\!/G$ by the universal mapping property of quotients. We denote the map from $Y/\!/G$ to $X/\!/G$ by ϕ_G. Explicitly, ϕ_G is given by $\phi_G(\pi_Y(y)) = \pi_X(\phi(y))$. Clearly this construction is functorial in the sense that $(\mathrm{id}_X)_G = \mathrm{id}_{X/\!/G}$ and if Z is a G-variety and $\psi\colon Z \to Y$ is a G-equivariant map then $(\phi \circ \psi)_G = \phi_G \circ \psi_G$.

For example, suppose a group Γ acts on X by automorphisms. Suppose the Γ-action commutes with the G-action (that is, $g \cdot (\gamma \cdot x) = \gamma \cdot (g \cdot x)$ for all $x \in X$, $g \in G$ and $\gamma \in \Gamma$). It follows from the universal mapping property that the Γ-action descends to a Γ-action on $X/\!/G$, given by $\gamma \cdot \pi_X(x) = \pi_X(\gamma \cdot x)$ for $x \in X$ and $\gamma \in \Gamma$.

Now suppose Y is a closed G-stable subset of X and let $i\colon Y \to X$ be the inclusion.

Lemma 3.6.1 ([132, Lem. 3.4.1, Rem. 3.4.2]) *Let $f \in k[Y]^G$.*

(a) *If $p = 0$ then f extends to an element of $k[X]^G$.*
(b) *If $p > 0$ then f^{p^r} extends to an element of $k[X]^G$ for some $r \in \mathbb{N}_0$.*

The geometric meaning of Lemma 3.6.1 is that i_G is a closed embedding when $p = 0$, while for $p > 0$, i_G is finite and injective; in particular, $\mathrm{Im}(i_G)$ is closed.

Remark 3.6.2 Here are two situations in which i_G is a closed embedding. First, suppose that Y is open as well as closed: that is, Y is a union of connected components of X. If $f \in k[Y]^G$ then we can extend f to an element of $k[X]^G$ just by, say, defining f to be 0 on $X \setminus Y$. Hence i_G is a closed embedding. Second, suppose instead that there is a G-equivariant map $j\colon X \to Y$ such that $j \circ i = \mathrm{id}_Y$. Then $j_G \circ i_G = \mathrm{id}_{Y/\!/G}$ by functoriality, so again i_G is a closed embedding.

The next result is [13, Prop. 3.10].

Lemma 3.6.3 *Let* $\phi\colon X \to Y$ *be a finite G-equivariant map of G-varieties. Then* $\phi_G\colon X/\!/G \to Y/\!/G$ *is finite.*

3.7 Visible G-Varieties and Stable Points

Let X be a G-variety. We define

$$X_s = \{x \in X \mid G \cdot x \text{ is closed and has maximal dimension among the closed orbits}\}.$$

By Remark 2.7.5, there is at least one closed orbit, so X_s is non-empty.

Definition 3.7.1 We say that X is *G-visible* or just *visible* if X_s is dense in X.

Example 3.7.2 Suppose G is connected and let $X = G$ with the conjugation action of G. It follows from Example 3.3.4 that X_s is the set of regular semisimple elements. Hence X is visible.

Lemma 3.7.3 *Suppose X is visible. Then X_s is open and for any $x \in X_s$, $\pi_X^{-1}(\pi_X(x)) = G \cdot x$.*

Proof Let r be the maximum orbit dimension and let r' be the maximal dimension of a closed orbit. Since X_s is dense, Lemma 2.7.4(i) implies that $r = r'$. The result now follows from [132, Prop. 3.8]. $\qquad\qquad\square$

Proposition 3.7.4 ([13, Lem. 2.7]) *Let X be an irreducible visible G-variety. Then* $\pi_X\colon X \to X/\!/G$ *is separable.*

Remark 3.7.5 In [8, 2.1.9(b)] it is claimed that if X is an irreducible G-variety then $\pi_X\colon X \to X/\!/G$ is separable. This is false without the assumption of visibility: see [122, Thm. 1.8] for a counter-example.

Lemma 3.7.6 *Let X be a visible G-variety. Suppose G permutes the irreducible components of X transitively. Let m be the maximal orbit dimension and let n be the minimal stabiliser dimension. Then*

$$\dim(X/\!/G) = \dim(X) - m = \dim(X) - \dim(G) + n.$$

Proof Since G acts transitively on the components of X, they all have the same dimension, so it makes sense to talk about $\dim(X)$. Similarly, since G is permuting these components, the quotient $X/\!/G$ is irreducible: each G-orbit meets every component, and so $X/\!/G$ is the image of any one of the components under the map $\pi_X\colon X \to X/\!/G$. If $x \in X_s$

then $\pi_X^{-1}(\pi_X(x)) = G \cdot x$ has dimension m, by Lemma 3.7.3. Now X_s is dense in X by hypothesis, so $\dim(X /\!\!/ G) = \dim(X) - m$ by Lemma 2.1.1. The second equality follows because $m = \dim(G) - n$, by Lemma 2.7.6. $\square$

Definition 3.7.7 An orbit $G \cdot x$ is called *stable* if it is closed and G_x is a finite extension of the kernel of the action; we call x a *stable point*.

Lemma 3.7.8 *Suppose X is an irreducible G-variety and X admits a stable point. Then X is visible, and X_s is precisely the set of stable points.*

Proof This is clear, since a stable point has the smallest possible stabiliser dimension, and hence the largest possible orbit dimension. $\square$

Note that the set of stable points might be empty: just let $G = \mathbb{G}_m = k^*$ act on $X = k$ by multiplication. Then the kernel of the action is the trivial group, but the only closed orbit is the orbit of 0. Another example: if G is connected but not a torus and G acts on $X = G$ by conjugation as in Example 3.7.2 then X is visible, but there are no stable points. On the other hand, if G is a subgroup of an algebraic group M and G acts on M by $g \cdot m = mg^{-1}$ then every stabiliser is trivial and all the orbits are pure-dimensional of dimension $\dim(G)$, so every orbit is closed (Remark 2.7.5). Hence every point of M is stable.

Remark 3.7.9 In [132] Newstead defines the terms *semistable* and *unstable* for G-orbits, and these terms are often used elsewhere in the literature too (not always consistently). For example, if V is a G-module and $v \in V$ then v is often said to be *unstable* if $0 \in \overline{G \cdot v}$ and *semistable* otherwise, but in other places the term semistable is used to simply mean that a point has a closed orbit. We have chosen not to use these terms to avoid confusion, and because they do not sit well with the terminology used in Sect. 8.6. For a G-module V we will use the term 0-*unstable* for a point with 0 in the orbit closure and 0-*stable* for other points.

3.8   Richardson's Tangent Space Argument

The following fact (and variations) is known in the literature as Richardson's tangent space argument; see [143, §3, §4] for Richardson's original formulation. We give a proof since the ideas are useful later on, especially in our discussion of reductive pairs in Sect. 10.5.

Theorem 3.8.1 *Let K be an algebraic group acting on the algebraic variety X. Let M be a subgroup of K and let Y be a locally closed M-stable subset of X. Suppose*

$$T_y(K \cdot y) \cap T_y(Y) = T_y(M \cdot y) \tag{3.8.1}$$

for each $y \in Y$. Then for any K-orbit O in X which meets Y, the intersection $O \cap Y$ is a finite union of M-orbits. To be precise, each irreducible component of $O \cap Y$ is a single M^0-orbit. Moreover, each M-orbit in $O \cap Y$ is closed in $O \cap Y$.

Proof There is no harm in assuming that M is connected. Let $y \in O \cap Y$, so that $O = K \cdot y$. Because Y is M-stable, we have $M \cdot y \subseteq O \cap Y$, and consequently, $T_y(M \cdot y) \subseteq T_y(O \cap Y)$. Since $T_y(O \cap Y) \subseteq T_y(O) \cap T_y(Y)$, it follows from (3.8.1) that

$$T_y(M \cdot y) = T_y(O \cap Y). \tag{3.8.2}$$

Now let Z be an irreducible component of $O \cap Y$ containing $M \cdot y$. Then

$$\dim(M \cdot y) \leq \dim Z \leq \dim T_y(O \cap Y). \tag{3.8.3}$$

Since $M \cdot y$ is smooth, we have $\dim(M \cdot y) = \dim T_y(M \cdot y)$. We infer from (3.8.2) and (3.8.3) that $\dim(M \cdot y) = \dim Z$. So the closure of $M \cdot y$ inside $O \cap Y$ coincides with Z. As $M \cdot y$ is open in its closure, $M \cdot y$ is open in Z. By Remark 2.7.5, the complement of $M \cdot y$ in Z is a union of M-orbits, each of which is also open in Z, by the same argument. It follows that $M \cdot y$ is closed in Z as well. Consequently, $M \cdot y = Z$. It follows also that $M \cdot y$ is closed in $O \cap Y$. $\qquad \square$

The following is immediate from Theorem 3.8.1.

Corollary 3.8.2 *In the setting of Theorem 3.8.1, if K acts on X with only a finite number of orbits, then so does M on Y.*

3.9 Behaviour Under Algebraically Closed Field Extensions

We briefly record the behaviour of some of the constructions discussed thus far under algebraically closed field extensions. Let H be an algebraic k-group and let X be an H-variety. Let E/k be an algebraically closed field extension. Base change gives an action of H_E on X_E, so X_E becomes an H_E-variety. It is clear that for any $x \in X$ we have

$$(H_E)_x = (H_x)_E. \tag{3.9.1}$$

Hence the dimensions of $H \cdot x$ and $H_E \cdot x$ coincide by Lemma 2.7.6(i).

Lemma 3.9.1 *Let $x \in X$. Set $C = \overline{H \cdot x}$ and $D = \overline{H_E \cdot x}$. Then:*

(i) $C_E = D$.

(ii) $H \cdot x$ *is closed if and only if* $H_E \cdot x$ *is closed.*

Proof (i). Since C contains $\mathrm{Im}(\kappa_x)$, C_E contains $\mathrm{Im}((\kappa_x)_E) = (\mathrm{Im}(\kappa_x))_E = H_E \cdot x$. Hence C_E contains D, since C_E is closed. Conversely, $\mathrm{Im}(\kappa_x)$ is dense in C, so $\mathrm{Im}(\kappa_x)_E$ is dense in C_E; it follows that D contains C_E.

(ii). This follows from (i), as for any map $f \colon Y \to Z$ of k-varieties, f is surjective if and only if f_E is surjective.

$\square$

Now suppose X is a G-variety. The ring of invariants $E[X_E]^{G_E}$ has a canonical k-structure; using this, we may identify $X_E /\!\!/ G_E$ with $(X /\!\!/ G)_E$ and the canonical projection $\pi_{X_E} \colon X_E \to X_E /\!\!/ G_E$ with $(\pi_X)_E$. For details, see [8, §2.2]. It follows from Lemma 3.9.1(ii) and (3.9.1) that X_E is G_E-visible if and only if X is G-visible.

3.10 Historical Remarks and References

Much of the development of the theory of quotients and associated invariant rings in this chapter is based on the excellent set of notes by Newstead [132], which explore a great deal more material than we have included here. We have supplemented this with some further results from [8]. For the key result Proposition 3.1.5, see [132, Ch. 3, §3] (and [132, Thm. 3.5], in particular). More material on GIT may be found in the celebrated book of Mumford-Kirwan-Fogarty [130].

The Hilbert-Mumford Theorem 3.3.2 has a long history. We have cited Kempf's paper [89] for the proof because he states a version which is most suited to our purposes. Note, however, that Kempf's proof is based on that given in [130, Thm. 2.1]. As explained in [130, Ch. 2, §1] a version of the Hilbert-Mumford Theorem in the case that $G = \mathrm{SL}_m$ acts on projective space was indeed known to Hilbert in the 1890s [76], and Kempf also credits Birkes [30] and Raghunathan [139] for proving cases of this result over particular types of fields.

Kempf [89, Lem. 1.1] gives a proof of Lemma 3.2.1; see also [32, Prop. 1.10, Prop. 1.12]. The limit calculations in Sect. 3.2 and the important result Theorem 3.3.6 to which they contribute are all contained in the authors' paper [28] with R. Tange, where the notion of cocharacter-closed orbit was also introduced. The notion of accessibility is from [17].

The terminology "visible" seems to be standard: see [130, App. 4A] and [142], for example. Luna uses the terminology "de bonne dimension" [112].

3.11 Exercises

Exercise 3.1 Show that if a categorical quotient as in Definition 3.1.1 exists, then it is unique: that is, show that if (Y, ϕ) and (Y', ϕ') are two categorical quotients of an H-variety X, then there exists a unique isomorphism $\psi : Y \to Y'$ such that $\psi \circ \phi = \phi'$.

Exercise 3.2 Let X be a G-variety.

(i) Show that if C is a connected component of X then $\pi_X(C)$ is a connected component of $X /\!\!/ G$ and $\pi_X^{-1}(\pi_X(C)) = G \cdot C$.

(ii) Suppose X is visible. Show that the irreducible components of $X /\!\!/ G$ are precisely the sets of the form $\pi_X(C)$, where C is an irreducible component of X, and show that $\pi_X^{-1}(\pi_X(C)) = G \cdot C$.

(iii) Give an example to show that (ii) can fail without the assumption that X is visible.

Exercise 3.3 (i) Let G_1 and G_2 be reductive groups and let X_i be a G_i-variety for $i = 1, 2$. Recall from Sect. 2.7 that we have the product action of $G_1 \times G_2$ on $X_1 \times X_2$. Show that we can identify $(X_1 \times X_2) /\!\!/ (G_1 \times G_2)$ with $(X_1 /\!\!/ G_1) \times (X_2 /\!\!/ G_2)$ and $\pi_{X_1 \times X_2, G_1 \times G_2}$ with $\pi_{X_1, G_1} \times \pi_{X_2, G_2}$.

(ii) Let X be a G-variety and let Y be a variety. Define an action of G on $Y \times X$ by $g \cdot (y, x) = (y, g \cdot x)$. Show that $(Y \times X) /\!\!/ G \cong Y \times X /\!\!/ G$.

(iii) Suppose G is a subgroup of an algebraic group M. Show that the action of M on M by left multiplication gives rise to an action of M on M / G.

Exercise 3.4 Let V be a G-module, let T be a maximal torus of G, let $\lambda \in Y_T$ and let $\Phi_T(V)$ be the set of weights of V with respect to T (as in Example 2.8.4). Let $\chi \in \Phi_T(V)$ such that $\langle \lambda, \chi \rangle = 0$ and let $v \in V_\chi$. Use Lemma 3.2.7 to prove that if $u \in R_u(P_\lambda)$ then $u \cdot v - v \in V_{\lambda, >0}$. (This is a special case of Lemma 3.2.6.)

Exercise 3.5 Suppose X is a G-variety, and $x \in X$ is such that $G \cdot x$ is not closed. By the Hilbert-Mumford Theorem 3.3.2, there exists $\lambda \in Y_G$ such that $\lim_{a \to 0} \lambda(a) \cdot x$ exists and is not G-conjugate to x. Suppose H is a subgroup of G such that λ evaluates in H (i.e., $\lambda \in Y_H$). Show that $H \cdot x$ is not closed.

Exercise 3.6 Recall that a nilpotent $m \times m$ matrix is a matrix X such that $X^r = 0$ for some r. The set $\mathcal{N}$ of all nilpotent $m \times m$ matrices is a closed subvariety of the set of all $m \times m$ matrices, and $G = \mathrm{GL}_m$ acts on $\mathcal{N}$ by conjugation.

(i) Show that the G-orbits in $\mathcal{N}$ are parametrised by partitions of n.

(ii) Show that for every $X \in \mathcal{N}$ there exists $\lambda \in Y_G$ such that $\lim_{t \to 0} \lambda(t) \cdot X = 0$. Deduce that $\{0\}$ is the unique closed G-orbit in $\mathcal{N}$.

(iii) Let $m = 4$ and let X and Y denote the following matrices:

$$X = \begin{pmatrix} 0 & 1 & 0 & 0 \\ 0 & 0 & 1 & 0 \\ 0 & 0 & 0 & 0 \\ 0 & 0 & 0 & 0 \end{pmatrix}, \quad Y = \begin{pmatrix} 0 & 1 & 0 & 0 \\ 0 & 0 & 0 & 0 \\ 0 & 0 & 0 & 1 \\ 0 & 0 & 0 & 0 \end{pmatrix}.$$

Show that there exist $g \in G$ and a diagonal cocharacter λ such that

$$\lim_{t \to 0} \lambda(t) \cdot (gXg^{-1}) = Y.$$

(iv) Explain why it is impossible to find μ and h such that $\lim_{t \to 0} \mu(t) \cdot (hYh^{-1}) = X$.

(v) Suppose X is a nilpotent matrix with partition π_X and Y is a nilpotent matrix with partition π_Y. Show that there exists $\lambda \in Y_G$ with $\lim_{a \to 0} \lambda(a) \cdot X \in G \cdot Y$ if and only if $\pi_X \geq \pi_Y$ (here we use the *dominance order* on partitions: for two partitions of n we define $[p_1, \ldots, p_s] \geq [q_1, \ldots, q_s]$ if and only if $\sum_{i=1}^{r} p_i \geq \sum_{i=1}^{r} q_i$ for every $1 \leq r \leq s$).

Exercise 3.7 Let V be a G-module and let $k[V]_+^G$ denote the ideal of $k[V]$ generated by the homogeneous G-invariant elements of positive degree. Show that the nullcone $\mathscr{N}(V)$ is the vanishing set in V of $k[V]_+^G$.

G-Complete Reducibility: First Definitions and Properties

4

In this chapter we introduce the central idea of this book—Serre's notion of G-complete reducibility for subgroups of a reductive group G—and set up some of the questions which motivate us for the rest of the book. In particular, we give a first discussion of the ideas of *ascent* and *descent* of complete reducibility from and to a reductive subgroup M of G, Questions 4.1.8 and 4.1.9. We also collect some of the basic properties of completely reducible subgroups, including the important fact that a G-completely reducible subgroup of G is reductive (Lemma 4.1.6) and record some aspects of the behaviour of G-complete reducibility in the presence of an algebraically closed extension field.

4.1 Definitions and Motivating Questions

Here is the main definition of the book.

Definition 4.1.1 Suppose H is a subgroup of G. We say H is *G-completely reducible (G-cr)* if whenever $H \subseteq P$ for an R-parabolic subgroup P of G, there exists a R-Levi subgroup L of P with $H \subseteq L$. Further, H is *G-irreducible (G-ir)* if H is not contained in any proper R-parabolic subgroup of G, and H is *G-indecomposable (G-ind)* if H is not contained in any R-Levi subgroup of any proper R-parabolic subgroup of G.

Remark 4.1.2 Definition 4.1.1 makes sense for an arbitrary abstract subgroup of G. Since R-parabolic subgroups and R-Levi subgroups are closed, an abstract subgroup H of G is G-cr (resp., G-ir, resp., G-ind) if and only if $\overline{H}$ is.

Remark 4.1.3 It is immediate that a G-irreducible subgroup is G-completely reducible. The notion of G-indecomposability does not play a large role in what follows.

© The Author(s), under exclusive license to Springer Nature Switzerland AG 2026

M. Bate et al., *G-Complete Reducibility, Geometric Invariant Theory and Spherical Buildings*, Oberwolfach Seminars 57, https://doi.org/10.1007/978-3-032-08866-6_4

Example 4.1.4 The terminology in Definition 4.1.1 is chosen for the following reason: if $G = \mathrm{GL}(V)$ or $\mathrm{SL}(V)$ (or even $\mathrm{SO}(V)$ or $\mathrm{Sp}(V)$ when the characteristic is not 2), then H is G-completely reducible (resp., G-irreducible, G-indecomposable) if and only if the natural module V is completely reducible (irreducible, indecomposable) as an H-module. See Exercises 4.1 and 4.9–4.11.

We introduce the following terminology.

Definition 4.1.5 Let H be a subgroup of G. We call an R-parabolic subgroup P of G a *witness to the non-G-complete reducibility of H* if H is contained in P but H is not contained in any R-Levi subgroup of P.

By definition, if H is not G-cr then there is a witness to the non-G-complete reducibility of H, and vice versa.

Lemma 4.1.6 *If H is G-completely reducible, then H is reductive.*

Proof This is a straightforward application of Theorem 2.9.23. Suppose H is a non-reductive subgroup of G. Then applying Theorem 2.9.23 to $U = R_u(H) \subseteq G^0$ we obtain an R-parabolic subgroup P of G with $R_u(H) \subseteq R_u(P)$ and $H \subseteq P$. Since $R_u(H)$ is non-trivial, H intersects $R_u(P)$ non-trivially, and so we cannot have $H \subseteq L$ for any R-Levi subgroup L of P. Therefore P is a witness to the fact that H is not G-cr. $\square$

Remarks 4.1.7 (i). In characteristic zero the converse of Lemma 4.1.6 holds (this follows from Lemma 4.2.1 below). However, in positive characteristic the converse to Lemma 4.1.6 is not true. To see this, note that in positive characteristic every connected reductive group H which is not a torus has a representation $\rho : H \to \mathrm{GL}(V)$ which is not completely reducible. Then the image of ρ is a reductive but non-G-cr subgroup of $G = \mathrm{GL}(V)$. For a concrete example of this, take the adjoint representation of $H = \mathrm{SL}_2$ in characteristic 2.

(ii). If U is any unipotent subgroup of G^0, then Theorem 2.9.23 shows that we can find an R-parabolic subgroup P of G with $U \subseteq R_u(P)$, so U cannot be contained in any R-Levi subgroup of P. Thus U is *not* G-cr; we use this frequently below. Note that this doesn't hold without the stipulation that $U \subseteq G^0$, for if the characteristic is $p > 0$ we can let $G = M \times U$ be the direct product of a connected reductive group M and a cyclic group $U \cong C_p$ of order p. Then U is unipotent since it is a finite p-group, but U is also G-cr since it is centralised by all cocharacters of G.

We now set up some basic questions which serve as motivation for much of what follows. A key goal of the theory is to find results which allow us to pass G-complete reducibility down to subgroups or up from subgroups; we informally refer to such results as providing criteria for "descent" and "ascent" of complete reducibility. Specifically, suppose M is a reductive subgroup of G, and H is a subgroup of M. It is natural to pose the following questions:

Question 4.1.8 (Ascent of Complete Reducibility) If H is M-completely reducible, must H be G-completely reducible?

Question 4.1.9 (Descent of Complete Reducibility) If H is G-completely reducible, must H be M-completely reducible?

Clearly, the answer to Question 4.1.8 is "no" in general. For instance, if H is a reductive but non-G-cr subgroup of G and $M = H$ then H is M-cr but H is not G-cr; see Remark 4.1.7(i) for examples of such H in every positive characteristic. Question 4.1.9 is a bit more subtle, and we will see that examples exist only when the characteristic is small; see Chap. 10 for much more on this. Here is an important example due to Liebeck where both implications fail; we revisit this example several times in the book.

Example 4.1.10 Suppose $p = 2$. Let $m \geq 2$ be even. The symplectic group $M := \mathrm{Sp}_{2m}$ is a subgroup of $G := \mathrm{SL}_{2m}$, and $K := \mathrm{Sp}_m \times \mathrm{Sp}_m$ is a subgroup of Sp_{2m}. Let H be Sp_m diagonally embedded in $\mathrm{Sp}_m \times \mathrm{Sp}_m$. We have a chain of inclusions

$$H \leq K \leq M \leq G.$$

It can be shown that H is K-cr—this follows from Lemma 4.2.7—and G-cr, but not M-cr (see Exercise 4.10). So, Question 4.1.8 has a negative answer for the chain of inclusions $H \subseteq K \subseteq M$, and Question 4.1.9 has a negative answer for the chain of inclusions $H \subseteq M \subseteq G$.

For a more complicated example see Sect. 10.1 and Examples 10.4.5 and 10.5.11.

On the positive side, the answer to both questions is "yes" in some very natural situations, for example when M is an R-Levi subgroup of G (see Theorem 5.4.4). See also Example 4.1.4 above, where we noted that if $p > 2$ and M is either the special orthogonal group $\mathrm{SO}(V)$ or the symplectic group $\mathrm{Sp}(V)$ with its natural embedding in $G = \mathrm{GL}(V)$ then H is G-cr if and only if H is M-cr. Serre's Theorem 10.3.1(ii) below shows that if $\mathrm{char}(k)$ is sufficiently large and $G = \mathrm{GL}(V)$, then descent holds. Several of the results in Chaps. 9 and 10 provide criteria to ensure that ascent or descent holds: for instance, see Corollary 9.2.4, Theorems 10.4.2 and 10.5.21, and Corollary 10.5.24.

Another motivating theme of the development of the theory of G-complete reducibility over the past 30 years is the question of finding conditions for the converse to Lemma 4.1.6 to hold in positive characteristic. We shall see that the geometric approach to G-complete reducibility which we will develop in the next chapters helps to get a handle on this. For now, we note the following.

Remark 4.1.11 The main result in [99, Thm. 1] asserts that if $p > 7$ and H is any connected reductive subgroup of the simple exceptional group G, then H is G-cr; see Sects. 10.2 and 12.1 for a discussion of related results. In fact, the bounds established in *loc. cit.* are much more detailed and smaller depending on the types of G and H. Using this result, Liebeck and Seitz proceed in [99, Thm. 2] to show that the centraliser $C_G(H)$ for each such H is again reductive. Theorem 5.4.1 below gives a short proof of this fact for any reductive G and any G-cr subgroup H (connected or not), without any characteristic restrictions.

The counterparts of Question 4.1.9 for G-irreducibility and G-indecomposability are easily settled, as shown by the next result, which follows from Lemma 2.9.8.

Corollary 4.1.12 *Let $H \subseteq M$ be subgroups of G with M reductive.*

 (i) *If H is G-irreducible, then it is M-irreducible.*
(ii) *If H is G-indecomposable, then it is M-indecomposable.*

Remark 4.1.13 If H is not G-irreducible, then taking $M = H$ gives a trivial counterexample to the converse of Corollary 4.1.12(i). Example 10.3.4 below (taking $H = \mathrm{PGL}_p$, $M = \mathrm{SL}_{p^2-2}$, and $G = \mathrm{GL}(\mathfrak{m})$) shows that the converse of Corollary 4.1.12(i) can fail even when H is G-ir. Likewise, if H is not G-ind, taking $M = H$ again gives a trivial counterexample to the converse of Corollary 4.1.12(ii).

We finish this section with a basic result which is used a lot when addressing Questions 4.1.8 and 4.1.9.

Lemma 4.1.14 *Suppose H is a subgroup of G and $\lambda \in Y_G$ is such that P_λ is a witness to the fact that H is not G-completely reducible. Then for any reductive subgroup M of G with $H \subseteq M$ and $\lambda \in Y_M$, $P_\lambda(M)$ is a witness to the fact that H is not M-completely reducible.*

Proof We have $H \subseteq P_\lambda \cap M = P_\lambda(M)$. Every R-Levi subgroup of $P_\lambda(M)$ is of the form $uL_\lambda(M)u^{-1}$ for some $u \in R_u(P_\lambda(M)) = R_u(P_\lambda) \cap M$. Since P_λ is a witness to the fact that H is not G-cr, there is no $u \in R_u(P_\lambda)$ with $H \subseteq uL_\lambda u^{-1}$ and hence, in particular, there is no R-Levi subgroup of $P_\lambda(M)$ containing H. $\qquad\square$

4.2   Initial Results

In this section we collect together some initial results which can be proved with the minimal amount of technology available to us at this point. Many of these results will be generalised in later chapters.

For the first two results, recall that in positive characteristic a linearly reductive group is a finite extension of a torus by a group of order coprime to the characteristic, and such a group consists entirely of semisimple elements. Conversely, since a connected group consisting of semisimple elements is a torus, in *any* characteristic a group consisting only of semisimple elements is linearly reductive, and is a finite extension of a torus.

Lemma 4.2.1 *Let S be a linearly reductive subgroup of G. Then S is G-completely reducible.*

Proof Let P be an R-parabolic subgroup of G containing S, let L be an R-Levi subgroup of P and let $U = R_u(P)$. Let $\rho : S \to P$ be the embedding of S into P. Since S is linearly reductive and $c_L : P \to L$ has kernel U, $\sigma := c_L \circ \rho$ is an injective homomorphism of S into L. By Theorem 2.10.3, $H^1(S, U)$ is trivial, and thus, by Lemma 2.10.4(c), S is U-conjugate to $\sigma(S) \subseteq L$, so S is contained in a U-conjugate of L in P. Consequently, S is G-cr, as desired. $\qquad\square$

A straightforward application of Theorem 2.9.23 gives the following result.

Lemma 4.2.2 *If $H \subseteq G$ is G-irreducible, then:*

(i) $C_{G^0}(H)$ *is linearly reductive, hence G-completely reducible;*
(ii) $C_G(H)^0 = Z(G)^0$.

Proof (i). If K is any subgroup centralising a non-trivial unipotent element u of G^0, then K centralises the subgroup of G^0 generated by u. Hence K is contained in a proper R-parabolic subgroup of G by Theorem 2.9.23. We deduce that since H is G-irreducible, all elements of $C_{G^0}(H)$ must be semisimple, and hence $C_{G^0}(H)$ is linearly reductive.

(ii). Since $C_G(H)^0 \subseteq C_{G^0}(H)$, we see from (i) that $C_G(H)^0$ is a torus: call it S. Then $L = C_G(S)$ is an R-Levi subgroup of G containing H, so $L = G$, again by the irreducibility of H. Thus $S = Z(G)^0$, as required.

$\qquad\square$

Remarks 4.2.3   (i). It is not true in general that the full centraliser of a G-ir subgroup H is linearly reductive. For a trivial example, let $G = G^0 \times C_p$ where C_p is the cyclic group of order $p = \mathrm{char}(k)$ and G^0 is an adjoint simple group. Then $H = G$ is G-ir and $C_G(H) = C_p$ is not linearly reductive.

(ii). Even though $C_G(H)$ may not be linearly reductive, it is in fact true that $C_G(H)$ is G-cr, which strengthens the conclusion of part (i) of the lemma, but we do not yet have the technology to prove this. Indeed, below we show that if H is G-cr then $C_G(H)$ is G-cr: see Corollary 9.2.2.

(iii). Suppose $G = G^0$ is connected. Then Lemma 4.2.2 says that the centraliser of any G-irreducible subgroup of G is linearly reductive with identity component central in G. If, on the other hand, H is a subgroup of G such that $C_G(H)$ is G-ir, then $H \subseteq C_G(C_G(H))$, so we deduce that H is linearly reductive with central identity component.

Remark 4.2.4 Suppose G^0 is a torus. Then G^0 contains no unipotent elements so $R_u(P_\lambda) = 1$ for every $\lambda \in Y_G$. Hence $P_\lambda = L_\lambda$ for all $\lambda \in Y_G$. It follows that every subgroup of G is G-completely reducible.

Lemma 4.2.5 *Let H be a subgroup of G and let P be an R-parabolic subgroup of G such that $H \subseteq P$. Let L be an R-Levi subgroup of P.*

(a) *$c_L(H)$ is L-irreducible if and only if H is not contained in any R-parabolic subgroup Q of G with Q properly contained in P.*
(b) *$c_L(H)$ is L-completely reducible if and only if for every R-parabolic subgroup Q of G such that $H \subseteq Q \subseteq P$, there is an R-parabolic subgroup R of G such that $H \subseteq R \subseteq P$ and $R \cap L$ is opposite to $Q \cap L$.*

Proof Part (a) follows from Lemma 2.9.10(iv), and part (b) from Lemma 2.9.12. □

Another basic result is the following. Again, it is a special case of a much more general result we can only prove later on (it is the case $H = G$ of Theorem 9.1.1).

Lemma 4.2.6 *Suppose N is normal in G. Then N is G-completely reducible.*

Proof Since N^0 is a normal subgroup of G^0, it is reductive. Let $P = P_\lambda$ be an R-parabolic subgroup of G containing N. Then $\lambda(a)n\lambda(a)^{-1} \in N$ for all $n \in N$, so $c_\lambda(N) \subseteq N$ because N is closed. Now consider the normal unipotent subgroup $N \cap R_u(P)$ of N; we claim it is connected, and hence trivial since N is reductive. Let $u \in N \cap R_u(P)$. Then $\lambda(a)u\lambda(a)^{-1} \in N \cap R_u(P)$ for all $a \in k^*$, since N is normal in G, and $\lim_{a \to 0} \lambda(a)u\lambda(a)^{-1} = 1 \in N \cap R_u(P)$ also. Therefore there is a morphism $\mathbb{A}^1 \to N \cap R_u(P)$ whose image contains 1 and u, so u lies in the identity component of $N \cap R_u(P)$, as required. Since we've shown that $N \cap R_u(P)$ is trivial and $c_\lambda(N) \subseteq N$, so we have equality $N = c_\lambda(N)$. Since $c_\lambda(N) \subseteq L_\lambda$, we're done. □

The next result shows that the concept of complete reducibility behaves well with respect to certain homomorphisms. A homomorphism of algebraic groups $f : G_1 \to G_2$ is *non-degenerate* provided $(\ker f)^0$ is a torus. In particular, f is non-degenerate if f is an isogeny, and the corresponding special case of Lemma 4.2.7(ii)(b) is one we use frequently below.

Lemma 4.2.7 *Let G_1 and G_2 be reductive groups.*

(i) *Let H be a subgroup of $G_1 \times G_2$. Let $\pi_i : G_1 \times G_2 \to G_i$ be the canonical projection for $i = 1, 2$. Then H is $(G_1 \times G_2)$-cr $((G_1 \times G_2)$-ir, $(G_1 \times G_2)$-ind) if and only if $\pi_i(H)$ is G_i-cr $(G_i$-ir, G_i-ind) for $i = 1, 2$.*

(ii) *Let $f : G_1 \to G_2$ be an epimorphism. Let H_1 and H_2 be subgroups of G_1 and G_2, respectively.*

  (a) *If H_1 is G_1-cr $(G_1$-ir, G_1-ind), then $f(H_1)$ is G_2-cr $(G_2$-ir, G_2-ind).*

  (b) *If f is non-degenerate, then H_1 is G_1-cr $(G_1$-ir, G_1-ind) if and only if $f(H_1)$ is G_2-cr $(G_2$-ir, G_2-ind), and H_2 is G_2-cr $(G_2$-ir, G_2-ind) if and only if $f^{-1}(H_2)$ is G_1-cr $(G_1$-ir, G_1-ind).*

Proof   (i). Since every $\lambda \in Y_G$ has the form (λ_1, λ_2) with $\lambda_i \in Y_{G_i}$ for $i = 1, 2$, every R-parabolic subgroup of G has the form $P_1 \times P_2$ for R-parabolics P_1, P_2 of G_1, G_2, and similarly for R-Levi subgroups. Now the results follow easily.

(ii)(a). Let $N = \ker(f)$. Let M be the subgroup provided by Lemma 2.4.6 with $MN = G_1$. Suppose $\tau \in Y_{G_2}$. The restriction of f to M gives an isogeny from M to G_2. By Remark 2.9.22 there is $\sigma \in Y_M$ such that $P_\sigma(M) = f^{-1}(P_\tau) \cap M$ and $L_\sigma(M) = f^{-1}(L_\tau) \cap M$. Clearly P_σ contains N, so $P_\tau(G_2) = P_\sigma(G_1)/N$ and $L_\tau(G_2) = L_\sigma(G_1)/N$. Thus, for every R-parabolic subgroup Q of G_2 containing $f(H_1)$ there is an R-parabolic subgroup P of G_1 such that $H_1 \subseteq P$, N is contained in *every* R-Levi subgroup of P, and $Q = P/N$. Further, for each R-Levi subgroup K of Q there is an R-Levi subgroup L of P with $K = L/N$. Hence if H_1 is G-cr, so is $f(H_1)$. The other results follow in a similar fashion.

(ii)(b). Let $S = \ker(f)^0$, which is a torus by assumption. Then S is central in G^0, so is contained in every parabolic subgroup and Levi subgroup of G^0, so is contained in every R-parabolic and R-Levi subgroup of G. It follows easily that the R-parabolic subgroups of G/S are precisely the subgroups P/S where P is an R-parabolic subgroup of G, and the same for R-Levi subgroups. We see that H is G-cr if and only if HS/S is G/S-cr. Therefore we may factor out S and assume that f is an isogeny. We can now use Lemma 2.9.21 to deduce the result.

$\square$

Remark 4.2.8 This lemma allows us to reduce some questions about complete reducibility in connected reductive groups to questions about complete reducibility in simple groups. For suppose G is connected reductive, with central torus $Z = Z(G)^0$ and commuting simple factors $G_1, \ldots, G_r$. The multiplication map $Z \times G_1 \times \cdots \times G_r \to G$ is a surjective non-degenerate homomorphism, so Lemma 4.2.7(ii)(b) applies, and then we can project from the direct product onto each G_i using part (i) of Lemma 4.2.7. This is not a hugely important tool in this book, but it is a key idea in some of the wider theory; see Sect. 12.1 for example.

A particular case of Lemma 4.2.7(i) is when $G = G_1 \times G_2$ and H is a subgroup of one of the factors, say $H \subseteq G_1$. Then we see that H is G-completely reducible if and only if H is G_1-completely reducible. We can extend this result to cover the case when H lies in a normal subgroup which is not necessarily a direct factor of G.

Proposition 4.2.9 *Let $H \subseteq N \subseteq G$ be subgroups of G with N normal in G. Then H is G-completely reducible if and only if H is N-completely reducible.*

Proof Suppose H is G-cr. Let $\lambda \in Y_N$ such that $H \subseteq P_\lambda(N)$. As H is G-cr, there exists $u \in R_u(P_\lambda)$ such that $H \subseteq uL_\lambda u^{-1} = L_{u \cdot \lambda}$. Since N is normal in G, $u \cdot \lambda \in Y_N$, so $L_{u \cdot \lambda}(N)$ is an R-Levi subgroup of $P_\lambda(N)$ containing H. So H is N-cr.

Conversely, suppose H is N-cr. Let M be as in Lemma 2.4.6. Let $\lambda \in Y_G$ such that $H \subseteq P_\lambda$. The multiplication map $M^0 \times N^0 \to G^0$ is an isogeny, so we can write $n\lambda = \mu + \nu$ for some $n \in \mathbb{N}$ by (2.5.2), where $\mu \in Y_M$ and $\nu \in Y_N$. Since μ centralises N, μ centralises H, so $H \subseteq P_\nu \cap N = P_\nu(N)$. As H is N-cr, there exists $u \in R_u(P_\nu(N)) = R_u(P_\nu) \cap N = R_u(P_\lambda) \cap N$ such that $uHu^{-1} \subseteq L_\nu(N) = L_\lambda \cap N$. Thus H is G-cr, as required. $\qquad\square$

Remark 4.2.10 Note that the previous result applies in particular when $N = G^0$: so a subgroup H of G^0 is G-completely reducible if and only if it is G^0-completely reducible.

Corollary 4.2.11 *Suppose $g \in G^0$ and let H be the subgroup of G generated by g. Then H is G-completely reducible if and only if g is semisimple.*

Proof We may assume that $G = G^0$ by Remark 4.2.10. Let $g = g_s g_u$ be the Jordan decomposition of G, and note that g_s and g_u both lie in the subgroup H. If $g_u \neq 1$ then since g centralises g_u there is a parabolic subgroup P of G with $H \subseteq P$ and $g_u \in R_u(P)$, by Theorem 2.9.23. But then $g_u \in H$ is not contained in any Levi subgroup of P, so H is not G-cr. Conversely, if $g_u = 1$ so that $g = g_s$ is semisimple, then H is linearly reductive, and hence H is G-cr by Lemma 4.2.1. $\qquad\square$

For the following result note that a finite-index subgroup of a reductive group is reductive.

Proposition 4.2.12 *Let $H \subseteq M$ be subgroups of G, with M of finite index in G. Then H is M-completely reducible if and only if H is G-completely reducible.*

Proof We have $M^0 = G^0$, so $Y_M = Y_G$. Then for $\lambda \in Y_G$, $H \subseteq P_\lambda$ if and only if $H \subseteq P_\lambda(M)$ and the same for L_λ and $L_\lambda(M)$. The result follows. $\qquad\square$

Corollary 4.2.13 *Let N be a subgroup of G such that the number of G-conjugates of N is finite, and let H be a subgroup of N. Then H is G-completely reducible if and only if H is N-completely reducible.*

Proof The subgroup $G_1 := N_G(N)$ has finite index in G. Then G_1 is reductive, so its normal subgroup N is also reductive. The result now follows from Propositions 4.2.9 and 4.2.12. $\qquad\square$

Remark 4.2.14 Let H, M be subgroups of G with M of finite index in G. It need not be true that if H is G-completely reducible then $H \cap M$ is M-completely reducible. For example, let $p = 2$, let ϕ be an irreducible embedding of the symmetric group S_3 in SL_2, let $G = S_3 \times \mathrm{SL}_2$ and let $H = \{g\phi(g) \mid g \in S_3\}$. Let C_2 be a cyclic subgroup of S_3 of order 2 and let $M = C_2 \times \mathrm{SL}_2 \subseteq G$. It is easily checked using Lemma 4.2.7(i) that H is G-ir but $H \cap M$ is not M-cr.

4.3   Algebraically Closed Field Extensions

Recall from Sect. 2.6 that if E/k is an algebraically closed extension of k, and G is a reductive group over k, then we obtain by base change a reductive group G_E over E. In this situation we have the following.

Proposition 4.3.1 *Suppose H is a subgroup of G. Then H is G-completely reducible if and only if H_E is G_E-completely reducible.*

Proof Let P be an R-parabolic subgroup of G. Then the set $\{g \in G_E \mid H \subseteq gP_Eg^{-1}\} \subseteq G_E$ is k-defined (since P and H are), and so it contains a k-point if and only if it contains an E-point, by the Nullstellensatz. Similarly, for an R-Levi subgroup L (of some R-parabolic subgroup) of G the set $\{g \in G_E \mid H_E \subseteq gL_Eg^{-1}\} \subseteq G_E$ contains a k-point if and only if it contains an E-point. Since every R-parabolic subgroup of G_E is conjugate to some P_E, where P is an R-parabolic subgroup of G, and likewise for R-Levi subgroups, this is enough to derive the result. $\qquad\square$

4.4   Historical Remarks and References

Definition 4.1.1 is due to Serre, cf. [157], [158], and [159]. Most of the other material in this chapter is taken from [18] and [19], but some of the arguments have been shortened or otherwise adapted for this book.

There was a historically separate development by Richardson of some of the ideas in this chapter and the next. A subgroup H of G is said to be *strongly reductive in G* provided H is not contained in any proper parabolic subgroup of $C_G(S)$, where S is a maximal torus of $C_G(H)$ [147, Def. 16.1] (this does not depend on the choice of S because all such S are $C_G(H)$-conjugate). The authors showed in [18, Thm. 3.1] that a subgroup is strongly reductive if and only if it is G-cr—see Sect. 5.7—which tied these two strands of work together and provided the inspiration for much of the work collected in this book.

Examples of what Richardson proved include the following: if the subgroup H is topologically generated by $h_1, \ldots, h_n$, then H is strongly reductive in G if and only if the G-orbit of the n-tuple $(h_1, \ldots, h_n)$ is closed in G^n, [147, Thm. 16.4]; compare our Theorem 5.3.3 below.

In the special case $G = \mathrm{GL}(V)$, Corollary 4.1.12(i) is due to Slodowy [163, Lem. 11].

4.5   Exercises

Exercise 4.1 Let H be a subgroup of $\mathrm{GL}(V)$. To prove the claim in Example 4.1.4 that H is $\mathrm{GL}(V)$-cr if and only if H acts completely reducibly on V, we need to show the following: if $H \subseteq \mathrm{GL}(V)$ acts completely reducibly on V, then for every H-stable flag of subspaces in V there is an opposite flag which is H-stable. Write down the details.

Exercise 4.2 Let $G = \mathrm{GL}_m$ and let P be a maximal parabolic subgroup of standard type: so P is block upper triangular with blocks of size $r \times r$ and $(m - r) \times (m - r)$ on the diagonal. Let L be the corresponding block diagonal Levi subgroup with blocks of size $r \times r$ and $(m - r) \times (m - r)$ on the diagonal. Let $e_1, \ldots, e_m$ be the standard basis for k^m and let $W = \mathrm{span}(e_1, \ldots, e_r)$. Now let $H \leq P$ (so that H stabilises W).

(i) Prove that if H stabilises a subspace of k^m that is complementary to W then H is $R_u(P)$-conjugate to a subgroup of L.

   [*Hint*: Choose a change of basis matrix carefully.]

(ii) Find an example where H is G-conjugate to some subgroup of L but H is **not** $R_u(P)$-conjugate to any subgroup of L.

   [*Hint*: Try taking $p > 0$, $m = 3$, $r = 2$ and H a finite cyclic group.]

Exercise 4.3 Recall that an algebraic group acts linearly on its Lie algebra via the adjoint representation. Let $G = \mathrm{SL}_2$. Show that if $p \neq 2$ then this adjoint module is simple as a G-module, but if $p = 2$ then $\mathfrak{g}$ is indecomposable and not simple. Deduce that the corresponding image of G in GL_3 is GL_3-ir when $p \neq 2$ but is not GL_3-cr when $p = 2$.

Exercise 4.4 Prove that if $\mathrm{char}(k) = p$ then $\mathrm{SL}_2(k)$ contains infinitely many conjugacy classes of subgroups isomorphic to $C_p \times C_p$.

Exercise 4.5 Let H be an abstract group and suppose V is a finite-dimensional H-module. Suppose for every *simple* submodule U of V there exists a submodule W of V such that $V = U \oplus W$. Show that V is completely reducible (semisimple).

Exercise 4.6 Let J be the following $2m \times 2m$ anti-diagonal matrix:

$$J = \begin{pmatrix} & & & & & 1 \\ & & & & \cdot^{\cdot^{\cdot}} & \\ & & & 1 & & \\ & & -1 & & & \\ & \cdot^{\cdot^{\cdot}} & & & & \\ -1 & & & & & \end{pmatrix}.$$

Then $G := \{g \in \mathrm{GL}_{2m} \mid g^\top J g = J\}$ is the symplectic group.

(i) Show that with this set-up, G has a diagonal maximal torus.
(ii) For some small values of m, determine the parabolic and Levi subgroups of G arising from diagonal cocharacters.
(iii) Do the same for orthogonal groups in characteristic not 2 (you can again define the group as the stabiliser of a form given by an antidiagonal matrix).

Exercise 4.7 Suppose G is connected. For a subgroup H of G, let

$$\mathcal{M}(H) := \{P \mid H \subseteq P \text{ and } P \text{ is a maximal parabolic subgroup of } G\}.$$

(i) Show that if Q is any parabolic subgroup of G, then $Q = \bigcap_{P \in \mathcal{M}(Q)} P$.
(ii) For a subgroup H of G, show that H is G-cr if and only if $\bigcap_{P \in \mathcal{M}(H)} P$ is G-cr.

Exercise 4.8 Let H be a subgroup of the connected reductive group G. Consider the following condition:

($\star$) for every maximal parabolic P of G with $H \subseteq P$, there is a Levi subgroup L of P with $H \subseteq L$.

Show ($\star$) holds if and only if H is G-cr.

    [*Hint*: For the forward implication, suppose P is maximal containing H (if none exists, you're done!), and let L be a Levi of P containing H. Try to show that ($\star$) passes down to L.]

Note to reader: The next three exercises explain how to relate complete reducibility in symplectic groups to complete reducibility of the natural module. A similar sequence of exercises could be completed for orthogonal groups, with appropriate changes being made.

Exercise 4.9 (Complete Reducibility in Symplectic Groups I) Suppose that $G \subseteq \mathrm{GL}(V)$ is the symplectic group corresponding to an alternating form $(\cdot\,,\cdot) : V \times V \to k$. Show that if $H \subseteq G$ is G-cr, then H is also $\mathrm{GL}(V)$-cr.

[*Hint*: By Exercise 4.5, it is enough to show the following: given a simple H-submodule U of V, there is an H-submodule W of V with $V = U \oplus W$. By considering the restriction of the form to U, prove that this does indeed hold.]

Exercise 4.10 (Complete Reducibility in Symplectic Groups II) We fill in the details of Liebeck's Example 4.1.10. This shows the converse to the result in the previous exercise is not always true when $p = 2$.

Let $m \geq 4$ be even, and consider $\mathrm{Sp}(W)$ with W of dimension m. Let W' denote another copy of W, and let $V = W \oplus W'$ with the inherited alternating form. For concreteness, denote the form-preserving isomorphism $W \to W'$ by ϕ. This set-up gives a natural embedding

$$K := \mathrm{Sp}(W) \times \mathrm{Sp}(W') \subseteq \mathrm{Sp}(V) =: M,$$

with V the natural module for M. Let H denote the diagonal copy of the symplectic group of an m-dimensional space inside the direct product $\mathrm{Sp}(W) \times \mathrm{Sp}(W')$.

(i) Show that the only H-stable isotropic m-dimensional subspace of V is

$$U = \{v + \phi(v) \mid v \in W\}.$$

Deduce that H is not M-cr.

(ii) Show that H *is* $\mathrm{GL}(V)$-cr (and hence also $\mathrm{SL}(V)$-cr), as claimed in Example 4.1.10.

(iii) Choose $a \neq b \in k \setminus \{0\}$ and let $z \in \mathrm{End}(V)$ be the element which acts on W as multiplication by a and on W' as multiplication by b. Show that $z \in \mathrm{Lie}(M)$ and $K = M_z$.

[*Hint*: For parts (i) and (ii), it helps first to show that $V = W \oplus W'$ is a decomposition of V into simple H-submodules.]

Exercise 4.11 (Complete Reducibility in Symplectic Groups III) To complete the story for symplectic groups, suppose that $p \neq 2$ and $G \subseteq \mathrm{GL}(V)$ is the symplectic group corresponding to an alternating form $f : V \times V \to k$. We show that if $H \subseteq G$ is $\mathrm{GL}(V)$-cr then it is G-cr as well. Suppose $U \subseteq V$ is a simple H-submodule which is totally isotropic. We aim to show the existence of a second isotropic H-submodule U' such that $V = U \oplus (U')^{\perp}$.

(i) Let W be an H-module complement to $U^{\perp}$ in V (why does W exist?). Show that W is also simple, and that the restriction of the form to $X := U \oplus W$ is non-degenerate.

We can now work inside the symplectic space X: for if we can find an isotropic subspace U' of X with $X = U \oplus U'$, then using the fact that X is non-degenerate, as subspaces of V we have $(U')^\perp = U' \oplus X^\perp$, and so U' is the subspace we are after.

(ii) Suppose W is not totally isotropic. Show that the restriction of the form to W is non-degenerate, and deduce that $W^\perp \cong U$ and $U \oplus W = W^\perp \oplus W$ inside X.

(iii) In the situation of (ii), given $u \in U$ we have a unique expression $u = w' + w$ with $w' \in W^\perp$ and $w \in W$. Let

$$U' = \{w' - w \mid w' + w \in U, w' \in W^\perp, w \in W\}.$$

Show that U' is a totally isotropic H-submodule, and that $X = U \oplus U'$.

With the above in hand, it should now be possible to prove that H is G-cr, using the characterisation of parabolic subgroups and opposition from Example 2.4.2. Note that the construction in part (iii) depends in an essential way on the characteristic being different from 2.

The Geometric Approach

5

The central idea in our geometric approach is to parametrise subgroups H of G using generating tuples $\mathbf{h} \in H^n$, allowing the use of tools from geometric invariant theory. The point is that orbits of tuples in H^n give us a way to describe conjugacy classes of subgroups of H. This idea goes back to work of R.W. Richardson [147].

In this chapter we begin by considering subgroups which are topologically finitely generated (i.e., Zariski closures of abstract finitely generated subgroups). In order to deal with arbitrary closed subgroups, we then introduce the notion of a *generic tuple* in Sect. 5.2 and collect some key properties. The main result of the chapter is our characterisation of closed orbits in G^n in terms of G-complete reducibility (Theorem 5.3.3), which in turn allows us to derive some important structural results about completely reducible subgroups, notably Theorems 5.4.1 and 5.4.4. Also in this chapter we introduce the notion of the *semisimplification* of a subgroup in Sect. 5.5 and collect some results about the quotient variety $G^n /\!/ G$ via a study of the invariant ring $k[G^n]^G$ in Sect. 5.6.

5.1 The Subgroup Generated by a Tuple

Definition 5.1.1 Given $\mathbf{h} = (h_1, \ldots, h_n) \in G^n$, we define $\mathcal{G}(\mathbf{h})$ to be the closed subgroup topologically generated by the h_i; that is, we take the Zariski closure of the abstract subgroup $\langle h_1, \ldots, h_n \rangle$ generated by the h_i. More generally, if $S \subseteq G$ then we define $\mathcal{G}(S)$ to be the closed subgroup topologically generated by S.

It is clear that

$$\mathcal{G}(g \cdot \mathbf{h}) = g\mathcal{G}(\mathbf{h})g^{-1} \text{ for any } g \in G, \text{ any } n \in \mathbb{N} \text{ and any } \mathbf{h} \in G^n. \tag{5.1}$$

© The Author(s), under exclusive license to Springer Nature Switzerland AG 2026
M. Bate et al., *G-Complete Reducibility, Geometric Invariant Theory and Spherical Buildings*, Oberwolfach Seminars 57, https://doi.org/10.1007/978-3-032-08866-6_5

We need some information about stabilisers and orbits. In what follows, for a subset S of G we define $C_{\mathfrak{g}}(S) := \{X \in \mathfrak{g} \mid \mathrm{Ad}(g)(X) = X \text{ for all } g \in S\}$.

Proposition 5.1.2 *Let $n \in \mathbb{N}$ and let $\mathbf{h} = (h_1, \ldots, h_n) \in G^n$.*

(i) *We have $G_{\mathbf{h}} = C_G(\mathcal{G}(\mathbf{h}))$ and $\mathfrak{g}_{\mathbf{h}} = C_{\mathfrak{g}}(\mathcal{G}(\mathbf{h}))$.*
(ii) $\dim(G \cdot \mathbf{h}) = \dim(G) - \dim(C_G(\mathcal{G}(\mathbf{h}))) \geq \dim(G) - \dim(Z(G))$.

Proof For the statement about $G_{\mathbf{h}}$ in part (i), the key is to note that for an abstract subgroup M of G we have $C_G(\overline{M}) = C_G(M)$ (the inclusion $C_G(\overline{M}) \subseteq C_G(M)$ is completely obvious, and the reverse inclusion follows since conjugation by an element of G sends closed subgroups to closed subgroups). Thus, if $g \in G$ then $g \cdot \mathbf{h} = \mathbf{h}$ if and only if $g h_i g^{-1} = h_i$ for $1 \leq i \leq n$ if and only if $g h g^{-1} = h$ for all $h \in \mathcal{G}(\mathbf{h})$. Now part (ii) follows by Lemma 2.7.6.

For the statement about the infinitesimal stabiliser $\mathfrak{g}_{\mathbf{h}}$, note that the differential of the orbit map $G \to G \cdot \mathbf{h}$ in this case is the map $\mathfrak{g} \to T_{\mathbf{h}}(G^n)$ given by

$$X \mapsto ((dR_{h_1})_1(\mathrm{Ad}(h_1)(X) - X), \ldots, (dR_{h_1})_1(\mathrm{Ad}(h_n)(X) - X))$$

by Exercise 2.1, and hence $\mathfrak{g}_{\mathbf{h}} = C_{\mathfrak{g}}(\{h_1, \ldots, h_n\})$ in the notation set up before the statement. It is now easy to see that $\mathfrak{g}_{\mathbf{h}} = C_{\mathfrak{g}}(\mathcal{G}(\mathbf{h}))$. $\square$

We have the following basic observations.

Lemma 5.1.3 *Suppose $\mathbf{h} \in G^n$ and $\lambda \in Y_G$. Then $\mathbf{h}' = \lim_{a \to 0} \lambda(a) \cdot \mathbf{h}$ exists if and only if $\mathcal{G}(\mathbf{h}) \subseteq P_\lambda$. In this situation, $c_\lambda(\mathcal{G}(\mathbf{h})) = \mathcal{G}(\mathbf{h}')$, and $\mathcal{G}(\mathbf{h}) \subseteq L_\lambda$ if and only if $\mathbf{h} = \mathbf{h}'$.*

Proof The given limit exists if and only if it exists for each component of $\mathbf{h}$, by Lemma 2.8.2(iii). Now the results claimed are clear from the definitions. $\square$

Lemma 5.1.4 *Suppose $\mathbf{h} \in G^n$ has a closed G-orbit. Then for every $\mathbf{m} \in G^n$ with $\pi_G(\mathbf{m}) = \pi_G(\mathbf{h})$ we have $\dim(\mathcal{G}(\mathbf{m})) \geq \dim(\mathcal{G}(\mathbf{h}))$.*

Proof Since $\pi_G(\mathbf{m}) = \pi_G(\mathbf{h})$ and $G \cdot \mathbf{h}$ is closed, we see that $G \cdot \mathbf{h}$ is the unique closed orbit in $\overline{G \cdot \mathbf{m}}$ (Proposition 3.1.5(v)). Since $\mathcal{G}(g \cdot \mathbf{m}) = g\mathcal{G}(\mathbf{m})g^{-1}$ by (5.1), we are free to replace $\mathbf{m}$ with some conjugate. Hence, by the Hilbert-Mumford Theorem 3.3.2, we may assume that there exists $\lambda \in Y_G$ with $\lim_{a \to 0} \lambda(a) \cdot \mathbf{m} = \mathbf{h}$. But then $\mathcal{G}(\mathbf{h}) = c_\lambda(\mathcal{G}(\mathbf{m}))$, and the result follows from Remark 2.9.7. $\square$

5.2 Generating Tuples and Generic Tuples

We need to look more closely at subgroups and generating tuples.

Definition 5.2.1 Let H be an algebraic group and let $n \in \mathbb{N}$. We say that $\mathbf{h} \in H^n$ is a *generating n-tuple for H* (or just a *generating tuple for H*) or that $\mathbf{h}$ *generates H* if $H = \mathcal{G}(\mathbf{h})$. We say that H is *topologically finitely generated* if H admits a generating tuple. We define

$$H_{\mathrm{gen}}^n = \{\mathbf{h} \in H^n \mid \mathbf{h} \text{ generates } H\}.$$

Lemma 5.2.2 *Let H be a topologically finitely generated subgroup of G and let $\mathbf{h} \in H_{\mathrm{gen}}^n$. Then $G \cdot \mathbf{h} \cap H^n = N_G(H) \cdot \mathbf{h}$.*

Proof Clearly $N_G(H){\cdot}\mathbf{h} \subseteq H^n$. Conversely, if $g \in G$ and $g{\cdot}\mathbf{h} \in H^n$ then $H \supseteq \mathcal{G}(g \cdot \mathbf{h}) = g\mathcal{G}(\mathbf{h})g^{-1} = gHg^{-1}$. Hence $gHg^{-1} = H$. $\qquad\square$

We want to parametrise and describe subgroups of G using generating tuples. Unfortunately not every algebraic group H is topologically finitely generated in general. For instance, if $p > 0$ then a finitely generated additive subgroup of the field k itself is finite; one can deduce easily by looking at a composition series that a positive-dimensional unipotent group cannot be topologically finitely generated. (On the other hand, if $p = 0$ then every unipotent group H is connected and torsion-free, and we see again by looking at a composition series that H is topologically finitely generated.) Further, if $p > 0$ and $k = \overline{\mathbb{F}_p}$ then any topologically finitely generated algebraic group H is finite: for in this case the construction of Proposition 2.6.3 exhibits $H(\overline{\mathbb{F}_p}) = H$ as a union of finite groups. In particular, if $k = \overline{\mathbb{F}_p}$ then a positive-dimensional subgroup H of G cannot be topologically finitely generated.

There are several ways to avoid this problem: see Sect. 5.7 for a brief discussion. In this section we present a solution using the notion of a *generic tuple*.

Definition 5.2.3 Let H be a subgroup of G and let $G \hookrightarrow \mathrm{GL}_m$ be an embedding of algebraic groups. Then $\mathbf{h} \in H^n$ is called a *generic tuple of H for the embedding $G \hookrightarrow \mathrm{GL}_m$* if $\mathbf{h}$ generates the associative subalgebra of Mat_m generated by H. We call $\mathbf{h} \in H^n$ a *generic tuple of H* if it is a generic tuple of H for some embedding $G \hookrightarrow \mathrm{GL}_m$.

Remarks 5.2.4   (i). If H is a subgroup of G which is topologically finitely generated by a tuple $\mathbf{h} = (h_1, \ldots, h_n) \in H^n$, then $\mathbf{h}$ is a generic tuple of H in the sense of Definition 5.2.3. To see this, consider an embedding $G \hookrightarrow \mathrm{GL}_m$. Since each h_i is invertible, the minimal polynomial of each h_i has nonzero constant term, and so we can express h_i^{-1} as a polynomial in h_i. Hence, if A is the associative subalgebra of

Mat$_m$ generated by $\mathbf{h}$, then A contains the inverses of each of the h_i, so it contains the subgroup of GL$_m$ generated by $\mathbf{h}$. But A is closed, so this means it contains H.

(ii). If $G_1 \subseteq G$ is a reductive subgroup with $H \subseteq G_1$, then it is clear that a generic tuple of H for an embedding $G \hookrightarrow \mathrm{GL}_m$ is also a generic tuple of H for the restriction $G_1 \hookrightarrow \mathrm{GL}_m$.

(iii). Given an embedding $G \hookrightarrow \mathrm{GL}_m$, the associative subalgebra of Mat$_m$ generated by H has dimension at most m^2, so a generic tuple exists for any subgroup H of G if $n \geq m^2$. Recall that for $n \in \mathbb{N}$ we also use the notation c_λ to denote the map $P_\lambda^n \to L_\lambda^n$, see Remark 2.9.7.

Lemma 5.2.5 *Let H be a subgroup of G and let $\mathbf{h} \in H^n$ be a generic tuple of H for some embedding $G \hookrightarrow \mathrm{GL}_m$. Then:*

(i) *$M_{\mathbf{h}} = C_M(\mathcal{G}(\mathbf{h})) = C_M(H)$ and $\mathfrak{m}_{\mathbf{h}} = C_{\mathfrak{m}}(\mathcal{G}(\mathbf{h})) = C_{\mathfrak{m}}(H)$ for any subgroup M of G.*

(ii) *$\mathcal{G}(\mathbf{h})$ is contained in the same R-parabolic and the same R-Levi subgroups of G as H.*

(iii) *If $H \subseteq P_\lambda$ for some $\lambda \in Y_G$, then $c_\lambda(\mathbf{h})$ is a generic tuple of $c_\lambda(H)$ for the given embedding $G \hookrightarrow \mathrm{GL}_m$.*

Proof Denote by A the associative subalgebra Mat$_m$ generated by H, so that $\mathbf{h}$ generates A. Recall the notation introduced in Example 2.8.5: for $\lambda \in Y_{\mathrm{GL}_m}$ we have the corresponding subalgebras $\mathcal{P}_\lambda$ and $\mathcal{L}_\lambda$ and the projection $c_\lambda : \mathcal{P}_\lambda \to \mathcal{L}_\lambda$. If $\lambda \in Y_G$ then we see that $P_\lambda(G) = G \cap \mathcal{P}_\lambda$ and $L_\lambda(G) = G \cap \mathcal{L}_\lambda$.

(i). If a subset S of Mat$_m$ generates the associative subalgebra E of Mat$_m$, then $C_M(S) = M \cap C_{\mathrm{Mat}_m}(E)$ and $C_{\mathfrak{m}}(S) = \mathfrak{m} \cap C_{\mathrm{Mat}_m}(E)$. Since each entry of $\mathbf{h}$ lies in $\mathcal{G}(\mathbf{h}) \subseteq H$, it is clear that $\mathbf{h}$, $\mathcal{G}(\mathbf{h})$ and H all generate A. The claimed equalities follow.

(ii). If a subset S of G generates the associative subalgebra E of Mat$_m$, then $S \subseteq P_\lambda(G)$ if and only if $E \subseteq \mathcal{P}_\lambda$, and $S \subseteq L_\lambda(G)$ if and only if $E \subseteq \mathcal{L}_\lambda$. This implies the assertion.

(iii). Since $c_\lambda : \mathcal{P}_\lambda \to \mathcal{L}_\lambda$ is a homomorphism of associative algebras, $c_\lambda(\mathbf{h})$ generates the image $c_\lambda(A)$ and this is also the associative subalgebra of Mat$_m$ generated by $c_\lambda(H)$.

$\square$

Lemma 5.2.5(i) shows that generic tuples behave well when we take centralisers of H. Normalisers, however, are not as well-behaved.

Example 5.2.6 We give an example of G, H, $\mathbf{h}$ and $g \in G$ such that $\mathbf{h}$ and $g \cdot \mathbf{h}$ are both generic tuples for H, but g does not normalise H. First observe that since k is algebraically closed, if M is any subgroup of GL$_m$ then M is GL$_m$-ir if and only if k^m is simple as an M-module if and only if M generates the full algebra Mat$_m$ [93, Ch. XVII, Cor. 3.6]. Hence if $n \in \mathbb{N}$ and $\mathbf{m} \in (\mathrm{GL}_m)^n$ such that $\mathcal{G}(\mathbf{m})$ is GL_m-ir then $\mathbf{m}$ is a generic tuple for *every* subgroup M of GL_m such that $\mathbf{m} \in M^n$. Now assume $p > 0$. Let $m \geq 2$, let V be the natural module for SL_m and set $G = \mathrm{GL}(V \otimes V)$. Let $F_q \colon \mathrm{SL}_m \to \mathrm{SL}_m$ be the q-Frobenius map, where q is a power of p. Define $\rho_1 \colon \mathrm{SL}_m \to G$ by $\rho_1(h) = h \otimes F_q(h)$, and set $H = \mathrm{Im}(\rho_1)$. Then ρ_1 is an irreducible representation of SL_m by the Steinberg tensor product theorem [82, Cor. 3.17], so H is a G-ir subgroup of G.

Let $g \in G$ be the involution that swaps the tensor factors of $V \otimes V$. Then $gHg^{-1} = \rho_2(\mathrm{SL}_m)$, where $\rho_2 \colon \mathrm{SL}_m \to G$ is given by $\rho_2(h) = F_q(h) \otimes h$. We claim that $M := \rho_1(\mathrm{SL}_m(q^2))$ is contained in $H \cap gHg^{-1}$. To see this, note that if $h \in \mathrm{SL}_m(q^2)$ then $(F_q)^2(h) = h$, so

$$\rho_1(h) = h \otimes F_q(h) = F_q(F_q(h)) \otimes F_q(h) = \rho_2(F_q(h)) \in gHg^{-1},$$

as required. One can check that M is G-ir, see [172, Thm. 43]. Now let $\mathbf{h} \in M^n$ be a tuple consisting of a list of the elements of M. Then $g \cdot \mathbf{h} \in (gMg^{-1})^n = M^n \subseteq H^n$, so $\mathbf{h}$ and $g \cdot \mathbf{h}$ are both generic tuples for H by the discussion in the first paragraph; but clearly $gHg^{-1} \neq H$.

It is easy to check that $g \cdot \mathbf{h} \notin N_G(H) \cdot \mathbf{h}$, since any element of G normalising H must actually normalise the two tensor factors, so this example also shows that Lemma 5.2.2 fails if we replace a generating tuple with a generic tuple.

5.3 The Geometric Characterisation of G-Complete Reducibility

The following result applies some aspects of the general theory from Chap. 3 in the specific case that H is a subgroup of G and G acts on G^n by simultaneous conjugation.

Lemma 5.3.1 *Let H be a subgroup of G, and let $\mathbf{h} = (h_1, \ldots, h_n) \in H^n$ be a generic tuple for H. Let $\lambda \in Y_G$, and suppose $H \subseteq P_\lambda$. Set $\mathbf{h}' = c_\lambda(\mathbf{h})$. Then the following are equivalent:*

(a) $uHu^{-1} \subseteq L_\lambda$ *for some* $u \in R_u(P_\lambda)$;
(b) $\mathbf{h}' \in R_u(P_\lambda) \cdot \mathbf{h}$;
(c) $\mathbf{h}' \in G \cdot \mathbf{h}$.

Proof Suppose (a) holds. Then λ fixes $u \cdot \mathbf{h}$, and hence $\mathbf{h}' = u \cdot \mathbf{h}$ by Lemma 3.2.8, so (b) holds. Conversely, if $\mathbf{h}' = u \cdot \mathbf{h}$ for $u \in R_u(P_\lambda)$, then $u^{-1} \cdot \lambda$ fixes $\mathbf{h}$ by Lemma 3.2.8, and hence λ fixes $u \cdot \mathbf{h}$. But this means that $uHu^{-1} \subseteq L_\lambda$, so (a) holds. It is clear that (b) implies (c). The implication (c) $\Longrightarrow$ (b) follows from Theorem 3.3.6. $\qquad\square$

Remark 5.3.2 Note that Exercise 4.2(ii) shows that it is not necessarily true that if $H \subseteq P_\lambda$ and $gHg^{-1} \subseteq L_\lambda$ for some $g \in G$ then $c_\lambda(\mathbf{h}) \in G \cdot \mathbf{h}$. So we cannot replace $R_u(P_\lambda)$ with G in Lemma 5.3.1(a).

The next result is the main result of the chapter, providing the bridge between complete reducibility and geometric invariant theory and allowing us to study subgroups and their properties via G-orbits in suitable varieties.

Theorem 5.3.3 *Let H be a subgroup of G and let $\mathbf{h} \in H^n$ be a generic tuple for H. Then H is G-completely reducible if and only if $G \cdot \mathbf{h}$ is closed in G^n.*

Proof By Lemma 5.2.5 H is G-cr if and only if $\mathcal{G}(\mathbf{h})$ is, so we can assume without loss that $H = \mathcal{G}(\mathbf{h})$. Suppose H is G-cr and let $\lambda \in Y_G$ such that $\mathbf{h}' := \lim_{a \to 0} \lambda(a) \cdot \mathbf{h}$ exists, so that $H \subseteq P_\lambda$. Since H is G-cr, there is an R-Levi subgroup of P_λ containing H, which is the same as saying that there exists $u \in R_u(P_\lambda)$ such that $uHu^{-1} \subseteq L_\lambda$. Thus $\mathbf{h}' \in G \cdot \mathbf{h}$ by Lemma 5.3.1. Since this holds for all such choices of λ, we see that $G \cdot \mathbf{h}$ is closed by the Hilbert-Mumford Theorem 3.3.2.

Conversely, suppose $G \cdot \mathbf{h}$ is closed. Then for every $\lambda \in Y_G$ with $H \subseteq P_\lambda$ we have $uHu^{-1} \subseteq L_\lambda$ for some $u \in R_u(P_\lambda)$, by Lemma 5.3.1. But then $H \subseteq u^{-1}L_\lambda u$, an R-Levi subgroup of P_λ, so H is G-cr. $\qquad\square$

Corollary 5.3.4 *Suppose H is a G-completely reducible subgroup of G and let P_λ be an R-parabolic subgroup of G containing H. Then there exists $u \in R_u(P_\lambda)$ such that for all $h \in H$, $c_\lambda(h) = uhu^{-1}$. In particular, $uHu^{-1} = c_\lambda(H)$.*

Proof Let $\mathbf{h} \in H^n$ be a generic tuple for H. Since H is G-cr, $G \cdot \mathbf{h}$ is closed in G^n by Theorem 5.3.3, and hence property (c) of Lemma 5.3.1 holds. Thus, by Lemma 5.3.1(a), there is some $u \in R_u(P)$ with $uHu^{-1} \subseteq L_\lambda$. But then for any $h \in H$, we have $c_\lambda(h) = c_\lambda(uhu^{-1}) = uhu^{-1}$. $\qquad\square$

Remark 5.3.5 Suppose G is connected. It follows from Corollary 4.2.11 that for $g \in G$, $\mathcal{G}(g)$ is G-cr if and only if g is semisimple. But g is semisimple if and only if $G \cdot g$ is closed (Example 3.3.4), so we see that this is a particular instance of Theorem 5.3.3.

We can now prove a variation on Lemma 5.2.2 for G-completely reducible subgroups which are topologically finitely generated. Recall from Example 3.1.6 that for the G-variety G^n, we denote the canonical projection $\pi_{G^n} : G^n \to G^n /\!/ G$.

Lemma 5.3.6 *Let H be a topologically finitely generated and G-completely reducible subgroup of G. Suppose $\mathbf{h} \in H^n_{\mathrm{gen}}$ and let $\mathbf{h}_1 \in H^n$ be such that $\pi_{G^n}(\mathbf{h}_1) = \pi_{G^n}(\mathbf{h})$. Then $\mathbf{h}_1$ is $N_G(H)$-conjugate to $\mathbf{h}$.*

Proof Given such a tuple $\mathbf{h}_1 \in H^n$, there exists $\lambda \in Y_G$ such that $\mathbf{h}_2 := \lim_{a \to 0} \lambda(a) \cdot \mathbf{h}_1$ exists and $G \cdot \mathbf{h}_2$ is closed, by Theorem 3.3.2. Since H is G-cr, $G \cdot \mathbf{h}$ is closed, by Theorem 5.3.3, and since $\pi_{G^n}(\mathbf{h}) = \pi_{G^n}(\mathbf{h}_1) = \pi_{G^n}(\mathbf{h}_2)$, so $G \cdot \mathbf{h}_2 = G \cdot \mathbf{h}$, by Proposition 3.1.5(vi). Since $\mathbf{h}_2 = \lim_{a \to 0} \lambda(a) \cdot \mathbf{h}_1$, we have $\dim(G \cdot \mathbf{h}_1) \geq \dim(G \cdot \mathbf{h}_2) = \dim(G \cdot \mathbf{h})$ by Corollary 2.7.7, and hence $\dim(G_{\mathbf{h}_1}) \leq \dim(G_{\mathbf{h}_2}) = \dim(G_{\mathbf{h}})$. But $\mathbf{h}_1 \in H^n$, so $G_{\mathbf{h}_1} = C_G(\mathcal{G}(\mathbf{h}_1)) \supseteq C_G(H) = G_{\mathbf{h}}$, by Proposition 5.1.2. We conclude that $\dim(G \cdot \mathbf{h}_1) = \dim(G \cdot \mathbf{h})$, so $G \cdot \mathbf{h}_1 = G \cdot \mathbf{h}$ by Proposition 3.1.5(v) and Remark 2.7.5. Hence $\mathbf{h}_1 = g \cdot \mathbf{h}$ for some $g \in G$. Now $gHg^{-1} = g\mathcal{G}(\mathbf{h})g^{-1} = \mathcal{G}(\mathbf{h}_1) \subseteq H$, which shows that $g \in N_G(H)$, and we are done. $\qquad\square$

Remark 5.3.7 Note that Example 5.2.6 above shows that the Lemma 5.3.6 fails if we replace generating tuples with generic tuples.

5.4   **Centralisers and Normalisers**

Now we can begin to reap the fruits of our geometric approach. We use the geometric characterisation from Theorem 5.3.3 to prove some results on G-complete reducibility. We will unlock further results once we have developed the machinery of optimality in Chap. 8. In this section we take a first look at how G-complete reducibility interacts with the process of taking centralisers and normalisers.

Theorem 5.4.1 *Let H be a G-completely reducible subgroup of G. Then $C_G(H)$ and $N_G(H)$ are reductive.*

Proof Pick a generic tuple $\mathbf{h} \in H^n$ for H. Then $G \cdot \mathbf{h}$ is closed by Theorem 5.3.3, so $G_{\mathbf{h}} = C_G(H)$ is reductive by Proposition 3.1.15. Since H is G-cr, it is also reductive by Lemma 4.1.6. Now, by Proposition 2.4.8, $N_G(H)$ is a finite extension of $HC_G(H)$, so $N_G(H)$ is reductive. $\qquad\square$

Remark 5.4.2 Recall that a linearly reductive subgroup S of G is G-cr (Lemma 4.2.1), so $C_G(S)$ and $N_G(S)$ are reductive, by Theorem 5.4.1. Thus Theorem 5.4.1 generalises Proposition 2.4.9.

The centraliser of an arbitrary reductive subgroup of G need not be reductive. For instance, suppose G is connected and let U be a maximal unipotent subgroup of G. Let F be a non-trivial finite subgroup of $Z(U)$. Then $F \subseteq U \subseteq C_G(F)^0$, so $C_G(F)^0$ cannot be reductive as it has a non-trivial normal unipotent subgroup F. Hence $C_G(F)$ cannot be reductive. We now give an example with a connected reductive subgroup H.

Example 5.4.3 Let $\mathrm{char}(k) = 2$ and consider SL_2, with natural module $V = k^2$ spanned by the standard basis vectors e and f. The module $V \otimes V$ is indecomposable and has a unique filtration with simple quotients (it is *uniserial*):

- the simple socle S consists of a copy of the trivial module, spanned by $e \otimes f + f \otimes e$;
- the radical R is three-dimensional and indecomposable, spanned by the basis vector in the socle together with $e \otimes e$ and $f \otimes f$. The quotient R/S is isomorphic to the simple module $L(2) = V^{F_q}$ of highest weight 2 (here F_q denotes standard Frobenius);
- the quotient V/R is another copy of the trivial module.

With respect to the basis $\{e \otimes f + f \otimes e, e \otimes e, f \otimes f, e \otimes f\}$ one can check that the representation is given by the following map on matrices:

$$
\begin{pmatrix} a & b \\ c & d \end{pmatrix} \mapsto
\begin{pmatrix}
1 & ac & bd & bc \\
0 & a^2 & b^2 & ab \\
0 & c^2 & d^2 & cd \\
0 & 0 & 0 & 1
\end{pmatrix}.
$$

The image H is a connected reductive subgroup of $G = \mathrm{GL}_4$ and we claim that its centraliser is not reductive. To see this, note that there are no nonzero SL_2-module maps from R to S since such a map would have to have a kernel of dimension 2 and R has no submodules of dimension 2, so the restriction of any endomorphism of V to R must either map R isomorphically to itself or kill R. Any endomorphism of V killing R factors through the quotient V/R, and so corresponds to a choice of isomorphism $V/R \to S$; that is, a scalar since S is simple. Finally, the only endomorphisms of the composition factors themselves are scalars, since they are simple, so we can deduce that with respect to our chosen basis,

$$
\mathrm{End}_{\mathrm{SL}_2}(V \otimes V) =
\left\{
\begin{pmatrix}
x & 0 & 0 & z \\
0 & x & 0 & 0 \\
0 & 0 & x & 0 \\
0 & 0 & 0 & x
\end{pmatrix}
\;\middle|\; x, z \in k
\right\}.
$$

So the centraliser $C_G(H)$ is the set of invertible matrices in the endomorphism ring, and we see that it contains a 1-dimensional normal unipotent subgroup consisting of the matrices with $x = 1$.

This sort of phenomenon cannot arise in characteristic 0 since connected reductive groups have semisimple representation theory, but does arise in *any* positive characteristic whenever H is a connected reductive group and V is an indecomposable H-module with the sort of structure seen above: if V has simple socle S and radical R with $V/R \cong S$, then we can construct a unipotent group of endomorphisms of the form $1 + \phi$ where ϕ is the composition of the map $V \to V/R$ together with a choice of isomorphism of V/R with S (i.e., a scalar), and this group ends up being normal as long as the other composition factors of V are different from S. For example, there are always tilting modules of SL_2 with this structure: see [60, Ex. 2].

Now we give our first proof of one of the fundamental results in the theory, which gives an instance where both Questions 4.1.8 and 4.1.9 have positive answers. That we can "drop into a Levi subgroup" is an extremely important tool for proofs using induction on dimension (recall from Corollary 2.9.15 that the centraliser of any subtorus of G is an R-Levi subgroup).

Theorem 5.4.4 *Let L be an R-Levi subgroup of G and let H be a subgroup of L. Then H is G-completely reducible if and only if H is L-completely reducible.*

Proof Pick a generic tuple $\mathbf{h} \in H^n$ for H in G. Then $\mathbf{h}$ is also a generic tuple for H in L by Remark 5.2.4(ii). If H is G-cr, then $G \cdot \mathbf{h}$ is closed by Theorem 5.3.3. We may write $L = C_G(S)$ for some torus S in G and note that when we do this we have $S \subseteq C_G(H) = G_{\mathbf{h}}$. Then $L \cdot \mathbf{h}$ is closed by Corollary 3.3.8, so H is L-cr.

For the converse, we may actually assume that $L = C_G(S)$ is minimal amongst the R-Levi subgroups of G containing H, by the previous paragraph, and hence H is L-ir. Let Q be an R-parabolic subgroup of G such that L is an R-Levi subgroup of Q; then we claim that Q is minimal amongst the R-parabolic subgroups of G containing H. For any proper sub-R-parabolic of Q containing H has the form $Q'R_u(Q)$ where Q' is a proper R-parabolic subgroup of L containing H, by Lemma 2.9.10(iv), and there are no such subgroups since H is L-ir.

Now let P be any R-parabolic subgroup of G containing H. Again, we may assume that P is minimal amongst the R-parabolic subgroups containing H: by Lemma 2.9.10(v), if we can put H inside an R-Levi subgroup of some sub-R-parabolic of P, then H automatically lies in an R-Levi subgroup of P. Now minimality of P and Q, together with Corollary 2.9.13(ii), implies that P and Q contain a common R-Levi subgroup, M say. Write $P = P_\lambda$ and $Q = P_\mu$ for some $\lambda, \mu \in Y_M$; then automatically $L_\lambda = M = L_\mu$. Since λ and μ commute, so do the projection maps c_λ and c_μ (Lemma 3.2.5), and since $M = L_\lambda = L_\mu$ we see that

$$c_\lambda(\mathbf{h}) = c_\lambda(c_\mu(\mathbf{h})) = c_\mu(c_\lambda(\mathbf{h})) = c_\mu(\mathbf{h}).$$

But H is contained in the R-Levi subgroup L of Q, so $uHu^{-1} \subseteq M$ for some $u \in R_u(Q)$, so $c_\lambda(\mathbf{h})$ is G-conjugate to $\mathbf{h}$ by Lemma 5.3.1. Since $c_\lambda(\mathbf{h}) = c_\mu(\mathbf{h})$, this implies $vHv^{-1} \subseteq M$ for some $v \in R_u(P)$, by Lemma 5.3.1 again, and hence $H \subseteq v^{-1}Mv$, an R-Levi subgroup of P. So H is G-cr. $\qquad\square$

Remarks 5.4.5 (i). For a sketch proof of this result using the framework of buildings, see Remark 7.3.10(i).

(ii). Once we have developed the optimality results in the following chapters, we can give another proof of this result in a more general form: see Corollary 9.2.4.

The following gives various useful characterisations of complete reducibility. One consequence is that G-complete reducibility and strong reductivity are equivalent (see Sect. 4.4 for a discussion of the latter).

Corollary 5.4.6 *Let H be a closed subgroup of G. Then the following are equivalent:*

(i)   *H is $C_G(S)$-irreducible, where S is some maximal torus of $C_G(H)$;*

(ii)   *H is $C_G(S)$-irreducible for every maximal torus S of $C_G(H)$;*

(iii)   *H is G-completely reducible;*

(iv)   *for every R-parabolic subgroup P of G which is minimal with respect to containing H, the subgroup H is L-irreducible for some R-Levi subgroup L of P;*

(v)   *there exists an R-parabolic subgroup P of G which is minimal with respect to containing H, such that H is L-irreducible for some R-Levi subgroup L of P.*

Proof The equivalence of (i) and (ii) follows since all the maximal tori of $C_G(H)$ are conjugate by elements of $C_G(H)$. That (i) and (iii) are equivalent follows from Theorem 5.4.4. To see that (iii) implies (iv) note that if H is G-cr and P is an R-parabolic subgroup of G which is minimal with respect to containing H, then there does exist an R-Levi subgroup L of P containing H, and H must be L-ir by the minimality of P and Lemma 2.9.10(iv). That (iv) implies (v) is clear, and that (v) implies (iii) is a case of Theorem 5.4.4. □

Corollary 5.4.7 *Suppose H is any subgroup of G and P is an R-parabolic subgroup of G which is minimal with respect to containing H. Then for any R-Levi subgroup L of P, $c_L(H)$ is L-irreducible, and hence G-completely reducible.*

Proof Given any R-parabolic subgroup Q of L containing $c_L(H)$ we get an R-parabolic subgroup $QR_u(P) \subseteq P$ from Lemma 2.9.10(iv). Since $QR_u(P)$ contains the kernel of $c_L : P \to L$ and Q contains $c_L(H)$, we see that $QR_u(P)$ contains H. But then minimality of P forces $QR_u(P) = P$ and hence $Q = L$, so $c_L(H)$ is L-ir, and hence is G-cr by Corollary 5.4.6. □

Remark 5.4.8 The subgroup $c_L(H)$ in the previous result can be thought of as a *semisimplification* of the subgroup H. We return to this idea and flesh it out in Sect. 5.5.

Remark 5.4.9 If H is a subgroup of G then there can exist a P and an L such that $H \subseteq P$ and $c_L(H)$ is G-cr, but P is not minimal among the R-parabolic subgroups containing H.

For instance, take $H = U_\alpha U_{\alpha+\beta}$ inside $G := \mathrm{SL}_3$ and P to be P_β. We look at examples like this in Sect. 8.9.

We finish the section with a version of a result which has appeared in several variant forms in the literature. We state and prove what is perhaps the most general version; note that in part (ii) we get positive answers to both Questions 4.1.8 and 4.1.9 in the presence of an "auxiliary" G-cr subgroup K.

Theorem 5.4.10 *Suppose $K \subseteq H \subseteq M$ are subgroups of G, with M reductive.*

(i) *Suppose M contains a maximal torus of $C_G(K)$. If H is G-completely reducible then M is G-completely reducible.*
(ii) *Suppose $C_G(K)^0 \subseteq M$.*
    (a) *If H is G-completely reducible, then H is M-completely reducible.*
    (b) *If K is G-completely reducible, then H is G-completely reducible if and only if H is M-completely reducible.*

Proof   (i). Suppose P is an R-parabolic subgroup of G with $M \subseteq P$. Then H is also in P, and hence H lies in an R-Levi subgroup L of P, since H is G-cr. This puts K inside L as well, so $Z(L)^0$ is a subtorus of $C_G(K)$. Let S be a maximal torus of $C_G(K)$ such that $S \subseteq M$. Then S is a maximal torus of $C_P(K)$, so, after conjugating L by some element of $C_P(K)$, we can actually assume that $Z(L)^0 \subseteq S$. Now we can find $\lambda \in Y_S \subseteq Y_M$ with $P = P_\lambda$. But then $P_\lambda(M) = M$, so λ is central in M by Lemma 2.9.9, so $M \subseteq L_\lambda$ and we're done.

(ii)(a). Let $\lambda \in Y_M$ be such that $H \subseteq P_\lambda(M)$. Then $H \subseteq P_\lambda$ and H is G-cr, so we can find $\mu \in Y_G$ such that $P_\lambda = P_\mu$ and $H \subseteq L_\mu$. Then $\mu \in Y_{C_G(H)^0}$ and since $C_G(H)^0 \subseteq C_G(K)^0 \subseteq M$ we have $\mu \in Y_M$. So $H \subseteq L_\mu(M)$ and we see that H is M-cr.

(ii)(b). The forward implication follows from (a). For the other direction, suppose that P is an R-parabolic subgroup of G containing H; then $K \subseteq P$ also. We are given that K is G-cr, so we can find $\lambda \in Y_{C_G(K)}$ such that $P = P_\lambda$. But then $\lambda \in Y_M$ as well, by hypothesis, and so $P_\lambda(M)$ is an R-parabolic subgroup of M containing H. Since H is M-cr, there exists $u \in R_u(P_\lambda(M)) = R_u(P_\lambda) \cap M$ such that $H \subseteq uL_\lambda(M)u^{-1}$. Hence $H \subseteq uL_\lambda u^{-1}$, an R-Levi subgroup of P. This shows that H is G-cr.

$\square$

Remark 5.4.11   (i). Note that in part (i) of Theorem 5.4.10 we only require M to contain a maximal torus of $C_G(K)$, rather than all of $C_G(K)^0$. Part (ii) fails under this weaker hypothesis, however. For example, we can take K to be $\{1\}$ and H to be a subgroup of M, where M contains a maximal torus of G. There exist examples in which H is M-cr but not G-cr, and others in which H is G-cr but not M-cr. For the former, see Example 4.1.10 and Exercise 4.10: we take H to be Sp_m embedded diagonally in $M := \mathrm{Sp}_m \times \mathrm{Sp}_m$ inside $G := \mathrm{Sp}_{2m}$, where $m \geq 4$ is even. For the latter, see Sect. 10.1.

(ii). Putting $H = K$ in Theorem 5.4.10, we obtain the following result: if H is a G-cr subgroup of G, then any reductive subgroup M of G containing $HC_G(H)^0$ is also G-cr. In particular, since $C_G(H)$ and $N_G(H)$ are reductive by Theorem 5.4.1, if H is G-cr then so are $HC_G(H)^0$, $HC_G(H)$ and $N_G(H)$.

(iii). Later on we will see that $C_G(H)$ itself is G-cr, but the proof of this relies on the optimality results developed in Chap. 8; see Corollary 9.2.2.

We have the following easy corollary.

Corollary 5.4.12 *Let H be a reductive subgroup of G. Suppose H contains a maximal torus T of G. Then H is G-completely reducible.*

Proof This follows from Theorem 5.4.10(i), applied to the inclusion of subgroups $\{1\} \subseteq \{1\} \subseteq H$. For H contains a maximal torus of $C_G(\{1\}) = G$, and $\{1\}$ is G-cr. $\qquad\square$

Remark 5.4.13 Of course, Corollary 5.4.12 is false without the assumption that H is reductive: any R-parabolic subgroup P of G with $R_u(P) \neq \{1\}$ gives an example.

5.5 The Semisimplification of a Subgroup

Let H be a subgroup of G. Recall that if P is minimal among the R-parabolic subgroups containing H and if L is an R-Levi subgroup of P then $c_L(H)$ is an L-ir subgroup of L and a G-cr subgroup of G, by Corollary 5.4.7.

Definition 5.5.1 Let H and H' be subgroups of G. We say that H' is a *semisimplification of H* if H' is G-completely reducible and there exists an R-parabolic subgroup P of G containing H and an R-Levi subgroup L of P with $H' = c_L(H)$. We also say the pair (P, L) *yields* the semisimplification H' of H.

Example 5.5.2 Suppose $G = \mathrm{GL}(V)$. In this case, a parabolic subgroup P of G minimal subject to containing H corresponds to a flag $(U_1 \subset \cdots \subset U_r)$ of H-submodules V such that U_1, V/U_r and the quotients U_{i+1}/U_i are simple as H-modules (for, if any quotient were not simple, then the flag could be refined, so P would not be minimal). In other words, we get a composition series for V as an H-module. Thus, the subgroup $c_L(H)$ is the *semisimplification* of H in the usual sense of representation theory: it yields an action of H on a semisimple module which is the sum of the composition factors of V as an H-module. Further, the Jordan-Hölder Theorem implies that any two such semisimplifications are isomorphic, which translates into any two such subgroups $c_L(H)$ being conjugate in G. We generalise this below to arbitrary G, see Proposition 5.5.8. See also Exercise 12.6.

Remarks 5.5.3 (i). Let H be a subgroup of G. If H is G-cr then clearly H is a semisimplification of itself, yielded by the pair (G, G).

(ii). By the remark made just before the definition, any pair (P, L), where P is a minimal R-parabolic subgroup containing H and L is an R-Levi subgroup of P, yields a semisimplification of H.

(iii). Suppose (P, L) yields a semisimplification H' of H. Let L_1 be another R-Levi subgroup of P. Then $L_1 = uLu^{-1}$ for some $u \in R_u(P)$, so $c_{L_1}(H) = uc_L(H)u^{-1}$. Hence (P, L_1) also yields a semisimplification of H. Because of this, we can say that P *yields a semisimplification of* H.

(iv). It is easily checked that if ϕ is an automorphism of G, H is a subgroup of G and (P, L) yields a semisimplification H' of H then $\phi(H')$ is a semisimplification of $\phi(H)$, yielded by $(\phi(P), \phi(L))$.

(v). It follows from Remark 2.9.7 that if H is a subgroup of G and H' is a semisimplification of H, then $\dim(H') \leq \dim(H)$.

Remark 5.5.4 Here is a geometric description of semisimplification. Let H be a subgroup of G and pick a generic tuple $\mathbf{h} \in H^n$ for H. By the Hilbert-Mumford Theorem 3.3.2, there exists $\lambda \in Y_G$ such that $\mathbf{h}' := \lim_{a \to 0} \lambda(a) \cdot \mathbf{h} = c_\lambda(\mathbf{h})$ exists and $G \cdot \mathbf{h}'$ is closed. Then $\mathbf{h}'$ is a generic tuple of $H' := c_\lambda(H)$ by Lemma 5.2.5(iii) and hence H' is G-cr, by Theorem 5.3.3. Thus H' is a semisimplification of H yielded by the pair (P_λ, L_λ). Conversely, suppose (P_λ, L_λ) yields a semisimplification $H' = c_\lambda(H)$ of H. Then $\mathbf{h}' := \lim_{a \to 0} \lambda(a) \cdot \mathbf{h}$ is a generic tuple for H'. As H' is G-cr, $G \cdot \mathbf{h}'$ is closed. This idea is related to the notion of *degeneration* for modules in representation theory from [91]; see Sect. 5.7 for some more details.

The following is a variation on Theorem 3.3.6 involving two orbits, but note that we do need to assume that both of the orbits are closed.

Lemma 5.5.5 *Let X be a G-variety and let $x \in X$. Let $\lambda_1, \lambda_2 \in Y_G$ such that $x_1 := \lim_{a \to 0} \lambda_1(a) \cdot x$ and $x_2 := \lim_{a \to 0} \lambda_2(a) \cdot x$ exist. Suppose $G \cdot x_1$ and $G \cdot x_2$ are closed. Then there exist $u_1 \in R_u(P_{\lambda_1})$ and $u_2 \in R_u(P_{\lambda_2})$ such that:*

(i) $u_1 \cdot \lambda_1$ *and* $u_2 \cdot \lambda_2$ *belong to* Y_T *for some maximal torus* T *of* G;
(ii) $\lim_{a \to 0}(u_1 \cdot \lambda_1)(a) \cdot x = \lim_{a \to 0}(u_2 \cdot \lambda_2)(a) \cdot x$.
(iii) $u_1 \cdot x_1 = u_2 \cdot x_2$.

Proof We can pick a maximal torus S of $P_{\lambda_1} \cap P_{\lambda_2}$ and then choose $w_i \in R_u(P_{\lambda_i})$ such that $\mu_i := w_i \cdot \lambda_i$ belongs to Y_S for $i = 1, 2$ (Corollary 2.9.14). Set $y_i = \lim_{a \to 0} \mu_i(a) \cdot x_i$. Since $\lim_{a \to 0} \lambda_i(a) \cdot x_i = x_i$ we have $y_i = w_i \cdot x_i$ for $i = 1, 2$ by Lemma 3.2.7. By

Lemma 3.2.5 the limits $\lim_{a\to 0}\mu_1(a)\cdot y_2$ and $\lim_{a\to 0}\mu_2(a)\cdot y_1$ exist and are equal; call this common value y.

Since $G\cdot y_1$ is closed and μ_1 fixes y_1, $L_{\mu_1}\cdot y_1$ is closed by Corollary 3.3.8. Hence $y = v_2\cdot y_1$ for some $v_2 \in R_u(P_{\mu_2}(L_{\mu_1})) \subseteq R_u(P_{\mu_2}) = R_u(P_{\lambda_2})$ by Theorem 3.3.6. Set $\tau_2 = v_2^{-1}\cdot\mu_2$ and $\tau_1 = v_2^{-1}\cdot\mu_1 = \mu_1$; then τ_2 fixes y_1 and $\tau_1,\tau_2 \in Y_{v_2^{-1}Sv_2}$. Now let $z_1 = y_1$ and $z_2 = v_2^{-1}\cdot y_2$. Since $v_2^{-1}w_2$ belongs to $R_u(P_{\lambda_2})$, applying Lemma 3.2.7 gives

$$\lim_{a\to 0}\tau_2(a)\cdot x = \lim_{a\to 0}(v_2^{-1}w_2\cdot\lambda_2)(a)\cdot x = v_2^{-1}w_2\cdot x_2 = z_2.$$

Moreover,

$$\lim_{a\to 0}\tau_1(a)\cdot x = \lim_{a\to 0}\mu_1(a)\cdot x = w_1\cdot x_1 = y_1 = z_1$$

by Lemma 3.2.7.

To finish, we apply a similar argument swapping the roles of 1 and 2. We have $\lim_{a\to 0}\tau_1(a)\cdot z_2 = \lim_{a\to 0}\tau_2(a)\cdot z_1 = z_1$ by Lemma 3.2.5 and we can find $w_3 \in R_u(P_{\tau_1}(L_{\tau_2}))$ such that $z_1 = w_3\cdot z_2$. Set $\sigma_1 = w_3^{-1}\cdot\tau_1$ and $\sigma_2 = w_3^{-1}\cdot\tau_2 = \tau_2$. Then

$$\lim_{a\to 0}\sigma_1(a)\cdot x = w_3^{-1}\cdot z_1 = z_2 = \lim_{a\to 0}\sigma_2(a)\cdot x,$$

and $\sigma_1,\sigma_2 \in Y_T$, where $T = w_3^{-1}v_2^{-1}Sv_2w_3$. Taking $u_1 = w_3^{-1}w_1$ and $u_2 = v_2^{-1}w_2$, we see that $\sigma_i = u_i\cdot\lambda_i$ for $i = 1,2$, so we are done for parts (i) and (ii). Part (iii) now follows from Lemma 3.2.7. $\square$

Remark 5.5.6 Exercise 5.7 shows that the above result doesn't hold if we replace the hypothesis that $G\cdot x_1$ and $G\cdot x_2$ are closed with the weaker hypothesis that x_1 and x_2 are conjugate. In that example, G is not connected; we do not know whether the result with weaker hypotheses holds for connected G.

We now apply Lemma 5.5.5 to generic tuples specifically.

Lemma 5.5.7 *Let H be a subgroup of G and let $\mathbf{h} = (h_1,\ldots,h_n) \in H^n$ be a generic tuple for H. Let $\lambda_1,\lambda_2 \in Y_G$ such that $H \subseteq P_{\lambda_1}\cap P_{\lambda_2}$. Let $H_i = c_{\lambda_i}(H)$, let $\mathbf{h}_i := \lim_{a\to 0}\lambda_i(a)\cdot\mathbf{h}$ and suppose $G\cdot\mathbf{h}_i$ is closed for $i = 1,2$. Then:*

(i) *There exist $u_1 \in R_u(P_{\lambda_1})$ and $u_2 \in R_u(P_{\lambda_2})$ such that $u_1\cdot\lambda_1$ and $u_2\cdot\lambda_2$ belong to Y_T for some maximal torus T of G, $u_1\cdot\mathbf{h}_1 = u_2\cdot\mathbf{h}_2$ and $u_1 c_{\lambda_1}(h)u_1^{-1} = u_2 c_{\lambda_2}(h)u_2^{-1}$ for all $h \in H$. In particular, $u_1 H_1 u_1^{-1} = u_2 H_2 u_2^{-1}$.*
(ii) *For any $g \in G$ such that $g\cdot\mathbf{h}_1 = \mathbf{h}_2$ we have $gH_1 g^{-1} = H_2$.*

Proof (i). Choose $u_1 \in R_u(P_{\lambda_1})$, $u_2 \in R_u(P_{\lambda_2})$ and a maximal torus T of G as in Lemma 5.5.5; then $u_1 \cdot \mathbf{h}_1 = u_2 \cdot \mathbf{h}_2$. Set $\mu_i = u_i \cdot \lambda_i \in Y_T$ for $i = 1, 2$. We have $c_{\mu_1}(\mathbf{h}) = c_{\mu_2}(\mathbf{h})$ by Lemma 5.5.5, so $c_{\mu_1}(h_j) = c_{\mu_2}(h_j)$ for any $1 \leq j \leq n$, so $c_{\mu_1}(h) = c_{\mu_2}(h)$ for any $h \in \mathcal{G}(\mathbf{h})$. Hence $c_{\mu_i}(\mathcal{G}(\mathbf{h})) \subseteq L_{\mu_1} \cap L_{\mu_2}$ for $i = 1, 2$. But $c_{\mu_i}(\mathbf{h})$ is a generic tuple for $c_{\mu_i}(H)$ for $i = 1, 2$, so $c_{\mu_1}(H), c_{\mu_2}(H) \subseteq L_{\mu_1} \cap L_{\mu_2}$. It follows that for any $h \in H$,

$$u_1 c_{\lambda_1}(h) u_1^{-1} = c_{\mu_1}(h) = c_{\mu_2}(c_{\mu_1}(h)) = c_{\mu_1}(c_{\mu_2}(h)) = c_{\mu_2}(h) = u_2 c_{\lambda_2}(h) u_2^{-1},$$

where the middle equality follows from Lemma 3.2.5, and the first and last equalities from Lemma 3.2.7. This proves part (i).

For part (ii), observe that $u_2^{-1} u_1 \cdot \mathbf{h}_1 = \mathbf{h}_2$ by (i), so if $g \in G$ and $g \cdot \mathbf{h}_1 = \mathbf{h}_2$ then $g = u_2^{-1} u_1 z$ for some $z \in G_{\mathbf{h}_1}$. But $\mathbf{h}_1$ is a generic tuple for H_1 by Lemma 5.2.5(iii), so $G_{\mathbf{h}_1} = C_G(H_1)$ by Lemma 5.2.5(i). Hence $g H_1 g^{-1} = u_2^{-1} u_1 H_1 u_1^{-1} u_2 = H_2$. $\square$

We can now prove the main result of this section, that any two semisimplifications of a subgroup H are in fact G-conjugate. As we remarked above, this is analogous to the Jordan-Hölder Theorem applied to representations in the case that $G = \mathrm{GL}(V)$.

Proposition 5.5.8 *Let H be a subgroup of G and let H_1, H_2 be semisimplifications of H. Then H_1 and H_2 are conjugate.*

Proof Let $\lambda_1, \lambda_2 \in Y_G$ be such that $H_1 = c_{\lambda_1}(H)$ and $H_2 = c_{\lambda_2}(H)$ are semisimplifications of H yielded by R-parabolic subgroups P_{λ_1}, P_{λ_2}, respectively. Pick a generic tuple $\mathbf{h}$ for H. Then $\mathbf{h}_1 := c_{\lambda_1}(\mathbf{h})$ and $\mathbf{h}_2 := c_{\lambda_2}(\mathbf{h})$ are generic tuples for H_1 and H_2, respectively (Lemma 5.2.5(iii) again). Since H_1 and H_2 are G-cr, $G \cdot \mathbf{h}_1$ and $G \cdot \mathbf{h}_2$ are closed by Theorem 5.3.3. The result now follows from Lemma 5.5.7. $\square$

Remark 5.5.9 It is false that if H' is a semisimplification of H then any conjugate of H' is a semisimplification of H. For instance, suppose H is G-ir. Clearly H is a semisimplification of itself, yielded by the pair (G, G). But H is the unique semisimplification of itself, because H is not contained in any proper R-parabolic subgroup of G.

Open Problem 5.5.10 Characterise which conjugates of a given semisimplification H' of H are also semisimplifications of H.

We will have more to say about semisimplification later. It is an ingredient in the optimality constructions of Chap. 8, and once those tools are available we can discuss how semisimplification interacts with normal subgroups, for example; see Sect. 9.4. We can also discuss semisimplification over non-algebraically closed fields; see Sect. 12.6. For now, we finish with a definition and a further remark.

Definition 5.5.11 We define $\mathcal{D}(H)$ to be the set of G-conjugates of any semisimplification of H. This is well-defined by Proposition 5.5.8.

Remark 5.5.12 Let H be a subgroup of G and let P be an R-parabolic subgroup of G with R-Levi subgroup L such that $H \subseteq P$. If $c_L(H)$ is G-cr, then $c_L(H)$ is reductive by Lemma 4.1.6, so we must have $R_u(H) \subseteq R_u(P)$. The converse is true in characteristic 0, by Remark 4.1.7(i): so in characteristic 0, $c_L(H)$ is a semisimplification of H if and only if $R_u(H) \subseteq R_u(P)$.

The previous observation is clearly not true in any positive characteristic, for we can take any situation where $H \subseteq G$ and H is reductive but not G-cr. Then G is an R-parabolic subgroup of itself, and $\{1\} = R_u(H) = R_u(G)$, but $c_G(H) = H$ is not a semisimplification of H.

5.6 The Ring of Invariants of G^n

In preparation for the next chapter, we finish this chapter with some further results on the G-variety G^n, the quotient $G^n /\!/ G$, and its coordinate ring $k[G^n]^G$. As usual, the action here is simultaneous conjugation. In the case $n = 1$, we note that the notation $G /\!/ G$ allows us to distinguish this quotient from the (trivial) quotient group G/G.

Let $n \in \mathbb{N}$ and let w be a word in the free group $\Gamma_n = \langle \gamma_1, \ldots, \gamma_n \rangle$. We obtain a map $\mathrm{ev}_w \colon G^n \to G$ as follows: if $w = \gamma_{i_1}^{\epsilon_1} \cdots \gamma_{i_s}^{\epsilon_s}$ with each $\epsilon_j = \pm 1$ then for each $(g_1, \ldots, g_n) \in G^n$ we define $\mathrm{ev}_w(g_1, \ldots, g_n) = g_{i_1}^{\epsilon_1} \cdots g_{i_s}^{\epsilon_s}$. For brevity we will sometimes write $w(g_1, \ldots, g_n)$ for $\mathrm{ev}_w(g_1, \ldots, g_n)$. Let $f \in k[G]^G$ and define $f_w \colon G^n \to k$ by

$$f_w(g_1, \ldots, g_n) = f(w(g_1, \ldots, g_n)).$$

It is clear that f_w belongs to $k[G^n]^G$.

Definition 5.6.1 We define $\mathcal{T}_n(G)$ to be the subalgebra of $k[G^n]^G$ generated by the functions of the form f_w.

It is natural to ask how large the subalgebra $\mathcal{T}_n(G)$ is inside $k[G^n]^G$. We give a result for general G in Chap. 6 (Theorem 6.9.1). For now we consider what happens when $G = \mathrm{GL}_m$ or $G = \mathrm{SL}_m$. We can give some kind of complete answer in the former case using two deep results: one due to Procesi and Sibirskiĭ in characteristic 0 and one due to Donkin in positive characteristic. Recall from Example 3.4.2 that we let $M_{m,n} = (\mathrm{Mat}_m)^n$ and we let GL_m act by simultaneous conjugation on $M_{m,n}$. We need an analogue of the functions f_w, and the construction above carries over as long as we avoid taking inverses. Let us call a word w in the γ_i *positive* if it involves only positive powers of the γ_i. Given a positive word w, we obtain a map $\mathrm{ev}_w \colon M_{m,n} \to \mathrm{Mat}_m$ as before. Now given $f \in k[\mathrm{Mat}_m]^{\mathrm{GL}_m}$ we define $f_w \in k[M_{m,n}]^{\mathrm{GL}_m}$ by

$$f_w(A_1, \ldots, A_n) = f(\mathrm{ev}_w(A_1, \ldots, A_n)).$$

We define $\mathcal{T}_n(\mathrm{Mat}_m)$ to be the subalgebra of $k[M_{m,n}]^{\mathrm{GL}_m}$ generated by the functions of the form f_w.

We start with characteristic 0, for which we cite [138, Thm. 1.3, Thm. 3.4].

Theorem 5.6.2 *Suppose $p = 0$. Then $k[M_{m,n}]^{\mathrm{GL}_m}$ is generated by the functions of the form tr_w, where w ranges over the positive words of length at most 2^m and tr is the usual trace function on matrices. In particular, $k[M_{m,n}]^{\mathrm{GL}_m} = \mathcal{T}_n(\mathrm{Mat}_m)$ and $\mathcal{T}_n(\mathrm{Mat}_m)$ is a finitely generated k-algebra.*

We can immediately deduce

Theorem 5.6.3 *Suppose $p = 0$, let $G = \mathrm{GL}_m$ and $H = \mathrm{SL}_m$.*

(i) *$k[H^n]^H$ is generated by the functions of the form tr_w, where w ranges over the positive words of length at most 2^m. In particular, $k[H^n]^H = \mathcal{T}_n(H)$ and $\mathcal{T}_n(H)$ is a finitely generated k-algebra.*

(ii) *$k[G^n]^G$ is generated by the functions of the form tr_w, where w ranges over the positive words of length at most 2^m and over the functions $\det_{\gamma_i^{-1}}$ for $1 \leq i \leq n$. In particular, $k[G^n]^G = \mathcal{T}_n(G)$ and $\mathcal{T}_n(G)$ is a finitely generated k-algebra.*

Proof (i). The G-action on $M_{m,n}$ restricts to give an H-action, and it is clear that $k[M_{m,n}]^H = k[M_{m,n}]^G$. The inclusion $i \colon H^n \to M_{m,n}$ is H-equivariant. Since $p = 0$, the map $i_H \colon H^n /\!\!/ H \to M_{m,n} /\!\!/ H$ is a closed embedding by Lemma 3.6.1. Hence any element of $k[H^n]^H$ extends to an element of $k[H^n]^H = k[M_{m,n}]^G$. Now Theorem 5.6.2 tells us that $k[M_{m,n}]^G$ is generated by the functions of the form tr_w, where w ranges over the positive words of length at most 2^m. But the restriction of tr_w (regarded as a function on $M_{m,n}$) to H^n is precisely tr_w (regarded as a function on H^n). The result now follows.

(ii). We may regard G^n as the open subset of $M_{m,n}$ given by the condition $\det_{\gamma_1} \cdots \det_{\gamma_n} \neq 0$. Hence elements of $k[G^n]$ can be written in the form $f \det_{\gamma_1}^{-t_1} \cdots \det_{\gamma_n}^{-t_n}$ for some $f \in k[M_{m,n}]$ and some $t_1, \ldots, t_n \in \mathbb{N}_0$. Such a function is G-invariant if and only if f is G-invariant, so elements of $k[G^n]^G$ can be written in the form $f \det_{\gamma_1}^{-t_1} \cdots \det_{\gamma_n}^{-t_n}$ for some $f \in k[M_{m,n}]^G$ and some $t_1, \ldots, t_n \in \mathbb{N}_0$. The result follows from Theorem 5.6.2.

$\square$

Now consider the positive characteristic case, for which we cite [59, Thm. 1].

Theorem 5.6.4 *Suppose $p > 0$. Then $k[M_{m,n}]^{\mathrm{GL}_m} = \mathcal{T}_n(\mathrm{Mat}_m)$.*

We deduce

Theorem 5.6.5 *Suppose $p > 0$, and let $G = \mathrm{GL}_m$. Then $k[G^n]^G = \mathcal{T}_n(G)$.*

Proof By an argument similar to the proof of Theorem 5.6.3(ii), $k[G^n]^G$ is generated by $k[M_{m,n}]^G$ together with the functions $\det_{\gamma_i}^{-1}$ for $1 \le i \le n$. The restriction of any element of $\mathcal{T}_n(\mathrm{Mat}_m)$ to G^n is an element of $\mathcal{T}_n(G)$. The result now follows from Theorem 5.6.4. $\qquad\square$

Remark 5.6.6 Let $H = \mathrm{SL}_m$. Recall the map $i_H \colon H^n /\!\!/ H \to M_{m,n} /\!\!/ H$ from the proof of Theorem 5.6.3(i). In positive characteristic we know that i_H is finite and injective (Lemma 3.6.1), and it follows that $k[H^n]^H$ is a finite extension of $\mathcal{T}_n(H)$. But we do not know an explicit description of $k[H^n]^H$ in this case.

The definition above of $\mathcal{T}_n(G)$ involves $k[G]^G$, and likewise the definition of $\mathcal{T}_n(\mathrm{Mat}_m)$ involves $k[\mathrm{Mat}_m]^{\mathrm{GL}_m}$. It is therefore instructive to consider the special case $n = 1$. Note that $\mathcal{T}_1(G) = k[G]^G$ and $\mathcal{T}_1(\mathrm{Mat}_m) = k[\mathrm{Mat}_m]^{\mathrm{GL}_m}$ by definition. Suppose G is connected reductive, let T be a maximal torus of G, and note that the action of $N_G(T)$ on T gives rise to an action of the Weyl group $W = N_G(T)/T$ on T. The inclusion of T in G is $N_G(T)$-equivariant, where $N_G(T)$ acts on G by conjugation. By the universal mapping property, we get a map $\psi \colon T /\!\!/ W \to G /\!\!/ G$, given by $\psi(\pi_W(t)) = \pi_G(t)$.

Lemma 5.6.7 *Let G, T, W and ψ be as above. Then ψ is an isomorphism.*

Proof This result is well known, but we give the proof as it illustrates the machinery that we have developed. Recall from Remark 5.3.5 that for any $g \in G$, the conjugacy class $G \cdot g$ is closed if and only if g is semisimple. So T meets every closed conjugacy class in G, and it follows that ψ is surjective since points in $G /\!\!/ G$ correspond to closed G-orbits. To prove injectivity, let $t_1, t_2 \in T$ such that $\pi_G(t_1) = \pi_G(t_2)$. Then $g \cdot t_1 = t_2$ for some $g \in G$. Since T is a maximal torus of both $C_G(t_1)$ and $C_G(t_2)$, we can choose g so that g normalises T. This shows that g_1 and g_2 are W-conjugate, so $\pi_W(t_1) = \pi_W(t_2)$.

Now we show that ψ is separable. Define $\beta \colon G \times T \to G$ by $\beta(g, t) = gtg^{-1}$. Now $\pi_G(T) = G /\!\!/ G$ from the argument above, and π_G is separable by Example 3.7.2 and Proposition 3.7.4. It follows from Lemma 2.1.5 that there is a non-empty open subset U of G such that $\pi_G(U)$ is smooth and for all $g \in U$, g is regular semisimple and $(d\pi_G)_g$ is surjective. Since π_G is G-invariant, we can assume without loss that U is stable under conjugation.

Pick any $t \in T \cap U$. Let $X \in \mathfrak{g}$ and let $Y \in \mathfrak{t}$. Then, by Exercise 2.1, we have $d\beta_{(1,t)}(X, 0) = (dR_t)_1(X - \mathrm{Ad}(t)(X))$ and $d\beta_{(1,t)}(0, (dR_t)_1(Y)) = (dR_t)_1(Y)$. Since t is regular semisimple, the subspace $\{X - \mathrm{Ad}(t)(X) \mid X \in \mathfrak{g}\}$ is the sum of the root spaces of $\mathfrak{g}$, so $d\beta_{(1,t)}$ is surjective. It follows from the definition of U that $d(\pi_G \circ \beta)_{(1,t)}$ is surjective. But $\pi_G \circ \beta$ is constant on $\{(g, t) \mid g \in G\}$, so $d(\pi_G \circ \beta)_{(1,t)}(X, 0) = 0$ for all $X \in \mathfrak{g}$, so $d(\pi_G \circ \beta)_{(1,t)}(0 \oplus (dR_t)_1(\mathfrak{t})) = T_{\pi_G(t)}(G /\!/ G)$. Untangling the definitions, we see that $d((\pi_G)|_T)_t$ is surjective. Hence

$$d\psi_{\pi_W(t)} \text{ is surjective.} \tag{$*$}$$

Now $T \cap U$ is a non-empty open subset of T. By Lemma 2.1.5 we can choose $t \in T \cap U$ such that $\pi_W(t)$ is smooth. It follows from (*) and Lemma 2.1.5 that ψ is separable.

We have shown that ψ is a separable bijective map of irreducible varieties. Since G is irreducible, normal and connected, $G /\!/ G$ is irreducible and normal by Proposition 3.1.9(i). It follows from Zariski's Main Theorem that ψ is an isomorphism. $\qquad\square$

There is an analogous result for Mat_m with the GL_m-action by conjugation. Let D_m be the subalgebra of diagonal matrices in Mat_m and let T_m be the usual diagonal maximal torus of GL_m. The group $N_{\mathrm{GL}_m}(T)$ acts on D_m by conjugation and T acts trivially, so $W = N_{\mathrm{GL}_m}(T)/T$ acts on D_m. The inclusion of D_m in Mat_m gives rise to an isomorphism ψ' from $D_m /\!/ W$ to $\mathrm{Mat}_m /\!/ \mathrm{GL}_m$. To see this, note first that for any $A \in \mathrm{Mat}_m$, $\mathrm{GL}_m \cdot A$ is closed if and only if A is semisimple (Exercise 5.9). The result now follows from an argument like the one in the proof of Lemma 5.6.7. Indeed, to prove that ψ' is separable, it is enough by Remark 3.1.8 to prove that the map $(D_m \cap \mathrm{GL}_m) /\!/ W \to \mathrm{GL}_m /\!/ G$ is separable, and this follows immediately from the separability argument of Lemma 5.6.7 applied to $G = \mathrm{GL}_m$.

The Weyl group W is isomorphic to S_m, and it acts on D_m by permuting the diagonal entries $a_1, \ldots, a_m$. It is well known that $k[D_m]^W$ is the polynomial ring generated by the elementary symmetric polynomials $f_1, \ldots, f_m$ in the a_i: explicitly,

$$f_t(a_1, \ldots, a_m) = \sum_{1 \le i_1 < \cdots < i_t \le m} a_{i_1} \cdots a_{i_t}.$$

Since $D_m /\!/ W$ is isomorphic to $\mathrm{Mat}_m /\!/ \mathrm{GL}_m$, it follows that every element of $k[D_m]^W$ has a unique extension to an element of $k[\mathrm{Mat}_m]^{\mathrm{GL}_m}$. Here is an explicit description. Let $f_A(\lambda)$ be the characteristic polynomial of $A \in \mathrm{Mat}_m$. Then $f_A(\lambda)$ has the form $\lambda^m + c_{m-1}(A)\lambda^{m-1} + \cdots + c_1(A)\lambda + c_0(A)$, where each c_i belongs to $k[\mathrm{Mat}_m]^{\mathrm{GL}_m}$, and c_i is the extension of f_{m-i} for $0 \le i \le m-1$: for instance, f_m and f_1 are the restrictions to D_m of $c_0 = \det$ and $c_{m-1} = \mathrm{tr}$, respectively, e.g., see [56, §1 (3)]. Define $f_j' \in k[D_m]^W$ for $1 \le j \le m$ by $f_j'(A) = \mathrm{tr}(A^j)$ for $1 \le j \le m$. If $p = 0$ then $k[D_m]^W$ is the polynomial algebra in the f_j', e.g., see [38, Ch. VIII, §8, Thm. 1], or [78, §23.1]; this explains why we need only functions involving tr to generate $\mathcal{T}_n(\mathrm{Mat}_m)$, and not arbitrary elements of

$k[\mathrm{Mat}_m]^{\mathrm{GL}_m}$. On the other hand, if $p \geq m$ then $f_1', \ldots, f_m'$ are not even algebraically independent, because $\mathrm{tr}(A^p) = \mathrm{tr}(A)^p$ for any $A \in \mathrm{Mat}_m$.

Remark 5.6.8 We can show when $p = 0$ that $k[M_{m,n}]^{\mathrm{GL}_m}$ is a finitely generated $\mathcal{T}(M_{m,n})$-module without using the results from [138] that we cited for the proof of Theorem 5.6.2, by applying Proposition 3.4.3. Let $(A_1, \ldots, A_n) \in M_{m,n}$. Recall from Example 3.4.2 that $(A_1, \ldots, A_n)$ belongs to the nullcone $\mathcal{N}(M_{m,n})$ if and only if the subalgebra A of Mat_m generated by the A_i consists of nilpotent elements.

By the discussion above, if $C \in \mathrm{Mat}_m$ then C is nilpotent if and only if $\mathrm{tr}(C^j) = 0$ for all $1 \leq j \leq m$. Suppose that $\mathrm{tr}_w(A_1, \ldots, A_n) = 0$ for all positive words w in the A_i. Then for any $C \in A$ and any $1 \leq j \leq m$ we have $\mathrm{tr}(C^j) = 0$ by the linearity of trace: for we can write C^j as a linear combination of positive words in the A_i, and the trace of each such term vanishes by hypothesis. Hence $\mathcal{N}(M_{m,n})$ is the set of zeroes of the functions tr_w, where w ranges over all the positive words. There is a finite set F of these functions such that $\mathcal{N}(M_{m,n})$ is the set of zeroes of F, and it follows from Proposition 3.4.3 that $k[M_{m,n}]^{\mathrm{GL}_m}$ is a finitely generated $\mathcal{T}(M_{m,n})$-module.

5.7　　Historical Remarks and References

The material in Sects. 5.1–5.4 of this chapter consists of updated versions of ideas we used in several of our earlier papers on G-complete reducibility, in particular [18], [19], [28]. We have also taken the opportunity to update several of the proofs. For example, the proof of Theorem 5.3.3 is quite different from our original one: the analogous result in our early work is [18, Cor. 3.7], which in turn uses our proof of the equivalence between G-complete reducibility and strong reductivity [18, Thm. 3.1] and Richardson's characterisation of closed orbits of tuples in terms of strong reductivity [147, Thm. 16.4]; see the historical remarks in Sect. 4.4.

The notion of a generic tuple was introduced in [28, Def. 5.4] to get around the failure of subgroups H of G to be topologically finitely generated. An earlier approach [18, Lem. 2.10] involved replacing H with a finitely generated subgroup Γ of H such that any R-parabolic subgroup (resp., R-Levi subgroup) which contains Γ also contains H; see Exercise 5.3. Proposition 2.6.3 also involves a variation on this idea. Another approach involves doing base change to a larger algebraically closed field: see Sect. 6.4.

We mention another very recent approach of Cotner to the problem of parametrising subgroups of G. Let H be an algebraic group and suppose H does not have the additive group k as a quotient. Cotner shows the set $\mathrm{Hom}(H, G)$ of algebraic group homomorphisms from H to G can be given the structure of a countable disjoint union of affine schemes of finite type over k. The group G acts on $\mathrm{Hom}(H, G)$ by conjugation in the obvious way. For more details and further references, see [49].

The material on semisimplification in Sect. 5.5 was implicit in much of our earlier work, but was pulled together and extended in various ways in [23]. The term is due to Serre: see his "G-analogue" of a semisimplification from [159, §3.2.4]. Definition 5.5.1 generalises the definition of $\mathcal{D}(H)$ following [119, Lem. 4.1] (in the algebraically closed case). In [23, Thm. 4.5], we proved Proposition 5.5.8 using a different argument involving a slight variation on the idea in Exercise 5.2. The new proof we give using Lemma 5.5.7 avoids us having to modify the generic tuple.

There is a notion of *degeneration* for representations of a finitely generated k-algebra A which is closely related to the notion of semisimplification, see [91], [42], [148], [194], and [195]. Given such an algebra A, say with generators $f_1, \ldots, f_n$, a d-dimensional representation of A over k is determined by assigning to each generator f_i an element $x_i \in \mathrm{Mat}_d$, such that the x_i satisfy the same relations as the f_i. We see that the space of all d-dimensional representations can be identified with a subvariety of $M_{d,n}$. Further, two representations are isomorphic if and only if the corresponding tuples are conjugate under the diagonal action of GL_d on tuples. The closed orbits in this situation correspond to d-dimensional *semisimple* representations, giving an exact analogue in this situation of the semisimplification of a subgroup of GL_d. Say a d-dimensional representation ρ *degenerates* to another d-dimensional representation τ if τ is in the closure of the GL_d-orbit of ρ. Then the main result of [195] is a very nice representation-theoretic description of this degeneration process in terms of exact sequences of A-modules: if ρ is afforded by the A-module M and τ by the A-module N, then ρ degenerates to τ if and only if there exists an A-module Z and a short exact sequence of A-modules $0 \to Z \to M \oplus Z \to N \to 0$.

For Theorem 5.6.2 (invariants of GL_m on tuples of matrices in characteristic 0) we have cited [138]; note that Sibirskiĭ proved a similar result independently [160]. Lemma 5.6.7 is well-known: see [171, §6]. In characteristic 0 there is a Lie algebra analogue of Lemma 5.6.7, known as the Chevalley Restriction Theorem: it states that if T is a maximal torus of G then the map $\mathfrak{t}/\!\!/W \to \mathfrak{g}/\!\!/G$ is an isomorphism. This result also holds in positive characteristic under some extra hypotheses, but fails in general. See [83, §7.12] for details. We may view the isomorphism of $D_m /\!\!/ W$ with $\mathrm{Mat}_m /\!\!/ \mathrm{GL}_m$ as the special case of the Chevalley Restriction Theorem with $G = \mathrm{GL}_m$ and $\mathfrak{g} = \mathrm{Lie}(\mathrm{GL}_m) \cong \mathrm{Mat}_m$.

5.8 Exercises

Exercise 5.1 In Exercise 4.2 we found an example of a reductive G and a subgroup H of a parabolic P of G such that H is G-conjugate to a subgroup of L but not $R_u(P)$-conjugate to a subgroup of L, where L is a Levi subgroup of P. Suppose we are in such a situation. Let $\mathbf{h}$ be a generic tuple for H, let $g \in G$ conjugate H to a subgroup $H' = gHg^{-1}$ of L. Then $\mathbf{h}' = g \cdot \mathbf{h}$ is a generic tuple for H', but $\mathbf{h}'$ cannot be $R_u(P)$-conjugate to $\mathbf{h}$ (for if it was, then H would be $R_u(P)$-conjugate to a subgroup of L). Explain why this does not contradict Lemma 5.3.1.

Exercise 5.2 Let H_1, H_2 be conjugate subgroups of G. Let $\mathbf{h}_1 = (h_1, \ldots, h_n)$ be a generic tuple for H_1. Show that we can extend $\mathbf{h}_1$ to be a possibly longer generic tuple $\mathbf{h}_1 = (h_1, \ldots, h_n, h_{n+1}, \ldots, h_{n+s})$ for H_1 with the following property: for any $g \in G$ such that $g \cdot \mathbf{h}_1$ belongs to H_2^{n+s}, we have $g H_1 g^{-1} = H_2$.

Exercise 5.3 Suppose H is a subgroup of G. Show that there exists a finitely generated subgroup Γ of H such that any R-parabolic subgroup (resp., R-Levi subgroup) of G which contains Γ also contains H, and $C_G(\Gamma) = C_G(H)$.

[*Hint*: This involves a descending chain argument similar to that in the solution of the previous exercise.]

Exercise 5.4 Recall from Remark 3.7.9 that if X is a G-variety and $Z := \bigcap_{x \in X} G_x$ is the kernel of the action of G on X, then a point $x \in X$ is a *stable point* if $G \cdot x$ is closed and G_x/Z is finite. Let H be a subgroup of G and let $\mathbf{h} \in G^n$ be a generic tuple for H. Show that H is G-irreducible if and only if $\mathbf{h}$ is a stable point for the G-action on G^n. (This result is [18, Prop. 2.13].) Deduce that the set

$$\{\mathbf{g} \in G^n \mid \mathcal{G}(\mathbf{g}) \text{ is } G\text{-irreducible}\}$$

is open.

Exercise 5.5 For each of the following examples of $H \subseteq G$, find a semisimplification of H in G. (Note that the answer in both cases depends on the characteristic.)

(i) H is the image of the symmetric group S_m in GL_m, where S_m acts by permuting the standard basis vectors in the obvious way.

(ii) H is the image of the adjoint representation of SL_2 inside GL_3.

Exercise 5.6 Let $p = 2$ and let V be a $(2m + 1)$-dimensional k-vector space equipped with the quadratic form $q = x_0^2 + x_1 x_2 + \cdots + x_{2m-1} x_{2m}$ and associated bilinear form $(v, w) = q(v + w) + q(v) + q(w)$. Let $H = \mathrm{SO}_{2m+1} \subseteq G = \mathrm{SL}_{2m+1}$, where H is the group of endomorphisms fixing the given form.

(i) Show that the form $(\cdot, \cdot)$ is also alternating, and deduce that it has a one-dimensional radical.

Fix a vector $0 \neq v \in V$ with $(v, v) = 0$ and let P be the parabolic subgroup of G which is the stabiliser of the space spanned by v (i.e., the stabiliser of the radical of the form). Choose a basis for V which includes the vector v, and such that the space spanned by the other $2m$ basis vectors is non-degenerate.

(ii) Show that the Levi subgroup L of P corresponding to our choice of basis is isomorphic to GL_{2m}.

(iii) Show that $H \subseteq P$ and that $c_L(H) \cong \mathrm{Sp}_{2m}$.

(iv) Deduce that H is not G-cr and that $c_L(H)$ is a semisimplification of H in G.

Exercise 5.7 Let $X = k^2$ be the natural module for GL_2 and let $G = N_{\mathrm{GL}_2}(T)$, where T is the diagonal maximal torus in GL_2. Let $x = \begin{bmatrix} 1 \\ 1 \end{bmatrix} \in X$. Define $\lambda_1, \lambda_2 \in Y_G$ by

$$\lambda_1(a) = \begin{pmatrix} a & 0 \\ 0 & 1 \end{pmatrix} \qquad \text{and} \qquad \lambda_2(a) = \begin{pmatrix} 1 & 0 \\ 0 & a \end{pmatrix}$$

for each $a \in k^*$.

 (i) Show that for $i = 1, 2$, $x_i := \lim_{a \to 0} \lambda_i(a) \cdot x$ exists.

 (ii) Show that x_1 and x_2 are G-conjugate.

(iii) Show that this gives a counterexample to the statement of Lemma 5.5.5 if we replace the hypothesis that $G \cdot x_1$ and $G \cdot x_2$ are closed with the weaker hypothesis that x_1 and x_2 are conjugate.

Exercise 5.8 Let H be a subgroup of G and let $\lambda \in Y_G$ such that $H \subseteq P_\lambda$. Show that $\mathcal{D}(c_\lambda(H)) = \mathcal{D}(H)$.

Exercise 5.9 Let $A \in \mathrm{Mat}_m$. Prove that $\mathrm{GL}_m \cdot A$ is closed if and only if A is semisimple. [*Hint*: Use the Jordan Normal Form for matrices.]

Exercise 5.10 Suppose G is connected. Let $n \in \mathbb{N}$. Show that $\pi_G(\{g \in G \mid g^n = 1\})$ is a finite subset of $G /\!/ G$. (In the language of Sect. 6.5, this says that the character variety $C(C_n, G)$ is finite, where C_n is the cyclic group of order n.)

Finiteness, Rationality and Rigidity Results

6

In this chapter we prove four important results which use the theory of G-complete reducibility and the geometric methods developed in the previous chapter. We focus on two related themes: understanding conjugacy classes of subgroups of G (or more generally conjugacy classes of homomorphisms into G) and understanding maps between quotients of the form $G^n /\!\!/ G$. The link to G-complete reducibility is Theorem 5.3.3, since the points of the quotient variety $G^n /\!\!/ G$ correspond to closed orbits in G^n, and these in turn correspond to conjugacy classes of G-cr subgroups of G. The main results are as follows. The Rationality Theorem 6.1.1 implies that if E/k is an extension of algebraically closed fields, then every conjugacy class of G_E-cr subgroups contains a k-defined subgroup; in particular, then, since G is always defined over the algebraic closure k_0 of the prime field (Theorem 2.6.2), in order to understand conjugacy classes of G-cr subgroups we can work over k_0. Since k_0 is countable, we can deduce Theorem 6.1.2, which says that there are only countably many conjugacy classes of G-cr subgroups of G over any algebraically closed field. Our main result on quotients is Theorem 6.2.1, which implies that if H is a reductive subgroup of G then the natural map of quotients $H^n /\!\!/ H \to G^n /\!\!/ G$ is finite. This is an extremely useful result, with applications to the study of character varieties, for example; see Sect. 6.9.

The chapter is laid out as follows. We state the results in the first two sections and give some brief discussion of their context. The proofs of the results are rather closely intertwined, so in Sect. 6.3 we give a detailed plan of how they fit together. We develop further material on generating tuples in Sect. 6.4 and introduce representation and character varieties in Sect. 6.5. We then embark on the proofs. The first stage is to deal with the case that $G = \mathrm{GL}_m$, which we do with a representation-theoretic argument (Sect. 6.6), and then we finish the proofs in Sects. 6.7 and 6.8. Finally, Sect. 6.9 gives some further consequences of our results in the study of character varieties.

© The Author(s), under exclusive license to Springer Nature Switzerland AG 2026

M. Bate et al., *G-Complete Reducibility, Geometric Invariant Theory and Spherical Buildings*, Oberwolfach Seminars 57, https://doi.org/10.1007/978-3-032-08866-6_6

6.1   The Rationality Theorem

Theorem 6.1.1 (Rationality Theorem) *Let E be an algebraically closed field extension of k. Let H be a G_E-completely reducible subgroup of G_E. Then there exists $g \in G_E$ such that gHg^{-1} is k-defined.*

Theorem 6.1.2 (Rigidity Theorem) *There are only countably many G-conjugacy classes of G-completely reducible subgroups of G.*

We give a brief justification of the terminology in Theorem 6.1.2. Ideally, we would like to endow the space of conjugacy classes of subgroups of G with a geometry and gain information about subgroups by studying this geometry. (As discussed in the Introduction and in Chap. 5, however, we parametrise subgroups indirectly by working with generating tuples instead.) For instance, one can try to give this space the structure of a variety over k, then ask whether there is a positive-dimensional space of deformations of a given subgroup. Theorem 6.1.2 shows that if we consider the space of conjugacy classes of G-cr subgroups of G then the answer is no: even when k is uncountable, there are only countably many conjugacy classes of G-cr subgroups. Hence we can view this result as a rigidity theorem.

Remark 6.1.3 Theorem 6.1.1 fails if we consider arbitrary reductive subgroups H of G. For instance, let $p > 0$, let $k = \overline{\mathbb{F}_p}$ and let E be an algebraically closed transcendental extension of k. Let $G = \mathrm{SL}_2$ (over k) and let H be the finite subgroup of G_E generated by $\begin{pmatrix} 1 & 1 \\ 0 & 1 \end{pmatrix}$ and $\begin{pmatrix} 1 & x \\ 0 & 1 \end{pmatrix}$, where $x \in E \backslash k$. Then H is reductive and it is easily checked that H is not G_E-conjugate to a subgroup of G.

We give a consequence of the Rationality Theorem.

Proposition 6.1.4 *Suppose $p > 0$, and let M be an $\overline{\mathbb{F}_p}$-descent of G. Let H be a finite subgroup of G.*

(a) *If H is G-completely reducible then H is conjugate to a subgroup of M.*
(b) *If $\dim(G) > 0$ then H is not maximal.*

Proof Recall that G admits an $\overline{\mathbb{F}_p}$-structure by Theorem 2.6.2, so the statement of the proposition makes sense. If H is G-cr then Theorem 6.1.1 implies that gHg^{-1} is $\overline{\mathbb{F}_p}$-defined for some $g \in G$: say, $gHg^{-1} = F_k$ for some subgroup F of M. But then F is finite, since H is, so $F_k = F$, so $gHg^{-1} = F \subseteq M$. This proves (a).

Now suppose $\dim(G) > 0$. If H is not G-cr then H is contained in a proper R-parabolic subgroup of G, so H is not maximal. Hence we can assume without loss that H is G-cr. By Proposition 2.6.3 (applied to G), there is an ascending chain $M_1 \subseteq M_2 \subseteq \cdots$ of finite subgroups of M such that $\bigcup_{n \in \mathbb{N}} M_n = M$. There exists $g \in G$ such that $gHg^{-1} \subseteq M$ by part (a). Since H is finite, gHg^{-1} is contained in M_n for some $n \in \mathbb{N}$. But $\dim(M) > 0$, so M is infinite. Hence gHg^{-1} is properly contained in $M_{n'}$ for some $n' \geq n$, which shows that H is not maximal. This proves (b). $\square$

6.2   The Finiteness Theorem

To state the next result, we need some notation. Let $\phi \colon H \to G$ be a homomorphism of reductive groups and let $n \in \mathbb{N}$. Let $\phi_n \colon H^n \to G^n$ be the Cartesian product $\phi \times \cdots \times \phi$. We have an action of H on G^n given by $h \cdot \mathbf{g} = \phi(h) \cdot \mathbf{g}$. Then ϕ_n is H-equivariant, so we get a map $(\phi_n)_H \colon H^n /\!/ H \to G^n /\!/ H$ (see Sect. 3.6). Composing with the map $G^n /\!/ H \to G^n /\!/ G$ from Example 3.1.7 gives a map

$$\Psi^n_{G,H,\phi} \colon H^n /\!/ H \to G^n /\!/ G.$$

Note that $\mathrm{Im}\left(\Psi^n_{G,H,\phi}\right) = \pi_{G^n}(\phi(H)^n)$. When H is a subgroup of G then we just write $\Psi^n_{G,H}$ instead of $\Psi^n_{G,H,i}$, where $i \colon H \to G$ is the inclusion map. If $\psi \colon M \to H$ is a homomorphism of reductive groups then

$$\Psi^n_{G,M,\phi \circ \psi} = \Psi^n_{G,H,\phi} \circ \Psi^n_{H,M,\psi}. \tag{6.2.1}$$

If E is an algebraically closed field extension of k then

$$\Psi^n_{G_E,H_E,\phi_E} = \left(\Psi^n_{G,H,\phi}\right)_E \tag{6.2.2}$$

by the results in Sect. 3.9.

Theorem 6.2.1 (Finiteness Theorem) *Let H be a reductive subgroup of G and let $n \in \mathbb{N}$. Then the map $\Psi^n_{G,H}$ is finite.*

If we are given an embedding $i \colon G \to \mathrm{GL}_m$ then Theorem 6.2.1 allows us to transfer information about completely reducible subgroups from GL_m (where much more is known) to G. This is striking given that we know that if H is a subgroup of G then H can be G-cr but not GL_m-cr, or vice versa: for the former, just take $H = G$ to be the image of a non-completely reducible representation of a reductive group, and for the latter see Example 4.1.10. Here is one important consequence which illustrates the power of the Finiteness Theorem 6.2.1.[1]

---

[1] We use Theorem 6.2.1 to prove Theorem 6.2.2 when $p = 0$, but for $p > 0$ we give an independent proof of the latter: see Sect. 6.3 for details.

Theorem 6.2.2 *Let F be a finite group. Then the character variety $C(F, G)$ is finite.*

Character varieties are defined in Sect. 6.5 below. Here is an equivalent formulation of the theorem, using Proposition 6.5.3 below: if F is a finite group then there are only finitely many conjugacy classes of homomorphisms from F to G with G-cr image.

6.3　Plan of the Proofs

To help the reader follow the arguments, we give a plan of the proofs of the main theorems. We need to make a distinction in various places between the characteristic 0 case and the positive characteristic case. We proceed as follows.

(1) We study generating sets in more detail and prove some density results about sets of generating tuples (Corollary 6.4.7 and Lemma 6.4.9). We also prove a result on generating sets for the derived subgroup $[G, G]$ (Proposition 6.4.13), using results of Selberg and Platonov on finitely generated linear groups.

(2) We review the theory of representation varieties and character varieties. This is a subject of considerable independent interest, so we present a range of material; for the purposes of this plan, the main result is Proposition 6.5.4, which shows that $C(F, \mathrm{GL}_m)$ is finite for all finite groups F.

(3) We show that the Rationality Theorem holds when $G = \mathrm{GL}_m$ (Lemma 6.6.1). This follows from some representation-theoretic results.

(4) We prove using Bruhat-Tits theory that if $p > 0$ then $C(F, G)$ is finite for every finite group F (Proposition 6.7.8).

(5) We deduce that the Rationality Theorem holds when $p > 0$, using (4) and Lemma 6.6.3; the idea is to approximate G with finite subgroups, in the sense of Proposition 2.6.3. We also prove the Rationality Theorem in characteristic 0, using an application of Richardson's tangent space argument from Sect. 10.5 (see the end of Sect. 6.7). This completes the proof of the Rationality Theorem.

(6) We state and prove two key technical results, Lemma 6.8.6 and Lemma 6.8.8. For the former we use Proposition 6.4.13, while for the latter we use the Rationality Theorem.

(7) We prove that the Finiteness Theorem holds using Lemma 6.8.8.

(8) We deduce from (2) and the Finiteness Theorem that $C(F, G)$ is finite for all finite groups F when $p = 0$ (see the text following Corollary 6.8.11). This completes the proof of Theorem 6.2.2.

(9) We prove that Theorem 6.1.2 follows from the Rationality Theorem (see the end of Sect. 6.8).

Steps (1) and (2) are covered in Sects. 6.4 and 6.5, respectively, step (3) is covered in Sect. 6.6 and steps (4) and (5) are covered in Sect. 6.7. Section 6.8 covers steps (6)–(9).

The hardest steps are (4), (6) and (7). We briefly explain the main ideas. For step (4) we pass to a suitable field extension K/k such that K admits a discrete valuation; this allows us to define a notion of boundedness on subsets of $X(K)$, where X is a variety over k. If $C(F, G)$ is infinite then $C(F, G)(K)$ must be unbounded; but it follows from Bruhat-Tits theory that any finite subgroup of $G(K)$ is $G(K)$-conjugate to a subgroup of a parahoric subgroup and we conclude that $C(F, G)(K)$ is bounded, a contradiction. This forces $C(F, G)$ to be finite.

Roughly speaking, Lemma 6.8.8 in step (6) says the following. Let D be a closed subset of $\overline{\Psi^n_{G,H}(H^n /\!/ H)}$ satisfying certain properties. Then there are reductive subgroups $M_1, \ldots, M_t$ of G for some $t \in \mathbb{N}$ such that $\dim(M_i) < \dim(H)$ for each i and $D \subseteq \bigcup_{i=1}^t \mathrm{Im}\left(\Psi^n_{G,M_i}\right)$. The proof uses the density results on generating tuples from step (1) and involves passing to an algebraically closed field extension E/k.

For the Finiteness Theorem in step (7), we prove first using step (1) that generic fibres of $\Psi^n_{G,H}$ are finite (Lemma 6.8.4), again using the density results from step (1). This takes some work but is reasonably straightforward. Next let D_1 be the set of points $y \in H^n /\!/ H$ such that $\left(\Psi^n_{G,H}\right)^{-1}(\Psi^n_{G,H}(y))$ is infinite. Using Lemma 6.8.8 (but applied to the inclusion of H in H), we show that there are reductive subgroups $M_1', \ldots, M_t'$ of H such that $D_1 \subseteq \bigcup_{i=1}^t \mathrm{Im}\left(\Psi^n_{H,M_i'}\right)$. But we can assume using induction on $\dim(H)$ that the maps $\Psi_{G,M_i'}$ are finite for each i, so the maps $\Psi_{H,M_i'}$ are finite as well. This forces D_1 to be empty, so $\Psi^n_{G,H}$ is quasi-finite.

If $\Psi^n_{G,H}$ is not finite then by some general algebraic geometry arguments, we obtain a closed subset D of codimension 1 in $\overline{\mathrm{Im}\left(\Psi^n_{G,H}\right)}$; D is the set of points over which $\Psi^n_{G,H}$ fails to be finite. By applying Lemma 6.8.8, we see that $D \subseteq \bigcup_{i=1}^t \mathrm{Im}\left(\Psi^n_{G,M_i}\right)$, where the M_i are as in Lemma 6.8.8. But by taking n sufficiently large and using Lemma 3.7.6 to bound $\dim\left((M_i)^n /\!/ M_i\right)$, we can show that D has codimension at least 2, a contradiction. We conclude that $\Psi^n_{G,H}$ is finite, as required.

6.4 Density of Generating Tuples

We saw in Sect. 5.2 that if $k = \overline{\mathbb{F}_p}$ then reductive groups need not be topologically finitely generated. We dealt with this in our application to G-complete reducibility by working with generic tuples. We can also avoid this problem by doing base change to a suitable algebraically closed extension field E of k, as described in Sect. 2.6. We develop these ideas now as we need them to establish our finiteness and rationality results.

Definition 6.4.1 Let k be an algebraically closed field. We call k *solid* if either (i) $\mathrm{char}(k) = 0$, or (ii) $\mathrm{char}(k) = p > 0$ and k is transcendental over $\mathbb{F}_p$.

Definition 6.4.2 We say a tuple $\mathbf{h} \in H^n$ *generates* H/H^0 if the restriction to $\mathcal{G}(\mathbf{h})$ of the quotient map $H \to H/H^0$ is surjective. Now define

$$H_0^n = \left\{ \mathbf{h} \in H^n \,\middle|\, \mathbf{h} \text{ generates } H/H^0 \right\}.$$

It is immediate that $H_{\text{gen}}^n \subseteq H_0^n$. Clearly H_0^n is an H-stable union of connected components of H^n, so H_0^n is a smooth pure-dimensional variety of dimension $n \dim(H)$, and the connected components and irreducible components of H^n coincide with each other (and likewise for H_0^n). It follows from Remark 3.6.2 that for reductive H we may identify $H_0^n /\!\!/ H$ with the closed subset $\pi_{H^n}(H_0^n)$ of $H^n /\!\!/ H$. By Exercise 3.2(i) we have

$$\pi_{H^n}^{-1}(H_0^n /\!\!/ H) = H_0^n.$$

It follows that the connected components and irreducible components of $H^n /\!\!/ H$ coincide with each other (and likewise for $H_0^n /\!\!/ H$).

Remark 6.4.3 Let C be a connected component of H^n. We can write $C = C_1 \times \cdots \times C_n$ for some connected components $C_1, \ldots, C_n$ of H. If $C \not\subseteq H_0^n$ then $C \subseteq M_0^n$, where M is the subgroup of H generated by the C_i; in this case M is a proper finite-index subgroup of H. It follows that if D is a connected component of $H^n /\!\!/ H$ and $D \not\subseteq H_0^n /\!\!/ H$ then $D \subseteq \pi_{H^n}(M_0^n)$ for some proper finite-index subgroup M of H.

Lemma 6.4.4 *Suppose k is solid and let T be a torus. Then T_{gen}^1 is dense in T.*

Proof We have $T \cong (k^*)^m$, where $m = \dim(T)$. Let S be a proper subgroup of T. Then T/S is a non-trivial torus, so there is a non-trivial character $\chi \colon T/S \to k^*$. Let $\chi' = \chi \circ \pi_S$, where $\pi_S \colon T \to T/S$ is the canonical projection. Then χ' is a non-trivial character of T, so χ' has the form $\chi'(t_1, \ldots, t_m) = t_1^{n_1} \cdots t_m^{n_m}$ where the n_i are integers, not all zero. But χ' kills S, so

$$s_1^{n_1} \cdots s_m^{n_m} = 1 \text{ for all } (s_1, \ldots, s_m) \in S. \tag{6.4.1}$$

Suppose $p = 0$. Let $p_1, \ldots, p_m \in \mathbb{N}$ be distinct primes and let $t = (p_1, \ldots, p_m)$. Then t cannot satisfy (6.4.1) for any integers n_i that are not all zero, so t cannot belong to any proper subgroup of T. Hence t topologically generates T. Now suppose $p > 0$; the proof is very similar. Choose $a \in k^*$ such that a is transcendental over $\mathbb{F}_p$. Choose pairwise coprime polynomials $f_1(x), \ldots, f_m(x) \in \mathbb{F}_p[x]$ and let $t = (f_1(a), \ldots, f_m(a))$. Then, again, t cannot satisfy (6.4.1) for any integers n_i that are not all zero, so t cannot belong to any proper subgroup of T. Hence t topologically generates T. Moreover, the set of elements $t \in T$ of the form given above is dense in T both when $p = 0$ and when $p > 0$ (Exercise 6.1), so T_{gen}^1 is dense in T. $\qquad\square$

Recall from Exercise 5.4 that the set of n-tuples that generate a G-ir subgroup of G is open in G^n. In contrast, G^n_{gen} is neither open nor closed: in fact, it is not even constructible in general (see Exercise 12.10).

Remark 6.4.5 If $p = 0$ then any algebraic group H is topologically finitely generated. For the unipotent group $R_u(H)$ is topologically finitely generated—see the paragraph following Lemma 5.2.2—and $H^0/R_u(H)$ is topologically finitely generated by Lemma 6.4.4, since any connected reductive group is generated by its maximal tori; it follows easily that H is topologically finitely generated. This need not be the case if $p > 0$, even when k is solid (see Exercise 6.2).

We also need to consider some obvious symmetries of our set-up, so we introduce some notation to deal with this. Let $n \in \mathbb{N}$. Let Γ_n denote the free group on generators $\gamma_1, \ldots, \gamma_n$ and let H be an algebraic group. We have an action of $\text{Aut}(\Gamma_n)$ on H^n given by

$$\phi \cdot (h_1, \ldots, h_n) = (w_1(h_1, \ldots, h_n), \ldots, w_n(h_1, \ldots, h_n)), \tag{6.4.2}$$

where $\phi^{-1}(\gamma_i) = w_i(\gamma_1, \ldots, \gamma_n)$ for each i (see Exercise 6.4). In particular, the symmetric group S_n acts by permuting the γ_i, and the corresponding action on H^n is given by $\sigma \cdot (h_1, \ldots, h_n) = (h_{\sigma^{-1}(1)}, \ldots, h_{\sigma^{-1}(n)})$. Clearly the action stabilises both H^n_{gen} and H^n_0. The $\text{Aut}(\Gamma_n)$-action commutes with the H-action by simultaneous conjugation, so for reductive H we obtain actions of $\text{Aut}(\Gamma_n)$ on $H^n /\!\!/ H$ and $H^n_0 /\!\!/ H$ as well by the constructions from Sect. 3.6. It is easy to check that $\Psi^n_{G,H}$ is $\text{Aut}(\Gamma_n)$-equivariant.

Proposition 6.4.6 *Suppose k is solid. Let $r \geq \kappa(G)$ and let $C_1, \ldots, C_r$ be connected components of G such that $C_1, \ldots, C_r$ generate G. Then the set*

$$\mathscr{E} := \left\{ \mathbf{g} = (g_1, \ldots, g_{r+1}) \in G^{r+1} \,\middle|\, g_i \in C_i \text{ for } 1 \leq i \leq r, g_{r+1} \in G^0, \, \mathcal{G}(\mathbf{g}) = G \right\}$$

is dense in $C_1 \times \cdots \times C_r \times G^0$.

Proof To ease notation in the proof, let $\mathscr{C} = C_1 \times \cdots \times C_r$. Fix a maximal torus T of G. By Lemma 2.4.10 there are only finitely many maximal closed subgroups of G that contain T and have dimension less than G; let D be their union. Then, for each i, $D \cap C_i$ is a proper closed subset of C_i for dimension reasons. Let

$$U = (C_1 \setminus D) \times \cdots \times (C_r \setminus D),$$

a non-empty open subset of $\mathscr{C}$. Let $\mathscr{E}_T = U \times T_{\text{gen}}$. Then $\mathscr{E}_T$ is dense in $\mathscr{C} \times T$ by Lemma 6.4.4. If $(\mathbf{g}, t) \in \mathscr{E}_T$ then $\mathcal{G}(t) = T \subseteq \mathcal{G}((\mathbf{g}, t))$, so $\mathcal{G}((\mathbf{g}, t)) \supseteq G^0$, by construction. But $C_1, \ldots, C_r$ are chosen to generate G, so any choice of $\mathbf{g} \in \mathscr{C}$ will generate the component group G/G^0, and hence $\mathcal{G}((\mathbf{g}, t)) = G$; that is, $\mathscr{E}_T \subseteq \mathscr{E}$. Finally,

$\mathscr{E}$ and $\mathscr{C}$ are stable under conjugation by G^0, since each component of G is, so

$$\overline{\mathscr{E}} \supseteq G^0 \cdot \overline{\mathscr{E}_T} = G^0 \cdot (\mathscr{C} \times T) = \mathscr{C} \times (G^0 \cdot T).$$

But $G^0 \cdot T$ is dense in G^0, so $\mathscr{C} \times (G^0 \cdot T)$ is dense in $\mathscr{C} \times G^0$. Hence $\mathscr{E}$ is dense in $\mathscr{C} \times G^0$, as required. $\square$

Corollary 6.4.7 *Let k be solid and let $n \geq \kappa(G) + 1$. Then G^n_{gen} is dense in G^n_0.*

Proof The components of G^n_0 have the form $\prod_{i=1}^n C_i$ where $C_1, \ldots, C_n$ are components of G which generate G/G^0. Since G has fewer than n components, there must be at least one repeat in the list $C_1, \ldots, C_n$. Suppose $C_{n-1} = C_n$, so that $C_1, \ldots, C_{n-1}$ still generate G/G^0. Let $\phi \in \mathrm{Aut}(\Gamma_n)$ be the element given by $\phi(\gamma_i) = \gamma_i$ for $1 \leq i \leq n-1$ and $\phi(\gamma_n) = \gamma_{n-1}\gamma_n$. Then

$$\phi \cdot \left(\prod_{i=1}^n C_i \right) = \left(\prod_{i=1}^{n-1} C_i \right) \times G^0.$$

By Proposition 6.4.6 (with $r = n-1$), $G^n_{\mathrm{gen}} \cap \left(\prod_{i=1}^{n-1} C_i \times G^0 \right)$ is dense in $\prod_{i=1}^{n-1} C_i \times G^0$. Since G^n_{gen} is $\mathrm{Aut}(\Gamma_n)$-stable, we can apply ϕ^{-1} to deduce that $G^n_{\mathrm{gen}} \cap \prod_{i=1}^n C_i$ is dense in $\prod_{i=1}^n C_i$. The analogous argument will work in all cases: when $C_i = C_j$ for $i < j$ we can replace C_j with G^0 by using an element of $\mathrm{Aut}(\Gamma_n)$, apply Proposition 6.4.6, and then move back to the original component. $\square$

Corollary 6.4.8 *Let $n \geq \kappa(G) + 1$. Then G^n_0 is visible and $G^n_0 /\!/ G$ is a pure-dimensional variety of dimension $(n-1)\dim(G) + \dim(Z(G))$. Moreover, the irreducible components of $G^n_0 /\!/ G$ are pairwise disjoint.*

Proof By the results in Sect. 3.9 we can do base change to a solid algebraically closed field, so we can assume that k is solid. Let C be any irreducible component of G^n_0. By Corollary 6.4.7, there exists some $\mathbf{g} \in G^n_{\mathrm{gen}} \cap C$. Then $\mathcal{G}(\mathbf{g}) = G$ is G-ir, so $G \cdot \mathbf{g}$ is closed by Theorem 5.3.3. Moreover, $G^0_{\mathbf{g}} = C_G(G)^0 = Z(G)^0$, by Lemma 4.2.2(ii), so $\mathbf{g}$ is a stable point for the action by Exercise 5.4. But C was an arbitrary irreducible component of G^n_0. It follows from Lemma 3.7.8 that G^n_0 is visible. Now let C_1 be a union of irreducible components of G^n_0 such that G permutes these components transitively. The maximal orbit dimension r for C_1 is $\dim(G) - \dim(Z(G))$ by the above argument. Recall that C_1 is pure-dimensional of dimension $n\dim(G)$ (Example 2.7.2), so the dimension formula follows from Lemma 3.7.6.

The final assertion follows from Exercise 3.2(ii). $\square$

Corollary 6.4.7 shows that G_{gen}^n is non-empty if k is solid and n is large enough. We finish the section with a further density result for elements of G_{gen}^n. In order to state and prove it we need some further notation. Let $s \in \mathbb{N}$. We have an inclusion i_s of G^n in G^{n+s} via $(g_1, \ldots, g_n) \mapsto (g_1, \ldots, g_n, 1, \ldots, 1)$. Define $j_s \colon G^{n+s} \to G^n$ by $(g_1, \ldots, g_{n+s}) \mapsto (g_1, \ldots, g_n)$. Then i_s and j_s are G-equivariant and $j_s \circ i_s = \mathrm{id}_{G^n}$, so $(i_s)_G$ is a closed embedding into $G^{n+s} /\!\!/ G$ by Remark 3.6.2. Hence we may identify $G^n /\!\!/ G$ with $\pi_{G^{n+s}}(i_s(G^n))$. Clearly i_s maps G_0^n into G_0^{n+s}, so we may identify $G_0^n /\!\!/ G$ with $\pi_{G^{n+s}}(i_s(G_0^n))$.

Given a subset C of G^n we define $\mathcal{R}_s(C)$ (or $\mathcal{R}_s^G(C)$) to be the closure in G^{n+s} of $\mathrm{Aut}(\Gamma_{n+s}) \cdot i_s(C)$. If $\mathbf{g} \in G^n$ then we write $\mathcal{R}_s(\mathbf{g})$ rather than $\mathcal{R}_s(\{\mathbf{g}\})$. Likewise, given a subset D of $G^n /\!\!/ G$, we define $Q_s(D)$ (or $Q_s^G(D)$) to be the closure of $\mathrm{Aut}(\Gamma_{n+s}) \cdot (i_s)_G(D)$ in $G^{n+s} /\!\!/ G$. If $C \subseteq G^n$ then $\pi_{G^n}(i_s(C)) = (i_s)_G(\pi_{G^n}(C))$, so $\pi_{G^n}(\Gamma_{n+s} \cdot i_s(C)) = \Gamma_{n+s} \cdot (i_s)_G(\pi_{G^n}(C))$, and hence

$$Q_s(\pi_{G^n}(C)) = \overline{\pi_{G^n}(\mathcal{R}_s(C))}. \tag{6.4.3}$$

Clearly G_0^{n+s} is $\mathrm{Aut}(\Gamma_{n+s})$-stable, so $\mathcal{R}_s(C) \subseteq G_0^{n+s}$ if $C \subseteq G_0^n$ and $Q_s(D) \subseteq G_0^{n+s} /\!\!/ G$ if $D \subseteq G_0^n /\!\!/ G$.

Lemma 6.4.9 *Suppose k is solid. Let $\mathbf{g} \in G_{\text{gen}}^n$ and let $s \geq \kappa(G)+1$. Then $\mathcal{R}_s(\mathbf{g}) = G_0^{n+s}$.*

Proof It is enough to show that $\Gamma_{n+s} \cdot \mathbf{g}$ is dense in G_0^{n+s}. Clearly $i_s(\mathbf{g})$ belongs to G_0^{n+s}. Since G_0^{n+s} is $\mathrm{Aut}(\Gamma_{n+s})$-stable, the closure of $\mathrm{Aut}(\Gamma_{n+s}) \cdot i_s(\mathbf{g})$ is contained in G_0^{n+s}. Now we prove the opposite inclusion. Let Λ be the abstract subgroup of G generated by $g_1, \ldots, g_n$, where $\mathbf{g} = (g_1, \ldots, g_n)$. Consider the subset of G_0^{n+s} consisting of elements obtained by applying to $(g_1, \ldots, g_n, 1, \ldots, 1)$ automorphisms of $\mathrm{Aut}(\Gamma_{n+s})$ of the form $\gamma_i \mapsto \gamma_i$ for $1 \leq i \leq n$ and $\gamma_j \mapsto \gamma_j w_j(\gamma_1, \ldots, \gamma_n)$ for $n + 1 \leq j \leq n + s$. This subset is $\{(g_1, \ldots g_n)\} \times \Lambda^s$. Since Λ is dense in G, $\overline{\mathrm{Aut}(\Gamma_{n+s}) \cdot i_s(\mathbf{g})}$ therefore contains $\{(g_1, \ldots g_n)\} \times G^s$.

Let $(g_{n+1}, \ldots, g_{n+s}) \in G_{\text{gen}}^s$, and let Λ' be the abstract subgroup of G generated by $g_{n+1}, \ldots, g_{n+s}$. The element $\mathbf{g}' := (g_1, \ldots, g_n, g_{n+1}, \ldots, g_{n+s})$ belongs to $\overline{\mathrm{Aut}(\Gamma_{n+s}) \cdot i_s(\mathbf{g})}$ by the previous paragraph. Consider the subset of G_0^{n+s} consisting of elements obtained by applying to $\mathbf{g}'$ automorphisms of $\mathrm{Aut}(\Gamma_{n+s})$ of the form $\gamma_i \mapsto \gamma_i w_i(\gamma_{n+1}, \ldots, \gamma_{n+s})$ for $1 \leq i \leq n$ and $\gamma_j \mapsto \gamma_j$ for $n + 1 \leq j \leq n + s$. This subset is $\{(g_1 \Lambda' \times \cdots \times g_n \Lambda') \times \{(g_{n+1}, \ldots, g_{n+s})\}$. Since Λ' is dense in G, the closure of $\mathrm{Aut}(\Gamma_{n+s}) \cdot i_s(\mathbf{g})$ therefore contains $G^n \times \{(g_{n+1}, \ldots, g_{n+s})\}$. Hence $\overline{\mathrm{Aut}(\Gamma_{n+s}) \cdot i_s(\mathbf{g})}$ contains $G^n \times G_{\text{gen}}^s$. But G_{gen}^s is dense in G_0^s by Corollary 6.4.7, so $\overline{\mathrm{Aut}(\Gamma_{n+s}) \cdot i_s(\mathbf{g})}$ contains $G^n \times G_0^s$.

Finally, let $\mathbf{h} = (h_1, \ldots, h_{n+s}) \in G_0^{n+s}$. There are distinct $i_1, \ldots, i_{\kappa(G)} \in \{1, \ldots, n+s\}$ such that the images of $h_{i_1}, \ldots, h_{i_{\kappa(G)}}$ in G/G^0 generate G/G^0. Choose a permutation

σ of $\{1, \ldots, n + s\}$ such that $\sigma^{-1}(i_j) = n + j$ for $1 \leq j \leq \kappa(G)$. Write $\sigma \cdot \mathbf{h} = (h'_1, \ldots, h'_{n+s})$. Then $h'_{n+j} = h_{i_j}$ for $1 \leq j \leq \kappa(G)$, so $(h'_{n+1}, \ldots, h'_{n+s}) \in G_0^s$. Hence $\sigma \cdot \mathbf{h}$ belongs to $\overline{\mathrm{Aut}(\Gamma_{n+s}) \cdot i_s(\mathbf{g})}$. But $\overline{\mathrm{Aut}(\Gamma_{n+s}) \cdot i_s(\mathbf{g})}$ is $\mathrm{Aut}(\Gamma_{n+s})$-stable, so $\mathbf{h}$ belongs to $\overline{\mathrm{Aut}(\Gamma_{n+s}) \cdot i_s(\mathbf{g})}$. We conclude that $G_0^{n+s} \subseteq \overline{\mathrm{Aut}(\Gamma_{n+s}) \cdot i_s(\mathbf{g})}$, as required. $\square$

Remark 6.4.10 Suppose C is a closed subset of G^n and let $s \in \mathbb{N}$. Let E/k be an algebraically closed field extension. Then C is dense in C_E, so $\Gamma_{n+s} \cdot i_s(C)$ is dense in $\Gamma_{n+s} \cdot i_s(C_E)$. It follows that $\mathcal{R}_s^{G_E}(C_E) = (\mathcal{R}_s^G(C))_E$. By a similar argument, if D is a closed subset of $G^n /\!\!/ G$ then $Q_s^{G_E}(D_E) = (Q_s^G(D))_E$.

We need some further results about finite generating sets. Below we use the following fact repeatedly: if H is a connected algebraic group and Γ is a dense subgroup of H then any finite-index subgroup of Γ is also dense in H. We also use some results about finitely generated linear groups: see [133] for a quick overview, including discussions of the results of Selberg and Platonov cited just below.

Proposition 6.4.11 *Let Γ be a finitely generated subgroup of G. Then there is a finite-index subgroup Γ_0 of Γ such that every non-trivial semisimple element of Γ_0 has infinite order. In particular, if $\dim(G) > 0$ and Γ is dense in G then Γ contains elements of infinite order.*

Proof We can embed Γ in GL_m for some m, so Γ is a linear group. If $p = 0$ then by a theorem of Selberg [155], Γ has a torsion-free finite-index subgroup Γ_0, and clearly this Γ_0 will do. So suppose $p > 0$. By a theorem of Platonov [137], Γ has a finite-index residually p-finite subgroup Γ_0. Now let $1 \neq g \in \Gamma_0$ be semisimple. By definition of residually p-finite, there is a homomorphism f from Γ_0 to a finite p-group such that $f(g) \neq 1$. If $|g| < \infty$ then $|g|$ is coprime to p as g is semisimple; but $f(g)$ has p-power order which forces $f(g) = 1$, a contradiction. Hence g has infinite order.

If Γ is dense in G then so is Γ_0. The second assertion follows because $\Gamma_0 \cap G^0$ must meet the non-empty open set of regular semisimple elements of G^0. $\square$

Lemma 6.4.12 *Let Γ be a dense subgroup of G. Then $[\Gamma, \Gamma]$ is dense in $[G, G]$.*

Proof Let $C = \overline{\{[g, g'] \mid g, g' \in G\}}$. Then the set $C' := \{[g, g'] \mid g, g' \in \Gamma\}$ is dense in C, since Γ is dense in G. Now $[G, G]$ is the subgroup of G generated by C, so the abstract subgroup generated by C' is dense in G. But the abstract subgroup generated by C' is $[\Gamma, \Gamma]$, so we are done. $\square$

We now prove a variation on Lemma 6.4.12 for **finitely generated** dense subgroups. This takes some work. Note that it is not true in general that if Γ is a dense subgroup of G then Γ contains a finitely generated subgroup which is also dense in G: for instance, take

G to be H_k and Γ to be H, where $p > 0$, H is a positive-dimensional reductive group over $\overline{\mathbb{F}_p}$ and k is any algebraically closed field extension of $\overline{\mathbb{F}_p}$.

Proposition 6.4.13 *Let Γ be a finitely generated dense subgroup of G. Then there is a finitely generated subgroup Γ' of $[\Gamma, \Gamma]$ such that Γ' is dense in $[G, G]$.*

Proof Let N be maximal among the connected subgroups of $[G, G]^0$ such that N is generated by a finite subset of $[\Gamma, \Gamma]$; clearly such an N exists for dimension reasons. We show that $N = [G, G]^0$.

We claim first that $N \trianglelefteq G$. Write $N = \mathcal{G}(g_1, \dots, g_r)$ for some $g_1, \dots, g_r \in [\Gamma, \Gamma]$. If $g \in \Gamma$ then $\mathcal{G}(g_1, \dots, g_r, gg_1g^{-1}, \dots, gg_rg^{-1})$ is the subgroup generated by $N \cup gNg^{-1}$, which is connected. Maximality forces this subgroup to be N, so $gNg^{-1} = N$. But Γ is dense in G by Lemma 6.4.12, so $N \trianglelefteq G$, as claimed.

Suppose for a contradiction that N is properly contained in $[G, G]^0$. Let $\widetilde{G} = G/N$ and let $f: G \to \widetilde{G}$ be the canonical projection. Set $\widetilde{\Gamma} = f(\Gamma)$. Then $\widetilde{\Gamma}$ is dense in $\widetilde{G}$, so $[\widetilde{\Gamma}, \widetilde{\Gamma}] = f([\Gamma, \Gamma])$ is dense in $[\widetilde{G}, \widetilde{G}]$. Hence $[\widetilde{\Gamma}, \widetilde{\Gamma}] \cap [\widetilde{G}, \widetilde{G}]^0$ is dense in $[\widetilde{G}, \widetilde{G}]^0$. By Proposition 6.4.11 there is a finite-index subgroup $\widetilde{\Gamma}_0$ of $\widetilde{\Gamma}$ such that every non-trivial semisimple element of $\widetilde{\Gamma}_0$ has infinite order. Now $\widetilde{\Gamma}_1 := ([\widetilde{\Gamma}, \widetilde{\Gamma}] \cap [\widetilde{G}, \widetilde{G}]^0) \cap \widetilde{\Gamma}_0$ is a finite-index subgroup of $[\widetilde{\Gamma}, \widetilde{\Gamma}] \cap [\widetilde{G}, \widetilde{G}]^0$, so $\widetilde{\Gamma}_1$ is dense in $[\widetilde{G}, \widetilde{G}]^0$. Choose $h \in \widetilde{\Gamma}_1$ such that h is regular semisimple in $[\widetilde{G}, \widetilde{G}]^0$. Since $[\widetilde{G}, \widetilde{G}]^0$ is non-trivial, h is non-trivial and so h has infinite order. Choose $g \in [\Gamma, \Gamma]$ such that $f(g) = h$. Then $\mathcal{G}(g)$ has positive dimension and $\mathcal{G}(g)^0$ is not contained in N. Choose $s \in \mathbb{N}$ such that $\mathcal{G}(g^s) = \mathcal{G}(g)^0$. The subgroup of $[G, G]^0$ generated by $N \cup \{g^s\}$ is connected and properly contains N, contradicting the maximality of N. We deduce that $N = [G, G]^0$.

Since $[\Gamma, \Gamma]$ is dense in $[G, G]$, $[\Gamma, \Gamma]$ meets every coset of $[G, G]^0$ in $[G, G]$, so we can therefore pick coset representatives $m_1, \dots, m_b$ for $[G, G]^0$ in $[G, G]$ such that each m_j belongs to $[\Gamma, \Gamma]$. To finish, we observe that $[G, G]$ is generated by the generators $g_1, \dots, g_r$ for N together with $m_1, \dots, m_b$. $\qquad\square$

6.5   Representation Varieties and Character Varieties

In this section we collect some material on representation varieties and character varieties. Although it is slightly tangential to the main development of this chapter, this material is closely related and also of interest in its own right. We look again at character varieties in Sect. 6.9.

Definition 6.5.1 Let Γ be a finitely generated group. Define

$$\mathrm{Hom}(\Gamma, G) = \{\rho: \Gamma \to G \mid \rho \text{ is a homomorphism}\}.$$

We call $\mathrm{Hom}(\Gamma, G)$ the *representation variety (for Γ and G)*, and we call elements of $\mathrm{Hom}(\Gamma, G)$ *representations*.

(We can define $\mathrm{Hom}(\Gamma, H)$ for non-reductive H, but we stick to the reductive case as this is all we need.) We now show that $\mathrm{Hom}(\Gamma, G)$ has the structure of a variety. Fix a finite set of generators $\{\gamma_1, \ldots, \gamma_n\}$ for Γ. We have an injective function $i\colon \mathrm{Hom}(\Gamma, G) \to G^n$ given by $i(\rho) = (\rho(\gamma_1), \ldots, \rho(\gamma_n))$. The image of i is closed. To see this, choose a presentation $\Gamma = \langle \gamma_1, \ldots, \gamma_t \mid \mathcal{R} \rangle$. Then $i(\mathrm{Hom}(\Gamma, G)) = \bigcap_{w \in \mathcal{R}} \mathrm{ev}_w^{-1}(1)$, where the maps ev_w are as defined at the start of Sect. 5.6.

We identify $\mathrm{Hom}(\Gamma, G)$ with the closed subvariety $i(\mathrm{Hom}(\Gamma, G))$ of G^n. In particular, if Γ is the free group on $\gamma_1, \ldots, \gamma_n$ then we identify $\mathrm{Hom}(\Gamma, G)$ with G^n. It can be shown that the structure of a variety inherited by $\mathrm{Hom}(\Gamma, G)$ doesn't depend on the choice of generating set; see Exercise 6.3. It follows from the construction that if A is an abstract subgroup of G and $(g_1, \ldots, g_n) \in A^n$ then $(g_1, \ldots, g_n)$ belongs to $\mathrm{Hom}(\Gamma, G)$ if and only if the assignment $\gamma_i \mapsto g_i$ gives a well-defined homomorphism from Γ to A. Hence we can identify the set of homomorphisms $\mathrm{Hom}(\Gamma, A)$ with a subset of the variety $\mathrm{Hom}(\Gamma, G)$.

Given $\gamma \in \Gamma$, define $\mathrm{ev}_\gamma\colon \mathrm{Hom}(\Gamma, G) \to G$ by $\mathrm{ev}_\gamma(\rho) = \rho(\gamma)$. We see that if w is a word in the γ_i and $w(\gamma_1, \ldots, \gamma_n) = \gamma$ then ev_γ is the restriction of ev_w to $\mathrm{Hom}(\Gamma, G)$. Hence ev_γ is a morphism.

The group G acts on $\mathrm{Hom}(\Gamma, G)$ by conjugation:

$$(g \cdot \rho)(\gamma) = g\rho(\gamma)g^{-1}$$

for $\rho \in \mathrm{Hom}(\Gamma, G)$ and $g \in G$. If $\{\gamma_1, \ldots, \gamma_n\}$ is a set of generators for Γ then we clearly get

$$((g \cdot \rho)(\gamma_1), \ldots, (g \cdot \rho)(\gamma_n)) = g \cdot (\rho(\gamma_1), \ldots, \rho(\gamma_n)).$$

Hence the map i is G-equivariant. We obtain a quotient variety $\mathrm{Hom}(\Gamma, G) /\!/ G$; we call this the *character variety (for Γ and G)*, and we denote it by $C(\Gamma, G)$. Let $f \in k[G]^G$ and let $\gamma \in \Gamma$. Define $f_\gamma\colon \mathrm{Hom}(\Gamma, G) \to k$ by

$$f_\gamma(\rho) = f(\rho(\gamma)) = f(\mathrm{ev}_\gamma(\rho)). \tag{6.5.1}$$

Clearly f_γ belongs to the ring of invariants $k[\mathrm{Hom}(\Gamma, G)]^G$; we study $k[\mathrm{Hom}(\Gamma, G)]^G$ in Sect. 6.9 in more detail. We see that if w is a word in the γ_i and $w(\gamma_1, \ldots, \gamma_n) = \gamma$ then f_γ is the restriction of f_w to $\mathrm{Hom}(\Gamma, G)$.

We denote the canonical projection from $\mathrm{Hom}(\Gamma, G)$ to $C(\Gamma, G)$ by π_Γ or $\pi_{\Gamma, G}$. The map i gives rise to a map

$$i_G\colon C(\Gamma, G) \to G^n /\!/ G; \tag{6.5.2}$$

i_G is finite and injective, and if $p = 0$ then i_G is a closed embedding (see Lemma 3.6.1). Now let $j: G \to M$ be a homomorphism of reductive groups. We have a map $\mathrm{Hom}(\Gamma, G) \to \mathrm{Hom}(\Gamma, M)$ given by $\rho \mapsto j \circ \rho$. It is easily checked that this descends to a map

$$j_\Gamma : C(\Gamma, G) \to C(\Gamma, M). \tag{6.5.3}$$

We say that $\rho \in \mathrm{Hom}(\Gamma, G)$ is *separable* if the orbit $G \cdot \rho$ is separable. Clearly $G \cdot \rho$ is separable if and only if $G \cdot (\rho(\gamma_1), \ldots, \rho(\gamma_n))$ is separable.

Remark 6.5.2 If G_1, G_2 are reductive groups then there is an obvious isomorphism of varieties from $\mathrm{Hom}(\Gamma, G_1 \times G_2)$ to $\mathrm{Hom}(\Gamma, G_1) \times \mathrm{Hom}(\Gamma, G_2)$, and this gives rise to an isomorphism of varieties from $C(\Gamma, G_1 \times G_2)$ to $C(\Gamma, G_1) \times C(\Gamma, G_2)$ by Exercise 3.3(i).

Proposition 6.5.3 *Let Γ be a finitely generated group and let $\rho \in \mathrm{Hom}(\Gamma, G)$. Let $H = \overline{\rho(\Gamma)}$. Then H is G-completely reducible if and only if $G \cdot \rho$ is closed.*

Proof Choose generators $\gamma_1, \ldots, \gamma_n$ for Γ. Recall from above that the map defined by $\rho \mapsto (\rho(\gamma_1), \ldots, \rho(\gamma_n))$ gives an embedding of $\mathrm{Hom}(\Gamma, G)$ as a closed subvariety of G^n, and this embedding maps $G \cdot \rho$ onto $G \cdot (\rho(\gamma_1), \ldots, \rho(\gamma_n))$. Clearly, $H = \mathcal{G}((\rho(\gamma_1), \ldots, \rho(\gamma_n)))$, so the result follows from Theorem 5.3.3. $\qquad\square$

Proposition 6.5.4 *Let F be a finite group and let $m \in \mathbb{N}$. Then $C(F, \mathrm{GL}_m)$ is finite.*

Proof The points of $C(F, \mathrm{GL}_m)$ correspond to the closed orbits in $\mathrm{Hom}(F, \mathrm{GL}_m)$, which correspond to the isomorphism classes of semisimple m-dimensional representations of F, by Proposition 6.5.3. It is standard representation theory that the finite group F has only finitely many isomorphism classes of simple modules: the group algebra kF is finite-dimensional, and acts on any simple module through the semisimple ring kF/R, where R is the Jacobson radical. But kF/R is a finite product of matrix rings over k, one for each isomorphism class of simple kF-modules, by the Artin-Wedderburn Theorem. Now any semisimple representation is built by taking direct sums of simple representations, and for a fixed dimension m there are only finitely many ways to do this up to isomorphism, so we are done. $\qquad\square$

Remark 6.5.5 Here is an alternative proof of Proposition 6.5.4 using some of our earlier results. Let $\gamma_1, \ldots, \gamma_n$ be generators for F. By Theorems 5.6.3 and 5.6.5 there exist words $w_1, \ldots, w_s$ in the γ_j and elements $f_1, \ldots, f_s \in k[\mathrm{GL}_m]^{\mathrm{GL}_m}$ such that the functions $(f_j)_{w_j}$ generate $k[(\mathrm{GL}_m)^n]^{\mathrm{GL}_m}$. Our choice of generators gives us an embedding i of $\mathrm{Hom}(F, \mathrm{GL}_m)$ in $(\mathrm{GL}_m)^n$, and we get a finite injective map $i_{\mathrm{GL}_m} : C(F, \mathrm{GL}_m) \to (\mathrm{GL}_m)^n /\!/ \mathrm{GL}_m$ as per (6.5.2). Hence to show that $C(F, \mathrm{GL}_m)$ is finite, it is enough to show that $(f_j)_{w_j} \circ i_{\mathrm{GL}_m}$ has finite image for each $1 \leq j \leq s$. It is clear that $(f_j)_{w_j} \circ i_{\mathrm{GL}_m} = (f_j)_{\delta_j}$, where $\delta_j := w_j(\gamma_1, \ldots, \gamma_n)$.

So fix j. Untangling the definitions, we see that $(f_j)_{\delta_j}$ factors as

$$C(F, \mathrm{GL}_m) \xrightarrow{\ (\mathrm{ev}_{\delta_j})_{\mathrm{GL}_m}\ } \mathrm{GL}_m /\!/ \mathrm{GL}_m \xrightarrow{\ f_j\ } k.$$

Now the cyclic subgroup of F generated by δ_j has order t_j for some $t_j \in \mathbb{N}$, so the image of ev_{δ_j} in $\mathrm{GL}_m /\!/ \mathrm{GL}_m$ is contained in $\pi_{\mathrm{GL}_m}(\{g \in \mathrm{GL}_m \mid g^{t_j} = 1\})$, which is finite by Exercise 5.10. Hence $C(F, \mathrm{GL}_m)$ is finite.

We finish by looking at how character varieties behave under an algebraically closed field extension E/k. Let Γ be a finitely generated group and fix generators $\gamma_1, \ldots, \gamma_n$ for Γ. We get embeddings $i \colon \mathrm{Hom}(\Gamma, G) \to G^n$ and $i' \colon \mathrm{Hom}(\Gamma, G_E) \to (G_E)^n$ as above. It is clear that $i(\mathrm{Hom}(\Gamma, G))_E = i'(\mathrm{Hom}(\Gamma, G_E))$, so $i' = i_E$. Hence we may identify $\mathrm{Hom}(\Gamma, G_E)$ with $\mathrm{Hom}(\Gamma, G)_E$ as E-varieties; we see that if $\rho \in \mathrm{Hom}(\Gamma, G_E)$ then ρ is a k-point if and only if $\rho(\gamma_i) \in G$ for $1 \leq i \leq n$ if and only if $\rho(\Gamma) \subseteq G$. By the results in Sect. 3.9, we may identify $C(\Gamma, G_E)$ with $C(\Gamma, G)_E$. The map j_Γ from (6.5.3) is compatible with extending fields from k to E.

6.6   Rationality for GL$_m$

In this section we prove that the Rationality Theorem holds for GL_m, where it reduces to fairly standard representation theory. We need this result only in characteristic 0 — later we give a proof of the Rationality Theorem which works for all G when $p > 0$ — but the arguments below hold in any characteristic.

If $H \subseteq \mathrm{GL}_m$ is either a finite GL_m-cr subgroup or a connected reductive GL_m-cr subgroup, then its action on the natural module k^m is semisimple, and Theorem 6.1.1 follows from the fact that every such representation is in fact defined over the algebraic closure of the prime field. In fact, such a statement is true for an arbitrary non-connected reductive H. This follows from work of Achar-Hardesty-Riche [2], who construct all simple modules for a non-connected reductive group over an algebraically closed field. We give some details of the construction from [2, §2]; we have chosen to adjust some of their notation slightly so as not to clash with the rest of the book.

Let H be a non-connected reductive group over the algebraically closed field k, and let $A = H/H^0$ denote the component group. Recall that the simple modules for the identity component H^0 are classified by *highest weight*: if we fix a maximal torus T of H^0 and a Borel subgroup B containing T then we can define a subset $X^+ \subseteq X(T)$ of *dominant weights* and construct for each $\chi \in X^+$ a simple H^0-module $L(\chi)$. Every simple module is isomorphic to precisely one $L(\chi)$. This theory is described in full detail in [82, Part II, Ch. 2], for example. We can make A act on X^+ in a fairly obvious way—given any

$a \in A$ we can find a representative $h \in H$ for a which fixes B and T. For each $\chi \in X^+$ we therefore obtain a stabiliser A_χ in A, and we can let H_χ denote the preimage in H of this subgroup. In [2, §2.4] it is explained how to construct from this data a twisted version of the group algebra of A_χ, which we denote here by $\mathscr{A}_\chi$. Briefly, for each $a \in A_\chi$ we obtain an isomorphism $\theta_a : L(\chi) \to L(\chi)$, and the difference between $\theta_a \circ \theta_b$ and θ_{ab} is given by a scalar $\alpha(a, b) \in k^*$, by Schur's Lemma. The function $\alpha : A_\chi \times A_\chi \to k^*$ is then a 2-cocycle which we can use to twist the usual multiplication in the group algebra of A_χ: we let $\mathscr{A}_\chi$ be the algebra with k-basis $\{\rho_a \mid a \in A\}$, and multiplication given by $\rho_a \rho_b := \alpha(a, b)\rho_{ab}$.

Finally, given a simple module V for $\mathscr{A}_\chi$, we obtain a module $V \otimes L(\chi)$ for H_χ and hence an H-module $L(V, \chi) := \mathrm{Ind}_{H_\chi}^{H}(V \otimes L(\chi))$; this is induction of modules in the algebraic groups setting (see [82, I.3.3]), although note that H_χ contains H^0 so has finite index in H. Then [2, Lem. 2.12] tells us that the $L(V, \chi)$ form a complete list of the simple H-modules up to isomorphism. (Note: there might be some repeats in this list; the question of isomorphisms between the $L(V, \chi)$ is also addressed in [2, §2.8], but that does not concern us here.)

Now let k_0 denote the algebraic closure of the prime field in k. Then by Theorem 2.6.2 we can fix a k_0-structure on H; note that then H^0 is also k_0-defined, and we may choose B and T in the previous paragraph to be k_0-defined also. With these choices made, all the objects in the previous paragraph are also k_0-defined. For the simple modules for the connected reductive group H^0, this follows from [82, Cor. II.2.9]. Given this, it is clear that the subgroup A_χ, its preimage H_χ, and the twisted group algebra $\mathscr{A}_\chi$ can be defined over k_0 (the last since the 2-cocycle α described above is defined over k_0). Now $\mathscr{A}_\chi$ is a finite-dimensional algebra over an algebraically closed field, so its simple modules are all k_0-defined by the Artin-Wedderburn Theorem. We deduce that every simple H-module over k has a k_0-structure, and we can now prove:

Lemma 6.6.1 *Theorem 6.1.1 holds if $G = \mathrm{GL}_m$.*

Proof It is obviously enough to prove the result when k is the algebraic closure of the prime field and E is any algebraically closed extension of that. Since a Levi subgroup of GL_m is a direct product of smaller general linear groups, it is enough, by Lemma 4.2.7(i), to deal with the case that H is GL_m-ir. But this happens precisely when the representation of H on E^m is simple. By the discussion before the statement, any such representation has a k-structure: that is, after adjusting by an isomorphism if necessary, we may assume that there is a compatible k-structure on H and on the module $E^m = k^m \otimes E$. In terms of H as a subgroup of GL_m, adjusting by an isomorphism is simply conjugating by an element of $\mathrm{GL}_m(E)$, and saying that we have compatible k-structures on H and E^m then implies that this conjugate of H is k-defined *as a subgroup of* GL_m. This completes the proof. $\qquad\square$

Remark 6.6.2 We can deduce from the discussion above that for a given non-connected reductive group H and a fixed $m \in \mathbb{N}$, there are only countably many conjugacy classes of homomorphisms $\rho : H \to \mathrm{GL}_m$ with GL_m-cr image. To see this, note first that the classification result described before the statement of the lemma implies that there are only countably many simple modules for H. For the set of dominant weights X^+ is countable, and for each $\chi \in X^+$ there are at most $|A_\chi| \le |A|$ simple modules for the twisted group algebra $\mathscr{A}_\chi$, for dimension reasons. Now a semisimple representation of H of dimension m is built by taking a partition of m and assigning to each part of the partition a simple module of the correct dimension. Since m has finitely many partitions, there are only countably many semisimple representations we can build, up to isomorphism, which gives the result. See Exercise 6.6 below for an analogous result for general G in place of GL_m.

Note that when $p = 0$ and H is semisimple we have a stronger result: the Weyl dimension formula for simple representations of a connected semisimple group (see [172, p. 228, Exercise (c)]) shows that there are only finitely many simple modules of dimension $\le m$. Hence there are only *finitely* many semisimple modules of dimension m (note also that in characteristic 0 every representation is semisimple, so has G-cr image). Obviously this stronger result fails when $p > 0$, even in the semisimple case: for if H has one simple module of a given dimension, then taking Frobenius twists yields infinitely many non-isomorphic simple modules of that dimension.

Let F be a finite group. Recall from Sect. 6.5 that we may identify $C(F, G)_E$ with $C(F, G_E)$. Now suppose that $C(F, G)$ is finite. Then $C(F, G) = C(F, G_E)$, so for each $x \in C(F, G)$, there exists $\rho_x \in \mathrm{Hom}(F, G)$ such that $\pi_F(\rho_x) = x$. We can assume that $G \cdot \rho_x$ is closed for each x by Proposition 3.1.5(v), so each ρ_x has G-cr image by Proposition 6.5.3, and hence has G_E-cr image by Proposition 4.3.1. We obtain a finite set $A_F = \{\rho_x \mid x \in C(F, G)\}$ with the property that any $\rho \in \mathrm{Hom}(F, G_E)$ with G_E-cr image is G_E-conjugate to some element of A_F.

Lemma 6.6.3 *Suppose $p > 0$ and*

$$C(F, G) \text{ is finite for every finite group } F.$$

Then Theorem 6.1.1 holds for G.

Proof We keep the notation of the paragraph preceding the statement of the lemma. Let H be a G_E-cr subgroup of G_E. Then H is reductive, so by Proposition 2.6.3, there is a family of finite subgroups $H_1 \subseteq H_2 \subseteq \ldots$ of H such that $\bigcup_{i=1}^\infty H_i$ is dense in H. Hence for i sufficiently large—say, for $i \ge i_0$—we see that $C_{G_E}(H_i) = C_{G_E}(H)$ and H_i is G_E-cr. By discarding $H_1, H_2, \ldots, H_{i_0-1}$, we may assume without loss that $i_0 = 1$. Let $\rho : H \to G_E$ be the embedding of H in G_E and let $\rho_i : H_i \to G_E$ be the restriction of ρ. We obtain subsets A_{H_i} for $i \in \mathbb{N}$ as above. For each $i > 1$, we can replace the elements of A_{H_i} by suitable G-conjugates in order to ensure that $\rho|_{H_1} \in A_{H_1}$ for each $\rho \in H_i$.

For each i, we can pick $g_i \in G_E$ such that $g_i \cdot \rho_i \in A_{H_i}$. By construction, $(g_i \cdot \rho_i)|_{H_1} = g_i \cdot \rho_1$ belongs to A_{H_1} for each $i \in \mathbb{N}$. Now A_{H_1} is finite, so replacing the sequences (H_i) and (g_i) by appropriate subsequences, we can assume that $g_1 \cdot \rho_1 = g_2 \cdot \rho_1 = \cdots$. But $C_G(H_1) = C_G(H)$, so $g_1 \cdot \rho = g_2 \cdot \rho = \cdots$. Thus

$$g_1 H_i g_1^{-1} = (g_1 \cdot \rho)(H_i) = (g_i \cdot \rho)(H_i) = (g_i \cdot \rho_i)(H_i) \subseteq G$$

for all i. Since $\bigcup_{i=1}^{\infty} g_1 H_i g_1^{-1}$ is dense in $g_1 H g_1^{-1}$, $g_1 H g_1^{-1}$ is k-defined. $\qquad\square$

Remark 6.6.4 When $p > 0$, Lemma 6.6.3 gives an alternative proof that Theorem 6.1.1 holds for GL_m: for $C(F, \mathrm{GL}_m)$ is finite for every finite group F, by Proposition 6.5.4.

6.7 Proof of the Rationality Theorem

In this section we complete the proof of the Rationality Theorem. For the characteristic 0 case, we use Lemma 6.6.1 and a result from Chap. 10; this argument is given at the end of the section. To tackle the positive characteristic case we apply some machinery from the theory of Bruhat-Tits buildings (see [84] for a recent and comprehensive account). Until further notice we assume that $k = \overline{\mathbb{F}_p}$. Recall that if K is a k-algebra and X is a variety then we denote by $X(K)$ the set of K-points of X; elements of $X(K)$ correspond to homomorphisms of k-algebras from $k[X]$ to K. If $f : X \to Y$ is a map of k-varieties then we obtain a function $f_K : X(K) \to Y(K)$. We may realise $X(K)$ more concretely as follows. Since X is affine by assumption, we may regard X as the subset of k^r for some $r \in \mathbb{N}$ defined by the vanishing of some polynomials $f_1, \ldots, f_s$ in r variables over k. The coordinate ring of k^r is the polynomial ring $k[T_1, \ldots, T_r]$ in indeterminates $T_1, \ldots, T_r$. We identify $X(K)$ with the subset of K^r defined by the vanishing of $f_1, \ldots, f_s$. We may view any $f \in k[T_1, \ldots, T_r]$ as a function from K^r to K.

Since $k = \overline{\mathbb{F}_p}$, we can pick a power q of p such that all the coefficients of the f_i belong to $\mathbb{F}_q$. For any $t \in \mathbb{N}$ we have a function $F_{q^t} : K^r \to K^r$ given by

$$F_{q^t}(a_1, \ldots, a_r) = ((a_1)^{q^t}, \ldots, (a_r)^{q^t}).$$

It follows from the choice of q that F_{q^t} maps $X(K)$ into $X(K)$. (In fact, the hypothesis on q implies that X is defined over $\mathbb{F}_q$ in the sense of Chap. 11, and F_{q^t} is the so-called *relative Frobenius map*.)

We need to recall some facts about discrete valuations. A *discrete valuation* on a field K is a group homomorphism $\omega : K^* \to \mathbb{Z}$ such that

$$\omega(a_1 + a_2) \geq \min\{\omega(a_1), \omega(a_2)\} \tag{6.7.1}$$

for all $a_1, a_2 \in K^*$ such that $a_1 + a_2 \neq 0$. We will always assume that valuations are non-trivial: that is, that ω is a non-trivial homomorphism. We extend ω to a function on K by defining $\omega(0) = \infty$. For background material on valuations, see [93, Ch. XII].[2] We define the *valuation ring* R of K by

$$R = \{a \in K \mid \omega(a) \geq 0\}.$$

Then R is a local ring with unique maximal ideal $\mathfrak{m} := \{a \in K \mid \omega(a) > 0\}$; we call $R/\mathfrak{m}$ the *residue field* of K.

We define an ultrametric d_ω on K by $d_\omega(a, b) = l^{-\omega(b-a)}$ if $a \neq b$ and $d_\omega(a, a) = 0$, where l is a fixed real number with $l > 1$ (it is conventional to take l to be the characteristic of the residue field if this is positive). We say that the valuation ω is *complete* if d_ω is a complete metric on K. For instance, let K be the field of Laurent series $k((T))$ and define $\omega(f) = n$ if $f = \sum_{i=n}^{\infty} a_i T^i$ with $a_n \neq 0$; then ω is a complete discrete valuation on K.

A discrete valuation ω on K is said to be *Henselian* if ω has a unique extension to a valuation on K' for every algebraic field extension K'/K (see [84, §2.1]). In this case the residue field of K' is an algebraic extension of the residue field of K: so if the latter is algebraically closed then these residue fields are equal. If ω is complete then ω is Henselian.

Now suppose that K is a field extension of k endowed with a Henselian discrete valuation $\omega \colon K \to \mathbb{Z} \cup \{\infty\}$ such that k is the residue field (for instance, we can take $K = k((T))$ and ω as above). We use d_ω to define an ultrametric on K^r, which we also denote by d_ω, given by $d_\omega((b_1, \ldots, b_r), (a_1, \ldots, a_r)) = \max_i d_\omega(b_i, a_i)$. The resulting topology on $X(K)$ does not depend on the choice of embedding of X in an affine space. If $C \subseteq X(K)$ and $f \in k[X]$ then we say that f is *bounded on C* if $\omega(f(C))$ is bounded below as a subset of $\mathbb{Z}$; otherwise we say that f is *unbounded on C*. It is clear that the sum and product of bounded functions is bounded. We say $C \subseteq X(K)$ is *bounded* if f is bounded on C for all $f \in k[X]$; otherwise we say that C is *unbounded*.

We will apply this formalism to $\mathrm{Hom}(F, G)$, where F is a finite group. Recall that if C is a subgroup of $G(K)$ then we define $\mathrm{Hom}(F, C) \subseteq \mathrm{Hom}(F, G(K))$ to be the set of homomorphisms from F to C. If C is bounded as a subset of $G(K)$ then $\mathrm{Hom}(F, C)$ is a bounded subset of $\mathrm{Hom}(F, G(K))$; to see this, observe that the functions $f \circ \mathrm{ev}_\gamma$ for $\gamma \in F$ and $f \in k[G]$ generate $k[\mathrm{Hom}(F, G)]$ as a k-algebra.

Lemma 6.7.1 *Let X be a k-variety. Suppose there exist $x \in X(K)$ and $f \in k[X]$ such that $\omega(f(x)) < 0$. Then f is unbounded on $X(K)$.*

---

[2] A different convention for valuations is adopted in [93, Ch. XII]: one reverses the direction of the inequality in (6.7.1) and defines $\omega(0) = 0$.

Proof Pick an embedding of X in k^r and let $T_1, \ldots, T_r$ be the coordinate functions, as above. We can identify x with a tuple $(a_1, \ldots, a_r) \in K^r$. We can write f as a polynomial in the T_i with coefficients in k; let S be the set of these coefficients. By picking a sufficiently large power q of p, we get functions $F_{q^t} : X(K) \to X(K)$ for all $t \in \mathbb{N}$ as described above. Increasing q if necessary, we can assume that $S \subseteq \mathbb{F}_q$. It follows that for every $t \in \mathbb{N}$ we have

$$f(F_{q^t}(x)) = f(F_{q^t}(a_1, \ldots, a_r)) = f((a_1)^{q^t}, \ldots, (a_r)^{q^t}) = f(a_1, \ldots, a_r)^{q^t} = f(x)^{q^t}.$$

Hence

$$\omega(f(F_{q^t}(x))) = \omega(f(x)^{q^t}) = q^t \omega(f(x)) \to -\infty \text{ as } t \to \infty.$$

This shows that f is unbounded on $X(K)$. $\qquad\square$

Lemma 6.7.2 *Suppose $C(F, G)$ is infinite. Then there exist $f \in k[\mathrm{Hom}(F, G)]^G$ and a field extension K of k admitting a Henselian discrete valuation ω with residue field k such that f is unbounded on $\mathrm{Hom}(F, G)(K)$.*

Proof Choose a non-trivial field extension K_0 of k such that K_0 admits a Henselian discrete valuation ω_0 with residue field k. Let $\overline{\omega_0}$ be the unique extension of ω_0 to a valuation on $\overline{K_0}$. Since $C(F, G)$ is infinite, there exists $f \in k[C(F, G)]$ such that f (regarded as a function from $C(F, G)(\overline{K_0})$ to $\overline{K_0}$) has cofinite image in $\overline{K_0}$. So there exist a finite field extension K_1/K_0 and a point $y \in C(F, G)(K_1)$ such that $\overline{\omega_0}(f(y)) < 0$. Then $\omega_1(f(y)) < 0$, where ω_1 is the restriction of $\overline{\omega_0}$ to K_1. Since the map

$$(\pi_F)_{\overline{K_0}} : \mathrm{Hom}(F, G)(\overline{K_0}) \to C(F, G)(\overline{K_0})$$

is surjective, there exist a finite field extension K/K_1 and a point $x \in \mathrm{Hom}(F, G)(K)$ such that $(\pi_F)_K(x) = y$. Let ω be the restriction of ω_0 to K; then ω is discrete because K/K_0 is finite, and clearly ω is Henselian. We may regard f as an element of $k[\mathrm{Hom}(F, G)]^G$. Since $\omega(f(x)) = \omega(f(y)) = \omega_1(f(y)) < 0$, we deduce from Lemma 6.7.1 that f is unbounded on $\mathrm{Hom}(F, G)(K)$. $\qquad\square$

Our proof of Theorem 6.2.2 rests on the next proposition, which comes out of the theory of Bruhat-Tits buildings associated to connected reductive groups over a field with a Henselian discrete valuation [84]. Let M be a connected reductive group over $\overline{\mathbb{F}_p}$. Let K be a field extension of k endowed with a Henselian discrete valuation ω with residue field k. We consider the group of K-points $M(K)$. Two important subgroups $M(K)^0$ and $M(K)^1$ of $M(K)$ are defined in [84, Def. 2.6.23, Notn. 2.6.15]. We assume now that M is semisimple and simply connected; this ensures that $M(K)^0 = M(K)^1 = M(K)$ [84, Rem. 4.1.13].

Associated to M we have a *Bruhat-Tits building* $\mathcal{B}(M, K)$ (see [84, §4.1]); this is (the geometric realisation of) a certain polysimplicial complex in the sense of [84, Def. 1.5.1]. We call the polysimplices of this complex *facets*. The conjugation action of M on itself gives rise to an action of $M(K)$ on $\mathcal{B}(M, K)$ by polysimplicial automorphisms [84, Prop. 7.9.5]. There is a canonical $M(K)$-invariant metric on $\mathcal{B}(M, K)$ (see [84, §4.2]). If $x \in \mathcal{B}(M, K)$ then x belongs to a facet $\mathcal{F}$, and the stabiliser $M(K)_x$ is equal to the stabiliser $M(K)_{\mathcal{F}}$ (see [84, Axiom 4.1.20]). We call a subgroup of the form $M(K)_{\mathcal{F}}$ a *parahoric subgroup* of $M(K)$. It follows from [84, Axiom 4.1.2] that parahoric subgroups are bounded in the sense we defined above. By [84, Axiom 4.1.9, §7.6], $\mathcal{B}(M, K)$ is the restricted building (in the sense of [84, Def. 1.5.20]) of the so-called *Iwahori-Tits system* $(M(K), \mathcal{I}, N, R)$ associated to M and K, and the parahoric subgroups are the parabolic subgroups of this Tits system [84, Rem. 4.1.11]. This implies that every parahoric subgroup is self-normalising in $M(K)$. Moreover, it turns out that R is finite (see [84, §1.3, §7.5]), so there are only finitely many parahoric subgroups up to $M(K)$-conjugacy.

We apply the above formalism to the identity component G^0 of our possibly non-connected reductive group G, assuming that G^0 is semisimple and simply connected. The group $G(K)$ acts on $G^0(K)$ by conjugation, and this gives rise to an action of $G(K)$ on $\mathcal{B}(G^0, K)$ [84, Prop. 7.9.5]. Let $P_1, \ldots, P_t$ be a set of representatives for the conjugacy classes of parahoric subgroups of $G^0(K)$, and set $Q_i = N_{G(K)}(P_i)$. Then each Q_i is a finite extension of $N_{G^0(K)}(P_i) = P_i$. It follows from [84, Lem. 2.2.6] (applied to the bounded subgroup P_i and a set of coset representatives for Q_i/P_i) that each Q_i is a bounded subgroup of $G(K)$.

Proposition 6.7.3 *Suppose G^0 is semisimple and simply connected. If F is a finite subgroup of $G(K)$ then F is $G^0(K)$-conjugate to a subgroup of Q_i for some $1 \leq i \leq t$.*

Proof Let F be any finite subgroup of $G(K)$. Pick any $y \in \mathcal{B}(G^0, K)$. The orbit $F \cdot y$ is finite. It follows from [84, Cor. 4.2.12] that the Bruhat-Tits fixed point theorem [84, Thm. 1.1.15] holds for $\mathcal{B}(G^0, K)$, so there exists x in the convex hull of $F \cdot y$ such that F fixes x. Now $P := G^0(K)_x$ is a parahoric subgroup of $G^0(K)$. Since F fixes x and stabilises $G^0(K)$, F is contained in $N_{G(K)}(P)$. We can choose $g \in G^0(K)$ and $1 \leq i \leq t$ such that $gPg^{-1} = P_i$. Then

$$gFg^{-1} \subseteq gN_{G(K)}(P)g^{-1} = N_{G(K)}(gPg^{-1}) = N_{G(K)}(P_i) = Q_i,$$

as required. $\qquad\square$

For the rest of the section we allow k to be an arbitrary algebraically closed field again.

Lemma 6.7.4 *Suppose $p > 0$, and suppose G^0 is semisimple and simply connected. Then for any finite group F, $C(F, G)$ is finite.*

Proof Recall that G admits an $\overline{\mathbb{F}_p}$-structure, so by the arguments on field extensions at the end of Sect. 6.5, it is enough to prove the result under the extra assumption that $k = \overline{\mathbb{F}_p}$. Suppose $C(F, G)$ is infinite. Choose K, f as in Lemma 6.7.2. Let $\rho \in \mathrm{Hom}(F, G)(K)$. Now $\rho(F)$ is finite, so by Proposition 6.7.3 there exists $g \in G^0(K)$ and i with $1 \le i \le t$ such that $g\rho(F)g^{-1} \subseteq Q_i$; that is, $g \cdot \rho$ belongs to $\mathrm{Hom}(F, Q_i)$. Since f is G-invariant we have $f(\rho) = f(g \cdot \rho)$. It follows that

$$f(\mathrm{Hom}(F, G)(K)) = \bigcup_{i=1}^{t} f(\mathrm{Hom}(F, Q_i)).$$

But $C := \bigcup_{i=1}^{t} \mathrm{Hom}(F, Q_i)$ is bounded since each Q_i is bounded, so f is bounded on C. Hence f is bounded on $\mathrm{Hom}(F, G)(K)$, a contradiction. We conclude that $C(F, G)$ is finite after all. $\qquad\square$

Lemma 6.7.5 *Let J be a finite group acting on a torus Z, and let F be a finite group. Then $C(F, J \ltimes Z)$ is finite.*

Proof This is immediate from Corollary 2.10.7. $\qquad\square$

Lemma 6.7.6 *Let $f\colon G_1 \to G_2$ be an isogeny of reductive groups. The following are equivalent.*

(i) *$C(F, G_1)$ is finite for every finite group F.*
(ii) *$C(F, G_2)$ is finite for every finite group F.*

Proof Given a finite group F, set $\mathrm{Hom}(F, G_i)_c = \{\rho \in \mathrm{Hom}(F, G_i) \mid G_i \cdot \rho \text{ is closed}\}$ for $i = 1, 2$. It suffices to prove that $\mathrm{Hom}(F, G_1)_c$ is a finite union of G_1-orbits for every finite group F if and only if $\mathrm{Hom}(F, G_2)_c$ is a finite union of G_2-orbits for every finite group F. For each $n \in \mathbb{N}$, the map $f_n := f \times \cdots \times f$ from $(G_1)^n$ to $(G_2)^n$ is finite. Hence if $\mathbf{g}_2 \in (G_2)^n$ then the irreducible components of $C := (f_n)^{-1}(G_2 \cdot \mathbf{g}_2)$ are precisely the $(G_1)^0$-orbits in C. It follows that $G_2 \cdot \mathbf{g}_2$ is closed if and only if $G_1 \cdot \mathbf{g}_1$ is closed for every $\mathbf{g}_1 \in C$.

Let F be a finite group with generators $\gamma_1, \ldots, \gamma_n$. We embed $\mathrm{Hom}(F, G_i)$ in $(G_i)^n$ in the usual way for $i = 1, 2$. It follows from the previous paragraph that $f_n(\mathrm{Hom}(F, G_1)_c) \subseteq f_n(\mathrm{Hom}(F, G_2)_c)$. Conversely, if $\rho \in \mathrm{Hom}(F, G_2)$ then $f^{-1}(\rho(F))$ has order at most $|F| \cdot |\ker(f)|$; hence $\mathrm{Hom}(F, G_2)_c \subseteq \bigcup_i f_n(\mathrm{Hom}(F_i, G_1)_c)$ by the previous paragraph, where the F_i are representatives for the isomorphism classes of finite groups of order at most $|F| \cdot |\ker(f)|$. This gives the result. $\qquad\square$

Lemma 6.7.7 *Suppose $p > 0$. Let M be a semisimple group and let J be a finite group acting on M. Then for any finite group F, $C(F, J \ltimes M)$ is finite.*

Proof Let $\widetilde{M}$ be the simply connected cover of M. Any automorphism of M lifts to a unique automorphism of $\widetilde{M}$ by [173, (9.16)], so we get an action of J on M, and this gives rise to an isogeny from $J \ltimes \widetilde{M}$ to $J \ltimes M$. The result now follows from Lemmas 6.7.4 and 6.7.6. $\square$

Proposition 6.7.8 *Suppose $p > 0$. Then for any finite group F, $C(F, G)$ is finite.*

Proof By Proposition 2.4.5 there is a finite subgroup J of G such that $G = JG^0$. The multiplication map gives an isogeny from $J \ltimes G^0$ to G. Let M be the adjoint form of $[G^0, G^0]$ and let $Z = Z(G^0)^0$. There is an isogeny from G^0 to $M \times Z$ given by factoring out the scheme-theoretic centre of $[G^0, G^0]$; since the scheme-theoretic centre is J-stable, the J-action on G^0 descends to a J-action on $M \times Z$, and we obtain an isogeny from $J \ltimes G^0$ to $J \ltimes (M \times Z)$. By Lemma 6.7.6, it suffices to prove that $C(F, J \ltimes (M \times Z))$ is finite for every finite group F.

So let F be a finite group. The J-action on $M \times Z$ stabilises both M and Z, so we get J-actions on M and Z and a resulting product action of $J \times J$ on $M \times Z$. Set $M_1 = J \ltimes M$ and $Z_1 = J \ltimes Z$. We have an embedding of $J \ltimes (M \times Z)$ in $(J \times J) \ltimes (M \times Z) = M_1 \times Z_1$ given by $(\gamma, m, z) \mapsto (\gamma, \gamma, m, z)$. Recall that points in the character variety correspond to closed conjugacy classes of homomorphisms. If $\rho \in \mathrm{Hom}(F, J \ltimes (M \times Z))$ then, since $J \ltimes (M \times Z)$ has finite index in $M_1 \times Z_1$, the $(M_1 \times Z_1)$-conjugacy class of ρ is a finite union of $(J \ltimes (M \times Z))$-conjugacy classes, so $(J \ltimes (M \times Z)) \cdot \rho$ is closed if and only if $(M_1 \times Z_1) \cdot \rho$ is closed. Hence it is enough to show that $C(F, M_1 \times Z_1)$ is finite. But $C(F, M_1 \times Z_1)$ is isomorphic to $C(F, M_1) \times C(F, Z_1)$ by Remark 6.5.2, which is finite by Lemmas 6.7.7 and 6.7.5. This proves the result. $\square$

Proof of Theorem 6.1.1 If $p > 0$ then the result follows from Proposition 6.7.8 and Lemma 6.6.3. So suppose $p = 0$. Let E be an algebraically closed field extension of k and let H be a G_E-cr subgroup of G_E. Below if $S \subseteq G$ then we write $\mathcal{G}(S)$ for the closure of $\langle S \rangle$ in G (as usual) and $\mathcal{G}^E(S)$ for the closure of $\langle S \rangle$ in G_E. Pick an embedding of G in some GL_m. This gives an embedding of G_E in $(\mathrm{GL}_m)_E$, and it follows from Remark 4.1.7(i) that H is $(\mathrm{GL}_m)_E$-cr. Applying Lemma 6.6.1, we obtain $g \in (\mathrm{GL}_m)_E$ such that gHg^{-1} is k-defined: so $gHg^{-1} = M_E$ for some subgroup M of GL_m. Since k is solid, we can choose $n \in \mathbb{N}$ and a generating tuple $\mathbf{m} \in M^n$ for M. Then $\mathbf{m}$ is also a generating tuple for M_E, so $g^{-1} \cdot \mathbf{m}$ is a generating tuple for H. Now M is reductive, since gHg^{-1} is, so M_E is reductive, so M_E is $(\mathrm{GL}_m)_E$-cr by Remark 4.1.7(i). Hence $(\mathrm{GL}_m)_E \cdot \mathbf{m}$ is closed, by Theorem 5.3.3.

Consider $C := (\mathrm{GL}_m)_E \cdot \mathbf{m} \cap (G_E)^n$; note that C is k-defined. It follows from Proposition 10.5.18 and Remark 10.5.19 that C is a finite union of G_E-orbits, each of which is closed. Hence the irreducible components of C are precisely the $(G_E)^0$-orbits, which implies that each $(G_E)^0$-orbit in C contains a k-point. In particular, there is a k-point $\mathbf{h}_1 := g_1 \cdot (g^{-1} \cdot \mathbf{m})$ for some $g_1 \in (G_E)^0$. Let $H_1 = \mathcal{G}(\mathbf{h}_1)$. Then

$$(H_1)_E = \mathcal{G}^E(\mathbf{h}_1) = \mathcal{G}^E(g_1 \cdot (g^{-1} \cdot \mathbf{m})) = g_1 \mathcal{G}^E(g^{-1} \cdot \mathbf{m})g_1^{-1} = g_1 H g_1^{-1},$$

so we are done. $\square$

6.8 Proof of the Finiteness Theorem

Before completing the proofs, we need some final pieces of notation and preliminary results. We assume in this section that H is a reductive subgroup of G. We define $\Xi^n_{G,H}$ to be the restriction of $\Psi^n_{G,H}$ to $H_0^n /\!/ H$. Define

$$X^n_{G,H} := \overline{\Psi^n_{G,H}(H^n /\!/ H)} \quad \text{and} \quad Y^n_{G,H} := \overline{\Xi^n_{G,H}(H_0^n /\!/ H)}.$$

It will follow from Theorem 6.2.1 that $X^n_{G,H} = \Psi^n_{G,H}(H^n /\!/ H)$ and $Y^n_{G,H} = \Xi^n_{G,H}(H_0^n /\!/ H)$ but we do not know this a priori—this is a cause of complications in some of the proofs below.

It is clear that

$$X^n_{G_E, H_E} = (X^n_{G,H})_E \quad \text{and} \quad Y^n_{G_E, H_E} = (Y^n_{G,H})_E. \tag{6.8.1}$$

It follows from Exercise 3.2(i) that $\left(\Psi^n_{G,H}\right)^{-1}(Y^n_{G,H}) = H_0^n /\!/ H$, so $\left(\Xi^n_{G,H}\right)^{-1}(D) = \left(\Psi^n_{G,H}\right)^{-1}(D)$ for any $D \subseteq Y^n_{G,H}$. It is not hard to check that

$$(i_s)_G \circ \Psi^n_{G,H} = \Psi^{n+s}_{G,H} \circ (i_s)_H$$

for any $s \in \mathbb{N}$, so $(i_s)_G(X^n_{G,H}) \subseteq X^{n+s}_{G,H}$ and $(i_s)_G(Y^n_{G,H}) \subseteq Y^{n+s}_{G,H}$. Since $\Psi^n_{G,H}$ is $\mathrm{Aut}(\Gamma_n)$-equivariant and H_0^n is $\mathrm{Aut}(\Gamma_n)$-stable, if D is a subset of $X^n_{G,H}$ (resp., of $Y^n_{G,H}$) then $Q^G_s(D)$ is a subset of $X^{n+s}_{G,H}$ (resp., of $Y^{n+s}_{G,H}$).

If M is a reductive subgroup of H then we have

$$\overline{\Psi^n_{G,H}\left(X^n_{H,M}\right)} = X^n_{G,M} \quad \text{and} \quad \overline{\Psi^n_{G,H}(Y^n_{H,M})} = Y^n_{G,M}. \tag{6.8.2}$$

(To see this, note that $\overline{\Psi^n_{G,H}\left(\overline{D}\right)} = \overline{\Psi^n_{G,H}(D)}$ for any $D \subseteq H^n /\!/ H$ since $\Psi^n_{G,H}$ is continuous, and apply (6.2.1).)

Remark 6.8.1 Let E be any irreducible component of $X^n_{G,H}$. There is a connected component D of $H^n /\!/ H$ such that $E = \overline{\Psi^n_{G,H}(D)}$. There is a connected component C of H^n such that $\pi_{H^n}(C) = D$. If $E \not\subseteq Y^n_{G,H}$ then $C \not\subseteq H_0^n$, so it follows from Remark 6.4.3 that $C \subseteq M_0^n$ for some proper finite-index subgroup M of H. Hence $D \subseteq \overline{\Psi^n_{H,M}(M_0^n)} = Y^n_{H,M}$. By (6.8.2),

$$E = \overline{\Psi^n_{G,H}(D)} \subseteq Y^n_{G,M}.$$

Lemma 6.8.2 *Let H be reductive and let $\phi\colon H \to G$ be an isogeny. Let $n \in \mathbb{N}$. Then $\Psi^n_{G,H}$ is finite.*

Proof The map $\phi_n\colon H^n \to G^n$ is an isogeny of groups, so ϕ_n is finite. We let H act on G^n via ϕ; then ϕ_n is H-equivariant. Since ϕ is surjective, we see that $k[G^n]^H = k[G^n]^G$, so the map $G^n /\!\!/ H \to G^n /\!\!/ G$ is an isomorphism. Hence it is enough to show that $(\phi_n)_H$ is finite. This follows from Lemma 3.6.3. $\qquad\qquad\square$

It is clear that if H is conjugate to H' then $X^n_{G,H'} = X^n_{G,H}$ and $Y^n_{G,H'} = Y^n_{G,H}$. If $\lambda \in Y_G$ and $H \subseteq P_\lambda$ then $\pi_{G^n}(c_\lambda(\mathbf{h})) = \pi_{G^n}(\mathbf{h})$ for all $\mathbf{h} \in H^n$, and it follows that

$$\Psi^n_{G,H} = \Psi^n_{G,c_\lambda(H)} \circ \Psi^n_{c_\lambda(H),H,c_\lambda}. \tag{6.8.3}$$

Since $c_\lambda(H^n) = c_\lambda(H)^n$ and $c_\lambda(H^n_0) \subseteq c_\lambda(H)^n_0$, it follows that

$$X^n_{G,c_\lambda(H)} = X^n_{G,H} \quad \text{and} \quad Y^n_{G,c_\lambda(H)} \supseteq Y^n_{G,H}. \tag{6.8.4}$$

It can happen that $Y^n_{G,H}$ is properly contained in $Y^n_{G,c_\lambda(H)}$. For instance, let $G = \mathrm{SL}_2$ and let $H = C_p \times C_p$ embedded as a subgroup of the upper unitriangular matrices. Choose $\lambda \in Y_G$ such that $c_\lambda(H) = 1$ and take $n = 1$: then $Y^n_{G,H}$ is empty but $Y^n_{G,c_\lambda(H)}$ is not.

Definition 6.8.3 Let $\mathscr{S}_G$ be the set of conjugacy classes of G-cr subgroups of G. Define a relation $\preceq$ on $\mathscr{S}_G$ as follows: $G \cdot H \preceq G \cdot K$ if there exists a subgroup M of K such that $\mathcal{D}(M) = G \cdot H$. Here $\mathcal{D}(M)$ is as in Definition 5.5.11.

We give some properties of $\preceq$ in Exercise 6.7. The reader can prove parts of the exercise by applying the Finiteness and Rigidity Theorems, but they should avoid the temptation to do this because we actually use the exercise in the proof of these theorems. We provide complete solutions to this exercise in Chap. 13.

Let $n \in \mathbb{N}$ and let $w_1, \ldots, w_r$ be words in $\gamma_1, \ldots, \gamma_n$. Define $f_H\colon H^n \to H^r$ by

$$f_H(h_1, \ldots, h_n) = (w_1(h_1, \ldots, h_n), \ldots, w_r(h_1, \ldots, h_n)).$$

Then f_H is H-equivariant, so it gives rise to a map $\widehat{f}_H\colon H^n /\!\!/ H \to H^r /\!\!/ H$. It follows easily from the definitions that the diagram

$$
\begin{array}{ccc}
H^n /\!\!/ H & \xrightarrow{\ \Psi^n_{G,H}\ } & G^n /\!\!/ G \\[4pt]
{\scriptstyle \widehat{f}_H}\Big\downarrow & & \Big\downarrow{\scriptstyle \widehat{f}_G} \\[4pt]
H^r /\!\!/ H & \xrightarrow[\ \Psi^r_{G,H}\ ]{} & G^r /\!\!/ G
\end{array}
\tag{6.8.5}
$$

commutes. Let M be a subgroup of H which contains the subgroup generated by $w_1(H^n) \cup \cdots \cup w_r(H^n)$. The map $f_H \colon H^n \to H^r$ factors as $H^n \to M^r \to H^r$, where the second map is the inclusion of M^r in H^r, so $\widehat{f}_H \colon H^n /\!/ H \to H^r /\!/ H$ factors as $H^n /\!/ H \to M^r /\!/ M \overset{\Psi^r_{H,M}}{\to} H^r /\!/ H$. It follows from (6.2.1) that $\operatorname{Im}\left(\Psi^r_{G,H} \circ \widehat{f}_H \right) \subseteq X^r_{G,M}$. We deduce from (6.8.5) that

$$\widehat{f}_G(X^n_{G,H}) \subseteq X^r_{G,M}. \tag{6.8.6}$$

Lemma 6.8.4 *Suppose k is solid and $n \ge \kappa(H) + 1$. Then there is an open dense subset U of $Y^n_{G,H}$ such that $\left(\Xi^n_{G,H} \right)^{-1}(y)$ is finite for all $y \in U$.*

Proof First suppose H is G-cr. We show that if $\mathbf{h} \in H^n_{\mathrm{gen}}$ then $\left(\Xi^n_{G,H} \right)^{-1}(\pi_{G^n}(\mathbf{h}))$ is finite. The result will then follow from Lemma 2.1.1, since H^n_{gen} is dense in H^n_0 (Corollary 6.4.7). Recall that $N_G(H)$ is a finite extension of $HC_G(H)$ by Proposition 2.4.8, so we can pick a finite set of coset representatives $g_1, \ldots, g_l \in N_G(H)$ for $HC_G(H)$ in $N_G(H)$. Let $\mathbf{h} \in H^n_{\mathrm{gen}}$. Take any $\mathbf{h}_1 \in H^n$ such that $\pi_{G^n}(\mathbf{h}_1) = \pi_{G^n}(\mathbf{h})$. By Lemma 5.3.6 there exists $g \in N_G(H)$ such that $\mathbf{h}_1 = g \cdot \mathbf{h}$. We have $g = hg_i z$ for some $h \in H$, some $z \in C_G(H)$ and some $1 \le i \le l$. So $\mathbf{h}_1 = g \cdot \mathbf{h} = hg_i \cdot \mathbf{h}$, which implies that $\pi_{H^n}(\mathbf{h}_1) = \pi_{H^n}(g_i \cdot \mathbf{h})$. Set $y = \pi_{G^n}(\mathbf{h})$. We deduce that $\left(\Xi^n_{G,H} \right)^{-1}(y) = \{\pi_{H^n}(g_1 \cdot \mathbf{h}), \ldots, \pi_{H^n}(g_l \cdot \mathbf{h})\}$, as required.

Now let H be arbitrary. Choose a cocharacter λ of G such that P_λ yields a semisimplification of H (see Remark 5.5.4). We have $\Psi^n_{G,H} = \Psi^n_{G,c_\lambda(H)} \circ \Psi^n_{c_\lambda(H),H,c_\lambda}$ by (6.8.3). Since H is reductive, c_λ gives an isogeny from H to $c_\lambda(H)$, so $\Psi^n_{H,c_\lambda(H),c_\lambda}$ is finite by Lemma 6.8.2. We have $Y^n_{G,H} \subseteq Y^n_{G,c_\lambda(H)}$ by (6.8.4), so the result follows from the case when H is G-cr. $\qquad\square$

Corollary 6.8.5 *Suppose k is solid and $n \ge \kappa(H) + 1$. Then $Y^n_{G,H}$ is pure-dimensional of dimension $(n-1)\dim(H) + \dim(Z(H))$.*

Proof The result follows from Lemma 6.8.4, Corollary 6.4.8 (applied to H) and Lemma 2.1.1. $\qquad\square$

Lemma 6.8.6 *Suppose k is solid and $n \ge \kappa(H) + 1$. Let $\mathbf{g} \in G^n$ such that $y := \pi_{G^n}(\mathbf{g})$ belongs to $Y^n_{G,H}$ and let $M = \mathcal{G}(\mathbf{g})$. Suppose M is reductive. Then:*

  (i) $\dim(M) \le \dim(H)$.
  (ii) $\dim([M,M]) \le \dim([H,H])$.
  (iii) *If M and H are G-completely reducible and $\dim(M) = \dim(H)$ then M is conjugate to H.*

Proof Pick $s \in \mathbb{N}$ such that $s \geq \max\{\kappa(H) + 1, \kappa(M) + 1\}$. Then $Q_s(y) \subseteq Y_{G,H}^{n+s}$ (see Section 6.4 for the definition of Q_s). We have

$$Q_s(y) = Q_s(\pi_{G^n}(\mathbf{g})) = \overline{\pi_{G^n}(\mathcal{R}_s(\mathbf{g}))} = \overline{\pi_{G^n}(M_0^n)} = Y_{G,M}^{n+s},$$

where the second equality is from (6.4.3) and the third equality is from Lemma 6.4.9 (applied to M). Hence

$$Y_{G,M}^{n+s} \subseteq Y_{G,H}^{n+s}.$$

Applying Corollary 6.8.5, $Y_{G,H}^{n+s}$ and $Y_{G,M}^{n+s}$ are pure-dimensional varieties of dimension $(n + s - 1)\dim(H) + \dim(Z(H))$ and $(n + s - 1)\dim(M) + \dim(Z(M))$, respectively. Hence

$$(n + s - 1)\dim(M) + \dim(Z(M)) \leq (n + s - 1)\dim(H) + \dim(Z(H)). \qquad (6.8.7)$$

We are free to take s as large as we like, so we deduce that $\dim(M) \leq \dim(H)$. This proves (i).

To prove (ii), we first show the following: there exist $r \in \mathbb{N}$ and $\mathbf{g}' \in G^r$ such that $\mathcal{G}(\mathbf{g}') = [M, M]$ and $y' := \pi_{G^r}(\mathbf{g}')$ belongs to $X_{G,[H,H]}^r$. We establish this using the construction preceding Lemma 6.8.4. By Proposition 6.4.13 there exist $r \in \mathbb{N}$ and words $w_1, \ldots, w_r$ in letters $\gamma_1, \ldots, \gamma_n$ such that $w_1(g_1, \ldots, g_n), \ldots, w_r(g_1, \ldots, g_n)$ generate $[M, M]$ and each w_i (regarded as a word in the γ_i) is a product of commutators. Set $\mathbf{g}' := (w_1(g_1, \ldots, g_n), \ldots, w_r(g_1, \ldots, g_n)) \in G^r$ and set $y' := \pi_{G^r}(\mathbf{g}')$. Regarding $\mathbf{g}$ and $\mathbf{g}'$ as elements of M^n and M^r, respectively, set $z := \pi_{M^n}(\mathbf{g}) \in M^n /\!\!/ M$ and $z' := \pi_{M^r}(\mathbf{g}') \in M^r /\!\!/ M$. Now $\mathbf{g}' = f_M(\mathbf{g})$, so $z' = \widehat{f}_M(z)$. Hence

$$y' = \pi_{G^r}(\mathbf{g}') = \Psi_{G,M}^r(z') = \Psi_{G,M}^r(\widehat{f}_M(z)) = \widehat{f}_G(\Psi_{G,M}^n(z)) = \widehat{f}_G(y)$$

by (6.8.5). Therefore $y' = \widehat{f}_G(y)$ belongs to $X_{G,H}^r$. The subgroup of H generated by $w_1(H^n) \cup \cdots \cup w_r(H^n)$ is contained in $[H, H]$, so in fact y' belongs to $X_{G,[H,H]}^r$ by (6.8.6). This gives the desired result. To finish the proof of (ii), observe from Remark 6.8.1 that there is a finite-index subgroup H' of $[H, H]$ such that $y' \in Y_{G,H'}^r$. Since $\mathcal{G}(\mathbf{g}') = [M, M]$, we deduce that $\dim([M, M]) \leq \dim(H') = \dim([H, H])$ by part (i) (applied to the subgroups $[H, H]$ and $[M, M]$ of G).

To prove (iii), suppose that M and H are G-cr and $\dim(M) = \dim(H)$. We have $\dim([M, M]) \leq \dim([H, H])$ by (ii), so $\dim(Z(M)) \geq \dim(Z(H))$ by Corollary 2.4.7. It follows from (6.8.7) that $\dim(Z(M)) = \dim(Z(H))$ and $\dim\left(Y_{G,M}^{n+s}\right) = \dim\left(Y_{G,H}^{n+s}\right)$. Since $Y_{G,M}^{n+s} \subseteq Y_{G,H}^{n+s}$, it follows that $Y_{G,M}^{n+s}$ contains a non-empty open subset U of $Y_{G,H}^{n+s}$. As $\mathrm{Im}\left(\Xi_{G,M}^{n+s}\right)$ and $\mathrm{Im}\left(\Xi_{G,H}^{n+s}\right)$ are dense constructible subsets of $Y_{G,M}^{n+s}$ and $Y_{G,H}^{n+s}$, respectively, we can assume that $U \subseteq \operatorname{Im}\left(\Xi_{G,M}^{n+s}\right) \cap \operatorname{Im}\left(\Xi_{G,H}^{n+s}\right)$. Now $\pi_{G^{n+s}}^{-1}(U) \cap M_0^{n+s}$ is a non-empty open subset of M_0^{n+s}, so by Corollary 6.4.7 there exists $\mathbf{m} \in M_{\mathrm{gen}}^{n+s}$ such that $\pi_{G^{n+s}}(\mathbf{m}) \in U$. Then $\pi_{G^{n+s}}(\mathbf{m}) = \pi_{G^{n+s}}(\mathbf{h})$ for some $\mathbf{h} \in H^{n+s}$, so $G \cdot M \preceq G \cdot H$ in the sense of Definition 6.8.3, by Exercise 6.7(v). Swapping the roles of M and H, we see that $G \cdot H \preceq G \cdot M$. Part (iii) now follows from Exercise 6.7(iii), so we are done. $\quad\square$

The key ingredient in the proof of the Finiteness Theorem is Lemma 6.8.8. First we need a preliminary result. Let $f\colon X \to Y$ be a map of varieties, with Y irreducible. We have an inclusion $k[Y] \to k(Y)$, where $k(Y)$ is the function field of Y. Therefore, if we let E be any algebraically closed extension field of $k(Y)$, we obtain an inclusion $k[Y] \to E$, which gives a point $y \in Y_E$.

Lemma 6.8.7 *With the notation as above, suppose there exists $x \in X_E$ such that $f(x) = y$. Then f is dominant.*

Proof The point x corresponds to a k-algebra homomorphism $j\colon k[X] \to E$. Since $f(x) = y$, the composition $k[Y] \xrightarrow{f^*} k[X] \xrightarrow{j} E$ is the inclusion of $k[Y]$ in E. Hence f^* is injective, which implies that f is dominant. $\quad\square$

Lemma 6.8.8 *Assume k is solid. Let H be a G-completely reducible subgroup of G and let $s \geq \kappa(H) + 1$. Let $n \in \mathbb{N}$ and let D be a closed subset of $Y_{G,H}^n$. Suppose that $Q_s^G(D)$ is a proper subset of $Y_{G,H}^{n+s}$. Then there are G-completely reducible subgroups $M_1, \ldots, M_t$ of G such that $\dim(M_i) < \dim(H)$ for $1 \leq i \leq t$ and $D \subseteq \bigcup_{i=1}^t \operatorname{Im}\left(\Psi_{G,M_i}^n\right)$.*

Proof By noetherian induction, we can assume the result holds for any proper closed subset D' of D (note that $Q_s^G(D') \subseteq Q_s^G(D)$, so D' satisfies the hypothesis of the lemma). Let D_0 be any irreducible component of D and let E be the algebraic closure of the function field $k(D_0)$. As above, the inclusion of $k[D_0]$ in E gives a point $y \in (D_0)_E \subseteq (Y_{G,H}^n)_E$, and there exists $\mathbf{g} \in (G_E)^n$ such that $\pi_{(G_E)^n}(\mathbf{g}) = y$ and $\mathcal{G}(\mathbf{g})$ is a G_E-cr subgroup of G_E. By Theorem 6.1.1, we can assume after conjugating by some element of G_E that $\mathcal{G}(\mathbf{g})$ is k-defined: say, $\mathcal{G}(\mathbf{g}) = M_E$ for some G-cr subgroup M of G. Set $C_0 = \pi_{G^n}^{-1}(D_0) \cap M^n$. Now $\mathbf{g} \in \pi_{(G_E)^n}^{-1}((D_0)_E) \cap (M_E)^n = (C_0)_E$ and $\pi_{(G_E)^n}(\mathbf{g}) = y$, so $\pi_{G^n}(C_0)$ contains a dense subset of D_0 by Lemma 6.8.7 (applied to π_{G^n}, regarded as a map from C_0 to D_0). Hence $\operatorname{Im}\left(\Psi_{G,M}^n\right)$ contains a non-empty open subset U of D. Set $D' = D \setminus U$. By noetherian induction, there exist G-cr subgroups $M_1, \ldots, M_r$ of G for some $r \in \mathbb{N}$ such that $\dim(M_i) < \dim(H)$ for $1 \leq i \leq r$ and $D' \subseteq \bigcup_{i=1}^r \operatorname{Im}\left(\Psi_{G,M_i}^n\right)$. So $D \subseteq \bigcup_{i=1}^t \operatorname{Im}\left(\Psi_{G,M_i}^n\right)$, where $t := r+1$ and $M_t := M$. To finish, therefore, it is enough to prove that $\dim(M) < \dim(H)$.

Since y belongs to Y_{G_E,H_E}^n and $\mathcal{G}(\mathbf{g}) = M_E$, we know from Lemma 6.8.6(i) that $\dim(M) = \dim(M_E) \le \dim(H_E) = \dim(H)$. Suppose $\dim(H) = \dim(M)$. Then $\dim(H_E) = \dim(M_E)$, so M_E and H_E are G_E-conjugate by Lemma 6.8.6(iii); in particular, $\kappa(M_E) = \kappa(H_E) = \kappa(H)$. Hence $\mathcal{R}_s^{G_E}(\mathbf{g}) = (M_E)_0^{n+s}$ by Lemma 6.4.9. So

$$Q_s^{G_E}(D_E) \supseteq Q_s^{G_E}(y) = Q_s^{G_E}(\pi_{(G_E)^n}(\mathbf{g})) = \overline{\pi_{G_E^{n+s}}(\mathcal{R}_s^{G_E}(\mathbf{g}))} = \overline{\pi_{G_E^{n+s}}\left((M_E)_0^{n+s}\right)}$$

$$= \overline{\pi_{G_E^{n+s}}\left((H_E)_0^{n+s}\right)} = Y_{G_E,H_E}^{n+s},$$

where the second equality is (6.4.3). Since $Q_s^{G_E}(D_E) \subseteq Y_{G_E,H_E}^{n+s}$, we deduce that $Q_s^{G_E}(D_E) = Y_{G_E,H_E}^{n+s}$. But this implies that $Q_s^G(D) = Y_{G,H}^{n+s}$ by Remark 6.4.10 and (6.8.1), a contradiction. We conclude that $\dim(M) < \dim(H)$, as required. $\qquad\square$

Let $f\colon X \to Y$ be a dominant quasi-finite map of irreducible varieties. By a version of Zariski's Main Theorem [67, Thm. 8.12.6], there exist a variety Z and maps $i\colon X \to Z$ and $h\colon Z \to Y$ such that i is an open embedding, h is finite and $h \circ i = f$. Replacing Z with $\overline{i(X)}$ if necessary, we can assume that $i(X)$ is dense in Z. Let $C = Z \setminus i(X)$. Now $i(X)$ is affine (since X is affine by assumption), so if C is non-empty then every irreducible component of C has codimension 1 in Z; this follows from [67, Cor. 21.12.7]. Since h is finite and dominant, $h(C)$ is a closed pure-dimensional subvariety of codimension 1 in Y.

Lemma 6.8.9 *With the notation as above, we have:*

(i) *for any $y \in Y$, $y \notin h(C)$ if and only if there is an open neighbourhood O of y such that $f|_{f^{-1}(O)}\colon f^{-1}(O) \to O$ is finite;*

(ii) *f is finite if and only if $h(C) = \varnothing$.*

Proof (i). Note that if O is any open neighbourhood of y then $f|_{f^{-1}(O)}$ factors as

$$f^{-1}(O) \xrightarrow{\ i|_{f^{-1}(O)}\ } h^{-1}(O) \xrightarrow{\ h|_{h^{-1}(O)}\ } O.$$

Suppose $y \notin h(C)$. Since $h(C)$ is closed, we can choose an open neighbourhood O of y such that $h^{-1}(O) \subseteq i(X)$. So i induces an isomorphism from $f^{-1}(O) = i^{-1}(h^{-1}(O))$ onto $h^{-1}(O)$. Since h is finite, $h|_{h^{-1}(O)}\colon h^{-1}(O) \to O$ is finite, so $f|_{f^{-1}(O)}\colon f^{-1}(O) \to O$ is finite. Conversely, suppose there is an open neighbourhood O of y such that $f|_{f^{-1}(O)}\colon f^{-1}(O) \to O$ is finite. Then the map $i|_{f^{-1}(O)}\colon f^{-1}(O) \to h^{-1}(O)$ is finite by Lemma 2.1.2(i) and hence is surjective as it has dense image. It follows that $y \notin h(C)$.

(ii). By [170, §29.44], f is finite if and only for all $y \in Y$ there is an open neighbourhood O of y such that $f|_{f^{-1}(O)}\colon f^{-1}(O) \to O$ is finite. The result now follows from (i). $\qquad\square$

It follows from Lemma 6.8.9 that $h(C)$ does not depend on the choice of Z, h and i. We call $h(C)$ the *non-finite locus* of f. More generally, suppose $f: X \to Y$ is a map of varieties and let $X_1, \ldots, X_t$ be the irreducible components of X. Set $Y_i = \overline{f(X_i)}$ and let $f_i: X_i \to Y_i$ be map arising from the restriction of f for $1 \leq i \leq t$. We define the non-finite locus of f to be the union (in Y) of the non-finite loci of the f_i. It is straightforward to check using Lemma 2.1.2 and Corollary 2.1.4 that Lemma 6.8.9(ii) still holds in this more general setting: that is, f is finite if and only if the non-finite locus of f is empty.

Proof of Theorem 6.2.1 Let H be a reductive subgroup of G. By (6.2.2) we can— after extending the field if necessary—assume that k is solid. Recall from the argument preceding Lemma 6.4.9 that if $n' \geq n$ then we may regard $H^n /\!/ H$ (resp., $G^n /\!/ G$) as a closed subset of $H^{n'} /\!/ H$ (resp., $G^{n'} /\!/ G$), and it is clear that the restriction of $\Psi_{G,H}^{n'}$ to $H^n /\!/ H$ is just $\Psi_{G,H}^n$. It follows from Lemma 6.8.9(i) and Lemma 2.1.2(i) that

(∗)   for any $y \in Y_{G,H}^n$, if $\left(\Psi_{G,H}^n\right)^{-1}(y)$ is infinite then $\left(\Psi_{G,H}^{n'}\right)^{-1}(y)$ is infinite and if y belongs to the non-finite locus of $\Psi_{G,H}^n$ then y belongs to the non-finite locus of $\Psi_{G,H}^{n'}$.

In particular, $\Psi_{G,H}^n$ is finite if $\Psi_{G,H}^{n'}$ is. Hence we can assume without loss that

$$n \geq \kappa(H) + 1 \quad \text{and} \quad n > \dim(H) + 1. \tag{6.8.8}$$

By noetherian induction we can assume that

(†)   if M is any reductive subgroup of G such that either (i) $\dim(M) < \dim(H)$, or (ii) $\dim(M) = \dim(H)$ and $\kappa(M) < \kappa(H)$, then $\Psi_{G,M}^n$ is finite.

We complete the proof in four steps. Recall that the connected components and the irreducible components of $H^n /\!/ H$ coincide.

(a) First we reduce to the case when H is G-cr; this will allow us to apply Lemma 6.8.8. Pick $\lambda \in Y_G$ such that (P_λ, L_λ) yields a semisimplification of H. Then c_λ gives rise to an isogeny ϕ from H to $c_\lambda(H)$, and $\Psi_{G,H}^n = \Psi_{G,c_\lambda(H)}^n \circ \Psi_{c_\lambda(H),H,\phi}^n$ by (6.8.3). But $\Psi_{c_\lambda(H),H,\phi}^n$ is finite by Lemma 6.8.2, so $\Psi_{G,H}^n$ is finite if $\Psi_{G,c_\lambda(H)}^n$ is. Hence we can assume without loss that H is G-cr.

(b) Let D be a connected component of $H^n /\!/ H$ such that $D \not\subseteq H_0^n /\!/ H$. Then there is a connected component C of H^n such that $C \not\subseteq H_0^n$ and $D = \pi_{H^n}(C)$. By Remark 6.4.3 there is a proper finite-index subgroup M of H such that C is a connected component of M^n. Let $D_1 = \pi_{M^n}(C)$, a connected component of $M^n /\!/ M$. Now $\Psi_{G,M}^n$ is finite by (†), so the restriction of $\Psi_{G,M}^n$ to D_1 is finite. Since $\Psi_{G,M}^n$ factors as $\Psi_{G,H}^n \circ \Psi_{H,M}^n$ by (6.2.1) and $\Psi_{H,M}^n(D_1) = \pi_{H^n}(C) = D$, the restriction of $\Psi_{G,H}^n$ to D is finite by Lemma 2.1.2(ii). Hence to complete the proof it is enough by Corollary 2.1.4 to show that $\Xi_{G,H}^n: H_0^n /\!/ H \to G^n /\!/ G$ is finite.

(c) Next we show that $\Xi_{G,H}^n$ is quasi-finite. Let

$$D_1 = \left\{ x \in H_0^n /\!/ H \;\middle|\; \left(\Xi_{G,H}^n\right)^{-1} \left(\Xi_{G,H}^n(x)\right) \text{ is infinite} \right\}$$

and set $D = \overline{D_1}$. We show that D satisfies the hypotheses of Lemma 6.8.8 (applied to the inclusion of H in H, rather than the inclusion of H in G; note that $Y_{H,H}^n = H_0^n /\!/ H$). Choose $s \geq \kappa(H) + 1$, let

$$D_1' = \left\{ x \in H_0^{n+s} /\!/ H \;\middle|\; \left(\Xi_{G,H}^{n+s}\right)^{-1} \left(\Xi_{G,H}^{n+s}(x)\right) \text{ is infinite} \right\}$$

and set $D' = \overline{D_1'}$. By Lemma 6.8.4, D and D' are proper subsets of $H_0^n /\!/ H = Y_{H,H}^n$ and $H_0^{n+s} /\!/ H = Y_{H,H}^{n+s}$, respectively. It follows from $(*)$ that we can regard D as a closed subset of D'. Now $\Xi_{G,H}^{n+s}$ is $\mathrm{Aut}(\Gamma_{n+s})$-equivariant, so D' is $\mathrm{Aut}(\Gamma_{n+s})$-stable. We deduce that $Q_s^H(D) \subseteq D'$, so $Q_s^H(D)$ is a proper subset of $H_0^{n+s} /\!/ H$. Hence we can apply Lemma 6.8.8. This tells us that there are H-cr subgroups $M_1, \ldots, M_t$ of H for some $t \in \mathbb{N}$ such that $\dim(M_i) < \dim(H)$ for each i and $D \subseteq \bigcup_{i=1}^t \mathrm{Im}\left(\Psi_{H,M_i}^n\right)$.

So let $x \in D_1$ and set $y = \Xi_{G,H}^n(x)$. Now $\left(\Psi_{G,H}^n\right)^{-1}(y) = \left(\Xi_{G,H}^n\right)^{-1}(y) \subseteq D_1$ by definition of D_1, and $\Psi_{G,M_i}^n = \Psi_{G,H}^n \circ \Psi_{H,M_i}^n$ for each $1 \leq i \leq t$ by (6.2.1). It follows that

$$\left(\Psi_{G,H}^n\right)^{-1}(y) = \bigcup_{i=1}^t \Psi_{H,M_i}\left(\Psi_{G,M_i}^n\right)^{-1}(y). \tag{6.8.9}$$

But each map Ψ_{G,M_i}^n is finite by $(\dagger)$, so the RHS of (6.8.9) is finite. We deduce that $\left(\Xi_{G,H}^n\right)^{-1}(y)$ is finite. This forces D_1 to be empty. Hence $\Xi_{G,H}^n$ is quasi-finite.

(d) Finally, we prove that $\Xi_{G,H}^n$ is finite. Let D be the non-finite locus of $\Xi_{G,H}^n$. We show that D satisfies the hypotheses of Lemma 6.8.8. Let $s \geq \kappa(H) + 1$ and let D' be the non-finite locus of $Y_{G,H}^{n+s}$. We can regard D as a closed subset of D' by $(*)$. By an argument like the one in part (c), $Q_s^G(D)$ is a proper subset of $Y_{G,H}^{n+s}$. Hence by Lemma 6.8.8 there are reductive subgroups $M_1, \ldots, M_t$ of G for some $t \in \mathbb{N}$ such that $\dim(M_i) < \dim(H)$ for each i and $D \subseteq \bigcup_{i=1}^t \mathrm{Im}\left(\Psi_{G,M_i}^n\right)$. Now each irreducible component of $Y_{G,H}^n$ has dimension $(n-1)\dim(H) + \dim(Z(H))$ by Corollary 6.8.5, and each irreducible component of $(M_i)^n /\!/ M_i$ has dimension at most $\dim((M_i)^n) = n\dim(M_i)$ since $\pi_{(M_i)^n}$ is surjective. Hence

$$\dim D \leq \max_{1 \leq i \leq t} n\dim(M_i)$$

$$\leq n(\dim(H) - 1)$$

$$= n \dim(H) - n$$

$$< (n-1) \dim(H) - 1$$

$$\leq \dim(Y^n_{G,H}) - 1,$$

where the second-last inequality is from (6.8.8). If $D \neq \varnothing$ then each irreducible component of D has codimension 1 in some irreducible component of $Y^n_{G,H}$, which is a contradiction. We deduce that $D = \varnothing$, so $\Xi^n_{G,H}$ is finite by Lemma 6.8.9. This completes the proof.

$\square$

Remark 6.8.10 We point out a serious mistake in the proof of Theorem 6.2.1 given in [117] when $p > 0$. Lemma 13.1 of [117]—which says that $\Xi^n_{G,H}$ is quasi-finite—has the extra assumption that our Theorem 6.2.2 holds for G, which implies that the Rationality Theorem holds for G. But in the proof of the lemma, we need the assumption that the Rationality Theorem holds for H, and we do not have this.[3] Our argument in step (c) of the proof of Theorem 6.2.1 follows that of [117, Lem. 13.1] closely, but we avoid the above problem because we have already proved that the Rationality Theorem holds in full generality. See Sect. 6.10 for further discussion.

Corollary 6.8.11 *Let* $j \colon G \to M$ *be an embedding of reductive groups and let* Γ *be a finitely generated group. Then the map* $j_\Gamma \colon C(\Gamma, G) \to C(\Gamma, M)$ *from (6.5.3) is finite.*

Proof Choose generators $\gamma_1, \ldots, \gamma_n$ of Γ for some $n \in \mathbb{N}$. We obtain finite maps $i_G \colon C(\Gamma, G) \to G^n /\!/ G$ and $i_M \colon C(\Gamma, M) \to M^n /\!/ M$, as described in Sect. 6.5. The composition

$$C(\Gamma, G) \xrightarrow{\ j_\Gamma\ } C(\Gamma, M) \xrightarrow{\ i_M\ } M^n /\!/ M$$

is equal to the composition

$$C(\Gamma, G) \xrightarrow{\ i_G\ } G^n /\!/ G \xrightarrow{\ \Psi^n_{M,G}\ } M^n /\!/ M.$$

The latter composition is a finite map by Theorem 6.2.1 and because i_G is finite, so j_Γ is finite by Lemma 2.1.2.

$\square$

---

[3] This problem does not apply to Vinberg's proof in [192] in characteristic 0, as he proves that $\Psi^n_{G,H}$ is quasi-finite in a different way.

Proof of Theorem 6.2.2 Let F be a finite group. Pick an inclusion j of G in some GL_m. Since $C(F, \mathrm{GL}_m)$ is finite (Proposition 6.5.4) and j_F is finite (Corollary 6.8.11), $C(F, G)$ must also be finite. $\qquad\square$

Remark 6.8.12 Recall that we proved Theorem 6.2.2 in positive characteristic earlier (Proposition 6.7.8), so we only needed to deal with the characteristic 0 case. The proof of Theorem 6.2.2 above appears to work in arbitrary characteristic, but we did, however, use Theorem 6.2.2 in the positive characteristic case to prove Theorem 6.2.1 in the positive characteristic case.

We show how to deduce Theorem 6.1.2 from Theorem 6.1.1.

Proof of Theorem 6.1.2 By Theorem 6.1.1, it is enough to prove the result when k is the algebraic closure of the prime field: so we can assume G is countable. Let S be a torus of G. If H is a G-cr subgroup of G having S as a maximal torus then H is reductive, so H is generated as an abstract group by S together with a finite set of elements $h_1, \ldots, h_r$ for some r: for instance, we can choose one non-trivial element of every root group of H with respect to S along with a set of representatives for $N_H(S)/S$. Since G is countable, there are only countably many possibilities for $h_1, \ldots, h_r$. But G contains only countably many conjugacy classes of tori. The result follows. $\qquad\square$

Remark 6.8.13 Let $n \in \mathbb{N}$ and let H be a reductive subgroup of G. It follows from Example 3.1.7 that the map $\Psi^n_{G,H}\colon H^n /\!/ H \to G^n /\!/ G$ can be factorised as $H^n /\!/ H \to H^n /\!/ N_G(H) \xrightarrow{f} G^n /\!/ G$. Suppose $p = 0$. Vinberg proved that if $n \geq 2$ and H is connected then $f\colon H^n /\!/ N_G(H) \to X^n_{G,H}$ is the normalisation map [192, Cor. 2a]; he also gives an example to show that in this case f need not be an isomorphism [192, §4].

6.9   Further Consequences of the Finiteness Theorem

Here are some more consequences of the Finiteness Theorem. Recall the definition of the subalgebra $\mathcal{T}_n(G)$ of $k[G^n]^G$ from Definition 5.6.1.

Theorem 6.9.1 *Let* $n \in \mathbb{N}$. *Then* $\mathcal{T}_n(G)$ *is a finitely generated k-algebra and $k[G^n]^G$ is a finitely generated* $\mathcal{T}_n(G)$*-module.*

Proof The first assertion follows from the second by a standard result [93, Ch. X, §1.4]. To prove the second, fix an embedding of G in some GL_m. Pick $t \in \mathbb{N}$ and pick words w_i and elements $f_i \in k[\mathrm{GL}_m]^{\mathrm{GL}_m}$ for $1 \leq i \leq t$ such that $(f_1)_{w_1}, \ldots, (f_t)_{w_t}$ generate $\mathcal{T}_n(\mathrm{GL}_m)$—this is possible by Theorems 5.6.3 and 5.6.5. Define $\psi\colon (\mathrm{GL}_m)^n \to (\mathrm{GL}_m /\!/ \mathrm{GL}_m)^t$ by

$$\psi(g_1, \dots, g_n) = \big(\pi_{\mathrm{GL}_m}(w_1(g_1, \dots, g_n)), \dots, \pi_{\mathrm{GL}_m}(w_t(g_1, \dots, g_n))\big).$$

Then ψ is GL_m-invariant, so it descends to a map

$$\psi_{\mathrm{GL}_m} : (\mathrm{GL}_m)^n /\!\!/ \mathrm{GL}_m \to (\mathrm{GL}_m /\!\!/ \mathrm{GL}_m)^t$$

by the universal mapping property of quotients. Likewise, the map $\phi \colon G^n \to (G /\!\!/ G)^t$ given by

$$\phi(g_1, \dots, g_n) = (\pi_G(w_1(g_1, \dots, g_n)), \dots, \pi_G(w_t(g_1, \dots, g_n)))$$

descends to a map $\phi_G \colon G^n /\!\!/ G \to (G /\!\!/ G)^t$.

Untangling the definitions, we get a commutative diagram

$$
\begin{array}{ccc}
G^n /\!\!/ G & \xrightarrow{\ \phi_G\ } & (G /\!\!/ G)^t \\[2pt]
\Psi^n_{\mathrm{GL}_m, G} \downarrow & & \downarrow \delta \\[2pt]
(\mathrm{GL}_m)^n /\!\!/ \mathrm{GL}_m & \xrightarrow[\psi_{\mathrm{GL}_m}]{} & (\mathrm{GL}_m /\!\!/ \mathrm{GL}_m)^t
\end{array}
$$

where δ is the product of t copies of the map $\Psi^1_{\mathrm{GL}_m, G} \colon G /\!\!/ G \to \mathrm{GL}_m /\!\!/ \mathrm{GL}_m$. We see from the construction that ψ_{GL_m} is a closed embedding, and $\Psi^n_{\mathrm{GL}_m, G}$ is finite by Theorem 6.2.1. Hence $\delta \circ \phi_G$ is finite, so ϕ_G is finite by Lemma 2.1.2(i). If $f \in k[G]^G$ then

$$\phi_G^*(1 \otimes \cdots \otimes 1 \otimes f \otimes 1 \otimes \cdots \otimes 1) = f_{w_i},$$

where the f appears in the ith position. It follows that $\phi^*(k[(G /\!\!/ G)^t])$ contains $\mathcal{T}_n(G)$, so we get the desired result. $\qquad\square$

Remark 6.9.2 We have shown that $\mathcal{T}_n(G)$ is a finitely generated k-algebra, so it corresponds to a variety $X_n(G)$. We can construct a closed embedding $\phi_G = \phi_{n,G}$ of $X_n(G)$ in $(G /\!\!/ G)^t$ for t sufficiently large, using the argument in the proof of Theorem 6.9.1; we may identify $X_n(G)$ with the image of $\phi_{n,G}$. The inclusion of $\mathcal{T}_n(G)$ in $k[G^n]^G$ gives a dominant map $\beta_{n,G}$ from $G^n /\!\!/ G$ to $X_n(G)$. By Theorem 6.9.1, $\beta_{n,G}$ is finite and hence surjective.

We can carry out these constructions more generally for character varieties. Let Γ be a finitely generated group. Fix generators $\gamma_1, \dots, \gamma_n$ for Γ; we get an embedding i of $\mathrm{Hom}(\Gamma, G)$ in G^n in the usual way. Given $f \in k[G]^G$ and $\gamma \in \Gamma$, we obtain $f_\gamma \in k[\mathrm{Hom}(\Gamma, G)]^G$ as in (6.5.1). We define $\mathcal{T}_\Gamma(G)$ to be the subalgebra of $k[\mathrm{Hom}(\Gamma, G)]^G$ generated by the functions of the form f_γ. Since every f_γ is the restriction to $\mathrm{Hom}(\Gamma, G)$ of the map $f_w \colon G^n \to G$ for some word w, we may regard $\mathcal{T}_\Gamma(G)$ as a quotient of $\mathcal{T}_n(G)$, so $\mathcal{T}_\Gamma(G)$ is a finitely generated k-algebra by Theorem 6.9.1. Thus $\mathcal{T}_\Gamma(G)$ is the coordinate ring of some variety $X(\Gamma, G)$. The inclusion of $\mathcal{T}_\Gamma(G)$ in $k[\mathrm{Hom}(\Gamma, G)]^G$ gives rise to a dominant map $\beta_{\Gamma,G}$ from $C(\Gamma, G)$ to $X(\Gamma, G)$.

Proposition 6.9.3 $k[\mathrm{Hom}(\Gamma, G)]^G$ *is a finitely generated* $\mathcal{T}_\Gamma(G)$-*module and* $\beta_{\Gamma,G}$ *is finite.*

Proof The first assertion follows from the second by [93, Ch. X, §1.4]. To prove the second, choose generators $\gamma_1, \ldots, \gamma_n$ for Γ, giving an embedding $i \colon \mathrm{Hom}(\Gamma, G) \to G^n$. Untangling the definitions, we get a commutative diagram

$$
\begin{array}{ccc}
C(\Gamma, G) & \xrightarrow{\;\; i_G \;\;} & G^n /\!\!/ G \\[2pt]
{\scriptstyle \beta_{\Gamma,G}} \downarrow & & \downarrow {\scriptstyle \beta_{n,G}} \\[2pt]
X(\Gamma, G) & \xrightarrow{\hspace{2em}} & X_n(G)
\end{array}
$$

where the bottom horizontal arrow is the map induced by the epimorphism $\mathcal{T}_n(G) \to \mathcal{T}_\Gamma(G)$. Now $\beta_{n,G}$ is finite by Theorem 6.9.1 and i_G is finite (see Sect. 6.5), so $\beta_{n,G} \circ i_G$ is finite. Hence $\beta_{\Gamma,G}$ is finite by Lemma 2.1.2(i), as required. $\qquad\qquad\square$

The precise structure of $k[\mathrm{Hom}(\Gamma, G)]^G$ is still rather mysterious, even when Γ is the free group Γ_n and we are able to identify $\mathrm{Hom}(\Gamma_n, G)$ with G^n.

Open Problem 6.9.4 Do there exist examples of Γ, G and n such that $i_G \colon C(\Gamma, G) \to G^n /\!\!/ G$ is not a closed embedding? (If this is to happen then we must have $p > 0$.)

Open Problem 6.9.5 Do there exist examples of G and n such that $k[G^n]^G$ properly contains $\mathcal{T}_n(G)$?

6.10 Historical Remarks and References

The Rationality Theorem and the Finiteness Theorem were proved by Vinberg in characteristic 0 [192]. These theorems were extended to positive characteristic by the second author in [117], and our presentation in Chap. 6 is based on that paper; we have, however, written the arguments in the language of G-complete reducibility rather than strong reductivity.[4] Much of the proof of the Finiteness Theorem follows Vinberg's closely, although we have added more detail in places: for instance, we have given a complete proof

---

[4] It was not known that these two notions are equivalent when [117] was written.

of Proposition 6.4.13. Some different techniques are needed when $p > 0$—for instance, the idea of approximating H with an increasing chain of finite subgroups, as is seen in the proof of Lemma 6.6.3—but many of the key ideas in Sect. 6.8 (including Lemma 6.8.8) are due to him.

During the preparation of this book, we realised that the proof of the Finiteness Theorem given in [117] is faulty: see Remark 6.8.10. Fortunately, Sean Cotner recently proved the theorem in a more general context (see [48, Thm. 1.1]); the proof of the Finiteness Theorem should therefore be credited to him. A crucial ingredient in his argument is the use of Bruhat-Tits theory to deduce boundedness results about subgroups of $G(K)$, where K is a Henselian field. We have adapted this idea for the proof of our Lemma 6.7.4; rather than using a version of the valuative criterion for properness, as is done in [48], we have given a longer but less technical argument. We are grateful to Sean for his comments on parts of a draft version of Sect. 6.7.

The term *solid* was introduced in [119]. Any algebraically closed field of characteristic 0 is solid. In positive characteristic, however, one must consider the non-solid fields $\overline{\mathbb{F}_p}$. One deals with this by passing to a solid algebraically closed extension field (see the proof of [147, Prop. 16.9], for example), or by working with generic tuples rather than generating tuples.

Vinberg obtained slightly stronger density results for generating tuples than the ones we give, but it is not clear whether his results can be extended to positive characteristic. For instance, he proved that if $n \geq 2$ and $G_0^n \neq \varnothing$ then G_{gen}^n is dense in G_0^n [192, Cor. 8a]; in particular, it follows that the set $\widetilde{G}^n$ from [192, §1] coincides with our G_0^n when $p = 0$ and $n \geq 2$ (although when $p > 0$ these sets are different). On the other hand, our proof of Corollary 6.4.7 requires n to be at least $\kappa(G) + 1$.

Open Problem 6.10.1 Suppose k is solid and $n \geq 2$. Is it true that if $G_0^n \neq \varnothing$ then G_{gen}^n is dense in G_0^n?

Burness, Gerhardt and Guralnick have proved density results for generating tuples in more general subsets of G^n, where G is simple [43].

The theory of representation varieties and character varieties is vast, and we have only just scratched the surface. For more on this topic, see, for example, [86], [94], [103], [111], [141], and [161]. There is a parallel theory of representation varieties and character varieties where we replace our reductive algebraic group G with a real or complex Lie group. There are many applications to low-dimensional topology and geometry [3], [53], [65], and [87]. Most of the work on representation varieties and character varieties in both senses has been over fields of characteristic 0, but there has been some recent interest in fields of positive characteristic and even more general rings: see [63], [64], and [135].

One important topic we did not mention in Sect. 6.5 is a cohomological description of the tangent space to a representation variety $\mathrm{Hom}(\Gamma, G)$ at a point ρ. The group Γ acts on $\mathfrak{g}$ by

$$\gamma \cdot X = \mathrm{Ad}(\rho(\gamma))(X).$$

This gives $\mathfrak{g}$ the structure of a Γ-module, which we denote by $\mathfrak{g}(\rho)$. It turns out that one may identify the tangent space $T_\rho(\mathrm{Hom}(\Gamma, G))$ with a subspace of the space $Z^1(\Gamma, \mathfrak{g}(\rho))$ of 1-cocycles.[5] If ρ is separable then $T_\rho(G \cdot \rho)$ corresponds to $B^1(\Gamma, \mathfrak{g}(\rho))$, and under further appropriate hypotheses we can identify $T_{\pi(\rho)}(C(\Gamma, G))$ with a subspace of the 1-cohomology $H^1(\Gamma, \mathfrak{g}(\rho))$. This idea is used, for example, by Goldman to define the symplectic form on the character variety of a surface group in [65]. Here is an application, which dates back to André Weil [193]: if $H^1(\Gamma, \mathfrak{g}(\rho)) = 0$ then one can deduce that ρ is *rigid*: that is, $G \cdot \rho$ is an open subset of $\mathrm{Hom}(\Gamma, G)$. We do not use this cohomology formalism explicitly but the tangent space calculations in the proof of Proposition 10.5.18 can be interpreted in terms of 1-cocycles, as can some of the work of Slodowy discussed in Sect. 12.7.

Many applications of character varieties involve the special case $G = \mathrm{SL}_2$ for $k = \mathbb{C}$. It follows then from Theorem 5.6.3(i) and Sect. 6.5 that for any finitely generated group Γ, $\mathcal{T}_\Gamma(\mathrm{SL}_2)$ and $\mathbb{C}[\mathrm{Hom}(\Gamma, \mathrm{SL}_2)]^{\mathrm{SL}_2}$ are equal, so $C(\Gamma, \mathrm{SL}_2)$ and $X(\Gamma, \mathrm{SL}_2)$ are canonically isomorphic. But if $p > 0$ then we do not know whether $\mathcal{T}_\Gamma(G)$ and $\mathbb{C}[\mathrm{Hom}(\Gamma, G)]^G$ are equal, not even when $\Gamma = \Gamma_n$ and $G = \mathrm{SL}_2$. Some variations on $\mathcal{T}_\Gamma(G)$ and $X(\Gamma, G)$ have been studied: see [96] and [162], for example.

6.11 Exercises

Exercise 6.1 Let $T \cong (k^*)^m$ be a torus of rank m over a solid field k. Complete the proof of Lemma 6.4.4 by showing the following:

(i) If $\mathrm{char}(k) = 0$ then the set

$$A := \{(p_1, \ldots, p_m) \mid p_i \in \mathbb{N} \text{ are distinct primes}\}$$

is dense in T.

(ii) If $\mathrm{char}(k) = p$ and $a \in k^*$ is transcendental over $\mathbb{F}_p$ then the set

$$\{(f_1(a), \ldots, f_m(a)) \mid f_i \in \mathbb{F}_p[x] \text{ are distinct coprime polynomials}\}$$

is dense in T.

Exercise 6.2 Suppose $p > 0$.

---

[5] Here we mean the usual (abelian) group cohomology.

(i) Show that if k is solid then $k^* \ltimes k$ is topologically finitely generated, where k^* acts on k by $a \cdot x = ax$.

(ii) Show that k and $k^* \times k$ are not topologically finitely generated, even when k is solid.

Exercise 6.3 Recall from Sect. 6.5 that for a group Γ generated by n elements $\gamma_1, \ldots, \gamma_n$, and an algebraic group H, we can identify $\mathrm{Hom}(\Gamma, H)$ with a closed subset of H^n, and this gives $\mathrm{Hom}(\Gamma, H)$ the structure of a variety. Show that this structure does not depend on the choice of generating set.

Exercise 6.4 Show that (6.4.2) does indeed define an action of $\mathrm{Aut}(\Gamma_n)$ on H^n.

[*Hint*: try using the previous exercise to identify H^n with $\mathrm{Hom}(\Gamma_n, H)$.]

Exercise 6.5 Let k_0 be the algebraic closure of the prime field and choose a k_0-descent G_0 of G. Suppose Γ is a finitely generated group such that $C(\Gamma, G)$ is finite. Show that for every $\rho \in \mathrm{Hom}(\Gamma, G)$ with G-cr image, there exists $g \in G$ such that $g\rho(\Gamma)g^{-1} \subseteq G_0$.

Exercise 6.6 Let H be a reductive group and let M be a subgroup of G. Show that there are only countably many M-conjugacy classes of homomorphisms $f \colon H \to G$ such that $f(H) = M$.

Exercise 6.7 Recall the definition of $\preceq$ from Definition 6.8.3. Prove the following (remembering not to apply the Finiteness Theorem or the Rigidity Theorem!).

(i) Show that $\preceq$ is well-defined.
(ii) Show that if $G \cdot H \preceq G \cdot K$ then either (a) $\dim(H) < \dim(K)$, or (b) $\dim(H) = \dim(K)$ and $\kappa(H) \leq \kappa(K)$.
(iii) Show that $\preceq$ is a partial ordering on $\mathscr{S}$.
(iv) Show that $\mathscr{S}$ satisfies the descending chain condition with respect to $\preceq$.
(v) Suppose k is solid. Show that the following are equivalent: (a) $H \preceq K$; (b) for all $n \in \mathbb{N}$ and all $\mathbf{h} \in H^n$, there exists $\mathbf{k} \in K^n$ such that $\pi_{G^n}(\mathbf{k}) = \pi_{G^n}(\mathbf{h})$; (c) there exist $n \in \mathbb{N}$, $\mathbf{k} \in K^n$ and $\mathbf{h} \in H^n_{\mathrm{gen}}$ such that $\pi_{G^n}(\mathbf{k}) = \pi_{G^n}(\mathbf{h})$.
(vi) Suppose k is solid. Let E be an algebraically closed field extension of k. Show that $H \preceq K$ if and only if $H_E \preceq K_E$.

(These results are used in Sect. 12.10.)

The Spherical Building of G 7

In this chapter we outline the connection between complete reducibility and spherical buildings, and the further link between buildings and our geometric invariant theory approach. This sequence of ideas is extremely important in the book: although most of what we do in the rest of the book could be achieved without explicit mention of buildings, the ideas in this section give a framework which brings together the various different points of view in a natural way. In particular, the material on the vector building in Sects. 7.5, 7.6 and 7.7 provides what we believe is the best setting within which to understand the optimality constructions of Kempf-Rousseau-Hesselink in Chap. 8 (indeed, this is precisely the point of view of Rousseau [150]). Much of the material in this chapter has been developed only for connected reductive groups (and some of it simply doesn't work in the same way for non-connected groups). We deal with this issue in a systematic way in the final section of the chapter, but for the majority of the chapter we enforce the following convention:

> unless otherwise indicated, in this chapter our reductive group G is assumed to be *connected.*

We may regard a spherical building Δ as a simplicial complex that satisfies certain axioms (see [1], [149], [183], and [185]). Rather than give a full overview of the theory of spherical buildings, which is a vast subject, we concentrate on describing the spherical building Δ_G of a (connected) reductive group G over an algebraically closed field k, and collecting together (mainly without proof) the key properties which allow us to study complete reducibility in this context; see Sects. 7.1, 7.2 and 7.3. Here we follow much of the development from the paper of Serre [159]. We then recall the important "Centre Conjecture" (now a theorem) of Tits in Sect. 7.4 and give a key example of how it can be used in the study of complete reducibility, Corollary 7.4.4. The rest of the chapter is devoted to the link between cocharacters and buildings. We introduce the vector building

© The Author(s), under exclusive license to Springer Nature Switzerland AG 2026   143
M. Bate et al., *G-Complete Reducibility, Geometric Invariant Theory and Spherical Buildings*, Oberwolfach Seminars 57, https://doi.org/10.1007/978-3-032-08866-6_7

in Sect. 7.5, explain how it connects with the theory of complete reducibility in Sect. 7.6, and elucidate the link with geometric invariant theory in Sect. 7.7. We consider a variation on the Centre Conjecture in Sect. 7.8. Finally, in Sect. 7.9, we indicate how to deal with the non-connected case.

There are various approaches to the theory of spherical buildings, see Sect. 7.10. We write "simplicial building" rather than "spherical building" to emphasise our simplicial approach, and to distinguish between this version of the building and others, such as the vector building that we introduce below and the Bruhat-Tits building which we encountered briefly in Sect. 6.7.

7.1   The Simplicial Building of a Reductive Group

We begin with the definition of the building of a connected reductive group.

Definition 7.1.1 Given a connected reductive group G, the *simplicial building of G* is the poset Δ_G formed by the parabolic subgroups of G, with inclusion reversed. That is, for each parabolic subgroup P of G, we have an element $\sigma_P \in \Delta_G$, and we write $\sigma_P \le \sigma_Q$ in Δ_G if and only if $P \supseteq Q$ in G. There is an action of G on Δ_G by poset automorphisms: for $g \in G$ and any parabolic subgroup P, we define $g \cdot \sigma_P := \sigma_{gPg^{-1}}$. We allow the case $P = G$ here, providing a minimal element σ_G in the poset. The maximal elements are those corresponding to the Borel subgroups of G; the conjugacy of Borel subgroups implies that the action of G is transitive on these maximal elements. We may regard Δ_G as a simplicial complex with set of simplices $\{\sigma_P \mid P$ is a parabolic subgroup of $G\}$ (see Sect. 7.2.1).

Example 7.1.2 Recall from Example 2.4.2 that if V is a finite-dimensional k-vector space, then for $G = \mathrm{GL}(V)$ or $\mathrm{SL}(V)$ the parabolic subgroups of G are the stabilisers of *flags* of subspaces of V. Such a flag $\mathcal{F}$ has the form

$$\mathcal{F} = (U_1 \subset U_2 \subset \cdots \subset U_r),$$

and we write $\mathcal{F}' \le \mathcal{F}$ if $\mathcal{F}'$ can be obtained from $\mathcal{F}$ by deletion; that is, if we have $1 \le i_1 < i_2 < \cdots < i_s \le r$ for some s such that

$$\mathcal{F}' = (U_{i_1} \subset U_{i_2} \subset \cdots \subset U_{i_s}).$$

Thus we see that the simplicial building Δ_G is isomorphic to the poset of flags of subspaces of V.

Suppose now that $H \subseteq G$ is a symplectic or special orthogonal group defined by a non-degenerate symmetric or alternating form on V. Looking again at Example 2.4.2, we recall that a subspace U of V is *isotropic* if the restriction of the form to U is 0. The parabolic subgroups of H are the stabilisers of flags of isotropic subspaces in V, and hence the simplicial building Δ_H is isomorphic to the poset of flags of isotropic subspaces in V.

7.2 Properties of Buildings

We collect together several properties of spherical buildings, and explain which properties of reductive groups and their parabolic subgroups they reflect.

7.2.1 Simplicial Structure

By definition, a simplicial building Δ is a simplicial complex satisfying certain axioms. For the building Δ_G, this follows from the characterisation of parabolic subgroups from the root system: a parabolic subgroup P is uniquely determined by the maximal parabolic subgroups containing it (see Exercise 4.7(i)), and so the vertex set of the simplicial complex Δ_G is the set of elements corresponding to the maximal parabolic subgroups of G. Note that in this way the minimal element σ_G of Δ_G corresponding to G itself identifies with the empty simplex $\varnothing$.

7.2.2 Axioms of a Building

The first axiom of a building Δ is that a building is the union of a system of subcomplexes called *apartments*, each isomorphic to the Coxeter complex of a fixed Coxeter group. For our $\Delta = \Delta_G$, we have one apartment $A_T = \{\sigma_P \mid T \subseteq P\}$ for each maximal torus T of G. By definition, the Coxeter group $W = N_G(T)/T$ is the Weyl group of G; since the maximal tori in G are all conjugate, this is independent of the choice of T. The minimal parabolic subgroups containing T are the Borel subgroups containing T, and these are in one-one correspondence with the elements of W, which induces the isomorphism of A_T with the Coxeter complex attached to W. (See Exercise 7.1 below for a definition of the Coxeter complex.)

Given the apartment system, the second building axiom states that any two simplices of a building must lie in a common apartment. For our Δ_G, this corresponds to the fact that the intersection of any two parabolic subgroups of G contains a maximal torus. We say that G satisfies the *common apartment property*.

The final building axiom states that given any two apartments A_1 and A_2 of a building, there is an isomorphism of simplicial complexes from A_1 to A_2 fixing $A_1 \cap A_2$ pointwise. For our Δ_G, given two apartments $A_1 = A_{T_1}$ and $A_2 = A_{T_2}$ corresponding to maximal tori T_1 and T_2, we can take the intersection

$$H = \bigcap_{\sigma_P \in A_1 \cap A_2} P.$$

This subgroup of G contains T_1 and T_2, and hence there is $h \in H$ such that $h T_1 h^{-1} = T_2$. Then $h \cdot A_1 = A_2$ and h fixes every element of the intersection $A_1 \cap A_2$.

7.2.3 The Geometric Realisation

Any abstract simplicial complex X has a *geometric realisation* as a complex living in some (possibly infinite-dimensional) Euclidean space. To construct this, one first constructs the abstract real vector space spanned by the vertices of X, and then attaches to each simplex $\sigma \in X$ the open simplex $|\sigma|$ which is the interior of the simplex spanned by its vertices. Then the geometric realisation $|X|$ is the union of these open simplices: see [1, §A.1.1].

We denote the geometric realisation of the simplicial building Δ by $|\Delta|$. In Sect. 7.5 below we show how to recover this geometric realisation using cocharacters in case $\Delta = \Delta_G$. There is a subtlety here which is important in the general theory but does not cause too many issues in the study of complete reducibility: since the centre $Z(G)$ is contained in every parabolic subgroup of G, the simplicial building of G is identical with the simplicial building of the semisimple group $[G, G]$, and hence they have the same geometric realisation. On the other hand, the geometric object we construct below does have a contribution from the centre $Z(G)$ whenever $Z(G)^0$ is non-trivial; see Sect. 7.5.8.

7.2.4 Spherical Buildings, the Type Function and Opposition

A building Δ is *spherical* if its associated Coxeter group is finite; in this case the Coxeter complex can be viewed geometrically as a triangulation of a sphere—that is, each simplex in the building gives rise to a spherical simplex in the geometric realisation. See Exercise 7.1 for some examples of this. Since the Weyl group W of G is finite, our building Δ_G is spherical.

We also see that a spherical building Δ is a *chamber complex*—the chambers are the maximal simplices in Δ—and it carries a *type function* in the sense of [1, §A.1.3]. The type function is a function from Δ to a set of subsets of a finite set I, such that the vertices of each chamber map bijectively to I. More generally, we define the *type* of an arbitrary simplex to be the set of types of its vertices. For our building $\Delta = \Delta_G$ we can take $I = \Pi$ to be the set of simple roots, as we now demonstrate.

Example 7.2.1 Suppose $G = \mathrm{GL}_m$; recall from Example 7.1.2[1] that simplices in Δ_G correspond to flags of subspaces of k^m. Let B be the standard upper triangular Borel subgroup of G. Then B is the stabiliser of a maximal flag $(U_1 \subset \cdots \subset U_{m-1})$ (see Example 2.4.2). There are $m - 1$ maximal parabolic subgroups of G containing B, which

---

[1] For more details on the structure of Δ_G when $G = \mathrm{GL}_m$, see [1, §4.3].

are the stabilisers $P_i := N_G(U_i)$ for $1 \le i \le m - 1$. There are also $m - 1$ simple roots $\Pi = \{\alpha_1, \ldots, \alpha_{m-1}\}$ corresponding to this choice of B, and we can label them so that $P_i = P_{\Pi_i}$, where $\Pi_i = \Pi \setminus \{\alpha_i\}$, and the notation for the standard parabolic subgroup P_{Π_i} is as in (2.4.1). This shows how to define the type function for Δ_G: every maximal parabolic is conjugate to a unique one of the P_i, and hence every vertex of Δ_G can be labelled by a unique simple root α_i (or equivalently a unique integer in the set $\{1, \ldots, m - 1\}$). Then every simplex gets labelled by the subset corresponding to the labels of its vertices.

The same scheme works for the building of any reductive group G: fix a Borel subgroup B and a corresponding set of simple roots. Then the maximal parabolics containing B can be labelled by the simple roots, and this extends to a function on all of Δ_G.

The spherical structure gives rise to a notion of *opposition* in a spherical building Δ: two simplices are opposite if they are opposite in some apartment (and hence all apartments) containing them both. In case $\Delta = \Delta_G$, the notion of opposition of simplices is precisely the notion of opposition of parabolic subgroups of G: σ_P and σ_Q are opposite in Δ_G if and only if P and Q are opposite in G if and only if $P \cap Q$ is a Levi subgroup of P and Q. In particular, if $G = \mathrm{GL}_m$ then two simplices are opposite if and only if the corresponding flags of subspaces are complementary.

7.2.5   Thickness

A *panel* in a spherical building Δ is a (non-empty) simplex of codimension 1; that is, it has type $I \setminus \{i\}$ for some $i \in I$, where I is the finite set of the type function. A building is *thick* if every panel is contained in at least 3 chambers. The building Δ_G of a reductive group G of semisimple rank at least 2 is thick:[2] this translates to the fact that away from type A_1, every parabolic subgroup which is not a Borel subgroup contains at least 3 Borel subgroups, since if P and B are a parabolic subgroup and a Borel subgroup with $B \subsetneq P$, then P/B is a connected projective variety of dimension at least 1, and hence is in fact infinite. Thus there are infinitely many distinct P-conjugates of B.

7.2.6   Convexity

There is a notion of *convexity* for subsets of a spherical building Δ (or, more precisely, for subsets of the geometric realisation), see [159, §2.1.4, §2.1.5]. For subcomplexes of the building, this notion can be captured combinatorially, and in case $\Delta = \Delta_G$ there is a nice interpretation in terms of parabolic subgroups due to Serre [159, Prop. 3.1]: a non-empty

---

[2] There is a converse result, proved by Tits: roughly speaking, it says that almost all thick buildings are the buildings of semisimple groups. For a precise statement, see [185, Thm. 41.23].

subset Σ of simplices of Δ_G is a convex subcomplex if and only if whenever P, Q, R are parabolic subgroups of G with $\sigma_P, \sigma_Q \in \Sigma$ and $P \cap Q \subseteq R$, then $\sigma_R \in \Sigma$. See also Sect. 7.6.

For the study of complete reducibility, the most important example of a convex subcomplex in the building Δ_G is the so-called *fixed-point subcomplex* for a subgroup H of G.

Definition 7.2.2 Let H be a subgroup of G. We define

$$(\Delta_G)^H := \{\sigma_P \mid H \subseteq P\};$$

this is the set of H-fixed points in Δ_G, where the action of G on Δ_G is as in Definition 7.1.1. It is clear from Serre's criterion above that $(\Delta_G)^H$ *is* a convex subcomplex.

7.2.7   Automorphisms of Δ

We define an *automorphism* of the spherical building Δ to be an automorphism of Δ as a simplicial complex. It can be shown that an automorphism maps apartments of Δ to apartments of Δ (see [1, Thm. 4.70]). The set of automorphisms of Δ forms a group under composition, which we denote by $\mathrm{Aut}(\Delta)$. Key examples of automorphisms in $\mathrm{Aut}(\Delta_G)$ are those afforded by the action of G by conjugation; that is, the maps given by $\sigma_P \mapsto \sigma_{gPg^{-1}}$ for $g \in G$.

An automorphism α of Δ is said to be *type-preserving* if $\alpha(\sigma)$ has the same type as σ for all simplices σ (it is clearly enough to check this on vertices). The action described in the previous paragraph of G on Δ_G is by type-preserving automorphisms.

Example 7.2.3 Examples of non-type-preserving automorphisms include those arising from outer automorphisms of G. For example, if $G = \mathrm{GL}_m$ and $\beta : G \to G$ is the inverse-transpose map, then β sends maximal parabolic subgroups to maximal parabolic subgroups, and hence induces an automorphism α, say, of Δ_G given by $\alpha(\sigma_P) = \sigma_{\beta(P)}$ for each parabolic P of G. But if $m \geq 3$ this map is *not* type-preserving. For example, in the notation introduced in Example 7.2.1 above, the maximal parabolic P_1 is the stabiliser of a 1-dimensional subspace of k^m (the span of the first standard basis vector), whereas the maximal parabolic $\beta(P_1)$ is the stabiliser of an $(m-1)$-dimensional subspace (the span of all the other standard basis vectors). This means that $\beta(P_1)$ is conjugate to P_{m-1}, and so α changes the type of the corresponding vertices in Δ_G. (It's not hard to see that the permutation of the simple roots induced by β is just the one given by the involution of the type A_{m-1} Dynkin diagram.)

When Δ_G is the simplicial building of a simple group G of rank at least 2, Tits showed in [183, Cor. 5.9, Cor. 5.10] that the automorphisms of Δ_G arise from a combination of inner automorphisms (i.e., those arising from the action of G itself), so-called field automorphisms (see Example 11.5.2), and certain special automorphisms arising from Frobenius automorphisms in characteristics 2 and 3. We observe also that a group of type-preserving automorphisms has a fixed-point set which is a subcomplex: if an automorphism fixes a simplex and preserves the type function, then it must also fix each face of that simplex, since the faces of a simplex all have distinct types.

7.3   Spherical Buildings and Complete Reducibility

Definition 7.3.1 Let Δ be a simplicial spherical building and Σ a convex subcomplex of Δ. Let $\sigma \in \Sigma$. We say that σ is *unopposed* in Σ if there does not exist $\tau \in \Sigma$ such that τ is opposite to σ.

Definition 7.3.2 ([159, §2.2]) Let Δ be a simplicial spherical building and Σ a convex subcomplex of Δ. We say that Σ is Δ-*completely reducible* (Δ-*cr* for short) if every $\sigma \in \Sigma$ has an opposite which also lies in Σ. So if a convex subcomplex Σ contains an unopposed simplex σ, then Σ is *not* Δ-cr.

Here is the observation allowing us to study G-complete reducibility via the building Δ_G.

Remark 7.3.3 Suppose that H is a subgroup of G. Then we claim that H is G-cr if and only if $(\Delta_G)^H$ is Δ_G-cr. Indeed, since $\sigma_P \in (\Delta_G)^H$ if and only if $H \subseteq P$, and H is contained in a Levi subgroup of P if and only if there is an opposite parabolic subgroup Q to P also containing H, this is immediate from Definitions 4.1.1 and 7.3.2.

The following two basic results are part of [159, Thm. 2.1].

Theorem 7.3.4 *Suppose Σ is a convex subcomplex of a spherical building Δ. Then Σ is Δ-completely reducible if and only if each vertex of Σ has an opposite in Σ.*

The proof of the preceding result is rather geometric in nature; for a proof using the subgroup structure of the reductive group G where $\Sigma = (\Delta_G)^H$ for a subgroup H of G, see Exercise 7.3 below.

Theorem 7.3.5 *Suppose Σ is a convex subcomplex of a spherical building Δ. Then at least one of the following holds: Σ is Δ-completely reducible, or the geometric realisation $|\Sigma|$ is contractible (homotopic to a point).*

Sketch Proof Given a convex subcomplex Σ, if it is not completely reducible, then there is a vertex $v \in \Sigma$ without an opposite in Σ, by Theorem 7.3.4. Now, in the geometric realisation $|\Sigma|$, given two non-opposite points there is a unique geodesic between them

[159, §2.1.4]; the vertex v corresponds to a point in the geometric realisation, and we can contract $|\Sigma|$ along geodesics issuing from this point. $\square$

Remark 7.3.6 A closely related argument can be used to show that the geometric realisation $|\Delta|$ of a spherical building Δ is homotopic to a bouquet (or wedge) of spheres; this is known as the Solomon-Tits Theorem. Fixing again a vertex $v \in \Delta$, every other simplex is contained in a common apartment with v. Thus every point of $|\Delta|$ is contained in a common sphere with the point corresponding to v, and one can then argue that $|\Delta|$ is homotopic to a bouquet of spheres. See [167] and [54] for more details.

The final result of this section is the analogue of Theorem 5.4.4 in this building-theoretic setting; indeed, this route via the building was the basis of Serre's original proof of Theorem 5.4.4 for connected G. In order to state the theorem we first need the concept of a *Levi sphere* in a spherical building Δ from [159, 2.1.6].

Definition 7.3.7 Suppose Δ is a spherical building. A *Levi sphere* in Δ is a subcomplex S which is contained in a single apartment A and for which, when the geometric realisation $|A|$ is viewed as a sphere in Euclidean space, $|S|$ is the intersection of $|A|$ with a vector subspace of the ambient Euclidean space.

Given a Levi sphere S in Δ, in [159, 2.1.8] Serre describes how to associate to S a spherical building Δ_S as follows: choose a maximal simplex $\sigma \in S$ and then consider the set of simplices $\Delta_\sigma = \{\tau \in \Delta \mid \sigma \leq \tau\}$. If σ' is another choice of maximal simplex in S, there is a canonical isomorphism $\Delta_\sigma \to \Delta_{\sigma'}$, which permits us to write Δ_S for this (abstract) simplicial complex. Further, if Σ is a convex subcomplex of Δ which contains the Levi sphere S, then for any choice of σ as above we may form the subset $\Sigma_\sigma = \{\tau \in \Sigma \mid \sigma \leq \tau\}$ and we obtain a subcomplex of Δ_σ. This again is independent of the choice of σ, so we can form a subcomplex Σ_S of Δ_S.

As usual, the rather geometric Definition 7.3.7 has a group-theoretic interpretation when $\Delta = \Delta_G$: in this case, given a Levi subgroup L of a parabolic subgroup P of G, we can form the fixed point subcomplex $S = (\Delta_G)^L = \{\sigma_P \mid L \subseteq P\}$. Then this is a Levi sphere in Δ_G [159, 3.1.7]. Moreover, the building $(\Delta_G)_S$ can be identified with Δ_L, as follows. A maximal simplex of S is of the form σ_P, where P is a parabolic subgroup such that L is a Levi subgroup of P. If we choose such a P then $(\Delta_G)_S = (\Delta_G)_{\sigma_P}$ consists of the simplices σ_Q such that $Q \subseteq P$. The map $Q \mapsto Q \cap L$ induces a bijection between the parabolic subgroups Q of G with $Q \subseteq P$ and the parabolic subgroups of L, with inverse $Q_L \mapsto Q_L R_u(P)$ (this is Lemma 2.9.12).

Example 7.3.8 We give details of the above construction for the two extreme cases $L = G$ and $L = T$, a maximal torus of G.

When $L = G$, we have $S = (\Delta_G)^G = \{\varnothing\}$ consists just of the empty simplex. In the geometric realisation $|\Delta_G|$ this corresponds to the (empty) intersection of any apartment with the zero subspace. The empty simplex itself is the maximal simplex of S, and $(\Delta_G)_S = (\Delta_G)_\varnothing$ is identified with the whole building Δ_G as intended.

When $L = T$, the subset $S = A_T$ is an apartment of Δ_G, which is the Levi sphere corresponding to this apartment itself. A maximal simplex of A_T is a simplex σ_B where B is some Borel subgroup containing T. Then the set of simplices $(\Delta_G)_S = (\Delta_G)_{\sigma_B} = \{\sigma_B\}$ is a singleton, and so we see that (abstractly) this does not depend on the choice of B, and $(\Delta_G)_S$ does indeed correspond to the building of the group T (which is just a single empty simplex, since T has no proper parabolic subgroups).

Now we can state [159, Prop. 2.5]:

Theorem 7.3.9 *Suppose Σ is a convex subcomplex of a spherical building Δ, and S is a Levi sphere contained in Σ. Then Σ is Δ-completely reducible if and only if Σ_S is Δ_S-completely reducible.*

Remarks 7.3.10   (i). To see how this implies Theorem 5.4.4 for connected G, suppose H is a subgroup of G, and let L be a Levi subgroup of G containing H. Then $(\Delta_G)^H \supseteq (\Delta_G)^L$, so $S = (\Delta_G)^L$ is a Levi sphere contained in $(\Delta_G)^H$. Further, under the identification of $(\Delta_G)_S$ with Δ_L, the subcomplex $((\Delta_G)^H)_S$ naturally identifies with $(\Delta_L)^H$. So, applying Theorem 7.3.9 to the fixed-point subcomplexes $(\Delta_G)^H$ and $(\Delta_L)^H$, we see that H is G-cr if and only if H is L-cr.

(ii). For connected G, Theorem 7.3.9 generalises Theorem 5.4.4, as it applies over arbitrary fields (see, for example, Remark 11.7.17(i) below). However, the proof does use in an essential way the fact that Σ is a subcomplex, whereas results analogous to Theorem 5.4.4 hold in much more generality: for example, for subsets of the building arising in a geometric invariant theory setting, as in Sect. 7.7.

7.4   The Tits Centre Theorem

Early in the development of the theory of buildings, Tits made a conjecture about the structure of convex subcomplexes which stood for many years. It was finally proven via a case-by-case analysis in three papers [97, 128, 140], leading to the following result.

Theorem 7.4.1 (Tits Centre Theorem) *Suppose Σ is a convex subcomplex of a thick spherical building Δ. Then one of the following is true:*

(i) *Σ is Δ-completely reducible;*

(ii) *there exists a non-empty simplex $\sigma \in \Sigma$ which is fixed by every building automorphism of Δ that stabilises Σ.*

Definition 7.4.2 A simplex σ satisfying property (ii) of Theorem 7.4.1 is called a *centre* for Σ.

Remark 7.4.3 Since a non-Δ-cr subcomplex is contractible, the Centre Theorem can be viewed as an analogue in the setting of spherical buildings of fixed point results due to Bruhat-Tits for affine buildings (or, more generally, complete CAT(0)-spaces, see [1, Thm. 11.23]). (Recall that we saw an application of the Bruhat-Tits fixed point theorem in Sect. 6.7.) It is also a "combinatorial analogue" of the optimality results described in Chap. 8 below. See Exercise 7.7 for the idea of the proof in type A, and Sect. 7.10 for some further discussion.

As an example of the use of the Tits Centre Theorem in the study of complete reducibility, we give a quick proof of one of the landmark results in the theory: an analogue for connected reductive groups of Clifford's Theorem in representation theory (for much more on this, including a version for non-connected G, see Sect. 9.1 below).

Corollary 7.4.4 *Suppose N and H are subgroups of G with N normal in H. If H is G-completely reducible then so is N.*

Proof Suppose that N is not G-cr. By using Theorem 5.4.4 (or Theorem 7.3.9), we may assume that H is not contained in any proper Levi subgroup of G. If N is not G-cr then the fixed point subcomplex $(\Delta_G)^N$ is not Δ_G-cr. Since N is normal in H it is clear that H stabilises the fixed point subcomplex $(\Delta_G)^N$. Hence we are in the situation of Theorem 7.4.1(ii); let $\sigma = \sigma_P$ be a non-empty H-fixed simplex of $(\Delta_G)^N$. Then P is a proper parabolic subgroup of G normalised by H, so $H \subseteq P$. But we reduced to the case that H is in no Levi subgroup of G in the first paragraph, so P witnesses the fact that H is also not G-cr. $\square$

Remark 7.4.5 Let Σ be a convex subcomplex of a thick spherical building Δ. Suppose Σ is not Δ-completely reducible. Then there is a centre Σ for Δ, by Theorem 7.4.1. By definition there is a an unopposed simplex in Σ. It is useful to know whether there is an unopposed centre: for instance, if G is connected, H is a subgroup of G and $\Sigma = (\Delta_G)^H$ then a simplex $\sigma_P \in \Sigma$ is unopposed if and only if H is not contained in a Levi subgroup of P (compare the proof of Corollary 7.4.4). It does not seem to be known in general whether a non-Δ-cr convex subcomplex of a thick spherical building Δ admits an unopposed centre; note, however, that if Δ has type A then the construction in [128] yields an unopposed centre. We discuss these matters further when we study the notion of centre for convex subsets of a vector building in Sect. 7.6.

7.5   Cocharacters and Vector Buildings

Since every parabolic subgroup of a connected reductive group has the form P_λ for a cocharacter λ, and the parabolics are identified with the simplices of the corresponding building, it is natural to ask whether we can make a link between the cocharacters of G and the building. We now answer this question. This section also lays the groundwork for our description of the optimality theory of Kempf-Rousseau-Hesselink in the next chapter.

7.5.1   Virtual Cocharacters of Maximal Tori

First let T be a maximal torus of G; then the set of cocharacters of T has the structure of an abelian group—we have $Y_T \cong \mathbb{Z}^r$, where r is the rank of T. Hence it makes sense to form the vector spaces

$$Y_T(\mathbb{Q}) := Y_T \otimes_{\mathbb{Z}} \mathbb{Q} \qquad \text{and} \qquad Y_T(\mathbb{R}) := Y_T \otimes_{\mathbb{Z}} \mathbb{R}.$$

The elements of $Y_T(\mathbb{R})$ are sometimes called *virtual cocharacters*. In what follows, we let $\mathbb{K}$ denote either one of the fields $\mathbb{Q}$ or $\mathbb{R}$. Note that Y_T can be identified as a subset of $Y_T(\mathbb{K})$ in the obvious way. For now we denote an element of $Y_T(\mathbb{K})$ as a pair (T, λ) so that we can keep track of the torus T as well as λ. (This temporary notation will be dropped at the end of the following subsection, once we have carried out a suitable gluing construction.)

To each element $(T, \lambda) \in Y_T(\mathbb{K})$ we can associate a parabolic subgroup P_λ and a Levi subgroup L_λ of P_λ. One way to do this is to observe that, given a root $\alpha \in \Phi(G, T)$, we can define $\langle \lambda, \alpha \rangle$ in the obvious way, and it still makes sense to ask whether $\langle \lambda, \alpha \rangle$ is positive or negative or 0. Hence we can define P_λ to be the subgroup generated by T and the root groups U_α with $\langle \lambda, \alpha \rangle \geq 0$, and L_λ to be the subgroup generated by T and the root groups U_α with $\langle \lambda, \alpha \rangle = 0$.

We now show that these constructions are compatible with the conjugation action of G. Suppose T is a maximal torus of G and let $g \in G$. Set $T' = gTg^{-1}$; then the map $\lambda \mapsto g \cdot \lambda$ gives an isomorphism of abelian groups $Y_T \to Y_{T'}$, which induces a $\mathbb{K}$-linear map $Y_T(\mathbb{K}) \to Y_{T'}(\mathbb{K})$. We denote this map by $(T, \lambda) \mapsto (gTg^{-1}, g \cdot \lambda)$ for $(T, \lambda) \in Y_T(\mathbb{K})$. Now given $\alpha \in \Phi(G, T)$, we have $g \cdot \alpha \in \Phi(G, T')$ and it is easy to see that

$$\langle g \cdot \lambda, g \cdot \alpha \rangle = \langle \lambda, \alpha \rangle. \tag{7.5.1}$$

(Note that the case when $\lambda \in Y_T$ is a genuine cocharacter is just (2.5.1), and the general case follows from this and the definition of $Y_T(\mathbb{K})$.) Hence $U_\alpha \subseteq P_\lambda$ if and only if $U_{g \cdot \alpha} \subseteq P_{g \cdot \lambda}$, and similarly for L_λ and $L_{g \cdot \lambda}$. This shows that for each $g \in G$ and $\lambda \in Y_T(\mathbb{K})$ we have

$$P_{g \cdot \lambda} = g P_\lambda g^{-1} \text{ and } L_{g \cdot \lambda} = g L_\lambda g^{-1}. \tag{7.5.2}$$

7.5.2   Virtual Cocharacters of G

Observe that the set Y_G of cocharacters of G can be constructed by "patching together" all the cocharacter groups Y_T as T varies over the maximal tori of G. To be more precise, it may happen that an element of Y_G arises as the composition of some $\lambda \in Y_T$ with the inclusion of T in G *and* as the composition of some $\lambda' \in Y_{T'}$ with the inclusion of T' in G, for two different maximal tori T and T' of G. However, whenever this happens, there will be some $g \in L_\lambda$ such that $(T', \lambda') = (gTg^{-1}, g \cdot \lambda)$. We turn this observation into a definition: given maximal tori T and T' of G, and virtual cocharacters $(T, \lambda) \in Y_T(\mathbb{K})$ and $(T', \lambda') \in Y_{T'}(\mathbb{K})$, write

$$(T, \lambda) \sim (T', \lambda') \iff \text{ there exists } g \in L_\lambda \text{ with } (T', \lambda') = (gTg^{-1}, g \cdot \lambda). \qquad (7.5.3)$$

To check that this is indeed an equivalence relation, note that reflexivity is obvious (take $g = 1$ in the definition). For symmetry and transitivity, the key observation is that if $(T, \lambda) \sim (T', \lambda')$ then $L_\lambda = L_{\lambda'}$ by (7.5.2). Note also that if $(T, \lambda) \sim (T, \lambda')$ then $P_\lambda = P_{\lambda'}$.

Now we may define

$$Y_G(\mathbb{K}) := \left(\coprod_T Y_T(\mathbb{K}) \right) \Big/ \sim$$

to be the quotient of the disjoint union of the spaces $Y_T(\mathbb{K})$ over all maximal tori T of G by the relation $\sim$. We allow ourselves to write simply λ for the equivalence class of each (T, λ) and note that for each maximal torus T of G, we have a natural embedding $(T, \lambda) \mapsto \lambda$ of $Y_T(\mathbb{K})$ in $Y_G(\mathbb{K})$. Using these embeddings, we identify $Y_T(\mathbb{K})$ with its image in $Y_G(\mathbb{K})$, and drop the (T, λ) notation for elements of $Y_T(\mathbb{K})$ when it is convenient to do so. We have a natural embedding of Y_G in $Y_G(\mathbb{K})$. The observations at the end of the previous paragraph also imply that it makes sense to write P_λ and L_λ for the corresponding parabolic and Levi subgroups.

7.5.3   The Vector Building

Given a simplex $\sigma = \sigma_P$ in the spherical building Δ_G, the apartments containing σ correspond to the maximal tori contained in P, and these tori are all conjugate in P. This motivates the following coarsening of the equivalence relation from the previous subsection. Given maximal tori T and T' of G, and $(T, \lambda) \in Y_T(\mathbb{K})$ and $(T', \lambda') \in Y_{T'}(\mathbb{K})$, write

$$(T, \lambda) \approx (T', \lambda') \iff \text{ there exists } g \in P_\lambda \text{ with } (T', \lambda') = (gTg^{-1}, g \cdot \lambda). \qquad (7.5.4)$$

That $\approx$ does define an equivalence relation is again an exercise using (7.5.2).

Now we may define

$$V_G(\mathbb{K}) := \left(\coprod_T Y_T(\mathbb{K}) \right) \Big/ \approx$$

to be the quotient of the disjoint union of the spaces $Y_T(\mathbb{K})$ over all maximal tori T of G by the relation $\approx$. It is clear that $V_G(\mathbb{K})$ is a quotient of $Y_G(\mathbb{K})$, and we denote the corresponding map by

$$\varphi_G : Y_G(\mathbb{K}) \to V_G(\mathbb{K}).$$

We let V_G denote the image of Y_G inside $V_G(\mathbb{K})$. Given $\zeta = \varphi_G(\lambda)$ we may write $P_\zeta = P_\lambda$ for the corresponding parabolic subgroup of G; doing so makes sense (i.e., is independent of the choice of λ with $\varphi_G(\lambda) = \zeta$) by (7.5.2). If T is a maximal torus of P_ζ then there exists $\lambda \in Y_T(\mathbb{K})$ such that $\zeta = \varphi_G(\lambda)$; this follows from the definition of $\approx$ and because any two maximal tori of P_ζ are P_ζ-conjugate.

We call $V_G(\mathbb{K})$ the *vector building of G*. In case we want to work specifically with $\mathbb{K} = \mathbb{R}$ we speak of the *real vector building of G* and we call $V_G(\mathbb{Q})$ the *rational vector building of G*. We again have a natural embedding of each $Y_T(\mathbb{K})$ inside $V_G(\mathbb{K})$, given by $(T, \lambda) \mapsto \varphi_G(\lambda)$. The images of the $Y_T(\mathbb{K})$ are denoted by $V_T(\mathbb{K})$ and are called the *apartments* of $V_G(\mathbb{K})$. Each apartment carries the structure of a $\mathbb{K}$-vector space and they are all naturally isomorphic to each other. The trivial cocharacters in each maximal torus all get mapped to the same point in $V_G(\mathbb{K})$, which we denote by $0 \in V_G(\mathbb{K})$; this is the common origin of all of these apartments. We define V_T to be the image of Y_T under our canonical identification of $Y_T(\mathbb{K})$ with $V_T(\mathbb{K})$. It follows from untangling the definitions that $V_T = V_G \cap V_T(\mathbb{K})$.

Given any two points $\zeta, \eta \in V_G(\mathbb{K})$, we may find an apartment $V_T(\mathbb{K})$ with $\zeta, \eta \in V_T(\mathbb{K})$. For $P_\zeta \cap P_\eta$ contains a maximal torus T of G, and there are elements $\lambda, \mu \in Y_T(\mathbb{K})$ such that $\zeta = \varphi_G(\lambda)$ and $\eta = \varphi_G(\mu)$. This allows us to define an addition operation on $V_G(\mathbb{K})$: we may define $\zeta + \eta$ to be the sum of ζ and η in any apartment containing them both. (That is, for λ, μ belonging to some $Y_T(\mathbb{K})$, we have $\varphi_G(\lambda) + \varphi_G(\mu) = \varphi_G(\lambda + \mu)$.) Note that if T' is another maximal torus contained in $P_\zeta \cap P_\eta$, then we can conjugate T to T' by an element $g \in P_\zeta \cap P_\eta$. But then it follows from the definitions that $\varphi_G(g \cdot \lambda) = \zeta$ and $\varphi_G(g \cdot \mu) = \eta$ and so $\zeta + \eta$ is independent of which apartment we choose containing ζ and η. Likewise, if $a \in \mathbb{K}_{\geq 0}$ then we can define $a\zeta \in V_G(\mathbb{K})$ by carrying out the scalar multiplication in any apartment to which ζ belongs: that is, we define $\varphi_G(a\lambda) = a\varphi_G(\lambda)$, and this does not depend on the choice of apartment. In particular, we obtain a well-defined ray $\mathbb{K}_+ \cdot \zeta = \{a\zeta \mid a \in \mathbb{K}_+\}$.

Care is needed, however, for a negative. For if $\zeta \in V_{T_1}(\mathbb{K}) \cap V_{T_2}(\mathbb{K})$ for T_1 and T_2 distinct maximal tori of G then $a\zeta$ taken in $V_{T_1}(\mathbb{K})$ can be different from $a\zeta$ taken in $V_{T_1}(\mathbb{K})$: hence we don't get a well-defined element $a\zeta$ of $V_G(\mathbb{K})$. On the other hand, this problem disappears at the level of cocharacters: if $\lambda \in Y_G(\mathbb{K})$ and $a < 0$ then $a\lambda$ does not depend on the choice of maximal torus T with $\lambda \in Y_T(\mathbb{K})$. Recall that the equivalence relation used to define $V_G(\mathbb{K})$ comes from the P_ζ-action, whereas the equivalence relation used to define $Y_G(\mathbb{K})$ comes from the L_λ-action: the difference in behaviour described above boils down to the fact that L_λ and $L_{-\lambda}$ are equal, but $R_u(P_\lambda)$ and $R_u(P_{-\lambda})$ are not.

Lemma 7.5.1 *Suppose $\lambda \in Y_G(\mathbb{Q})$. Then there exists $n \in \mathbb{N}$ such that $n\lambda \in Y_G$. Similarly, for any $\zeta \in V_G(\mathbb{Q})$ there exists $n \in \mathbb{N}$ such that $n\zeta \in V_G$.*

Proof Find some maximal torus T such that $\lambda \in Y_T(\mathbb{Q})$. Since $Y_T(\mathbb{Q}) = Y_T \otimes_{\mathbb{Z}} \mathbb{Q}$, a suitably large positive multiple of λ will in fact lie in Y_T, and this gives the result. For $\zeta \in V_G(\mathbb{Q})$ we may write $\zeta = \varphi_G(\lambda)$ for some $\lambda \in Y_G(\mathbb{Q})$, and then for every $n \in \mathbb{N}$ we have $n\zeta = n\varphi_G(\lambda) = \varphi_G(n\lambda)$; choosing n large enough so that $n\lambda \in Y_G$ gives the required result. $\square$

7.5.4 The Geometric Spherical Building

As noted above, there is a natural action of $\mathbb{K}_+$ on $Y_G(\mathbb{K})$ and $V_G(\mathbb{K})$. In particular, we can form the set of *rays* in $V_G(\mathbb{K})$:

$$\Delta_G(\mathbb{K}) := (V_G(\mathbb{K}) \setminus \{0\}) / \mathbb{K}_+.$$

We call $\Delta_G(\mathbb{K})$ the *geometric spherical building* of G; it is the object "at infinity" for the vector building $V_G(\mathbb{K})$. Since $P_\lambda = P_{a\lambda}$ for all $a \in \mathbb{K}_+$, we see that for any element $\zeta \in V_G(\mathbb{K})$ we also get $P_\zeta = P_{a\zeta}$, and hence it makes sense to write P_z for any $z \in \Delta_G(\mathbb{K})$.

7.5.5 Action of G

Recall from Sect. 7.5.1 that, given a maximal torus T and $g \in G$, conjugation by g induces a $\mathbb{K}$-linear isomorphism $Y_T(\mathbb{K}) \to Y_{gTg^{-1}}(\mathbb{K})$. It is straightforward to check that all the constructions above are compatible with these individual linear maps at the level of tori, and hence that the group G acts on the objects $Y_G(\mathbb{K})$, $V_G(\mathbb{K})$ and $\Delta_G(\mathbb{K})$. It is easy to see that the map $\varphi_G : Y_G(\mathbb{K}) \to V_G(\mathbb{K})$ is G-equivariant: if $(T, \lambda) \sim (T', \lambda')$ then $(gTg^{-1}, g \cdot \lambda) \sim (gT'g^{-1}, g \cdot \lambda')$, and similarly for $\approx$. Similarly, the projection from $V_G(\mathbb{K}) \setminus \{0\}$ to $\Delta_G(\mathbb{K})$ is equivariant. Clearly the embeddings of Y_G (resp., V_G) in $Y_G(\mathbb{K})$ (resp., $V_G(\mathbb{K})$) are G-equivariant, so Y_G and V_G are G-stable subsets of $Y_G(\mathbb{K})$ and $V_G(\mathbb{K})$, respectively. Moreover, it follows from (7.5.2) that

$$P_{g \cdot \zeta} = g P_\zeta g^{-1} \quad \text{and} \quad L_{g \cdot \lambda} = g L_\lambda g^{-1} \tag{7.5.5}$$

for all $\lambda \in Y_G(\mathbb{K})$ and all $\zeta \in V_G(\mathbb{K})$.

In the proof of the next result we use repeatedly the fact that for a parabolic subgroup P of G, we have $N_G(P) = P$; see Sect. 2.3.

Lemma 7.5.2 *For any $\lambda \in Y_G(\mathbb{K})$, we have $G_\lambda = L_\lambda$, where G_λ denotes the stabiliser of λ in G. Similarly, for any $\zeta \in V_G(\mathbb{K})$, we see that $G_\zeta = P_\zeta$.*

Proof For any $\lambda \in Y_G(\mathbb{K})$ and $g \in G$ such that $g \cdot \lambda = \lambda$, it follows from (7.5.5) that g normalises L_λ and P_λ. Since $P_\lambda = N_G(P_\lambda)$, and $N_{P_\lambda}(L_\lambda) = L_\lambda$, we have $G_\lambda \subseteq L_\lambda$. Conversely, if $g \in L_\lambda$ then it is clear by the definition of $Y_G(\mathbb{K})$ via the relation $\sim$ that $g \cdot \lambda = \lambda$, so we have $G_\lambda = L_\lambda$.

Now given $\zeta \in V_G(\mathbb{K})$, we see using (7.5.5) that $g P_\zeta g^{-1} = P_{g \cdot \zeta}$, for all $g \in G$. Hence any element of G stabilising ζ must normalise P_ζ and hence lie in P_ζ. The reverse inclusion follows from the definition of $V_G(\mathbb{K})$ via $\approx$. $\qquad\square$

7.5.6 Linear Maps

Suppose G, G' are connected reductive groups. A map $\kappa : V_G(\mathbb{K}) \to V_{G'}(\mathbb{K})$ is called a *linear map of vector buildings* if for each apartment $V_T(\mathbb{K})$ of $V_G(\mathbb{K})$ there is an apartment $V_{T'}(\mathbb{K})$ of $V_{G'}(\mathbb{K})$ such that the restriction of κ induces a $\mathbb{K}$-linear map $V_T(\mathbb{K}) \to V_{T'}(\mathbb{K})$. If κ is bijective and the inverse is also a linear map, then κ is called an *isomorphism of vector buildings*; in this case, if $G = G'$, then we say κ is an *automorphism* of the vector building $V_G(\mathbb{K})$. We define $\mathrm{Aut}(V_G(\mathbb{K}))$ to be the set of automorphisms of $V_G(\mathbb{K})$; it is easy to show that this is a group under composition.

Here are some important examples of linear maps:

Example 7.5.3 (i). The action of G on $V_G(\mathbb{K})$ from Sect. 7.5.5 is an action by automorphisms.

(ii). When $\phi : G \to \widetilde{G}$ is a homomorphism of reductive groups, then ϕ gives rise to a linear map $\kappa_\phi : V_G(\mathbb{K}) \to V_{\widetilde{G}}(\mathbb{K})$. This is the map induced by the map $\lambda \mapsto \phi \circ \lambda$ on cocharacters. To see that this induces a map on the vector building, the key case to consider is when T and T' are maximal tori of G, $\lambda \in Y_T$ and $\lambda' \in Y_{T'}$ are cocharacters, and $(T, \lambda) \approx (T', \lambda')$. We wish to show that for any choice of maximal tori S and S' of $\widetilde{G}$ with $\phi(T) \subseteq S$ and $\phi(T') \subseteq S'$ we have $(S, \phi \circ \lambda) \approx (S', \phi \circ \lambda')$. Since $(T, \lambda) \approx (T', \lambda')$, there is some $g \in P_\lambda$ with $gTg^{-1} = T'$ and $g \cdot \lambda = \lambda'$, and hence if we set $h = \phi(g)$ then $h\phi(T)h^{-1} = \phi(T')$ and $h \cdot (\phi \circ \lambda) = \phi \circ \lambda'$. But then hSh^{-1} is a maximal subtorus of $L_{\phi \circ \lambda'}$, and so we may adjust by an element of $L_{\phi \circ \lambda'}$ to ensure that $hSh^{-1} = S'$, and we're done. It is easy to check that $\kappa_\phi(g \cdot x) = \phi(g) \cdot \kappa_\phi(x)$ for all $x \in V_G(\mathbb{K})$ and all $g \in G$.

(iii). When $H \subseteq G$ is a reductive subgroup of G (or, more generally, if $\phi : H \to G$ is a homomorphism with finite kernel), the induced map on vector buildings allows us to identify $V_H(\mathbb{K})$ as a subset of $V_G(\mathbb{K})$. That we get a map from $V_H(\mathbb{K})$ to $V_G(\mathbb{K})$

follows from part (ii). That it is an inclusion amounts to showing that there is no fusion of equivalence classes when moving from the classes of the relation $\approx$ on virtual cocharacters of H to the classes for the same relation for G; this follows quite easily from Corollary 2.9.17.

(iv). If $\phi : G \to G'$ is an isogeny, then κ_ϕ is actually an isomorphism, since for each maximal torus T of G the induced map $\lambda \mapsto \phi \circ \lambda$ gives an isomorphism of vector spaces $Y_T(\mathbb{K}) \to Y_{\phi(T)}(\mathbb{K})$. This holds, in particular, when ϕ is a so-called Steinberg endomorphism (an isogeny with finite fixed point subgroup); see Sect. 11.8 for more on this.

An injective linear map $\kappa : V_G(\mathbb{K}) \to V_{G'}(\mathbb{K})$ of vector buildings descends to a map $\Delta_G(\mathbb{K}) \to \Delta_{G'}(\mathbb{K})$ on the geometric spherical buildings (injectivity is necessary here for well-definedness). In particular, automorphisms of $V_G(\mathbb{K})$ give rise to automorphisms of $\Delta_G(\mathbb{K})$. It is also the case that an automorphism of $V_G(\mathbb{K})$ induces an automorphism of the corresponding simplicial spherical building (the proof of this is rather involved, but the key idea is to show that a parabolic subgroup P of G is determined by its cocharacters Y_P: see [25, Lem. 4.22, Lem. 4.24]). We call an automorphism of $V_G(\mathbb{K})$ *type-preserving* if the induced automorphism of Δ_G is type-preserving. We give an important basic example of these ideas.

Example 7.5.4 Let $p > 0$, let $k = \overline{\mathbb{F}_p}$ and let $G = \mathrm{GL}_m$ over k. Let $F_p \colon G \to G$ be the standard Frobenius map raising matrix entries to the power p, and let $\kappa_p : V_G(\mathbb{K}) \to V_G(\mathbb{K})$ be the corresponding linear map κ_{F_p} of vector buildings. If we let T be the standard diagonal maximal torus in G, then we have $F_p \circ \lambda = p\lambda$ for each $\lambda \in Y_T$. For a general maximal torus T' of G, if we write $T' = gTg^{-1}$ for some $g \in G$, then the maximal torus $F_p(T')$ is equal to $F_p(g)TF_p(g)^{-1}$. Further, every $\mu \in Y_{T'}$ has the form $g \cdot \lambda$ for some $\lambda \in Y_T$ and $F_p \circ \mu = F_p(g) \cdot (p\lambda) = p(F_p(g) \cdot \lambda)$. Thus we see that the induced map κ_p permutes the apartments of $V_G(\mathbb{K})$ according to the action of F_p on the maximal tori of G, and also scales each element by p.

There's another closely related automorphism $\kappa : V_G(\mathbb{K}) \to V_G(\mathbb{K})$ which acts by permuting the apartments in the same way that κ_p does, but *without the scaling*. (This is actually the linear map arising from *base change* along the field automorphism $\alpha_p : k \to k, x \mapsto x^p$: see Example 11.5.3). On the other hand, the automorphisms of $\Delta_G(\mathbb{K})$ induced by κ_p and κ are identical: they are simply given by the permutation of the apartments, since $\Delta_G(\mathbb{K})$ does not "see" the scaling by p involved in κ_p. The same is true for the induced automorphisms of Δ_G.

7.5.7 $V_G(\mathbb{R})$ as a Metric Space

Recall the notion of a length function from Definition 2.5.1: such a function restricts to a positive-definite integer-valued bilinear form on Y_T for each maximal torus T of G. Note that, by extending the form to $Y_T(\mathbb{K})$, a length function on Y_G extends naturally to a G-invariant function $Y_G(\mathbb{K}) \to \mathbb{K}$ which we also denote by $\|\cdot\|$ and call a *length function on* $Y_G(\mathbb{K})$.

A choice of length function induces a metric on $V_G(\mathbb{R})$ in which all the apartments become isomorphic to Euclidean spaces. Given any two points $\zeta, \eta \in V_G(\mathbb{R})$ we can find a common apartment containing them both and use the length function to define the distance between them; this doesn't depend on the choice of apartment, by the G-invariance of the length function. Under such a metric, the building $V_G(\mathbb{R})$ is a complete metric space: see [1, Prop. 12.10] for G semisimple, and [25, Prop. 6.8] in general. For $V_G(\mathbb{Q})$ we also obtain a metric from a length function, the inclusion of $V_G(\mathbb{Q})$ in $V_G(\mathbb{R})$ is an isometry—so $V_G(\mathbb{Q})$ has the subspace topology—and $V_G(\mathbb{Q})$ is dense in $V_G(\mathbb{R})$. In particular, for any maximal torus T of G, the topology on $V_T(\mathbb{Q})$ is the subspace topology from the Euclidean topology on $V_T(\mathbb{R})$. It is not hard to see that any two length functions give equivalent metrics on $V_G(\mathbb{K})$—this follows because any two positive-definite bilinear forms on a finite-dimensional $\mathbb{K}$-vector space give equivalent metrics. Hence the topology on $V_G(\mathbb{K})$ induced by the metric does not depend on the choice of length function.

We may identify the geometric spherical building $\Delta_G(\mathbb{R})$ with the unit sphere in $V_G(\mathbb{R})$. The situation for $\mathbb{K} = \mathbb{Q}$ is slightly more complicated: we may identify $\Delta_G(\mathbb{Q})$ with the projection of $V_G(\mathbb{Q}) \setminus \{0\}$ onto the unit sphere in $V_G(\mathbb{R})$ (in general this properly contains the unit sphere in $V_G(\mathbb{Q})$).

7.5.8 Relationship with the Simplicial Building

Recall our comment in Sect. 7.2.3 above that the simplicial spherical building Δ_G (and hence its geometric realisation) cannot tell apart G and its derived group $[G, G]$. This is not a problem with the objects we have just defined. Let $Z = Z(G)^0$, so that $G = Z \cdot [G, G]$.

Suppose first that Z is trivial so that $G = [G, G]$ is semisimple. Then we can naturally identify $\Delta_G(\mathbb{R})$ with the geometric realisation $|\Delta_G|$. Given a parabolic subgroup P of G, if we let

$$\widetilde{\sigma}_P := \{z \in \Delta_G(\mathbb{R}) \mid P_z = P\},$$

then we recover an open spherical simplex for each parabolic subgroup of G. The resulting poset of subsets of $\Delta_G(\mathbb{R})$, given by $\widetilde{\sigma}_P \leq \widetilde{\sigma}_Q$ if and only if $\widetilde{\sigma}_P$ is contained in the closure of $\widetilde{\sigma}_Q$, is isomorphic to the simplicial building Δ_G.

Now, if Z has positive dimension, then $Y_Z(\mathbb{K})$ is a positive-dimensional $\mathbb{K}$-vector space. We have $Z \subseteq T$ for every maximal torus T of G, so we deduce a natural vector space inclusion $Y_Z(\mathbb{K}) \subseteq Y_T(\mathbb{K})$ for every maximal torus, and this descends to an injective linear map of vector buildings $V_Z(\mathbb{K}) \to V_G(\mathbb{K})$ (since for each $\lambda \in Y_Z(\mathbb{K})$ we have $P_\lambda = L_\lambda = G$). That is, all the apartments of the vector building share a common subspace, rather than just a common origin. When we project to the geometric spherical building

$\Delta_G(\mathbb{R})$, we still see this contribution from Z as a common sphere in each apartment, and we see that we cannot identify $\Delta_G(\mathbb{R})$ with $|\Delta_G|$. Nevertheless, if we define subsets $\widetilde{\sigma}_P$ as above, the corresponding poset of subsets of $\Delta_G(\mathbb{R})$ is still isomorphic to the simplicial building Δ_G, even though the individual subsets are not themselves spherical simplices (since each one contains a subsphere).

7.6   Convex Subsets of the Vector Building and Complete Reducibility

We introduce some notation and several further definitions:

Notation 7.6.1 *If T is a maximal torus of G and $C \subseteq V_G(\mathbb{K})$ then we define C_T to be $C \cap V_T(\mathbb{K})$.*

The following result was proved in [25, Prop. 6.33] in the more general context of vector edifices. The result takes some effort to establish, so we do not give the proof here.

Theorem 7.6.2 *Suppose we have chosen a length function on $Y_G(\mathbb{K})$, giving a metric on $V_G(\mathbb{K})$. Let $C \subseteq V_G(\mathbb{R})$. Then C is closed if and only if C_T is closed for every maximal torus T of G.*

Definition 7.6.3 Suppose $\zeta, \eta \in V_G(\mathbb{K})$. Then ζ and η lie in some apartment $V_T(\mathbb{K})$, and within that apartment it makes sense to consider the subset $[\zeta, \eta]$ as in (2.2.1). We say a subset $C \subseteq V_G(\mathbb{K})$ is *convex* if for every $\zeta, \eta \in C$ we have $[\zeta, \eta] \subseteq C$: that is, if C_T is convex for every maximal torus T of G. We call C a *cone (in $V_G(\mathbb{K})$)* if C_T is a cone for every maximal torus T of G. We call C a *convex cone* if C is a cone and C is convex. (Equivalently, C_T is a convex cone for every maximal torus T of G.)

The following definition and lemma give a large source of convex cones: those arising from convex subcomplexes of Δ_G.

Definition 7.6.4 Suppose $\Sigma \subseteq \Delta_G$ is a subset of the simplicial spherical building of G. Define

$$C_\Sigma := \{ \zeta \in V_G(\mathbb{K}) \mid \sigma_{P_\zeta} \in \Sigma \}.$$

Lemma 7.6.5 *For any $\Sigma \subseteq \Delta_G$, C_Σ is a cone in $V_G(\mathbb{K})$. If Σ is a convex subcomplex of Δ_G, then C_Σ is a convex cone.*

Proof The first statement follows from the fact that for any $a \in \mathbb{K}_+$ and $\zeta \in V_G(\mathbb{K})$ we have $P_\zeta = P_{a\zeta}$. We leave the proof of the second statement as Exercise 7.4, which uses Serre's criterion for recognising a convex subcomplex from Sect. 7.2.6. $\square$

There is a further discussion of properties of cones in Sect. 8.4, where we show that the cones C_Σ are also closed (with respect to the topology coming from any length function). For now, we move on to further analogues in the vector building of constructions introduced above for the simplicial building, leading to a characterisation of G-complete reducibility. Suppose $H \subseteq G$ is a subgroup of G. Then letting $\Sigma = (\Delta_G)^H$ be the fixed point subcomplex of G we can form the convex cone C_Σ as above, and we can also form the fixed point subset

$$V_G(\mathbb{K})^H := \{\zeta \in V_G(\mathbb{K}) \mid h \cdot \zeta = \zeta\}.$$

It follows quickly from Lemma 7.5.2 that $C_\Sigma = V_G(\mathbb{K})^H$ in this case.

Now we make the following definitions.

Definition 7.6.6 Given $\zeta, \eta \in V_G(\mathbb{K})$ we say ζ and η are *opposite* if $\zeta + \eta = 0$, where $0 \in V_G(\mathbb{K})$ is the common zero element of all the apartments of $V_G(\mathbb{K})$. Given a convex subset C of $V_G(\mathbb{K})$, we say C is $V_G(\mathbb{K})$-*completely reducible ($V_G(\mathbb{K})$-cr)* if whenever $\zeta \in C$, there exists some $\eta \in C$ with ζ and η opposite.

Note that ζ and η are opposite in $V_G(\mathbb{K})$ if and only if there is some maximal torus T of G and some $\lambda \in Y_T(\mathbb{K})$ such that $\zeta = \varphi_G(\lambda)$ and $\eta = \varphi_G(-\lambda)$. With this in hand, we can prove a vector building analogue of Remark 7.3.3.

Lemma 7.6.7 *Let H be a subgroup of G. Then H is G-completely reducible if and only if $V_G(\mathbb{K})^H$ is $V_G(\mathbb{K})$-completely reducible.*

Proof Set $C = V_G(\mathbb{K})^H$. Suppose H is G-cr, and let $\zeta \in C$. Then H fixes ζ, so $H \subseteq P_\zeta$, and we can find some Levi subgroup L of P_ζ with $H \subseteq L$. Choose a maximal torus T of L, and write $\zeta = \varphi_G(\lambda)$ with $\lambda \in Y_T(\mathbb{K})$, so that $L = L_\lambda$. Then we have the point $\eta = \varphi_G(-\lambda) \in V_G(\mathbb{K})$, and η is opposite ζ. But $P_\eta = P_{-\lambda}$ has the Levi subgroup $L_\lambda = L_{-\lambda}$, so $H \subseteq P_\eta$, so $\eta \in C$. This shows that C is $V_G(\mathbb{K})$-completely reducible.

For the converse, suppose that C is $V_G(\mathbb{K})$-completely reducible, and let $H \subseteq P$ for some parabolic subgroup P of G. Then we may write $P = P_\zeta$ for some $\zeta \in C$ and we may find some $\eta \in C$ such that η is opposite ζ. But this corresponds to finding a maximal torus T and $\lambda \in Y_T(\mathbb{K})$ such that $\zeta = \varphi_G(\lambda)$ and $\eta = \varphi_G(-\lambda)$, and then we see that $H \subseteq P_\lambda \cap P_{-\lambda} = L_\lambda$, which is a Levi subgroup of P. Hence H is G-cr. $\square$

7.7   Links to Geometric Invariant Theory

One of the main uses of the vector building is to allow the formulation of general geometric invariant theory problems involving orbits and cocharacters in the language of buildings, generalising the set-up which works so well in the case of G-complete reducibility. Suppose that G is a connected reductive group, now acting on a variety X. Given $x \in X$, we can form the subset

$$\Lambda_x = \left\{ \lambda \in Y_G \,\middle|\, \lim_{a \to 0} \lambda(a) \cdot x \text{ exists} \right\} \subseteq Y_G.$$

It is easy to "fill out" this set to give a subset of $Y_G(\mathbb{Q})$: by Lemma 7.5.1 for every element $\lambda \in Y_G(\mathbb{Q})$ there is some $n \in \mathbb{N}$ such that $n\lambda \in Y_G$, so it makes sense to define

$$\Lambda_x(\mathbb{Q}) = \{\lambda \in Y_G(\mathbb{Q}) \mid n\lambda \in \Lambda_x \text{ for some } n \in \mathbb{N}\}.$$

Let $\mathcal{D}_x = \varphi_G(\Lambda_x) \subseteq V_G$ and $\mathcal{D}_x(\mathbb{Q}) = \varphi_G(\Lambda_x(\mathbb{Q}))$ be the corresponding subsets of $V_G(\mathbb{Q})$. We refer to the subset $\Lambda_x(\mathbb{Q})$ (resp., $\mathcal{D}_x(\mathbb{Q})$) as the *destabilising locus for x inside* $Y_G(\mathbb{Q})$ (resp., *inside $V_G(\mathbb{Q})$*).

We have the following result, which is the generalisation in this setting of the results from the previous section.

Theorem 7.7.1 *Keeping the set-up above:*

(i) $\mathcal{D}_x(\mathbb{Q})$ *is a convex cone in $V_G(\mathbb{Q})$;*
(ii) *for any $\lambda \in \Lambda_x$, $\lim_{a \to 0} \lambda(a) \cdot x$ lies outside $G \cdot x$ if and only if $\zeta = \varphi_G(\lambda)$ has no opposite in $\mathcal{D}_x(\mathbb{Q})$;*
(iii) *the orbit $G \cdot x$ is closed if and only if $\mathcal{D}_x(\mathbb{Q})$ is $V_G(\mathbb{Q})$-completely reducible.*

Proof (i). That $\mathcal{D}_x(\mathbb{Q})$ is closed under taking positive multiples is obvious from the definition. For the other conclusion, it is therefore enough to show that for any maximal torus T of G and any $\lambda, \mu \in \Lambda_x \cap Y_T$, we have $\lambda + \mu \in \Lambda_x$. This follows from Lemma 3.2.5.

(ii). Suppose $\lambda \in \Lambda_x$ and $\zeta = \varphi_G(\lambda)$, and set $x' = \lim_{a \to 0} \lambda(a) \cdot x$. We prove the contrapositive in both directions. Theorem 3.3.6 implies that x' is G-conjugate to x if and only if there is some $u \in R_u(P_\lambda)$ with $x' = u \cdot x$ and this is true if and only if $\mu := u^{-1} \cdot \lambda$ fixes x by Lemma 3.2.8. Now μ fixes x if and only if μ and $-\mu$ lie in Λ_x, and we have $\zeta = \varphi_G(\mu)$ since μ and λ are $R_u(P_\lambda)$-conjugate, so $\eta = \varphi_G(-\mu)$ is an opposite to ζ in $\mathcal{D}_x(\mathbb{Q})$. Conversely, any opposite to ζ in $\mathcal{D}_x(\mathbb{Q})$ arises in such a way, so we see that x' is G-conjugate to x if and only if ζ has an opposite in $\mathcal{D}_x(\mathbb{Q})$. This gives the result.

(iii). Given any point $\zeta \in \mathscr{D}_x(\mathbb{Q})$ there is a suitably large $n \in \mathbb{N}$ such that $n\zeta = \varphi_G(\lambda)$ for some $\lambda \in \Lambda_x$, by Lemma 7.5.1. Since ζ has an opposite in $\mathscr{D}_x(\mathbb{Q})$ if and only if $n\zeta$ does, we may assume $\zeta \in \mathscr{D}_x$. Now the result follows by applying (ii) and the Hilbert-Mumford Theorem 3.3.2.

$\square$

7.8   The Strong Tits Centre Conjecture

Theorem 7.7.1 shows that the destabilising locus $\mathscr{D}_x(\mathbb{Q})$ can detect whether an orbit $G \cdot x$ is closed and this gives us a way to translate from a geometric invariant theory setting to a building setting (and vice versa). Note that $\mathscr{D}_x(\mathbb{Q})$ is not of the form C_Σ for a subcomplex Σ of the simplicial building in general—see [21, Rem. 5.10]—but, given that we have the Tits Centre Theorem for subcomplexes, it is natural to ask whether there is an analogue for subsets. In order to formulate a precise conjecture, we need to define a suitable notion of a centre. We mentioned in Sect. 7.5.6 that an automorphism of $V_G(\mathbb{K})$ gives rise to an automorphism of the simplicial building Δ_G, so it induces a permutation of the set of parabolic subgroups of G. However, the structure of $\mathrm{Aut}(V_G(\mathbb{K}))$ is more complicated than that of $\mathrm{Aut}(\Delta_G)$: see Example 7.5.4 for an example of how complications can arise. For this reason—and for reasons related to geometric invariant theory, which we will pick up in Sect. 8.9—it is convenient to consider centres with respect to a subgroup Γ of $\mathrm{Aut}(V_G(\mathbb{K}))$.

Definition 7.8.1 Let $C \subseteq V_G(\mathbb{K})$ be a convex cone and let $\zeta \in C$. We say that ζ is *unopposed* in C if there does not exist $\eta \in C$ such that η is opposite to ζ.

Definition 7.8.2 Let $C \subseteq V_G(\mathbb{K})$ be a convex cone, and let Γ be a subgroup of $\mathrm{Aut}(V_G(\mathbb{K}))$. We say that $\zeta \in C$ is a *Γ-centre* for C if P_ζ is fixed by every $\gamma \in \Gamma$ that stabilises C. We say that ζ is an *unopposed Γ-centre* if ζ is a Γ-centre for C and ζ is unopposed in C. By an *(unopposed) centre* we mean an (unopposed) $\mathrm{Aut}(V_G(\mathbb{K}))$-centre.

Note that in the above definition we don't require that ζ itself is fixed by $N_\Gamma(C)$, for reasons explained in Sect. 7.10. See also [25, §7], where the following appears as Conjecture 7.5.

Conjecture 7.8.3 (Strong Tits Centre Conjecture) Let C be a closed convex cone in $V_G(\mathbb{K})$. Suppose C is not $V_G(\mathbb{K})$-completely reducible. Then there exists an unopposed centre for C.

The Tits Centre Theorem 7.4.1 can be viewed as a special case of Conjecture 7.8.3, see [21, §2.6] and [25, §7].

A full resolution of Conjecture 7.8.3 is still well out of reach, but in Chap. 8 we show how one can proceed for destabilising loci at least. Note that we are asking for the existence of an **unopposed** centre, as this extra property is useful to have (compare the discussion in Remark 7.4.5). Moreover, the centres we obtain from geometric invariant theory (see Sect. 8.9) are unopposed. Note also that there is a closedness hypothesis on the cone C; again, for the cones we consider in Chap. 8, this hypothesis is automatic.

7.9   Non-Connected G

Our convention through this chapter has been that the reductive group G is connected. There are several reasons for this, but it is very important for applications that we can work with non-connected reductive G as well. We resolve this issue in this section. First we give a couple of examples to indicate some problems with the non-connected case.

Example 7.9.1 Suppose that G is reductive but not necessarily connected. Then we can form a poset from the R-parabolic subgroups of G as in Definition 7.1.1. However, it is not as well-behaved as in the connected case. For example, it doesn't form a simplicial complex in general; see [188, Thm. 1.8].

Example 7.9.2 Note that if G is a non-connected reductive group, then conjugation induces an action of G on Δ_{G^0}, since parabolic subgroups of G^0 get sent to parabolic subgroups of G^0 by the conjugation action of G on G^0. However, this action is not type-preserving in general: for an example, just take G^0 of the form $H \times H$ where H is a connected reductive group, and let G be generated by G^0 and the automorphism τ swapping the two factors. Then τ does not preserve the type of any σ_P where P is a parabolic subgroup of G^0 of the form $P = P_1 \times P_2$ with P_1 and P_2 being non-conjugate parabolic subgroups of H.

Note also that in this example the subgroup of G generated by τ is G-cr, since any R-parabolic of G containing τ has identity component of the form $P \times P$, and τ fixes the corresponding Levi subgroups of the form $L \times L$. However, the subset $(\Delta_{G^0})^\tau$ of τ-fixed simplices in Δ_{G^0} is not a subcomplex: for example, it contains all maximal simplices $\sigma_{B \times B}$ but not the simplices $\sigma_{B \times P}$ where P properly contains B.

Serre observes in [157, §2.3.1 Rem.] that the fixed point subset for a group of non-type preserving automorphisms of a spherical building will at least be a subcomplex of its *barycentric subdivision*, but we do not need to go down this route.

We now come to the key definition:

Definition 7.9.3 Suppose G is non-connected. Then by Δ_G, $Y_G(\mathbb{K})$, $V_G(\mathbb{K})$ and $\Delta_G(\mathbb{K})$ we mean the sets Δ_{G^0}, $Y_{G^0}(\mathbb{K})$, $V_{G^0}(\mathbb{K})$ and $\Delta_{G^0}(\mathbb{K})$, defined as previously in this chapter. We set $\varphi_G = \varphi_{G^0} : Y_G(\mathbb{K}) \to V_G(\mathbb{K})$.

Assumption Whenever we choose a length function on $Y_G = Y_{G^0}$—for instance, when we metrise $V_G(\mathbb{K})$—we always choose to do so using a G-invariant length function on Y_G (which is also a G^0-invariant length function). Such a length function exists by Remark 2.5.4.

It is very important to note that we are **not** defining new objects here: we still use only elements of L_λ^0 in the definition of $\sim$ in (7.5.3), and we use only elements of P_λ^0 in the definition of $\approx$ in (7.5.4). The new ingredient is that we consider an action by conjugation of the whole of G on $V_G(\mathbb{K}) = V_{G^0}(\mathbb{K})$, as we now explain.

The conjugation action of G on G^0 induces an action of G on Δ_G by automorphisms, but they may no longer be type-preserving: see Example 7.9.2. Similarly, we get actions on $Y_G(\mathbb{K})$, $V_G(\mathbb{K})$ and $\Delta_G(\mathbb{K})$, and the action on $V_G(\mathbb{K})$ is an action by automorphisms. The projection $\varphi_G : Y_G(\mathbb{K}) \to V_G(\mathbb{K})$ is still G-equivariant. However, the situation with stabilisers is a little more complex than in Lemma 7.5.2: the essential issue is that we can have $\lambda, \mu \in Y_G$ with $P_\lambda(G^0) = P_\mu(G^0)$ but $P_\lambda \neq P_\mu$. We also do not yet have definitions of L_λ and P_λ for G non-connected and $\lambda \in Y_G(\mathbb{K})$. We resolve both these issues in the next result. First we set up some notation, extending our previous notation for cocharacters: given $\lambda \in Y_G(\mathbb{K})$, we write $P_\lambda(G^0)$ and $L_\lambda(G^0)$ for the corresponding parabolic and Levi subgroups of G^0.

Lemma 7.9.4 *Let $\lambda \in Y_G(\mathbb{K})$ and $\zeta = \varphi_G(\lambda) \in V_G(\mathbb{K})$. Let G_λ be the stabiliser of λ in G and let G_ζ be the stabiliser of ζ in G. Then:*

(i) $G_\zeta^0 = P_\lambda(G^0)$ *and* $G_\lambda^0 = L_\lambda(G^0)$.
(ii) *Suppose* $\lambda \in Y_G$. *Then* $G_\zeta = P_\lambda$ *and* $G_\lambda = L_\lambda$ *(where P_λ and L_λ are as defined in Theorem/Definition 2.9.1).*
(iii) G_ζ *is an R-parabolic subgroup of G and G_λ is an R-Levi subgroup of G_ζ.*

Proof Part (i) follows from Lemma 7.5.2. For (ii), let $g \in L_\lambda$. We cannot conclude immediately that if T is a maximal torus of G such that $\lambda \in Y_T$ then $g \cdot (T, \lambda) \sim (gTg^{-1}, g \cdot \lambda)$, because we don't know that g belongs to L_λ^0. Instead we can proceed as follows. Recall from Sect. 7.5.2 that Y_G embeds G-equivariantly in $Y_G(\mathbb{K})$. We deduce that $G_\lambda = L_\lambda$. It now follows from Lemma 2.9.16 that $P_\lambda = G_\zeta$.

Part (iii) when $\mathbb{K} = \mathbb{Q}$ follows from part (ii) and Lemma 7.5.1. For (iii) in the case that $\mathbb{K} = \mathbb{R}$ we can use the fact that $Y_G(\mathbb{Q})$ is dense in $Y_G(\mathbb{R})$ to approximate λ arbitrarily closely by an element of $Y_G(\mathbb{Q})$, which allows us to deduce the result. (See [25, §3.2] for more on this idea of approximating real virtual cocharacters by rational ones.) $\square$

Given the lemma, for $\lambda \in Y_G(\mathbb{K})$ we can now define L_λ to be the stabiliser of λ in G, and P_λ to be the stabiliser of $\varphi_G(\lambda)$ in G. We set $P_\zeta = P_\lambda$ for any $\lambda \in Y_G(\mathbb{K})$ such that $\zeta = \varphi_G(\lambda)$; as before, this is well-defined. We emphasise once again that it is possible to have $P_\lambda^0 = P_\mu^0$ but $P_\lambda \neq P_\mu$.

Recall from Example 7.9.2 that if H is a subgroup of non-connected G then $(\Delta_G)^H$ need not be a subcomplex. Our next result shows that nevertheless we can capture G-complete reducibility using this fixed point set.

Lemma 7.9.5 *Suppose H is a subgroup of G, and let $(\Delta_G)^H \subseteq \Delta_G$ and $C_H = V_G(\mathbb{K})^H \subseteq V_G(\mathbb{K})$ be the corresponding sets of fixed points. Then:*

(i) *H is G-completely reducible if and only if every $\sigma \in (\Delta_G)^H$ has an opposite in $(\Delta_G)^H$;*

(ii) *C_H is a convex cone in $V_G(\mathbb{K})$ and H is G-completely reducible if and only if C_H is $V_G(\mathbb{K})$-completely reducible.*

Proof (i). Suppose H is G-cr. If $\sigma_P \in (\Delta_G)^H$, then $H \subseteq N_G(P)$, an R-parabolic subgroup of G by Lemma 2.9.18, and so we can find an R-Levi subgroup L of $N_G(P)$ with $H \subseteq L$. Let Q be the unique R-parabolic subgroup opposite to $N_G(P)$ with respect to L. Then Q^0 is opposite to P by Remark 2.9.5, H normalises Q^0, and hence σ_{Q^0} is a simplex of $(\Delta_G)^H$ opposite to σ_P. Conversely, if every simplex of $(\Delta_G)^H$ has an opposite in $(\Delta_G)^H$, then for each R-parabolic subgroup P of G containing H, we have $\sigma_{P^0} \in (\Delta_G)^H$ and hence we can find some Q normalised by H and opposite P^0 in G^0. It is not hard to see that $N_G(Q) \cap P$ is an R-Levi subgroup of P containing H, and hence H is G-cr.

(ii). The key here is that Lemma 7.9.4 allows us to identify C_H as precisely the set of $\zeta \in V_G(\mathbb{K})$ such that $H \subseteq P_\zeta$. This makes it easy to see that C_H is a cone and that it is convex (since if H fixes ζ and η then certainly H fixes all points of $[\zeta, \eta]$). The statement about complete reducibility can be proved by making the appropriate changes to the proof of Lemma 7.6.7.

$\square$

Remark 7.9.6 Note that the argument in part (ii) of the proof shows that if $\zeta \in V_G(\mathbb{K})^H$ and $\eta \in V_G(\mathbb{K})$ is such that $P_\eta = P_\zeta$, then $\eta \in V_G(\mathbb{K})^H$ also.

Finally, we outline how to generalise the material from Sect. 7.7 to the non-connected setting. Suppose X is a G-variety and $x \in X$. Then we can keep precisely the same definitions of the sets Λ_x, $\Lambda_x(\mathbb{Q})$, $\mathcal{D}_x$ and $\mathcal{D}_x(\mathbb{Q})$ as before, since all these sets are defined with reference to cocharacters (or virtual cocharacters) of G, and these are the same for G as for G^0. The proof of Theorem 7.7.1 also goes through without any problem; the essential fact is that if $x' := \lim_{a \to 0} \lambda(a) \cdot x$ exists then $x' \in G \cdot x$ if and only if $x' \in G^0 \cdot x$ (this follows from Theorem 3.3.6).

7.10 Historical Remarks and References

There is a vast literature on buildings, from all sorts of points of view. An excellent overview of the topic, covering many of the different ways to approach it, can be found in [1]. Tits' original notes on buildings [183] are also highly recommended, as are the books of Ronan [149] and Tits and Weiss [185]. As we mentioned in the introduction to the chapter, much of the development of Sects. 7.1–7.3 is based on Serre's Bourbaki Seminar [157]. The vector building was introduced by Rousseau in [150]; it also appears in a paper of Curtis, Lehrer and Tits [55] and in [130, Ch. 2, §2], among other places. The authors have extended the definition of the vector building to arbitrary smooth connected affine algebraic groups over arbitrary fields in [25], where we introduced the notion of *edifices*. Some of the ideas from Sects. 7.5 and 7.6 are taken from [25]. Many of the constructions presented in this chapter have generalisations to edifices, including the Strong Centre Conjecture 7.8.3.

There is an important approach to buildings which we haven't touched on above. It is possible to metrise (the geometric realisation of) a spherical building with an angular metric in such a way that it becomes a CAT(1) space. Very roughly speaking, these are metric spaces which locally look like some model space: for a CAT(1) space, that model is the unit sphere. To realise the geometric realisation $|\Delta|$ as a CAT(1) space we take the unit sphere S with its partition into spherical simplices according to the action of the Weyl group, and then give an atlas of maps from S to $|\Delta|$ with compatibility properties matching the building axioms; the images of these maps are then the apartments of the building. This "metric geometry" point of view is the one taken in the work of Leeb and Ramos-Cuevas cited below in their proof of various cases of the Tits Centre Theorem 7.4.1. For generalities on CAT(κ) spaces, see [39, Ch. II.1].

The formulation of the Tits Centre Conjecture (now Theorem 7.4.1) apparently goes back to early attempts to prove the Borel-Tits Theorem 2.9.23. Note that if U is a unipotent subgroup of a connected reductive group G, then the fixed point subcomplex $(\Delta_G)^U$ is not completely reducible, and hence Theorem 7.4.1 does give a parabolic subgroup P of G with $N_G(U) \subseteq P$. However, it is not immediately clear that $U \subseteq R_u(P)$: this is related to the question of whether or not one can guarantee an *unopposed* centre when applying that theorem. In any case, Theorem 2.9.23 was proved in [36] long before the Centre Conjecture was resolved.

The Centre Theorem itself is the result of work by Mühlherr and Tits [128] for the classical type buildings, Leeb and Ramos-Cuevas [97] for buildings of type F_4 and E_6, and Ramos-Cuevas [140] for the remaining cases of types E_7 and E_8. See also [136] for a very short proof in the classical types, but only for subcomplexes containing a chamber, and for a more uniform approach see [129] (again, this is only for chamber complexes, although some indications are given for how to proceed to the general case). We note that the fact that the proof of Theorem 7.4.1 is completed by dealing with each irreducible type is implicit in all these papers but not stated or proved explicitly; it does however follow by a relatively straightforward argument using the fact that the building of an (almost) direct product of groups is isomorphic to the *join* of the buildings of the factors. (The join is defined in [183, §1.1]; see also [1, p. 124].) We note also that the Mühlherr-Tits proofs for classical types are relatively straightforward and constructive; for the exceptional types the arguments become increasingly involved as the rank increases.

For the Strong Centre Conjecture (or variations of it), much less is known. Rousseau [150] essentially proves a variation of this conjecture by working in the vector building and exploiting the fixed point result of Bruhat-Tits we mentioned in Remark 7.4.3; similarly, we show in the next chapter how the results of Kempf can be viewed as special cases of the Strong Centre Conjecture. Balser and Lytchak [6] give a proof in the CAT(1) setting when the dimension of the subset is at most 2 and the group acting is acting by isometries. The authors proved some special cases in [21]. We note that the version of the Strong Centre Conjecture we gave in [21] was not the same as the one given in Conjecture 7.8.3 above. This is because we have realised that it is too much to expect a fixed point in general: if the group of automorphisms is not acting by isometries (with respect to the norm coming from some length function), then there can be instances where the version of the conjecture demanding a fixed point fails. We believe that the more complicated version of what a centre should be in Conjecture 7.8.3 is the correct formulation.

The result given in Corollary 7.4.4 was originally posed as a question by Serre in [158], and was proved by the authors using optimality methods (see Sect. 9.1) several years *before* the proof of the Tits Centre Theorem 7.4.1 had been completed. It is also worth re-emphasising that although the proof of Corollary 7.4.4 is very short, the only known proof of Theorem 7.4.1 is not. The proof of Theorem 9.1.1 below is short, importantly works for non-connected G, and it relies only on the optimality material in the next chapter (which is free of any case checks).

7.11　Exercises

Exercise 7.1 The Coxeter group W of type A_{n-1} for $2 \leq n \in \mathbb{N}$ is the symmetric group S_n with generating set $S = \{(1, 2), (2, 3), \ldots, (n - 1, n)\}$.

  (i) Recall that the Cayley graph of W with respect to S has vertices the elements $w \in W$ and edges $w \sim w s_i$ for $s_i \in S$. Draw a picture of the Cayley graph for A_2.

  (ii) The reflection representation (or canonical representation) of W is a real representation with basis a set of vectors e_s for $s \in S$, a bilinear form defined by $(e_s, e_t) = -\cos\left(\frac{\pi}{m_{st}}\right)$, where m_{st} is the order of st, and action given by $s(v) = v - 2(e_s, v)e_s$ for $s \in S$. Draw a picture of the reflection representation for A_2, labelling the two basis vectors and the hyperplanes corresponding to the reflections $(1, 2)$, $(2, 3)$ and $(1, 3)$.

(iii) The Coxeter complex for W consists of the cosets of its parabolic subgroups, ordered by reverse inclusion. (A parabolic subgroup of W is a subgroup of the form $W_I :=$ $\langle s \mid s \in I \rangle$, where $I \subseteq S$.) Work out the Coxeter complex for A_2.

(iv) Try to do the same for A_3.

Exercise 7.2 Let V be a vector space equipped with a non-degenerate symmetric or alternating bilinear form. Show that the set of flags of totally isotropic subspaces in V satisfies the building axioms.

Exercise 7.3 By translating Exercise 4.8 into the language of buildings, give a proof of Theorem 7.3.4 in the case that $\Delta = \Delta_G$ for a connected reductive group G and $\Sigma = \Delta^H$ for some subgroup $H \subseteq G$.

Exercise 7.4 Let $\Delta = \Delta_G$ be the spherical building of a reductive group G (thought of as a simplicial complex), and let Σ be a subcomplex. In Sect. 7.2.6, we gave a criterion of Serre for Σ to be convex. For parabolic subgroups P, Q, R of G, we have:

$$\sigma_P, \sigma_Q \in \Sigma \text{ and } R \supseteq P \cap Q \implies \sigma_R \in \Sigma. \tag{$\star$}$$

At the start of Sect. 7.6, we gave a definition of convexity for a subset C of the vector building. For points $\zeta, \eta \in V_G(\mathbb{K})$, we have:

$$\zeta, \eta \in C \implies [\zeta, \eta] \subseteq C. \tag{$\dagger$}$$

We show in this exercise that the two definitions are equivalent for complex subcomplexes.

(i) Let T be a maximal torus of G and let $\lambda, \mu \in Y_T(\mathbb{K})$. Show that for all non-negative $a, b \in \mathbb{K}$, we have $P_{a\lambda+b\mu} \supseteq P_\lambda \cap P_\mu$.

(ii) Let Σ be a subset of Δ_G, and let C_Σ be the corresponding cone in $V_G(\mathbb{K})$ (Definition 7.6.4 and Lemma 7.6.5). Show that Σ satisfies $(\star)$ if and only if C_Σ satisfies $(\dagger)$.

Exercise 7.5 Find an example of a connected reductive group G and a subgroup H of G such that $N_G(\Delta^H)$ properly contains $N_G(H)$.

[*Hint*: Pick any G you like and a G-irreducible subgroup H of G!]

Exercise 7.6 Suppose G is connected reductive, let H be a subgroup of G, and let $\widetilde{H} = \bigcap_{H \subseteq P} P$, where the intersection is over the parabolic subgroups P of G containing H. Let $\Sigma = \Delta^H$ be the fixed point subcomplex of H in $\Delta = \Delta_G$. Show that $\Sigma = \Delta^{\widetilde{H}}$ and $N_G(\Sigma) = N_G(\widetilde{H})$.

Exercise 7.7 We prove the Tits Centre Theorem 7.4.1 for fixed point subcomplexes in type A. Suppose $H \subseteq G = \mathrm{GL}(V)$. Recall that the building $\Delta = \Delta_G$ can be identified with the set of flags of subspaces in the space V. Let

$$m(V) = \{U \subseteq V \mid U \text{ is a simple } H\text{-submodule}\},$$

and

$$M(V) = \{U \subseteq V \mid U \text{ is a maximal } H\text{-submodule}\}.$$

 (i) Show that H is G-cr if and only if every element of $m(V)$ has a complement in $M(V)$ if and only if every element of $M(V)$ has a complement in $m(V)$.
    [*Hint*: Have another look at Exercise 4.5.]
 (ii) Suppose H is not G-cr, and let $m_0(V) \subseteq m(V)$ and $M_0(V) \subseteq M(V)$ denote the (non-empty, by (i)) subsets of elements without complements. Show that if $U \in m_0(V)$ and $W \in M_0(V)$, then $U \subseteq W$.
 (iii) Let $U_0 = \Sigma_{U \in m_0(V)} U$ and $W_0 = \bigcap_{W \in M_0(V)} W$. Show that the flag $(U_0 \subseteq W_0) \in \Delta^H$ is an unopposed centre for the subcomplex Δ^H.

Remark: This same basic argument works for symplectic and orthogonal groups too, and a version of it works for subcomplexes which are not fixed point subcomplexes.

Exercise 7.8 Let $G = \mathrm{GL}_2$ and let T be the usual diagonal maximal torus. Let $\lambda, \mu \in Y_T$ be the following two cocharacters:

$$\lambda(a) = \mathrm{diag}(a, 1), \qquad \mu(a) = \mathrm{diag}(1, a)$$

for each $a \in k^*$. Then we have an isomorphism $Y_T \cong \mathbb{Z}^2$ via the map $a\lambda + b\mu \mapsto (a, b)$, which induces an isomorphism $Y_T(\mathbb{R}) \cong \mathbb{R}^2$. Let $\| \cdot \|$ be the length function from Example 2.5.2: so $(a_1, b_1) \cdot (a_2, b_2) = a_1 a_2 + b_1 b_2$. There are three parabolic subgroups of G containing T: the upper triangular Borel B^+, the lower triangular Borel B^-, and G itself.

Draw a picture of $\mathbb{R}^2$ which shows the following features:

 (i) the image 0 of the trivial cocharacter;
 (ii) the images of the rays $\mathbb{R}_+\lambda$ and $\mathbb{R}_+\mu$ and the correct angle between them according to the dot product;
 (iii) the image of the set $\{\lambda \in Y_T(\mathbb{R}) \mid P_\lambda = B^+\}$ and similarly for B^- and G;
 (iv) the subspace $Y_S(\mathbb{R})$ where S is the diagonal maximal torus of $\mathrm{SL}_2 \subseteq \mathrm{GL}_2$.

Exercise 7.9 Let $V_G(\mathbb{K})$ be equipped with the topology arising from a length function on $Y_G(\mathbb{K})$. Suppose G is connected. Prove that the map $\zeta \mapsto P_\zeta$ is semi-continuous in the following sense: for each $\zeta \in V_G(\mathbb{K})$ there is an open neighbourhood U of ζ such that for all $\eta \in U$ we have $P_\eta \subseteq P_\zeta$.

The Optimality Formalism

8

This chapter is devoted to a complete proof of the so-called optimality results of Kempf [89], Rousseau [150] and Hesselink [75], which give a strengthening of the Hilbert-Mumford Theorem 3.3.2 and form a central plank of our programme of applying geometry to the theory of G-complete reducibility. We have chosen to give a detailed treatment, since we can introduce some generalisations of the original results which are particularly well suited to the study of complete reducibility (particularly the notion of *uniform S-instability*, Definition 8.6.2). We are also able to develop the results within the framework of vector buildings introduced in the previous chapter, which makes the links between the different areas we have so far studied much more transparent, and allows us to deal with non-connected reductive groups without too much extra effort. All of this means that our results are much easier to apply in practice, making many of our subsequent proofs disarmingly straightforward! (See Chap. 9, for example.)

Recall the concept of an R-parabolic subgroup *witnessing* the fact that a subgroup H of G is not G-cr (Definition 4.1.5). The main idea of the chapter is to prove the existence of a witness that is canonical in a suitable sense, as we explain in Sect. 8.1. There we give the main theorem for the study of G-complete reducibility (Theorem 8.1.1) whilst also giving some details about the proof. Section 8.2 addresses the key example of instability in a G-module; the calculations here are particularly transparent, but essential for the general argument. Sections 8.3–8.5 build up the notation and terminology required to give a description of optimality in the context of vector buildings. With these ideas made concrete, we can prove the main optimality theorem (Theorem 8.5.5) and then apply it to prove our version of the "Optimal Hilbert-Mumford Theorem" (Theorem 8.6.7). We make some further observations about how our constructions depend on the choice of length function in Sect. 8.7 and then complete the proof of Theorem 8.1.1 in Sect. 8.8. Section 8.9 explains how the ideas in this chapter can be viewed through the lens of the Strong Tits Centre Conjecture 7.8.3. The results in this chapter do not rely on any of the

© The Author(s), under exclusive license to Springer Nature Switzerland AG 2026   171
M. Bate et al., *G-Complete Reducibility, Geometric Invariant Theory and Spherical Buildings*, Oberwolfach Seminars 57, https://doi.org/10.1007/978-3-032-08866-6_8

main results from Chap. 6 (Theorems 6.1.1, 6.1.2, 6.2.1 and 6.2.2), so this chapter can be read independently of Chap. 6.

8.1   An Optimal Witness for Non-G-Complete Reducibility

We start with a motivating example. Suppose U is a non-trivial unipotent subgroup of G^0. Theorem 2.9.23 yields a witness $P_{\mathrm{BT}}(U)$ to the non-G-complete reducibility of U with the added property that $N_G(U) \subseteq P_{\mathrm{BT}}(U)$. In fact, since U is contained in $R_u(P_{\mathrm{BT}}(U))$, $P_{\mathrm{BT}}(U)$ yields a semisimplification of U. We extend this result to any non-G-cr subgroup of G.

Theorem 8.1.1 *Let H be a subgroup of G such that H is not G-completely reducible. Let*

$$\Gamma = N_G((\Delta_G)^H) = \left\{ g \in G \,\middle|\, g \cdot (\Delta_G)^H = (\Delta_G)^H \right\}.$$

Then there is an R-parabolic subgroup $\mathcal{P}(H)$ of G such that

(i) $\mathcal{P}(H)$ yields a semisimplification of H, and
(ii) $\Gamma \subseteq \mathcal{P}(H)$.

In particular, $\mathcal{P}(H)$ is a witness that H is not G-completely reducible: for $H \subseteq \mathcal{P}(H)$ but H is not contained in any R-Levi subgroup of $\mathcal{P}(H)$ since H is not conjugate to its semisimplification. In applications of Theorem 8.1.1 one usually does not need that $\Gamma \subseteq \mathcal{P}(H)$; it is enough to know that $N_G(H) \subseteq \mathcal{P}(H)$. It is clear that $N_G(H) \subseteq \Gamma$, so the second condition follows from the first. We give the stronger condition here because it has a building-theoretic interpretation (see Sect. 8.9) and it is not much harder to prove.

   There may exist several R-parabolic subgroups of G with the desired properties (see Example 8.9.2), but we have a particular construction in mind. We take $\mathcal{P}(H)$ to be a so-called "optimal destabilising R-parabolic subgroup" obtained by applying a theorem from geometric invariant theory, which we describe now. Let X be a G-variety and let $x \in X$ such that $G \cdot x$ is not closed. Let O be the unique closed G-orbit contained in $\overline{G \cdot x}$ (Proposition 3.1.5(v)). By the Hilbert-Mumford Theorem 3.3.2, there exists $\lambda \in Y_G$ such that λ destabilises x into O: that is, such that $x' := \lim_{a \to 0} \lambda(a) \cdot x$ exists and $x' \in O$. The key insight of Kempf[1] is that one can construct a "best" cocharacter λ_{opt} with this property; roughly speaking, we can think of λ_{opt} as the cocharacter that takes x outside the orbit $G \cdot x$ and into the closed orbit O "as quickly as possible". This is the content of what

---

[1] Kempf and Rousseau independently came up with similar ideas but in slightly different contexts, and also Hesselink developed Kempf's work; see Sect. 8.10 for details.

we call the Optimal Hilbert-Mumford Theorem 8.6.7. To prove Theorem 8.1.1 we apply this machinery with $X := G^n$ and $x := \mathbf{h} \in H^n$ a generic tuple for H.

Here is an outline of how we prove the Optimal Hilbert-Mumford Theorem 8.6.7. Recall the destabilising locus $\mathscr{D}_x(\mathbb{Q})$ from Sect. 7.7. We define a real-valued function $f \colon \mathscr{D}_x(\mathbb{Q}) \setminus \{0\} \to \mathbb{R}$ such that $f(\varphi_G(\lambda))$ measures how quickly a cocharacter λ takes x into O. The function f is concave, so we can optimise it over $\mathscr{D}_x(\mathbb{Q})$; by "optimise" we mean find a point $x \in \mathscr{D}_x(\mathbb{Q})$ at which f attains its maximum value. We show that under certain hypotheses on f, there is a unique ray $\mathcal{R}$ in $\mathscr{D}_x(\mathbb{Q})$ such that f attains its maximal value on $\mathcal{R}$. We take λ_{opt} to be the unique cocharacter of G of minimal length such that $\varphi_G(\lambda_{\mathrm{opt}})$ belongs to $\mathcal{R}$, and we set $\zeta_{\mathrm{opt}} = \varphi_G(\lambda_{\mathrm{opt}})$. It turns out that λ_{opt} is not quite unique, but ζ_{opt} and the associated R-parabolic subgroup $P_{\mathrm{opt}} := P_{\lambda_{\mathrm{opt}}}$ are unique. We call λ_{opt} an *optimal destabilising cocharacter* for x, and we call P_{opt} the *optimal destabilising parabolic subgroup*. The uniqueness implies that G_x fixes ζ_{opt}, and we deduce that G_x is contained in P_{opt}.

There are two parts to the proof. The first is a local calculation:[2] for each maximal torus T of G, we optimise f over a subset of $\mathscr{D}_x(\mathbb{Q})_T$ and obtain an optimal ray $R_T \subseteq \mathscr{D}_x(\mathbb{Q})_T$.[3] The second is a global argument: we show that one of the rays R_T yields our optimal ray $\mathcal{R}$. In fact this process works for a more general class of functions f, the so-called "stately" functions. We give a precise statement of our optimality result in Theorem 8.5.5, which we state and prove in Sects. 8.3, 8.4 and 8.5.

In Sect. 8.2 we consider the special case when X is a G-module and $O = \{0\}$. In Sect. 8.6 we apply Theorem 8.5.5 to prove Theorem 8.6.15—using results from Sect. 8.2— and Theorem 8.6.7 then follows. We use the constructions from Sect. 8.6 to prove Theorem 8.1.1 in Sect. 8.8. Finally we give an interpretation of Theorem 8.6.15 in Sect. 8.9 in terms of the Strong Tits Centre Conjecture 7.8.3.

Throughout this chapter $\mathbb{K}$ is $\mathbb{Q}$ or $\mathbb{R}$. We fix a length function $\| \cdot \|$ on Y_G (see Definition 2.5.1, and see also Sect. 7.5.7). We warn the reader that some of our constructions below can depend on the choice of length function: see Sect. 8.7 for details.

8.2   0-Unstable Points in a Rational G-Module

We present a special case which both illustrates the main idea in the proof of Theorem 8.6.7 and forms a vital ingredient in that proof (see Sect. 8.6). In this section we take $\mathbb{K} = \mathbb{Q}$. Let T be a torus and let V be a T-module. We can write $V = \bigoplus_{\chi \in \Phi_T(V)} V_\chi$: see (2.8.1). Write $\Phi_T(V) = \{\chi_1, \ldots, \chi_l\}$ for some $l \in \mathbb{N}$. Let $0 \neq v \in V$. Then we write $v = v_1 + \cdots + v_l$ where $v_i \in V_{\chi_i}$ for each i. Now let $\lambda \in Y_T$. Then for each $a \in k^*$ we have

---

[2] Here "local" means the calculation is carried out within a single apartment.

[3] Recall that if $C \subseteq V_G(\mathbb{K})$ then we write C_T for $C \cap V_T(\mathbb{K})$ (Notation 7.6.1).

$$\lambda(a) \cdot v = a^{n_1} v_1 + \cdots + a^{n_l} v_l,$$

where $n_i := \langle \lambda, \chi_i \rangle$. Define $\mu_{T,v} \colon Y_T \to \mathbb{Z}$ by

$$\mu_{T,v}(\lambda) = \min_{\chi_i \in \mathrm{supp}_T(v)} \langle \lambda, \chi_i \rangle. \tag{8.2.1}$$

We see from Example 2.8.4 that

$$\lim_{a \to 0} \lambda(a) \cdot v \text{ exists if and only if } \mu_{T,v}(\lambda) \geq 0$$

and

$$\lim_{a \to 0} \lambda(a) \cdot v = 0 \text{ if and only if } \mu_{T,v}(\lambda) > 0. \tag{8.2.2}$$

We regard λ as taking v to 0 quickly if $\mu_{T,v}(\lambda)$ is large. We extend $\mu_{T,v}$ to a map from $Y_T(\mathbb{Q})$ to $\mathbb{Q}$ in the obvious way. It is also convenient to define $\mu_{T,v}$ when $v = 0$ as well: we set $\mu_{T,0}(\lambda) = \infty$ for all $\lambda \in Y_G$. (Recall our convention that the infimum of the empty set is ∞, so this agrees with the prescription in (8.2.1) since $\mathrm{supp}_T(0) = \varnothing$.)

If $r \in \mathbb{Q}$ then

$$\langle r\lambda, \chi_i \rangle = r \langle \lambda, \chi_i \rangle \tag{8.2.3}$$

for each i, so $\mu_{T,v}(r\lambda) = r\mu_{T,v}(\lambda)$ if $r > 0$. Thus in order to use the idea of maximising the speed at which λ takes v to 0, we need to normalise our cocharacters. We define

$$f_{T,v} \colon Y_T(\mathbb{Q}) \backslash \{0\} \to \mathbb{R}$$

by

$$f_{T,v}(\lambda) = \frac{\mu_{T,v}(\lambda)}{\|\lambda\|}.$$

The plan is to show the following: if $0 \in \overline{T \cdot v}$ then there is a unique ray $\mathcal{R}$ in Y_T such that $f_{T,v}$ attains its maximal value on $\mathcal{R}$.

Now let V be a rational G-module and let $0 \neq v \in V$. Suppose v is 0-unstable: recall this means that $0 \in \overline{G \cdot v}$. We want to extend the above ideas to this setting. The extra complication is that we have to consider not just a single maximal torus of G, but all of them. To compare different maximal tori we need some further results. First observe that given $g \in G$, we have

$$\mu_{gTg^{-1}, g \cdot v}(g \cdot \lambda) = \mu_{T,v}(\lambda). \tag{8.2.4}$$

We leave the proof of this as Exercise 8.1, noting that by (8.2.3) we can replace $\mu_{T,v}$ with $r\mu_{T,v}$ for sufficiently large $r \in \mathbb{N}$, so we can assume that $\lambda \in Y_T$.

Lemma 8.2.1 *Let V and v be as above, let T be a maximal torus of G, let $\lambda \in Y_T(\mathbb{Q})$ and let $g \in L_\lambda$. Then $\mu_{T,v}(\lambda) = \mu_{gTg^{-1},v}(\lambda)$.*

Proof Choose $n \in \mathbb{N}$ such that $n\lambda \in Y_T$. Set $T_\lambda = \mathrm{Im}(n\lambda)$ (this does not depend on the choice of n). We can write V as a sum of T_λ-weight spaces $V = \sum_{\eta \in \Phi_{T_\lambda}(V)} V_\eta$. Note that each V_η is the sum of certain T-weight spaces of V. It is not hard to see that

$$\mu_{T,v}(\lambda) = \min_{\eta \in \mathrm{supp}_{T_\lambda}(v)} \langle \lambda, \eta \rangle.$$

Clearly, g stabilises V_η for each $\eta \in \Phi_{T_\lambda}(V)$, so $\mu_{T,v}(\lambda) = \mu_{T,g^{-1}.v}(\lambda)$. So

$$\mu_{gTg^{-1},v}(\lambda) = \mu_{T,g^{-1}.v}(g^{-1} \cdot \lambda) = \mu_{T,g^{-1}.v}(\lambda) = \mu_{T,v}(\lambda),$$

where the first equality follows from (8.2.4). $\qquad\square$

Lemma 8.2.2 *Let V and v be as above, let T be a maximal torus of G, let $\lambda \in Y_T(\mathbb{Q})$ and let $u \in R_u(P_\lambda)$. Then $\mu_{T,v}(\lambda) = \mu_{uTu^{-1},v}(u \cdot \lambda)$.*

Proof As above we can assume that $\lambda \in Y_T$. By (8.2.4), $\mu_{uTu^{-1},v}(u \cdot \lambda) = \mu_{T,u^{-1}.v}(\lambda)$, so it's enough to show that $\mu_{T,u^{-1}.v}(\lambda) = \mu_{T,v}(\lambda)$. Let $n = \mu_{T,v}(\lambda)$; then there exists at least one $\widetilde{\chi} \in \mathrm{supp}_T(v)$ such that $\langle \lambda, \widetilde{\chi} \rangle = n$, and $\langle \lambda, \chi \rangle \geq n$ for all $\chi \in \mathrm{supp}_T(v)$. Lemma 3.2.6 implies that $u^{-1} \cdot v = v + w$ for some $w \in V$ such that $\langle \lambda, \chi' \rangle > n$ for all $\chi' \in \mathrm{supp}_T(w)$. It follows that $\mu_{T,u^{-1}.v}(\lambda) = \mu_{T,v}(\lambda)$, as required. $\qquad\square$

Recall from Sect. 7.5.3 that we form $V_G(\mathbb{Q})$ by taking the quotient of $\bigsqcup_T Y_T(\mathbb{Q})$ by the equivalence relation $(T, \lambda) \approx (gTg^{-1}, g \cdot \lambda)$ for $g \in P_\lambda$. By applying Lemmas 8.2.1 and 8.2.2 we get the following result.

Lemma 8.2.3 *The maps $\mu_{T,v} \colon V_T(\mathbb{Q}) \to \mathbb{Q}$ paste together across all maximal tori T of G to give a well-defined map $\mu_v \colon V_G(\mathbb{Q}) \to \mathbb{Q}$. We call μ_v the numerical function associated to v.*

We define $f_v \colon V_G(\mathbb{Q})\backslash\{0\} \to \mathbb{R}$ by

$$f_v(\zeta) = \frac{\mu_v(\zeta)}{\|\zeta\|}.$$

It follows from (8.2.4) that

$$\mu_{g \cdot v}(g \cdot \zeta) = \mu_v(\zeta) \tag{8.2.5}$$

for all $\zeta \in V_G(\mathbb{Q})$ and all $g \in G$, and

$$f_{g \cdot v}(g \cdot \zeta) = f_v(\zeta) \tag{8.2.6}$$

for all $\zeta \in V_G(\mathbb{Q}) \setminus \{0\}$ and all $g \in G$.

Remark 8.2.4 Our setting in this section is a special case of the setting of Sect. 8.6 with $X = V$, $A = \{x\}$ and $S = \{0\}$, and it follows from the results of Sect. 8.6 that f_v attains its maximum value on a unique ray $\mathcal{R}$ in $V_G(\mathbb{Q}) \setminus \{0\}$. Note, however, that we use the results of the current section to help establish the results in Sect. 8.6—including Theorem 8.6.7—by showing that the hypotheses of Theorem 8.5.5 hold.

8.3   Convex Cones and Optimality in Finite-Dimensional Vector Spaces

Now we lay the groundwork which will allow us to state and prove Theorem 8.5.5 below. Recall from Sect. 8.1 that the proof of Theorem 8.5.5 has a local part and a global part; here we deal with the local part.

In this section we fix a finite-dimensional vector space E over $\mathbb{K}$ and a positive-definite $\mathbb{K}$-valued bilinear form $(\cdot, \cdot)$ on E. We will optimise certain functions defined on polyhedral cones in E (we gave an overview of cones in Sect. 2.2). We describe these functions now.

Definition 8.3.1 Let $C \subseteq E$ be convex. A function $f : C \to \mathbb{R}$ is *concave* if for all $v, w \in E$ and all $t \in [0, 1] \cap \mathbb{K}$,

$$f(tv + (1 - t)w) \geq tf(v) + (1 - t)f(w).$$

We say that f is *strictly concave* if for all $v, w \in E$ such that $v \neq w$ and all $t \in (0, 1) \cap \mathbb{K}$,

$$f(tv + (1 - t)w) > tf(v) + (1 - t)f(w).$$

Remark 8.3.2 It is immediate from the definition that if f is concave then $\{v \in C \mid f(v) \geq 0\}$ is convex. Note that a concave function need not be continuous: see [85, §1.2] for a simple example.

Lemma 8.3.3 *Let $C \subseteq E$ be a convex cone, let $r \in \mathbb{N}$ and let $f_1, \ldots, f_r : C \to \mathbb{R}$ be concave. Define $f : C \to \mathbb{R}$ by $f(v) = \min_{1 \leq i \leq r} f_i(v)$. Then f is concave.*

Proof Since $\min_{1 \le i \le r} f_i(v) = \min \left\{ \min_{1 \le i \le r-1} f_i(v), f_r(v) \right\}$, we can reduce to the case $r = 2$ using induction on r. So let $v, w \in E$ and let $t \in [0, 1] \cap \mathbb{K}$. Now

$$f_1(tv + (1 - t)w) \ge tf_1(v) + (1 - t)f_1(w),$$

$$f_2(tv + (1 - t)w) \ge tf_2(v) + (1 - t)f_2(w)$$

and

$$f(tv + (1 - t)w) = \min\{f_1(tv + (1 - t)w), f_2(tv + (1 - t)w)\},$$

so

$$f(tv + (1 - t)w) \ge \min\{tf_1(v) + (1 - t)f_1(w), tf_2(v) + (1 - t)f_2(w)\}$$
$$\ge t \min\{f_1(v), f_2(v)\} + (1 - t)\min\{f_1(v), f_2(v)\}$$
$$= tf(v) + (1 - t)f(w),$$

as required. $\qquad\qquad\square$

The following is the key result for the local part of the construction.

Proposition 8.3.4 *Suppose* $\mathbb{K} = \mathbb{R}$. *Let* $C \subseteq E$ *be a closed convex cone and let* $\mu \colon C \to \mathbb{R}$ *be a continuous concave function such that*

$$\mu(av) = a\mu(v) \text{ for all } a \in \mathbb{R}_+ \text{ and all } v \in C. \qquad (8.3.1)$$

Define $f \colon C \setminus \{0\} \to \mathbb{R}$ *by* $f(v) = \frac{\mu(v)}{\|v\|}$. *Suppose* $\mu(v) > 0$ *for some* $v \in C$. *Then there is a unique ray* $\mathcal{R}$ *in* E *such that* $\mathcal{R} \subseteq C \setminus \{0\}$ *and* f *attains its maximum value on* $\mathcal{R}$.

Proof The set $C' := \{v \in C \mid \mu(v) \ge 0\}$ is a convex cone, by (8.3.1) and the concavity of μ. The intersection of convex cones is a convex cone, so without loss we can replace C with $C \cap C'$. We can therefore assume μ is non-negative on C. Let $\mathbb{S}$ be the unit sphere in E. Then $\mathbb{S} \cap C = \mathbb{S} \cap C \setminus \{0\}$ is a closed bounded subset of E, so it is compact. Hence the continuous function f attains its maximum value on $\mathbb{S} \cap C$ at some $v \in \mathbb{S} \cap C$: call this maximum value M. Let $\mathcal{R} = \mathbb{R}_+ \cdot v$. By (8.3.1), $f(w) = M$ for all $w \in \mathcal{R}$. Now take any $w \in C \setminus \{0\}$ such that $f(w) = M$. We prove that $w \in \mathcal{R}$.

Replacing w with $\frac{w}{\|w\|}$, we can assume without loss that $\|v\| = \|w\| = 1$: so $\mu(v) = f(v) = M$ and $\mu(w) = f(w) = M$. Suppose $v \ne w$. Choose any $t \in (0, 1)$ and set $w' = tv + (1 - t)w$. If $w \ne v$ then $\|w'\| < 1$ (see Exercise 8.2), and we have

$$f(w') = \frac{\mu(w')}{\|w'\|} > \mu(w') \ge t\mu(v) + (1 - t)\mu(w)$$

$$= tf(v) + (1 - t)f(w) = tM + (1 - t)M = M,$$

by the concavity of μ. But this contradicts the maximality. We deduce that $w = v$, so we are done. $\square$

Remark 8.3.5 One can define the notion of a *convex* function and *strictly convex* function by reversing the direction of the inequality signs in Definition 8.3.1. Much of optimisation theory is concerned with the problem of minimising a convex real-valued function defined on a convex domain: see [85, Ch. 3], for example. If C is a convex cone in E and $f : C \to \mathbb{R}$ then f is concave if and only if $-f$ is convex. Hence minimising a convex function f is equivalent to maximising the concave function $-f$.

Let F be a non-empty finite subset of E^*. Define $\mu_F : E \to \mathbb{R}$ by

$$\mu_F(v) = \min_{\chi \in F} \chi(v). \tag{8.3.2}$$

For example, if $\mathbb{K} = \mathbb{Q}$, T is a torus, V is a T-module, $0 \neq v \in V$ and $E = Y_T(\mathbb{Q})$ then

$$\mu_{T,v} = \mu_{\operatorname{supp}_T(v)},$$

where $\mu_{T,v}$ is as in (8.2.1). (We alert the reader that there is a small notational clash here which we have chosen to accept since it is very localised: in (8.3.2) the vector v is a vector in the $\mathbb{K}$-space E; in (8.2.1) the vector v is a vector in the k-space V.) It is convenient to define $\mu_F(v)$ to be ∞ for all $v \in E$ if $F = \varnothing$, following the convention that the minimum of the empty set is ∞.

For any $\chi \in E^*$, the function $E \to \mathbb{R}$, $v \mapsto \chi(v)$ is linear and hence concave. It follows from Lemma 8.3.3 that μ_F is concave. It is easily checked that μ_F is continuous. We have

$$\mu_F(av) = a\mu_F(v)$$

for all $v \in E$ and all $a \in \mathbb{K}_+$. We define $f_F : E \setminus \{0\} \to \mathbb{R}$ by $f_F(v) = \frac{\mu_F(v)}{\|v\|}$.

We need a version of Proposition 8.3.4 when $\mathbb{K} = \mathbb{Q}$, where μ is of the form μ_F for some F as above.

Proposition 8.3.6 *Let $\mathbb{K} = \mathbb{Q}$. Let F be a finite subset of E^* and let C be a polyhedral cone in E. Regard μ_F and f_F as maps from C to $\mathbb{R}$, and suppose $\mu_F(v) > 0$ for some $v \in C$. Then there is a unique ray $\mathcal{R}$ contained in $C \setminus \{0\}$ such that f_F attains its maximum value M on $\mathcal{R}$.*

Proof Our first step is to extend scalars from $\mathbb{Q}$ to $\mathbb{R}$, so that we can apply Proposition 8.3.4. Let $E' = E \otimes_{\mathbb{Q}} \mathbb{R}$. We extend our given $\mathbb{Q}$-valued bilinear form $(\cdot, \cdot)$ on E to an $\mathbb{R}$-valued bilinear form $(\cdot, \cdot)'$ on E' in the obvious way. By definition, there exist $\xi_1, \ldots, \xi_t \in E^*$ such that

$$C = \{v \in E \mid \xi_j(v) \geq 0 \text{ for } 1 \leq j \leq t\}.$$

Each ξ_j extends to an element ξ'_j of $(E')^*$ in the obvious way. We define

$$C' = \{v' \in E' \mid \xi'_j(v') \geq 0 \text{ for } 1 \leq j \leq t\};$$

we have $C = E \cap C'$. Likewise, each element χ of F extends to an element χ' of $(E')^*$, so we get a finite subset F' of $(E')^*$ and a function $\mu_{F'} \colon E' \to \mathbb{R}$. We regard $\mu_{F'}$ and $f_{F'}$ as functions from C' to $\mathbb{R}$. It follows from the definitions that the restriction of $\mu_{F'}$ to C is μ_F.

Since $\mu_{F'}$ is concave and continuous, we can now apply Proposition 8.3.4 to conclude that there is a unique ray $\mathcal{R}'$ in $C' \setminus \{0\}$ such that $f_{F'}$ attains its maximum value M' on $\mathcal{R}'$ (note that the polyhedral cone C' is closed in E'). To complete the proof, it suffices to show that $\mathcal{R}' \cap E \neq \varnothing$: for then M' is the maximum value of f_F on $C \setminus \{0\}$, and clearly if $v_1, v_2 \in \mathcal{R}' \cap E$ then $v_2 = a v_1$ for some $a \in \mathbb{Q}$.

Pick $v' \in \mathcal{R}'$. We want to reduce to the case when $\mu_{F'}$ is given by a single linear function and C' contains an open neighbourhood of v'. To achieve this, we pass to a smaller subspace W' of E' as follows. Choose $\chi \in F$ such that $\chi'(v') = M'\|v'\|$. Define $F_0 = \{\tau \in F \mid \tau'(v') = \chi'(v')\}$. Now set

$$W'_1 = \{w' \in E' \mid \xi'_j(w') = 0 \text{ for all } 1 \leq j \leq t \text{ such that } \xi'_j(v') = 0\}$$

and

$$W'_2 = \{w' \in E' \mid \tau'(w') = \chi'(w') \text{ for all } \tau \in F_0\}.$$

Set

$$W' = W'_1 \cap W'_2.$$

By construction, if $1 \leq j \leq t$ then either $\xi'_j(v') > 0$ or $\xi'_j|_{W'}$ is the zero function, and if $\tau \in F \setminus F_0$ then $\tau'(v') > \chi'(v')$. Hence there is an open neighbourhood U' of v' in W' such that $U' \subseteq C'$ and $\chi'(w') = \mu_{F'}(w')$ for all $w' \in U'$.

Let $H' \subseteq W'$ be the hyperplane $\{w' \in W' \mid \chi'(w') = 0\}$. Let L' be the orthogonal complement of H' in W' with respect to $(\cdot, \cdot)'$. We claim that $v' \in L'$. Suppose not. Write $v' = v'_1 + v'_2$, where $v'_1 \in L'$ and $0 \neq v'_2 \in H'$. Let $w' = v' - \epsilon v'_2$, where $\epsilon \in (0, 1)$ is small enough to ensure that w' belongs to U'. Then $\|w'\| < \|v'\|$ and

$$\mu'(w') = \chi'(w') = \chi'(v') - \epsilon \chi'(v'_2) = \chi'(v'),$$

so $f_{F'}(w') > f_{F'}(v')$. But this contradicts maximality. We deduce that $v' \in L'$ and $L' = \mathbb{R} \cdot \mathcal{R}'$.

To finish the proof, we observe that L' contains $\mathbb{Q}$-points, because it is defined over $\mathbb{Q}$. To be precise, define

$$W_1 = \{w \in E \mid \xi_j(w) = 0 \text{ for all } 1 \leq j \leq t \text{ such that } \xi'_j(v') = 0\}$$

and

$$W_2 = \{w \in E \mid \tau(w) = \chi(w) \text{ for all } \tau \in F_0\}.$$

Set $W = W_1 \cap W_2$ and $H = \{w \in W \mid \chi(w) = 0\}$, and let L be the orthogonal complement of H in W with respect to $(\cdot, \cdot)$. We may identify W' (resp., H', resp., L') with the images in E' of $W \otimes_{\mathbb{Q}} \mathbb{R}$ (resp., $H \otimes_{\mathbb{Q}} \mathbb{R}$, resp., $L \otimes_{\mathbb{Q}} \mathbb{R}$). The result follows. $\square$

Example 8.3.7 We show that the conclusion of Proposition 8.3.6 can fail if C is not polyhedral. Let $E = \mathbb{Q}^2$ with the usual positive-definite bilinear form (i.e., the usual dot product) and let C be the closed subset of $\mathbb{Q}^2$ defined by the conditions $y \geq 0$ and $y \leq \sqrt{2}x$. Let $u = (0, 1) \in E$ and define $\xi \in E^*$ by $\xi(v) = u \cdot v$. Set $F = \{\xi\}$. Then f_F does not attain a maximum value on $C \setminus \{0\}$: for if $(x, y) \in C$ then we can find $x' \in \mathbb{Q}$ such that $x' < x$ and $(x', y) \in C$, and we have $f_F(x', y) > f_F(x, y)$. Observe that C is of the form $C' \cap \mathbb{Q}^2$, where C' is a polyhedral cone in $\mathbb{R}^2$.

8.4   Polyhedral Cones in the Vector Building

In this section we describe the first ingredient of the "global" part of our proof by introducing the sets that we wish to optimise over. Recall our convention that if G is a non-connected reductive group we use the notation $V_G(\mathbb{K})$ to denote the vector building $V_{G^0}(\mathbb{K})$ (but with the action of all of G and, where relevant, the norm coming from a G-invariant length function on Y_G; see Sect. 2.5).

Definition 8.4.1 Let $C \subseteq V_G(\mathbb{K})$. We call C a *polyhedral cone (in $V_G(\mathbb{K})$)* if C_T is a polyhedral cone in $V_T(\mathbb{K})$ for every maximal torus T of G.

Definition 8.4.2 A subset C of $V_G(\mathbb{K})$ is *of finite type* or *has finite type* if the following holds: for some maximal torus T_0 of G, the set

$$\{g^{-1} \cdot C_{gT_0g^{-1}} \mid g \in G\} \tag{8.4.1}$$

is a finite set of subsets of $V_{T_0}(\mathbb{K})$.

It is not hard to show that if the condition in Definition 8.4.2 holds for one maximal torus of G then it holds for every maximal torus of G (Exercise 8.3). Note also that since

G^0 has finite index in G, the set in (8.4.1) is finite where g ranges over all elements of G if and only if it is finite where g ranges over all elements of G^0. Hence in many arguments below we can reduce to the case that G is connected.

Lemma 8.4.3 *The intersection and the union of finitely many subsets of finite type have finite type.*

Proof In both cases, by induction on the number of subsets, we can reduce to the case of two subsets. So suppose that $C, C' \subseteq V_G(\mathbb{K})$ have finite type. Fix a maximal torus T_0 of G. Let $g \in G$. We have $g^{-1} \cdot (C \cap C')_{gT_0g^{-1}} = g^{-1} \cdot C_{gT_0g^{-1}} \cap g^{-1} \cdot C'_{gT_0g^{-1}}$. But there are only finitely many possibilities for $g^{-1} \cdot C_{gT_0g^{-1}}$ and $g^{-1} \cdot C'_{gT_0g^{-1}}$ as g ranges over the elements of G, so there are only finitely many possibilities for $g^{-1} \cdot (C \cap C')_{gT_0g^{-1}}$. Hence $C \cap C'$ has finite type. The argument for $C \cup C'$ is very similar. $\qquad\square$

We can now show that parabolic subgroups of G^0 and subcomplexes of Δ_G give rise to polyhedral cones of finite type in $V_G(\mathbb{K})$.

Lemma 8.4.4 *Suppose P is a parabolic subgroup of G^0 and define*

$$C(P) := \{\zeta \in V_G(\mathbb{K}) \mid P_\zeta \supseteq P\}.$$

Then:

(i) for every maximal torus T of P, $C(P)$ is a polyhedral cone in $V_T(\mathbb{K})$;
(ii) $C(P)$ is a polyhedral cone of finite type in $V_G(\mathbb{K})$.

Proof Since P is a parabolic subgroup of G^0, we have $P_\zeta \supseteq P$ if and only if $P_\zeta^0 \supseteq P$, so we can assume that G is connected.

Let T be a maximal torus of P. Then $T \subseteq P_\zeta$ for all $\zeta \in C(P)$, so $\zeta \in V_T(\mathbb{K})$. Hence $C(P) \subseteq V_T(\mathbb{K})$. Now $\Phi(P, T) \subseteq \Phi(G, T)$ is a subset of the root system, hence finite, and $\zeta \in C(P)$ if and only if $\langle \zeta, \alpha \rangle \geq 0$ for all $\alpha \in \Phi(P, T)$, so we see that $C(P)$ is a polyhedral cone in $V_T(\mathbb{K})$, which completes the proof of (i).

For (ii), we need to understand the intersection $C(P)_T = C(P) \cap V_T(\mathbb{K})$ for each maximal torus T of G. Note that $\zeta \in C(P)_T$ if and only if $P_\zeta \supseteq P$ and $P_\zeta \supseteq T$. But this implies that if $\zeta \in C(P)_T$ then $C(P_\zeta) \subseteq C(P)_T$. Thus we see that $C(P)_T$ is the union of the $C(P_\zeta)$ for $\zeta \in C(P)$ such that $T \subseteq P_\zeta$. There may be infinitely many such ζ, but there are only finitely many P_ζ appearing, since there are only finitely many parabolic subgroups of G containing P. Thus $C(P)_T$ is a polyhedral cone by Lemma 2.2.2, and we see that $C(P)$ is a polyhedral cone in $V_G(\mathbb{K})$.

To show that $C(P)$ has finite type, we need to fix a maximal torus T_0 of G and consider all the cones $g^{-1} \cdot C(P)_{gT_0g^{-1}}$ as g runs over the elements of G. By Exercise 8.3, we are free to choose T_0, so let us pick $T_0 \subseteq P$ again. By the previous paragraph, $C(P)_{gT_0g^{-1}}$ is a finite union of cones $C(Q)$ where Q is a parabolic subgroup containing P and $gT_0g^{-1} \subseteq Q$. Since there are only finitely many parabolic subgroups of G containing P, there are only finitely many such $C(Q)$, so to complete the proof that $C(P)$ has finite type it is enough to show that for a given $Q \supseteq P$, there are only finitely many possibilities for $g^{-1} \cdot C(Q)$ where $g \in G$ is such that $gT_0g^{-1} \subseteq Q$. Given such a $g \in G$, note that since $T_0 \subseteq P \subseteq Q$, there is some $q \in Q$ such that $gT_0g^{-1} = qT_0q^{-1}$. But we also have $q \in P_\zeta$ for all $\zeta \in C(Q)$ since $P_\zeta \supseteq Q$, and hence $q \cdot C(Q) = C(Q)$, and hence $g^{-1}q \cdot C(Q) = g^{-1} \cdot C(Q)$. Since $g^{-1}q \in N_G(T_0)$, we see that there are only finitely many different $g^{-1} \cdot C(Q)$ as g ranges over the elements of G with $gT_0g^{-1} \subseteq Q$, as claimed. This completes the proof of (ii). $\qquad\square$

Corollary 8.4.5 *Suppose C is a convex subset of $V_G(\mathbb{K})$ such that for all $\zeta \in C$, $C((P_\zeta)^0) \subseteq C$. Then C is a polyhedral cone of finite type in $V_G(\mathbb{K})$. In particular, the convex cone C_Σ corresponding to a convex subcomplex Σ of Δ_G (Definition 7.6.4) is a polyhedral cone of finite type.*

Proof Again, we can easily reduce to the case that G is connected. Now C is convex by hypothesis, and C is clearly closed under multiplication by $\mathbb{K}_+$, since $P_{a\zeta} = P_\zeta$ for all $\zeta \in V_G(\mathbb{K})$ and $a \in \mathbb{K}_+$, so C is a convex cone. It is now clear that $C = \bigcup_{\zeta \in C} C(P_\zeta)$ is the union of certain $C(P)$. Since for each maximal torus T of G, there are only finitely many parabolic subgroups containing T, we deduce from Lemma 2.2.2 and Lemma 8.4.4 that C_T is a polyhedral cone in $V_T(\mathbb{K})$. To see that C has finite type, fix T_0 and let $g \in G$. Then $C_{gT_0g^{-1}}$ is a union of certain $C(Q)$ for Q containing gT_0g^{-1}, and since it is clear that $g^{-1} \cdot C(Q) = C(g^{-1}Qg)$, $g^{-1} \cdot C_{gT_0g^{-1}}$ is a finite union of certain $C(P)$ for P containing T_0. But there are only finitely many such P, and hence only finitely many possibilities for $g^{-1} \cdot C_{gT_0g^{-1}}$ as g ranges over the elements of G, as required.

Now let Σ be a subcomplex of Δ_G. By definition, $\zeta \in C_\Sigma$ if and only if $\sigma_{P_\zeta} \in \Sigma$, and since Σ is a subcomplex, for any $P_\eta \supseteq P_\zeta$ we have $\sigma_{P_\eta} \in \Sigma$; we deduce that $\eta \in C_\Sigma$, and hence $C(P_\zeta) \subseteq C_\Sigma$. We deduce from the first part of the corollary that C_Σ is a polyhedral cone of finite type. $\qquad\square$

Corollary 8.4.6 *If C is a polyhedral cone in $V_G(\mathbb{K})$ then C is closed. In particular, the convex cone corresponding to any convex subcomplex of Δ_G is closed.*

Proof It is enough to consider the case that G is connected. When $\mathbb{K} = \mathbb{R}$ the first assertion follows immediately from Theorem 7.6.2, and the second from Corollary 8.4.5. Now for the rational case, let $\overline{C}$ denote the closure of C in $V_G(\mathbb{R})$. Given a maximal torus T of G, let $\overline{C_T}$ denote the closure of $C_T = C \cap V_T(\mathbb{Q})$ in $V_T(\mathbb{R})$. Set $D = \bigcup_T \overline{C_T}$. We prove that $D_T = \overline{C_T}$. It then follows from Theorem 7.6.2 that D is closed, so $D = \overline{C}$, which implies that

$$\overline{C} \cap V_G(\mathbb{Q}) = D \cap V_G(\mathbb{Q}) = \bigcup_T (\overline{C_T} \cap V_G(\mathbb{Q})) = \bigcup_T (\overline{C_T} \cap V_T(\mathbb{Q})) = \bigcup_T C_T = C,$$

as required.

We claim that if $\zeta \in \overline{C_T}$, then for any open ball B in $V_G(\mathbb{R})$ around ζ there is some $\eta \in C_T \cap B$ such that $P_\zeta = P_\eta$. Given this claim, the result follows quickly: for if $\zeta \in V_{T'}(\mathbb{R})$ for some other maximal torus T', then $P_\zeta = P_\eta$ implies that $\eta \in V_{T'}(\mathbb{Q})$. Since $\eta \in C$ also, we get $\eta \in C_{T'}$ and hence deduce that $\zeta \in \overline{C_{T'}}$. This shows that $\overline{C_T} \cap V_{T'}(\mathbb{R}) \subseteq \overline{C_{T'}}$, and we deduce that $D_{T'} = \overline{C_{T'}}$.

It remains to prove the claim of the previous paragraph. So suppose T is a maximal torus of G, and $\zeta \in \overline{C_T}$. Let $P = P_\zeta$, and let L be the Levi subgroup of P containing T. Write $\Phi(P, T) = \Phi_0 \cup \Phi_+$, where Φ_0 consists of the roots for L, and Φ_+ consists of the roots for $R_u(P)$ with respect to T. Then Φ_0 gives a subspace $S \subseteq V_T(\mathbb{Q})$:

$$S := \{\eta \in V_T(\mathbb{Q}) \mid \langle \eta, \alpha \rangle = 0 \text{ for all } \alpha \in \Phi_0\}.$$

Let $\overline{S}$ be the subspace of $V_T(\mathbb{R})$ formed by closing up S. Since C_T is a polyhedral cone in $V_T(\mathbb{Q})$, it is finitely generated by Lemma 2.2.2 and its closure $\overline{C_T}$ is just the polyhedral cone in $V_T(\mathbb{R})$ with the same finite set of generators. Now by hypothesis, we have $\zeta \in \overline{S} \cap \overline{C_T}$, and since this is the intersection of a $\mathbb{Q}$-defined subspace and a $\mathbb{Q}$-defined polyhedral cone, this is enough to deduce that the intersection $S \cap C_T$ is non-empty and

$$\overline{S \cap C_T} = \overline{S} \cap \overline{C_T}.$$

(This final equality follows since each side is a polyhedral cone in $V_T(\mathbb{R})$, and they are defined by the same sets of linear inequalities.) Now ζ belongs to the open subset of $\overline{S} \cap C_T$ defined by the extra conditions $\langle \eta, \alpha \rangle > 0$ for $\alpha \in \Phi_+$, and this open subset contains rational points arbitrarily close to ζ, completing the proof. $\square$

8.5   Optimality in the Vector Building

Now we move on to the second global part of our recipe for proving Theorem 8.5.5 by describing the types of functions that we want to optimise. After that we can prove our main result Theorem 8.5.5. Ultimately we are concerned with the case $\mathbb{K} = \mathbb{Q}$, but some of the constructions need to be carried out over $\mathbb{R}$. First note that the action of G on $\bigsqcup_T V_T(\mathbb{K}) = \bigsqcup_T Y_T(\mathbb{K})$ gives rise to an action of G on $\bigsqcup_T V_T(\mathbb{K})^*$ in a natural way: explicitly, $g \in G$ maps $V_T(\mathbb{K})^*$ to $V_{gTg^{-1}}(\mathbb{K})^*$, and we define

$$(g \cdot \chi)(\zeta) = \chi(g^{-1} \cdot \zeta)$$

for $\chi \in V_T(\mathbb{K})^*$ and $\zeta \in V_{gTg^{-1}}(\mathbb{K})$. Since X_T is the dual of Y_T as a $\mathbb{Z}$-module, we may identify $V_G(\mathbb{K})^* = (Y_T \otimes_{\mathbb{Z}} \mathbb{K})^*$ with $X_T \otimes_{\mathbb{Z}} \mathbb{K}$; the action of G on $\coprod_T V_T(\mathbb{K})^*$ is then compatible with the action of G on $\coprod_T X_T$ defined in Sect. 2.5.

Definition 8.5.1 Let C be a convex cone in $V_G(\mathbb{K})$. A function $\mu \colon C \to \mathbb{K}$ is *stately* if there exists a finite subset F_T of $V_T(\mathbb{K})^*$ for each maximal torus T of G such that the following hold:

(i) $\mu|_{C_T} = \mu_{F_T}$ for every maximal torus T of G.

(ii) For some maximal torus T_0 of G, $\bigcup_{g \in G} g^{-1} \cdot F_{gT_0g^{-1}}$ is a finite subset of $V_{T_0}(\mathbb{K})^*$.

(Here μ_{F_T} is as defined in (8.3.2); the subsets F_T are necessarily non-empty since μ does not take the value ∞.) We may regard any $\chi \in X_T$ as an element of $V_T(\mathbb{K})^* \cong (Y_T \otimes_{\mathbb{Z}} \mathbb{K})^*$ via the pairing between X_T and Y_T.

It is not hard to show that if the condition in Definition 8.5.1(ii) holds for one maximal torus of G then it holds for every maximal torus of G (Exercise 8.4). Note that the sets F_T are not uniquely determined by μ. For instance, suppose $\mu|_{C_T} \geq 0$. Suppose F is a finite subset of $V_T(\mathbb{K})^*$ such that $\mu|_{C_T} = \mu_F$, and choose any $\chi \in F$. Fix $a \in \mathbb{K}$ such that $a \geq 1$ and set $F' = F \cup \{a\chi\}$. Then $\mu|_{C_T} = \mu_{F'}$.

Example 8.5.2 Let V be a G-module. Suppose we are given a subset F_T of $\Phi_T(V)$ for each maximal torus T of G. Let C be a convex cone in $V_G(\mathbb{K})$ and suppose $\mu \colon C \to \mathbb{K}$ is a function such that $\mu|_{C_T} = \mu_{F_T}$ for every maximal torus T of G (i.e., such that condition (i) of Definition 8.5.1 holds). Then μ is stately. To see this, fix a maximal torus T_0 of G. Let $T = gT_0g^{-1}$ be another maximal torus of T and let $\chi \in F_T$. Then $\chi \in \Phi_T(V)$, so $g^{-1} \cdot \chi \in \Phi_{T_0}(V)$. Since $\Phi_{T_0}(V)$ is finite, it follows that there are only finitely many possibilities for $g^{-1} \cdot F_T$ as g ranges over the elements of G, so condition (ii) of Definition 8.5.1 holds.

In particular, if $0 \neq v \in V$ then the function μ_v from Lemma 8.2.3 is stately; we take F_T to be $\mathrm{supp}_T(v)$.

Remark 8.5.3 We cannot expect to obtain a well-defined function on $V_G(\mathbb{K})$ by choosing an arbitrary finite subset F_T of $V_T(\mathbb{K})$ for each maximal torus T of G. The point is that if $\zeta \in V_G(\mathbb{K})$ then ζ can belong to $V_{T_1}(\mathbb{K})$ and to $V_{T_2}(\mathbb{K})$ for different maximal tori T_1 and T_2, and we need to know that $\mu_{F_{T_1}}(\zeta) = \mu_{F_{T_2}}(\zeta)$. This places strong constraints on the choice of the F_T.

Lemma 8.5.4 *Let C be a polyhedral cone of finite type in $V_G(\mathbb{K})$ and let $\mu \colon C \to \mathbb{K}$ be stately. Define*

$$N(\mu) := \{\zeta \in C \mid \mu(\zeta) \geq 0\}. \tag{8.5.1}$$

Then $N(\mu)$ is a polyhedral cone of finite type in $V_G(\mathbb{K})$.

Proof By definition of stately, there is a finite subset F_T of $V_T(\mathbb{K})^*$ for each maximal torus T of G such that $\mu|_{V_T(\mathbb{K})} = \mu_{F_T}$. It follows from (8.2.3) that $N(\mu)$ is a cone, and $N(\mu)$ is convex because checking convexity boils down to checking that each $N(\mu)_T$ is convex; this follows from Remark 8.3.2, since each μ_{F_T} is a concave function. Fix a maximal torus T_0 of G. For each maximal torus T of G, fix $g_T \in G$ such that $g_T T_0 g_T^{-1} = T$. By definition, there exist $r \in \mathbb{N}$ and polyhedral cones $C_1, \ldots, C_r$ in $V_{T_0}(\mathbb{K})$ such that for every maximal torus T of G, there exists i_T with $1 \le i_T \le r$ such that $g_T \cdot C_{i_T} = C_T$. Choose finite $S_i \subseteq V_T(\mathbb{K})^*$ such that for each $1 \le i \le r$ we have $C_i = \{\zeta \in V_{T_0}(\mathbb{K}) \mid \xi(\zeta) \ge 0 \text{ for all } \xi \in S_i\}$. Set $F_T' = F_T \cup g_T \cdot S_{i_T}$. For each maximal torus T of G we have

$$N(\mu)_T = N(\mu) \cap V_T(\mathbb{K}) = \{\zeta \in V_T(\mathbb{K}) \mid \xi(\zeta) \ge 0 \text{ for all } \xi \in F_T'\}$$

$$= \{\zeta \in V_T(\mathbb{K}) \mid \mu_{F_T'}(\zeta) \ge 0\},$$

so $N(\mu)$ is a polyhedral cone and

$$g_T^{-1} \cdot N(\mu)_T = \{\zeta \in V_{T_0}(\mathbb{K}) \mid \mu_{g_T^{-1} \cdot F_T'}(\zeta) \ge 0\}.$$

But $g_T^{-1} \cdot F_T' = g_T^{-1} \cdot F_T \cap S_{i_T}$ and the set $\{g^{-1} \cdot F_{gT_0g^{-1}} \mid g \in G\}$ is finite, so there are only finitely many possibilities for $g_T^{-1} \cdot F_T'$, so there are only finitely many possibilities for $g_T^{-1} \cdot N(\mu)_T$: say, $D_1, \ldots, D_s$.

To finish, let $n_1, \ldots, n_m \in N_G(T_0)$ be a set of representatives for the Weyl group. Pick any $g \in G$ and set $T := gT_0g^{-1}$. We can write $g = g_T n_j$ for some $1 \le j \le m$. We have

$$g^{-1} \cdot N(\mu)_T = n_j^{-1} g_T^{-1} \cdot N(\mu)_T \in \{n_j^{-1} \cdot D_i \mid 1 \le i \le s, 1 \le j \le m\}.$$

It follows that $N(\mu)$ is a polyhedral cone of finite type, as required. $\qquad\square$

Now we can state and prove our main result.

Theorem 8.5.5 *Let C be a polyhedral cone of finite type in $V_G(\mathbb{Q})$ and let $\mu \colon C \to \mathbb{Q}$ be stately. Define $f \colon C \setminus \{0\} \to \mathbb{R}$ by $f(\zeta) = \frac{\mu(\zeta)}{\|\zeta\|}$. Suppose there exists $\zeta \in C \setminus \{0\}$ such that $f(\zeta) > 0$. Then there is a unique ray $\mathcal{R}_{\mathrm{opt}}$ in $C \setminus \{0\}$ such that f attains its maximum value on $\mathcal{R}_{\mathrm{opt}}$.*

Proof For each maximal torus T of G, pick F_T as in Definition 8.5.1 with respect to μ. Choose a maximal torus T_0 of G such that $\zeta \in C_{T_0}$. Let $\mathcal{D}$ be the set of maximal tori T of G such that $f(\eta) > 0$ for some $\eta \in C_T$: note that $T_0 \in \mathcal{D}$ by hypothesis. If $T \in \mathcal{D}$ then $V_T(\mathbb{Q})$, C_T and μ_T satisfy the hypotheses of Proposition 8.3.6, so $f|_{C_T \setminus \{0\}} = f_{F_T}$ attains its maximum value M_T on a unique ray $\mathcal{R}_T$ in $C_T \setminus \{0\}$.

We claim that if $g \in G$ and $g T_0 g^{-1} \in \mathcal{D}$ then the maximum value of $\mu_{g^{-1} \cdot F_{g T_0 g^{-1}}}$ on $g^{-1} \cdot C_{g T_0 g^{-1}}$ is $M_{g T_0 g^{-1}}$. To see this, let $\zeta \in C_{g T_0 g^{-1}}$ and $\chi \in F_{g T_0 g^{-1}}$. Then we have $(g^{-1} \cdot \chi)(g^{-1} \cdot \zeta) = \chi(\zeta)$, and so

$$\mu_{g^{-1} \cdot F_{g T_0 g^{-1}}}(g^{-1} \cdot \zeta) = \min_{\chi \in F_{g T_0 g^{-1}}} \chi(\zeta) = \mu_{F_{g T_0 g^{-1}}}(\zeta).$$

The claim now follows. By Definition 8.5.1(ii) there are only finitely many possibilities for $\mu_{g^{-1} \cdot F_{g T_0 g^{-1}}}$ as g runs over the elements of G, and since C has finite type there are only finitely many possibilities for $g^{-1} \cdot C_{g T_0 g^{-1}}$ as g runs over the elements of G. This implies that $\{M_T \mid T \in \mathcal{D}\}$ is finite.

So let $M = \max\{M_T \mid T \in \mathcal{D}\} > 0$ and pick $\zeta_1 \in C \backslash \{0\}$ such that $f(\zeta_1) = M$. By construction, f attains its maximal value at ζ_1. Set $\mathcal{R}_{\mathrm{opt}} = \mathbb{Q}_+ \cdot \zeta_1$. Choose any $\zeta_2 \in C \backslash \{0\}$ such that $f(\zeta_2) = M$. By the common apartment property there exists a maximal torus T of G such that $\zeta_1, \zeta_2 \in C_T$. Clearly $T \in \mathcal{D}$ and $M = M_T$, so ζ_1, ζ_2 belong to $\mathcal{R}_T$. But then $\mathcal{R}_T = \mathcal{R}_{\mathrm{opt}}$, so ζ_2 belongs to $\mathcal{R}_{\mathrm{opt}}$. This completes the proof. $\qquad\square$

Definition 8.5.6 Under the hypotheses of Theorem 8.5.5, we define $P_{\mathrm{opt}} = P_\zeta$ for any $\zeta \in \mathcal{R}$, and we call P_{opt} the *optimal destabilising R-parabolic subgroup associated to C and f*. We define ζ_{opt} to be the unique element of $\mathcal{R} \cap V_G$ of minimal length. To see that this makes sense, choose a maximal torus T of G such that $\mathcal{R} \subseteq V_T(\mathbb{Q})$. Now V_T is a lattice inside $V_T(\mathbb{Q})$ and $\mathcal{R} \cap V_T \neq \varnothing$ by Lemma 7.5.1, so $\mathcal{R}$ contains an element of V_T of minimal length. Often we abuse notation and write λ_{opt} for any element of Y_G such that $\varphi_G(\lambda_{\mathrm{opt}}) = \zeta_{\mathrm{opt}}$ and call λ_{opt} *"the" optimal destabilising cocharacter associated to C and f*, even though λ_{opt} is only defined up to $R_u(P_{\mathrm{opt}})$-conjugacy. Note that if T is any maximal torus of P_{opt} then we can choose λ_{opt} to belong to Y_T by Corollary 2.9.14.

We write $\mathcal{R}_{\mathrm{opt}}(f, C)$, $\zeta_{\mathrm{opt}}(f, C)$, $P_{\mathrm{opt}}(f, C)$ and $\lambda_{\mathrm{opt}}(f, C)$ if we want to make the choice of f and C explicit.

Remark 8.5.7 Both the cocharacter λ_{opt} and the associated R-parabolic subgroup can depend on the choice of length function: see Examples 8.7.1 and 8.7.2.

We need to investigate the behaviour of these constructions under the action of G. Let C be a polyhedral cone of finite type in $V_G(\mathbb{Q})$ and let $\mu \colon C \to \mathbb{K}$ be stately. Given $g \in G$, define $g \cdot \mu \colon g \cdot C \to \mathbb{Q}$ by

$$(g \cdot \mu)(g \cdot \zeta) = \mu(\zeta).$$

It is easy to check that $g \cdot C$ is a polyhedral cone of finite type and $g \cdot \mu$ is stately (Exercise 8.5). Define $g \cdot f \colon g \cdot C \backslash \{0\} \to \mathbb{R}$ by

$$(g \cdot f)(g \cdot \zeta) = \frac{(g \cdot \mu)(g \cdot \zeta)}{\|g \cdot \zeta\|} = f(\zeta),$$

where the second equality follows from the conjugation-invariance of $\| \cdot \|$. It follows from the constructions that

$$\mathcal{R}_{\mathrm{opt}}(g \cdot f, g \cdot C) = g \cdot \mathcal{R}_{\mathrm{opt}}(f, C),$$

$$\zeta_{\mathrm{opt}}(g \cdot f, g \cdot C) = g \cdot \zeta_{\mathrm{opt}}(f, C) \text{ and} \qquad (8.5.2)$$

$$P_{\mathrm{opt}}(g \cdot f, g \cdot C) = g \cdot P_{\mathrm{opt}}(f, C).$$

In particular, if $g \cdot C = C$ and $g \cdot f = f$ then

$$\mathcal{R}_{\mathrm{opt}}(f, C) = g \cdot \mathcal{R}_{\mathrm{opt}}(f, C),$$

$$\zeta_{\mathrm{opt}}(f, C) = g \cdot \zeta_{\mathrm{opt}}(f, C) \text{ and} \qquad (8.5.3)$$

$$P_{\mathrm{opt}}(f, C) = g \cdot P_{\mathrm{opt}}(f, C).$$

and it follows from Lemma 7.9.4(ii) that

$$g \in P_{\mathrm{opt}}(f, C). \qquad (8.5.4)$$

8.6   The Hesselink-Kempf-Rousseau Optimality Construction

We can now state and prove a rigorous formulation of the optimality results sketched in Sect. 8.1. We take $\mathbb{K}$ to be $\mathbb{Q}$ for the rest of the chapter. For our application to Theorem 8.1.1, we consider destabilising cocharacters not just for single points of a G-variety but for subsets. We are concerned below mainly with uniform $\mathcal{S}$-instability rather than uniform instability, but we define both notions and give examples to show that some care is needed when working with uniform instability.

Definition 8.6.1 Let X be a G-variety and let $\mathcal{A} \subseteq X$. We say that $\lambda \in Y_G$ *uniformly destabilises* $\mathcal{A}$ if $x' := \lim_{a \to 0} \lambda(a) \cdot x$ exists for all $x \in \mathcal{A}$. We say that $\mathcal{A}$ is *uniformly unstable* if there exists $\lambda \in Y_G$ such that λ uniformly destabilises $\mathcal{A}$. We say that $\lambda \in Y_G$ *properly uniformly destabilises* $\mathcal{A}$ if λ destabilises $\mathcal{A}$ and for all $u \in R_u(P_\lambda)$, there exists $x \in \mathcal{A}$ such that $\lim_{a \to 0} \lambda(a) \cdot x \neq u \cdot x$. We say that $\mathcal{A}$ is *properly uniformly unstable* if there exists $\lambda \in Y_G$ such that λ properly uniformly destabilises $\mathcal{A}$.

Definition 8.6.2 Let X be a G-variety and let $\mathcal{S}$ be a closed G-stable subvariety of X. Let $\mathcal{A} \subseteq X$. We say that $\lambda \in Y_G$ *uniformly destabilises* $\mathcal{A}$ *into* $\mathcal{S}$ if $x' := \lim_{a \to 0} \lambda(a) \cdot x$ exists and belongs to $\mathcal{S}$ for all $x \in \mathcal{A}$. We say that $\mathcal{A}$ is *uniformly $\mathcal{S}$-unstable* if there exists

$\lambda \in Y_G$ such that λ uniformly destabilises $\mathcal{A}$ into $\mathcal{S}$. We say that $\mathcal{A}$ is *properly uniformly $\mathcal{S}$-unstable* if $\mathcal{A}$ is uniformly $\mathcal{S}$-unstable and $\mathcal{A} \nsubseteq \mathcal{S}$.

Remark 8.6.3 If $\mathcal{A}$ is properly uniformly $\mathcal{S}$-unstable then there exists $\lambda \in Y_G$ and $x \in \mathcal{A}$ such that λ destabilises $\mathcal{A}$ into $\mathcal{S}$ and $x \notin \mathcal{S}$. Then $x' := \lim_{a \to 0} \lambda(a) \cdot x \in \mathcal{S}$ cannot be $R_u(P_\lambda)$-conjugate to x since $\mathcal{S}$ is G-stable. Hence $\mathcal{A}$ is properly uniformly unstable.

Example 8.6.4 Let $X = G$ with the conjugation action and let $\mathcal{S} = \{1\}$. Let $\mathcal{A} \subseteq X$. Then $\mathcal{A}$ is uniformly $\mathcal{S}$-unstable if and only if $G(\mathcal{A})$ is unipotent: this follows from a slight adaptation of the arguments in Example 3.3.3. Note that if G is connected and semisimple then there exists a subset $\mathcal{A}$ consisting of unipotent elements such that $G(\mathcal{A}) = G$, so it is not enough to assume that every element of $\mathcal{A}$ is unipotent.

We show how these definitions relate to our earlier notions when $\mathcal{A} = \{x\}$.

Example 8.6.5 Let X be a G-variety and let $x \in X$. Take $\mathcal{A}$ to be $\{x\}$ and take $\mathcal{S}$ to be the unique closed G-orbit contained in $\overline{G \cdot x}$. Suppose $G \cdot x$ is not closed. By the Hilbert-Mumford Theorem 3.3.2, there exists $\lambda \in Y_G$ such that $x' := \lim_{a \to 0} \lambda(a) \cdot x$ belongs to $\mathcal{S}$; hence λ uniformly destabilises $\mathcal{A}$ into $\mathcal{S}$, and $\mathcal{A}$ is uniformly $\mathcal{S}$-unstable. Since $x \notin \mathcal{S}$, $\mathcal{A}$ is properly uniformly $\mathcal{S}$-unstable and hence is properly uniformly unstable by Remark 8.6.3. Next suppose that $\mathcal{A}$ is properly uniformly unstable. There exists $\lambda \in Y_G$ such that $x' := \lim_{a \to 0} \lambda(a) \cdot x$ belongs to $\mathcal{S}$: this follows from the Hilbert-Mumford Theorem if $x \notin \mathcal{S}$ and is immediate if $x \in \mathcal{S}$ (although we see in a moment that this second possibility cannot occur). Hence $\mathcal{A}$ is uniformly $\mathcal{S}$-unstable. By hypothesis, x' is not $R_u(P_\lambda)$-conjugate to x, so $x' \notin G \cdot x$ by Theorem 3.3.6, so $x \notin \mathcal{S}$. It follows that $G \cdot x$ is not closed and $\mathcal{A}$ is properly uniformly $\mathcal{S}$-unstable. Finally, if $\mathcal{A}$ is properly uniformly $\mathcal{S}$-unstable then $\mathcal{A}$ is properly uniformly unstable by Remark 8.6.3, so $G \cdot x$ is not closed by the above argument. We have shown:

$$\mathcal{A} \text{ is properly uniformly } \mathcal{S}\text{-unstable} \iff \mathcal{A} \text{ is properly uniformly unstable}$$

$$\iff G \cdot x \text{ is not closed.}$$

It is natural to ask whether there is a converse to Remark 8.6.3 in the following sense: if $\mathcal{A}$ is a properly uniformly unstable subset of X, does there exists a closed G-stable subset $\mathcal{S}$ of X such that $\mathcal{A}$ is uniformly $\mathcal{S}$-unstable? This is the case if $\mathcal{A} = \{x\}$ by Example 8.6.5. The next example shows, however, that the answer is no in general.

Example 8.6.6 Suppose G is connected. Let X be G with the conjugation action and let P be a proper parabolic subgroup of G. Set $\mathcal{A} = P$. We can write $P = P_\lambda$ for some $\lambda \in Y_G$. If $1 \neq u \in R_u(P_\lambda)$ then $\lim_{a \to 0} \lambda(a) \cdot u = 1$ is not $R_u(P_\lambda)$-conjugate to u. So λ properly uniformly destabilises $\mathcal{A}$ and $\mathcal{A}$ is properly uniformly unstable.

Now let S be any closed G-stable subset of X such that $\mathcal{A}$ is uniformly S-unstable, and let $\mu \in Y_G$ such that μ uniformly destabilises $\mathcal{A}$ into S. Fix a maximal torus T of P. If $x \in T$ then $x' := \lim_{a \to 0} \mu(a) \cdot x$ belongs to S. But semisimple conjugacy classes are closed (Remark 5.3.5), so x' is conjugate to x. Since S is G-stable, S must therefore contain $G \cdot T$, and it follows that $S = G$ since S is closed. Hence $\mathcal{A}$ is not properly uniformly S-unstable.

Examples like this one explain why we can't just take $X = G$ and $\mathcal{A} = H$ in the proof of Theorem 8.1.1 (see Sect. 8.8).

We spend the rest of the section proving the following result.

Theorem 8.6.7 (Optimal Hilbert-Mumford Theorem) *Let X be a G-variety, let S be a closed G-stable subset of X and let $\mathcal{A} \subseteq X$. Suppose $\mathcal{A}$ is properly uniformly S-unstable. Then there exists $\lambda_{\mathcal{A}} \in Y_G$ such that $\lambda_{\mathcal{A}}$ uniformly destabilises $\mathcal{A}$ into S and $N_G(\mathcal{A}) \subseteq P_{\lambda_{\mathcal{A}}}$.*

The idea is to use Theorem 8.5.5 to associate an optimal destabilising cocharacter λ_{opt} and the corresponding optimal destabilising R-parabolic subgroup P_{opt} to $\mathcal{A}$ and S. To do this we need a stately function defined on a convex polyhedral cone C of finite type, which we construct using ingredients from Sect. 8.2. In our proof of Theorem 8.6.7 we just take $C = \mathscr{D}_{\mathcal{A}}(\mathbb{Q})$, but we formulate our arguments in terms of a more general C—see Theorem 8.6.15—which is useful later on.

To apply our machinery we need to consider G-equivariant maps φ from X to a G-module V. Let $x \in X$ and let $\lambda \in Y_G$. If $x' := \lim_{a \to 0} \lambda(a) \cdot x$ exists then $\lim_{a \to 0} \lambda(a) \cdot \varphi(x)$ exists, and the latter limit equals $\varphi(x')$, by Lemma 2.8.2(i). Conversely, if $\lim_{a \to 0} \lambda(a) \cdot \varphi(x)$ exists then $\lim_{a \to 0} \lambda(a) \cdot x$ exists under the assumption that φ is a closed embedding (Lemma 2.8.2(ii)), but this is false for general φ.

We need to extend the definition of destabilising locus from Sect. 7.7 to the setting of uniform instability.

Definition 8.6.8 Let $\mathcal{A} \subseteq X$. Define

$$\Lambda_{\mathcal{A}} = \left\{ \lambda \in Y_G \;\middle|\; \lim_{a \to 0} \lambda(a) \cdot x \text{ exists for all } x \in \mathcal{A} \right\}$$

and

$$\Lambda_{\mathcal{A}}(\mathbb{Q}) = \{\lambda \in Y_G(\mathbb{Q}) \mid n\lambda \in \Lambda_{\mathcal{A}} \text{ for some } n \in \mathbb{N}\}.$$

We call $\Lambda_{\mathcal{A}}$ the *destabilising locus of* $\mathcal{A}$. Define $\mathscr{D}_{\mathcal{A}} = \varphi_G(\Lambda_{\mathcal{A}})$ and $\mathscr{D}_{\mathcal{A}}(\mathbb{Q}) = \varphi_G(\Lambda_{\mathcal{A}}(\mathbb{Q}))$. We refer to the subset $\Lambda_{\mathcal{A}}(\mathbb{Q})$ (resp., $\mathscr{D}_{\mathcal{A}}(\mathbb{Q})$) as the *destabilising locus for* $\mathcal{A}$ *inside* $Y_G(\mathbb{Q})$ (resp., *inside* $V_G(\mathbb{Q})$).

Remark 8.6.9 (a). The destabilising locus defined in Sect. 7.7 is the special case of this construction with $\mathcal{A} = \{x\}$.

(b). It follows easily from the definitions that for any $g \in G$ we have

$$\Lambda_{g \cdot \mathcal{A}} = g \cdot \Lambda_{\mathcal{A}}, \quad \Lambda_{g \cdot \mathcal{A}}(\mathbb{Q}) = g \cdot \Lambda_{\mathcal{A}}(\mathbb{Q}), \quad \mathscr{D}_{g \cdot \mathcal{A}} = g \cdot \mathscr{D}_{\mathcal{A}}, \quad \mathscr{D}_{g \cdot \mathcal{A}}(\mathbb{Q}) = g \cdot \mathscr{D}_{\mathcal{A}}(\mathbb{Q}).$$

In Exercise 8.8 we give a criterion for $\mathscr{D}_{\mathcal{A}}(\mathbb{Q})$ to be $V_G(\mathbb{Q})$-cr. It follows from the proof of Theorem 7.7.1(i) with some minor modifications that $\mathscr{D}_{\mathcal{A}}(\mathbb{Q})$ is a convex cone in $V_G(\mathbb{Q})$, but we can prove something stronger (see Proposition 8.6.11).

Below we let X, S and $\mathcal{A}$ be as above; we don't assume for now that $\mathcal{A}$ is properly uniformly S-unstable. Suppose V is a G-module and $\varphi \colon X \to V$ is a G-equivariant map. Recall that for any $0 \neq v \in V$ there is a map $\mu_v \colon V_G(\mathbb{Q}) \to \mathbb{Q}$ as described in Lemma 8.2.3. As before, it is convenient to extend the definition and set $\mu_v(\zeta) = \infty$ for all $\zeta \in V_G(\mathbb{Q})$ if $v = 0$. Let T be a maximal torus of G. For $x \in \mathcal{A}$, set $F_{\varphi,T,x} = \mathrm{supp}_T(\varphi(x))$: then $\mu_{T,\varphi(x)} = \mu_{F_{\varphi,T,x}}$ (see (8.2.1)). Define $F_{\varphi,T,\mathcal{A}} = \bigcup_{x \in \mathcal{A}} F_{\varphi,T,x}$. We define $\mu_{\varphi,\mathcal{A}} \colon V_G(\mathbb{Q}) \to \mathbb{R} \cup \{\infty\}$ by

$$\mu_{\varphi,\mathcal{A}}(\zeta) = \min_{x \in \mathcal{A}} \mu_{\varphi(x)}(\zeta). \tag{8.6.1}$$

Untangling the definitions, we see that for any maximal torus T of G and any $\zeta \in V_T(\mathbb{Q})$,

$$\mu_{\varphi,\mathcal{A}}(\zeta) = \mu_{F_{\varphi,T,\mathcal{A}}}(\zeta) = \min_{\chi \in F_{\varphi,T,\mathcal{A}}} \langle \lambda, \chi \rangle. \tag{8.6.2}$$

Note that $F_{\varphi,T,\mathcal{A}}$ is finite because it is contained in $\Phi_T(V)$, so taking the minimum in (8.6.1) and (8.6.2) makes sense (as usual, we interpret the minimum as ∞ if $F_{\varphi,T,x}$ is empty). We have

$$\mu_{\varphi,\mathcal{A}}(c\zeta) = c\mu_{\varphi,\mathcal{A}}(\zeta) \tag{8.6.3}$$

for any $c \in \mathbb{Q}_{\geq 0}$ and any $\zeta \in V_T(\mathbb{Q})$.

Lemma 8.6.10 *Suppose $\varphi(\mathcal{A}) \neq \{0\}$. Then $\mathrm{Im}(\mu_{\varphi,\mathcal{A}}) \subseteq \mathbb{Q}$—that is, $\infty \notin \mathrm{Im}(\mu_{\varphi,\mathcal{A}})$—and $\mu_{\varphi,\mathcal{A}}$ is stately.*

Proof By assumption there exists $x \in \mathcal{A}$ such that $\varphi(x) \neq 0$; then $\mathrm{supp}_T(\varphi(x)) \neq \varnothing$, so $F_{T,\mathcal{A}}$ is non-empty. Hence $\mathrm{Im}(\mu_{\varphi,\mathcal{A}}) \subseteq \mathbb{Q}$. It follows from Example 8.5.2 that $\mu_{\varphi,\mathcal{A}}$ is stately. $\qquad\square$

Proposition 8.6.11 *Let φ be a closed G-equivariant embedding of X in a G-module V, and suppose $\varphi(\mathcal{A}) \neq \{0\}$. Then, using the notation from (8.5.1) and (8.6.1), we have:*

(i) $\mathscr{D}_{\mathcal{A}}(\mathbb{Q}) = N(\mu_{\varphi,\mathcal{A}})$.

(ii) $\mathscr{D}_{\mathcal{A}}(\mathbb{Q})$ *is a polyhedral cone of finite type.*

Proof Let $\zeta \in Y_G(\mathbb{Q})$. We claim that $\zeta \in \mathscr{D}_{\mathcal{A}}(\mathbb{Q})$ if and only if $\mu_{\varphi,\mathcal{A}}(\zeta) \geq 0$. We are free by (8.6.3) to multiply ζ by a positive integer, so we can assume without loss that $\zeta = \varphi_G(\lambda)$ for some $\lambda \in Y_T$. We see from Sect. 8.2 that for any $x \in \mathcal{A}$, $\lim_{a \to 0} \lambda(\varphi(x))$ exists if and only if $\mu_{\varphi(x)}(\zeta) \geq 0$, so λ uniformly destabilises $\mathcal{A}$ if and only if $\min_{x \in \mathcal{A}} \mu_{\varphi(x)}(\zeta) \geq 0$. But $\min_{x \in \mathcal{A}} \mu_{\varphi(x)}(\zeta) = \mu_{\varphi,\mathcal{A}}(\zeta)$, so the claim follows. This proves (i). Part (ii) now follows from Lemmas 8.6.10 and 8.5.4. $\qquad\square$

Lemma 8.6.12 *Let V be a G-module and let $\psi\colon X \to V$ be a G-equivariant map such that S is the scheme-theoretic preimage $\psi^{-1}(0)$. The restriction of $\mu_{\psi,\mathcal{A}}$ to $\mathscr{D}_{\mathcal{A}}(\mathbb{Q})$ does not depend on the choice of V and the map ψ.*

Proof Let $\zeta \in \mathscr{D}_{\mathcal{A}}(\mathbb{Q})$. Let $x \in \mathcal{A}$; it is enough to show that the value of $\mu_{\psi(x)}(\zeta)$ does not depend on the choice of V and ψ. Recall that if $v = 0$ then $\mu_v(\lambda) = \infty$ for all $\lambda \in Y_G$, so we get $\mu_{\psi(x)}(\lambda) = \infty$ for all $\lambda \in G$ if $x \in S$. So without loss let us assume that $x \notin S$. Choose $\lambda \in Y_G(\mathbb{Q})$ such that $\varphi_G(\lambda) = \zeta$. By (8.6.3) there is no harm in multiplying λ by a positive integer, so without loss we can assume that $\lambda \in Y_G$. Set $x' := \lim_{a \to 0} \lambda(a) \cdot x$. Then $\psi(x') = \lim_{a \to 0} \lambda(a) \cdot \psi(x)$ by Lemma 2.8.2(ii) and the map $k^* \to X$, $a \mapsto \lambda(a) \cdot \psi(x)$ extends to a map $\phi_\lambda\colon k \to X$. If we choose a suitable basis of weight vectors for V with respect to T then ϕ_λ is given by the map $a \mapsto (a^{n_1}, \ldots, a^{n_t}, 0, \ldots, 0)$ for some t; each n_i belongs to $\mathbb{N}_0$, and $\psi(x') = 0$ if and only if none of the n_i is 0. Moreover, the n_i are precisely the numbers $\langle \lambda, \chi \rangle$ as χ ranges through the elements of $\mathrm{supp}_T(\psi(x))$, so we have $\mu_{\psi(x)}(\zeta) = \min_{1 \leq i \leq t} n_i$. Hence $\mu_{\psi(x)}(\zeta) = 0$ if and only if $\psi(x') \neq 0$ if and only if $x' \notin S$, and this condition does not depend on the choice of V and ψ.

So we can assume without loss that $x' \in S$. The scheme-theoretic preimage $\phi_\lambda^{-1}(S)$ is therefore a non-empty closed subscheme of k, and 0 is the unique closed point of $\phi_\lambda^{-1}(S)$ because $\psi(x) \neq 0$. Hence the coordinate ring of $\phi_\lambda^{-1}(S)$ has the form $k[Z]/(Z^n)$ for some $n \in \mathbb{N}$. (Here Z is an indeterminate and (Z^n) denotes the ideal generated by Z^n.) By construction, n does not depend on the choice of V and ψ. Set $\phi'_\lambda = \psi \circ \phi_\lambda$. Then $(\phi'_\lambda)^{-1}(0) = (\phi_\lambda)^{-1}(\psi^{-1}(0)) = \phi_\lambda^{-1}(S)$, so $(\phi'_\lambda)^{-1}(0)$ has coordinate ring $k[Z]/(Z^n)$. We leave it as an exercise (Exercise 8.9) to show that this positive integer n is the minimum of the n_i. Hence $\mu_{\psi(x)}(\zeta) = n$, so $\mu_{\psi(x)}(\zeta)$ does not depend on the choice of V and ψ. This completes the proof. $\qquad\square$

Notation 8.6.13 *We write $\mu_{\mathcal{A}}\colon \mathscr{D}_{\mathcal{A}}(\mathbb{Q}) \to \mathbb{Q}$ for the restriction of $\mu_{\psi,\mathcal{A}}$ to $\mathscr{D}_{\mathcal{A}}(\mathbb{Q})$, where V and ψ are as in Lemma 8.6.12; note that such a V and ψ exist by Lemma 3.2.1. We define*

$$f_{\mathcal{A}}\colon \mathscr{D}_{\mathcal{A}}(\mathbb{Q}) \setminus \{0\} \to \mathbb{R} \quad by \quad f_{\mathcal{A}}(\zeta) = \frac{\mu_{\mathcal{A}}(\zeta)}{\|\zeta\|}.$$

(We interpret $f_{\mathcal{A}}(\zeta)$ as ∞ if $\mu_{\mathcal{A}}(\zeta) = \infty$.)

By (8.2.5), (8.2.6) and the definition of $\mu_{\mathcal{A}}$, we have

$$\mu_{g\cdot\mathcal{A}}(g \cdot \zeta) = \mu_{\mathcal{A}}(\zeta) = (g \cdot \mu_{\mathcal{A}})(g \cdot \zeta)$$

and

$$f_{g\cdot\mathcal{A}}(g \cdot \zeta) = f_{\mathcal{A}}(\zeta) = (g \cdot f_{\mathcal{A}})(g \cdot \zeta) \tag{8.6.4}$$

for all $\zeta \in \mathscr{D}_{\mathcal{A}}(\mathbb{Q})$ and all $g \in G$.

To apply our optimality construction to $f_{\mathcal{A}}$ we need a positivity condition to be satisfied. The next result gives a criterion for this to hold.

Lemma 8.6.14 *Let V and ψ be as in Lemma 8.6.12. Let $\lambda \in Y_G$ and set $\zeta = \varphi_G(\lambda)$. Suppose λ uniformly destabilises $\mathcal{A}$. Then $f_{\mathcal{A}}(\zeta) > 0$ if and only if λ uniformly destabilises $\mathcal{A}$ into S.*

Proof If λ destabilises $\mathcal{A}$ into S then $\lim_{a \to 0} \lambda(a) \cdot \psi(x) = 0$ for each $x \in \mathcal{A}$, so $\mu_{T,\psi(x)}(\lambda) > 0$ for each $x \in \mathcal{A}$ and each T such that $\lambda \in Y_T$ by (8.2.2), so $\mu_{\mathcal{A}}(\zeta) > 0$, so $f_{\mathcal{A}}(\zeta) > 0$. The opposite implication follows by reversing this argument. (Note that the argument works even if $\mathcal{A} \subseteq S$.) $\square$

We now have everything we need to prove Theorem 8.6.7. In fact, we prove something slightly more general. Let C be a polyhedral cone of finite type such that $C \subseteq \mathscr{D}_{\mathcal{A}}(\mathbb{Q})$. We abuse notation slightly and denote the restriction of $\mu_{\mathcal{A}}$ to C by $\mu_{\mathcal{A}}$ (and likewise for $f_{\mathcal{A}}$). Suppose $\mathcal{A} \nsubseteq S$. Note that $\mu_{\mathcal{A}}$ is a stately function, by Lemma 8.6.10. Suppose there exists $\zeta \in C \setminus \{0\}$ such that $f_{\mathcal{A}}(\zeta) > 0$. Then the hypotheses of Theorem 8.5.5 hold, so we obtain $\zeta_{\mathrm{opt}} \in C \setminus \{0\}$ and a destabilising R-parabolic subgroup P_{opt} of G as in Definition 8.5.6. For any $g \in G$ we have $f_{g\cdot\mathcal{A}} = g \cdot f_{\mathcal{A}}$ by (8.6.4), so

$$\zeta_{\mathrm{opt}}(f_{g\cdot\mathcal{A}}, g \cdot C) = g \cdot \zeta_{\mathrm{opt}}(f_{\mathcal{A}}, C),$$

by (8.5.2). If $g \cdot \mathcal{A} = \mathcal{A}$ and $g \cdot C = C$ then

$$\zeta_{\mathrm{opt}}(f_{\mathcal{A}}, C) = g \cdot \zeta_{\mathrm{opt}}(f_{\mathcal{A}}, C), \tag{8.6.5}$$

by (8.5.3) and

$$g \in P_{\mathrm{opt}}(f_{\mathcal{A}}, C), \tag{8.6.6}$$

by (8.5.4).

Theorem 8.6.15 *Let X be a G-variety, let S be a closed G-stable subset of X and let $\mathcal{A} \subseteq X$. Let $C \subseteq \mathcal{D}_{\mathcal{A}}(\mathbb{Q})$ be a polyhedral cone of finite type. Suppose there exists $\lambda \in Y_G$ such that λ destabilises $\mathcal{A}$ into S and $\varphi_G(\lambda) \in C$. Suppose also that $\mathcal{A} \not\subseteq S$. Then there exists $\lambda_{\mathcal{A}} \in Y_G$ such that $\lambda_{\mathcal{A}}$ destabilises $\mathcal{A}$ into S, $N_G(\mathcal{A}) \cap N_G(C) \subseteq P_{\lambda_{\mathcal{A}}}$ and $\varphi_G(\lambda_{\mathcal{A}}) \in C$.*

Proof Set $\zeta = \varphi_G(\lambda)$. Then $f_{\mathcal{A}}(\zeta) > 0$ by Lemma 8.6.14, so we obtain $\zeta_{\mathrm{opt}} \in C \setminus \{0\}$ and P_{opt} by the argument above. Pick any $\lambda_{\mathcal{A}} \in Y_G$ such that $\varphi_G(\lambda_{\mathcal{A}}) = \zeta_{\mathrm{opt}}$. Then $\lambda_{\mathcal{A}}$ destabilises $\mathcal{A}$ into S by Lemma 8.6.14, since $f_{\mathcal{A}}(\zeta_{\mathrm{opt}}) > 0$. It follows from (8.6.6) that $N_G(\mathcal{A}) \cap N_G(C) \subseteq P_{\lambda_{\mathcal{A}}}$. $\qquad\square$

Proof of Theorem 8.6.7 This follows from Theorem 8.6.15, taking $C = \mathcal{D}_{\mathcal{A}}(\mathbb{Q})$. $\qquad\square$

Notation 8.6.16 *We write $\lambda_{\mathrm{opt}}(\mathcal{A}, S, C)$ for the cocharacter $\lambda_{\mathcal{A}}$ constructed in the proof of Theorem 8.6.15, or just $\lambda_{\mathrm{opt}}(\mathcal{A})$ or $\lambda_{\mathrm{opt}}(S)$ or $\lambda_{\mathrm{opt}}(C)$ if the other arguments are understood. Likewise we write $\zeta_{\mathrm{opt}}(\mathcal{A}, S, C)$ or $\zeta_{\mathrm{opt}}(\mathcal{A})$ or $\zeta_{\mathrm{opt}}(S)$ or $\zeta_{\mathrm{opt}}(C)$, and $P_{\mathrm{opt}}(\mathcal{A}, S, C)$ or $P_{\mathrm{opt}}(\mathcal{A})$ or $P_{\mathrm{opt}}(S)$ or $P_{\mathrm{opt}}(C)$.*

Remark 8.6.17 Under the hypotheses of Theorem 8.6.15 we have

$$\zeta_{\mathrm{opt}}(g \cdot \mathcal{A}, S, g \cdot C) = g \cdot \zeta_{\mathrm{opt}}(\mathcal{A}, S, C)$$

and

$$P_{\mathrm{opt}}(g \cdot \mathcal{A}, S, g \cdot C) = g \cdot P_{\mathrm{opt}}(\mathcal{A}, S, C)$$

for all $g \in G$ by (8.5.2).

Remark 8.6.18 Let $X, \mathcal{A}, S$ and C be as in Theorem 8.6.15. Suppose G is a finite-index subgroup of a reductive group G_1. Suppose the G-action on X extends to a G_1-action, and suppose that S is G_1-stable and the length function $\| \cdot \|$ is invariant under conjugation by G_1. (For instance, if G is connected and semisimple then $\mathrm{Out}(G)$ is finite, so $\mathrm{Aut}(G)$ has the structure of an algebraic group; if moreover G is adjoint then we may regard G as a finite-index subgroup of $\mathrm{Aut}(G)$, and we could take G_1 to be a subgroup of $\mathrm{Aut}(G)$ containing G.) Since $Y_{G_1} = Y_G = Y_{G^0}$, the notions of uniformly unstable, uniformly S-unstable, etc., are unchanged if we pass from G to G_1; likewise, each of $\lambda_{\mathrm{opt}}(\mathcal{A}, S, C)$ and $\zeta_{\mathrm{opt}}(\mathcal{A}, S, C)$ is the same for G_1 as for G. Hence the formulas in Remark 8.6.17 hold for $g \in G_1$.

Remark 8.6.19 We took C to be $\mathcal{D}_{\mathcal{A}}(\mathbb{Q})$ in the proof of Theorem 8.6.7. It can be useful to allow C to be a proper subset of $\mathcal{D}_{\mathcal{A}}(\mathbb{Q})$ (see Sect. 11.6). Note that $P_{\mathrm{opt}}(C)$ does depend on the choice of C. For instance, suppose $\lambda \in Y_G$ is indivisible and λ destabilises $\mathcal{A}$ into

S. Set $\zeta = \varphi_G(\lambda)$ and $C = \mathbb{Q}_{\geq 0} \cdot \zeta$. It is clear that $\zeta_{\mathrm{opt}}(C) = \zeta$, and that this need not be equal to $\zeta_{\mathrm{opt}}(\mathscr{D}_{\mathscr{A}}(\mathbb{Q}))$.

Remark 8.6.20 Let X, S, $\mathscr{A}$ and C be as in Theorem 8.6.15. Let H be a reductive subgroup of G. Recall from Example 7.5.3(ii) that we have an H-equivariant inclusion of $V_H(\mathbb{Q})$ in $V_G(\mathbb{Q})$. Suppose the optimal point $\zeta = \zeta_{\mathrm{opt}}(\mathscr{A}, S, C)$ for the G-action belongs to $V_H(\mathbb{Q})$. We can pick $\lambda \in Y_H$ such that $\varphi_H(\lambda) = \zeta$. It is clear from the constructions that ζ is also the optimal point $\zeta_{\mathrm{opt}}(\mathscr{A}, S, C \cap V_H(\mathbb{Q}))$ for the H-action. In fact, if S' is a closed H-stable subset of S and λ uniformly destabilises $\mathscr{A}$ into S' then $\zeta = \zeta_{\mathrm{opt}}(\mathscr{A}, S', C \cap V_H(\mathbb{Q}))$. This set-up was studied by the second author in the special case when $\mathscr{A} = \{x\}$ for some $x \in X$, S is the unique closed G-orbit contained in $\overline{G \cdot x}$ and $C = \mathscr{D}_x(\mathbb{Q})$, and applied in particular to the study of nilpotent orbits (see [9]). Following *op. cit.*, we say that H is an *optimal subgroup for* $X, S, \mathscr{A}$ and C if $\zeta_{\mathrm{opt}}(\mathscr{A}, S, C)$ belongs to $V_H(\mathbb{Q})$.

We now give some remarks on how to calculate optimal destabilising cocharacters. To simplify our discussion we restrict attention to the case that X is a G-module and $S = \{0\}$; this is the standing assumption in the next couple of paragraphs. This is no real loss, since given general X, $\mathscr{A}$, S and C the first step of the optimality construction is to fix a G-equivariant map from X to some G-module such that S is the scheme-theoretic preimage of 0, and then work with points in that module.

Let P be any R-parabolic subgroup of G. There is a maximal torus T of G contained in $P \cap P_{\mathrm{opt}}$, and there exists an optimal destabilising cocharacter λ_{opt} in Y_T. In particular, take P to be a Borel subgroup B of G. The maximal tori of B are all $R_u(B)$-conjugate and so there exists $\tilde{u} \in R_u(B)$ such that $\zeta_{\mathrm{opt}} \in V_{\tilde{u}T\tilde{u}^{-1}}$. Now suppose that C is $R_u(B)$-stable (this is the case when $C = \mathscr{D}_{\mathscr{A}}(\mathbb{Q})$, for example). Let $u \in R_u(B)$ be any element. Let ζ_u be the unique indivisible element of $C_T \cap V_T$ such that $f_{u^{-1} \cdot \mathscr{A}}$ attains its maximum value M_u on C_T at ζ_u. So if $\zeta \in C_T \cap V_T$ and ζ is indivisible then

$$f_{u^{-1} \cdot \mathscr{A}}(\zeta) \leq f_{u^{-1} \cdot \mathscr{A}}(\zeta_u) = f_{\mathscr{A}}(u \cdot \zeta_u) \leq f_{\mathscr{A}}(\zeta_{\mathrm{opt}}) = f_{u^{-1} \cdot \mathscr{A}}(u^{-1} \cdot \zeta_{\mathrm{opt}}) \qquad (8.6.7)$$

using (8.6.5). Equality in (8.6.7) holds if and only if $\zeta = \zeta_u$ and $u \cdot \zeta_u = \zeta_{\mathrm{opt}}$; the second equality holds if and only if ζ_{opt} belongs to $C_{uTu^{-1}}$.

Hence to find ζ_{opt} we can proceed as follows. For each $u \in R_u(B)$, find the unique indivisible $\zeta_u \in V_T$ such that $f_{u^{-1} \cdot \mathscr{A}}$ attains its maximum value M_u on C_T at ζ_u. Since $F_{T, u^{-1} \cdot \mathscr{A}} \subseteq \Phi_T(X)$ (recall that we are assuming X is a G-module), there are only finitely many possibilities for ζ_u and M_u as u ranges over the elements of $R_u(B)$, so $\{M_u \mid u \in R_u(B)\}$ has a maximum M. Choose any $u \in R_u(B)$ such that $M_u = M$. Then $\zeta_{\mathrm{opt}} = u \cdot \zeta_u \in V_{uTu^{-1}}$, and if we wish we can choose an optimal destabilising cocharacter λ_{opt} from $Y_{uTu^{-1}}$. Note that if $\mathrm{supp}_T(u_1^{-1} \cdot \mathscr{A}) \subseteq \mathrm{supp}_T(u_2^{-1} \cdot \mathscr{A})$ then $f_{u_2^{-1} \cdot \mathscr{A}}(\zeta) \leq f_{u_1^{-1} \cdot \mathscr{A}}(\zeta)$

for all $\zeta \in C_T$, so we need only consider those $u \in R_u(B)$ such that $\operatorname{supp}_T(u^{-1} \cdot \mathcal{A})$ is minimal among the set of all such supports.

Remark 8.6.21 Suppose moreover that $\mathcal{A}$ is $R_u(B)$-stable. Let T be any maximal torus of B. Then $\zeta_{\mathrm{opt}} = \zeta$, where ζ is the unique indivisible element of C_T where $f_{\mathcal{A}}$ attains its maximum value. (Note that $P_{\mathrm{opt}} \supseteq R_u(B)$ by Theorem 8.6.15, so $u \cdot \zeta = \zeta = \zeta_{\mathrm{opt}}$ for any $u \in R_u(B)$.)

Here is an extended example.

Example 8.6.22 Let $G = X = \mathrm{SL}_3$ with the action of G by conjugation, and let $S = \{1\}$. Let B be the usual Borel subgroup of upper triangular matrices, let T be the usual diagonal maximal torus and let $U = R_u(B)$. Let α and β be the usual labelling of the simple roots for this choice of B and T. Concretely, let e_α denote the 3×3 matrix with a 1 in the $(1, 2)$-position and 0 elsewhere, so the root homomorphism $\mathbb{G}_a \to U_\alpha$ is the map $a \mapsto 1 + a e_\alpha$ (where the 1 on the RHS is the identity matrix). Similarly, we have elements e_β (nonzero in the $(2, 3)$-position) and $e_{\alpha+\beta}$ (nonzero in the $(1, 3)$-position) in Mat_3 giving the usual parametrisation of the positive root groups.

We may identify $Y_T \cong \{(r, s, t) \in \mathbb{Z}^3 \mid r + s + t = 0\}$, where a triple $(r, s, t) \in \mathbb{Z}^3$ represents the cocharacter $a \mapsto \operatorname{diag}(a^r, a^s, a^t)$. Letting $\|\cdot\|$ be the length function from Example 2.5.2, we then have $\|\lambda\| = \sqrt{r^2 + s^2 + t^2}$. We have $\langle(r, s, t), \alpha\rangle = r - s$ and $\langle(r, s, t), \beta\rangle = s - t$.

A subset $\mathcal{A}$ of X is uniformly S-unstable if and only if $\mathcal{A}$ is contained in a maximal unipotent subgroup of G if and only if $G(\mathcal{A})$ is unipotent (see Example 8.6.4). We calculate λ_{opt} for some uniformly S-unstable subsets of X (taking $C = \mathcal{D}_{\mathcal{A}}(\mathbb{Q})$). Since any unipotent subgroup of G is conjugate to a subgroup of U, it is enough by Remark 8.6.17 to consider subsets of U.

The first step for all the cases below is to linearise the action, which we can do every time with the map $\psi : X \to \mathrm{Mat}_3, g \mapsto g - 1$. This does indeed satisfy $\psi^{-1}(0) = S$, and ψ is G-equivariant, where G acts on Mat_3 also by conjugation.

Case 1 Let $x_1 = 1 + e_{\alpha+\beta}$, so that $\psi(x_1) = e_{\alpha+\beta}$. Set $\mathcal{A}_1 = \{x_1\}$. Then $\mathcal{A}_1$ is $R_u(B)$-stable since $x_1 \in Z(R_u(B))$, so there is an optimal destabilising cocharacter λ_{opt} in Y_T by Remark 8.6.21. Now $\operatorname{supp}_T(\psi(\mathcal{A}_1)) = \{\alpha + \beta\}$, so in order to find $\lambda_1 = \lambda_{\mathrm{opt}}(\mathcal{A}_1)$, we have to maximise the function

$$f(r, s, t) = \frac{r - t}{\sqrt{r^2 + s^2 + t^2}}.$$

Clearly $s = 0$ at the maximum. The condition $r + s + t = 0$ therefore implies that $t = -r$, so $\lambda_1 = (1, 0, -1)$ and $P_{\mathrm{opt}}(\mathcal{A}_1) = B$.

Case 2 Let $x_2 = 1 + e_\alpha$. Set $\mathcal{A}_2 = \{x_2\}$ and note that $\psi(\mathcal{A}_2) = \{e_\alpha\}$. This time $\mathcal{A}_2$ is not $R_u(B)$-stable, so following the procedure described just before Remark 8.6.21 we need to consider the supports $\operatorname{supp}_T(u^{-1} \cdot \{e_\alpha\})$ for $u \in R_u(B)$. But actually it is clear that $\{\alpha\}$ is the unique minimal element in this set of supports (e.g., take $u = 1$), so again we see that we can find an optimal destabilising cocharacter $\lambda_2 = \lambda_{\mathrm{opt}}(\mathcal{A}_2)$ evaluating in T. This time we need to maximise the function

$$f(r, s, t) = \frac{r - s}{\sqrt{r^2 + s^2 + t^2}},$$

giving $\lambda_2 = (1, -1, 0)$, so $P_{\mathrm{opt}}(\mathcal{A}_2)$ is a different Borel subgroup B' of G (it is the conjugate $n_\beta B n_\beta^{-1}$, where $n_\beta \in N_G(T)$ is any representative of the reflection s_β in the Weyl group of G).

Note that we could instead have argued that the sets $\mathcal{A}$ from (i) and (ii) are conjugate to each other and applied Remark 8.6.17: the reflection s_β swaps the roots α and $\alpha + \beta$, so a suitable choice for n_β conjugates x_1 to x_2 and vice versa.

Case 3 Let $x_3 = 1 + e_\alpha + e_{\alpha+\beta}$. Set $\mathcal{A}_3 = \{x_3\}$, so that $\psi(\mathcal{A}_3) = \{e_\alpha + e_{\alpha+\beta}\}$. As in Case 2, to do a direct calculation we need to consider the supports $\operatorname{supp}_T(u^{-1} \cdot \psi(\mathcal{A}_3))$ for $u \in R_u(B)$. A quick calculation (either with matrices or with commutator relations) shows that if we take $u = 1 - e_\beta$ then $u^{-1} \cdot (e_\alpha + e_{\alpha+\beta}) = e_\alpha$, so $\operatorname{supp}_T(u^{-1} \cdot \psi(\mathcal{A}_3)) = \{\alpha\}$. Again, it is clear that this must be the unique minimal support, so we can take $\lambda_3 = \lambda_{\mathrm{opt}}(\mathcal{A}_3) \in Y_{uTu^{-1}}$ to be $u \cdot \lambda_2$, where $\lambda_2 \in Y_T$ is the cocharacter from Case 2. Hence $P_{\mathrm{opt}}(\mathcal{A}_3) = uB'u^{-1}$.

Again, we could instead have argued using the fact that the elements x_1 and x_3 are conjugate and applying Remark 8.6.17.

Case 4 Let $x_4 = 1 + e_\alpha + e_\beta$, and set $\mathcal{A}_4 = \{x_4\}$. Then $\operatorname{supp}_T(\psi(\mathcal{A}_4)) = \{\alpha, \beta\}$, and this is the unique minimal support (e.g., take $u = 1$). So we can find an optimal destabilising cocharacter $\lambda_4 = \lambda_{\mathrm{opt}}(\mathcal{A}_4)$ evaluating in T by maximising the function

$$f(r, s, t) = \frac{\min\{r - s, s - t\}}{\sqrt{r^2 + s^2 + t^2}}. \tag{8.6.8}$$

This is slightly less straightforward to calculate, but note straight away that it must be the case that $P_{\lambda_4} = B$. To see this, we can argue that both $\langle \lambda_4, \alpha \rangle > 0$ and $\langle \lambda_4, \beta \rangle > 0$, so $\langle \lambda_4, \alpha + \beta \rangle > 0$, so $R_u(B) \subseteq R_u(P_{\lambda_4})$, which forces $P_{\lambda_4} = B$. (An alternative argument is that x_4 in this case is a *regular* unipotent element contained in B, and since $x_4 \in R_u(P_{\lambda_4})$ this forces $P_{\lambda_4} = B$; see Sect. 12.2 for more on regular elements, and in particular Lemma 12.2.2(ii).)

Now to maximise the function in (8.6.8), observe that since $P_{\lambda_4} = B$, we must have $\lambda_4 = (r, s, t)$ with $r > s > t$. Since $r + s + t = 0$, this means that $r > 0$ and $t < 0$. We claim that in fact we must have $r - s = s - t$, which combined with $r + s + t = 0$ and the indivisibility of λ_4 forces $\lambda_4 = (1, 0, -1)$. Suppose for a contradiction that $r - s < s - t$ (the other case is similar). Then we can choose small positive $\delta, \epsilon \in \mathbb{Q}$ such that $r - s < (r + \delta) - s < s - (t + \epsilon)$ and $\sqrt{(r + \delta)^2 + s^2 + (t + \epsilon)^2} < \sqrt{r^2 + s^2 + t^2}$. But then $f(r + \delta, s, t + \epsilon) > f(r, s, t)$, contradicting maximality (recall that λ_4 maximises the value of f as we range over the elements of $V_T(\mathbb{Q})$). This means we do have $\lambda_{\mathrm{opt}}(\mathcal{A}_4) = (1, 0, -1)$ and $P_{\mathrm{opt}}(\mathcal{A}_4) = B$.

For further examples along these lines, see Exercise 8.11.

For convenience, we state a special case of the Optimal Hilbert-Mumford Theorem 8.6.7 which is often used in practice.

Theorem 8.6.23 *Let X be a G-variety and let $x \in X$ such that $G \cdot x$ is not closed. Then there exists $\lambda \in Y_G$ such that $x' := \lim_{a \to 0} \lambda(a) \cdot x$ exists, $G \cdot x'$ is closed and $G_x \subseteq P_\lambda$.*

Proof This follows from Theorem 8.6.7, taking $\mathcal{A} = \{x\}$ and S to be the unique closed G-orbit contained in $\overline{G \cdot x}$. $\qquad\qquad\square$

Now we use optimality to prove a counterpart of Levi descent for closed orbits (Corollary 3.3.8).

Corollary 8.6.24 *Suppose X is a G-variety, $x \in X$, and H is a subgroup of G_x such that H is G-completely reducible. If $C_G(H) \cdot x$ is closed then $G \cdot x$ is closed.*

Proof We prove the contrapositive. Suppose $G \cdot x$ is not closed. By Theorem 8.6.23 there exists $\lambda \in Y_G$ such that $x' := \lim_{a \to 0} \lambda(a) \cdot x$ exists, $G \cdot x'$ is closed and $H \subseteq P_\lambda$. Now H is G-cr, so there exists $u \in R_u(P_\lambda)$ such that $u \cdot \lambda$ centralises H. This means that $u \cdot \lambda \in Y_{C_G(H)}$. We have $\lim_{a \to 0}(u \cdot \lambda)(a) \cdot x = u \cdot x'$ by Lemma 3.2.7, so $u \cdot x' \in \overline{C_G(H) \cdot x}$. However, since $u \cdot x'$ is not G-conjugate to x, it is not $C_G(H)$-conjugate to x, and hence $C_G(H) \cdot x$ is not closed. $\qquad\qquad\square$

We immediately deduce the following counterparts of the descent results Corollaries 3.3.7 and 3.3.8. Both the results are special cases of the previous one, using the fact that a linearly reductive subgroup of G is G-cr (Lemma 4.2.1, which applies in particular to a torus).

Corollary 8.6.25 *Let X be a G-variety and let $x \in X$. Suppose S is a linearly reductive subgroup of G_x. If $C_G(S) \cdot x$ is closed then $G \cdot x$ is closed.*

Corollary 8.6.26 (Levi Ascent for Closed Orbits) *Let X be a G-variety and let $x \in X$. Suppose $L = C_G(S)$ with S a subtorus of G_x. If $L \cdot x$ is closed then $G \cdot x$ is closed.*

Corollary 8.6.27 *Let X be a G-variety. Suppose $x \in X$ is fixed by a maximal torus T of G. Then $G \cdot x$ is closed.*

Proof Since $T \cdot x = \{x\}$ is closed and $C_G(T)^0 = T$, we see that $C_G(T) \cdot x$ is closed. Now apply Corollary 8.6.26. $\qquad\qquad\square$

8.7   Dependence on the Length Function

Often when the optimality formalism is described in the literature, the length function is fixed early on and then all further mention of it is suppressed. We have also suppressed the mention of $\|\cdot\|$ to avoid cluttering up our notation: but we warn the reader that the optimal destabilising cocharacter and the optimal destabilising R-parabolic subgroup can depend on the choice of length function. So strictly speaking we should, for example, have written $\lambda_{\mathrm{opt}}(f, C, \|\cdot\|)$ instead of $\lambda_{\mathrm{opt}}(f, C)$ at the end of Sect. 8.5.

Suppose we replace $\|\cdot\|$ with another length function $\|\cdot\|'$ of the form $\|\cdot\|' = a\|\cdot\|$ for some $a > 0$. The effect of this for a fixed stately function μ is to scale the corresponding function f from Theorem 8.5.5 by the constant factor a^{-1}, so λ_{opt}, ζ_{opt}, etc., do not change; in particular, this is the case if G is simple, by Lemma 2.5.3. But if G has more than one simple factor then we are free to scale $\|\cdot\|$ on the different simple factors independently, and this can make a difference. We give two examples.

Example 8.7.1 Let $G_1 = G_2 = \mathrm{SL}_2$. Let $V_i = k^2$ be the natural module for G_i, let T_i be the usual diagonal maximal torus of G_i and let B_i be the usual upper triangular Borel subgroup of G_i. Define $\lambda_i \in Y_{T_i}$ by $\lambda_i(a) = \mathrm{diag}(a, a^{-1})$. Fix $n_1, n_2 \in \mathbb{N}$. Choose a $N_{G_i}(T_i)$-invariant positive-definite bilinear form $(\cdot\,,\cdot)_i$ on Y_{T_i} such that $(\lambda_i, \lambda_i)_i = n_i$, and let $\|\cdot\|_i$ be the corresponding length function on Y_{G_i}; here we have scaled the length function on Y_{G_i} from Example 2.5.2 by a factor of n_i for $i = 1, 2$.

Let $G = G_1 \times G_2$ with maximal torus $T = T_1 \times T_2$ and Borel subgroup $B = B_1 \times B_2$. Let $(\cdot\,,\cdot)$ be the orthogonal direct sum of $(\cdot\,,\cdot)_1$ and $(\cdot\,,\cdot)_2$ on $Y_{T_1 \times T_2}$ and let $\|\cdot\|$ be the corresponding length function on G. Regard λ_1 and λ_2 as elements of Y_T in the obvious way.

Let $V = V_1 \times V_2$ with the product action of G, and $v = (v_1, v_2) \in V$, where $v_i = \begin{pmatrix} 1 \\ 0 \end{pmatrix} \in V_i$ for $i = 1, 2$. Then $\mathrm{supp}_T(v) = \{\alpha_1, \alpha_2\}$ where $\langle \lambda_1, \alpha_1 \rangle = \langle \lambda_2, \alpha_2 \rangle = 1$ and $\langle \lambda_1, \alpha_2 \rangle = \langle \lambda_2, \alpha_1 \rangle = 0$.

We see that $\{v\}$ is uniformly $\{0\}$-unstable: for instance, $\lambda_1 + \lambda_2$ destabilises v to 0. We find the optimal destabilising cocharacter and parabolic subgroup, and we investigate the effect of changing the length function by varying n_1 and n_2. Since $R_u(B)$ fixes v, it suffices by Remark 8.6.21 to find $\lambda \in Y_T$ such that $f_{T,v}$ attains its maximum value at λ. We can write any element λ of Y_T as $m_1\lambda_1 + m_2\lambda_2$ for some $m_1, m_2 \in \mathbb{Z}$. Since we want λ to destabilise v to 0, we can assume without loss that $m_1, m_2 > 0$. We have

$$f_T(m_1\lambda_1 + m_2\lambda_2) = \frac{\min\{\langle m_1\lambda_1 + m_2\lambda_2, \alpha_1\rangle, \langle m_1\lambda_1 + m_2\lambda_2, \alpha_2\rangle\}}{\|m_1\lambda_1 + m_2\lambda_2\|} = \frac{\min\{m_1, m_2\}}{\sqrt{m_1^2 n_1^2 + m_2^2 n_2^2}}.$$

This value is maximised when $(m_1, m_2) = c(n_2, n_1)$ for some $c \in \mathbb{Q}_+$. Hence we can take λ_{opt} to be $n_2\lambda_1 + n_1\lambda_2$. This depends on the choice of n_1 and n_2. However, $P_{\mathrm{opt}} = B$ does not depend on n_1 and n_2.

Example 8.7.2 Now we consider an example where the optimal destabilising parabolic subgroup does depend on the choice of length function. Let G_i, T_i, B_i, n_i, λ_i, α_i, for $i = 1, 2$ and G, T, B be as in Example 8.7.1. Let V be the Weyl module of G with highest weight $\alpha_1 + \alpha_2$ and let $0 \neq v \in V$ be a highest weight vector. Then $\mathrm{supp}_T(v) = \{\alpha_1 + \alpha_2\}$ and v is fixed by $R_u(B)$. We find the optimal destabilising cocharacter and parabolic subgroup for $\mathcal{A} = \{v\}$ and $\mathcal{S} = \{0\}$. As in Example 8.7.1, it suffices to find $m_1, m_2 \in \mathbb{Z}$ such that $f_{T,v}(m_1\lambda_1 + m_2\lambda_2)$ is maximal. We have

$$f_{T,v}(m_1\lambda_1 + m_2\lambda_2) = \frac{\langle m_1\lambda_1 + m_2\lambda_2, \alpha_1 + \alpha_2\rangle}{\|m_1\lambda_1 + m_2\lambda_2\|} = \frac{m_1 + m_2}{\sqrt{m_1^2 n_1^2 + m_2^2 n_2^2}}.$$

When $n_1 < n_2$ this is maximised when $m_2 = 0$; when $n_1 > n_2$ we take $m_1 = 0$; when $n_1 = n_2$ we take $m_1 = m_2$. Hence we can take λ_{opt} to be λ_1 when $n_1 < n_2$, λ_2 when $n_1 > n_2$ and $\lambda_1 + \lambda_2$ when $n_1 = n_2$. We see that P_{opt} is $B_1 \times G_1$ in the first case, $G_1 \times B_2$ in the second and $B_1 \times B_2$ in the third.

8.8   Proof of Theorem 8.1.1

Now that we have our optimality machine up and running, the obvious way to prove Theorem 8.1.1 is to apply these results directly to the G-variety G^n, taking $\mathcal{A}$ to be H^n and $\mathcal{S}$ to be a suitable subset of G^n; the optimal destabilising subgroup that we obtain will contain $N_G(H)$, since $N_G(H)$ stabilises H^n. There is, however, a small issue remaining: the subgroup Γ in the theorem (the stabiliser in G of the fixed-point subcomplex $(\Delta_G)^H$) might be larger than $N_G(H)$; see Example 8.8.4. Because of this, it is convenient to be able to pass from H to a possibly larger subgroup K with better properties, so we begin the section by setting up the tools to do this.

Let K be any non-G-cr subgroup of G. Pick $\lambda \in Y_G$ such that (P_λ, L_λ) yields a semisimplification $M := c_\lambda(K)$ of K. Pick $n \in \mathbb{N}$ and $\mathbf{k} \in K^n$ such that $\mathbf{k}$ is a generic tuple for K. Then $\mathbf{m} := c_\lambda(\mathbf{k})$ is a generic tuple for M by Lemma 5.2.5(iii); in particular, $G \cdot \mathbf{m}$ is closed but $G \cdot \mathbf{k}$ is not.

We apply the optimality construction to $X := G^n$ with G acting by simultaneous conjugation. Set $\mathcal{S}_n = \overline{G \cdot M^n}$ and $\mathcal{A}_n = K^n$. Note that λ uniformly destabilises $\mathcal{A}_n$

into $\mathcal{S}_n$, by construction, and $\mathcal{S}_n$ does not depend on the choice of semisimplification of K by Proposition 5.5.8.

Lemma 8.8.1 *Let* $\mu \in Y_G$ *such that* $\mathbf{k}' := \lim_{a \to 0} \mu(a) \cdot \mathbf{k}$ *exists and belongs to* $\mathcal{S}_n$. *Then* $G \cdot \mathbf{k}'$ *is closed.*

Proof Set $r = \dim(C_G(M))$. Then $\dim(G_{\mathbf{g}}) \geq r$ for every $\mathbf{g} \in G \cdot M^n$, so $\dim(G_{\mathbf{g}}) \geq r$ for every $\mathbf{g} \in \overline{G \cdot M^n}$ by Lemma 2.7.4(i). Now suppose $G \cdot \mathbf{k}'$ is not closed. There exists $\nu \in Y_G$ such that $\mathbf{k}'' := \lim_{a \to 0} \nu(a) \cdot \mathbf{k}'$ exists and $G \cdot \mathbf{k}''$ is closed. Now $\mathbf{k}' \in \overline{G \cdot \mathbf{k}}$, so $G \cdot \mathbf{k}'' = G \cdot \mathbf{m}$, since $\overline{G \cdot \mathbf{k}}$ contains exactly one closed G-orbit. Hence

$$\dim(G_{\mathbf{k}'}) < \dim(G_{\mathbf{k}''}) = \dim(G_{\mathbf{m}}) = \dim(C_G(M)) = r,$$

where the inequality is from Lemma 2.7.6(ii). But $\dim(G_{\mathbf{k}'}) \geq r$ by the argument at the start of the proof, a contradiction. We conclude that $G \cdot \mathbf{k}'$ is closed after all. $\square$

Since $G \cdot \mathbf{k}$ is not closed, we deduce from Lemma 8.8.1 (taking $\mu = 0$) that $\mathbf{k} \notin \mathcal{S}_n$. This shows that $\mathcal{A}_n \not\subseteq \mathcal{S}_n$, so it makes sense to define $\widehat{\mathcal{P}}(K)$ to be the optimal destabilising R-parabolic subgroup $P_{\mathrm{opt}}(\mathcal{A}_n, \mathcal{S}_n, C_n)$ from Theorem 8.6.15, taking C_n to be $\mathcal{D}_{\mathcal{A}_n}(\mathbb{Q})$. Clearly $N_G(K) \subseteq N_G(\mathcal{A}_n) \cap N_G(C_n)$, so $N_G(K) \subseteq \widehat{\mathcal{P}}(K)$.

Lemma 8.8.2 $\widehat{\mathcal{P}}(K)$ *yields a semisimplification of* K.

Proof Set $\mu = \lambda_{\mathrm{opt}}(\mathcal{A}_n, \mathcal{S}_n, C_n)$. Then $\mathbf{k}' := \lim_{a \to 0} \mu(a) \cdot \mathbf{k}$ belongs to $\mathcal{S}_n$, so $G \cdot \mathbf{k}'$ is closed by Lemma 8.8.1. It follows from Theorem 5.3.3 that $\mathcal{G}(\mathbf{k}')$ is G-cr, as required. $\square$

Proof of Theorem 8.1.1 Let H be as in the statement of the theorem, and recall that the group Γ in the statement of the theorem is the stabiliser in G of $(\Delta_G)^H$. As explained above, we aim to replace H with a possibly larger group K such that $\Gamma = N_G(K)$. Then we can take $\mathcal{P}(H)$ to be $\widehat{\mathcal{P}}(K)$. Set

$$K := \bigcap_{\sigma_P \in (\Delta_G)^H} N_G(P),$$

a subgroup of G which clearly contains H. Since H is not G-cr, there is some R-parabolic subgroup Q of G witnessing that fact. Note that Q^0 is a parabolic subgroup of G^0 normalised by H, so $\sigma_{Q^0} \in (\Delta_G)^H$, so $K \subseteq N_G(Q^0)$. But H is contained in no R-Levi subgroup of Q, so is contained in no R-Levi subgroup of $N_G(Q^0)$, by Lemma 2.9.20. Thus K is contained in no R-Levi subgroup of $N_G(Q^0)$, so $N_G(Q^0)$ witnesses the fact that K is not G-cr.

If P is a parabolic subgroup of G^0 then $H \subseteq N_G(P)$ if and only if $K \subseteq N_G(P)$. Hence $(\Delta_G)^H = (\Delta_G)^K$, and it follows easily that $N_G(K) = \Gamma$ (cf. Exercise 7.6);

in particular, Γ is closed. Set $\mathcal{P}(H) = \widehat{\mathcal{P}}(K)$. Then $\Gamma = N_G(K) \subseteq \mathcal{P}(H)$, so we are done if we can show that $\mathcal{P}(H)$ gives a semisimplification of H. Fix an R-Levi subgroup L of $\mathcal{P}(H)$. Then $c_L(K)$ is G-cr by Lemma 8.8.2. We claim that $c_L(H)$ is also G-cr. By Theorem 5.4.4, it suffices to show that $c_L(H)$ is L-cr. Suppose P is an R-parabolic subgroup of L containing $c_L(H)$. Then $c_L(H)$ normalises P^0, so H normalises $P^0 R_u(\mathcal{P}(H))$ since $H \subseteq c_L(H) R_u(\mathcal{P}(H))$. Since $P^0 R_u(\mathcal{P}(H))$ is a parabolic subgroup of G^0 by Lemma 2.9.12 (applied to G^0), this implies that K normalises $P^0 R_u(\mathcal{P}(H))$ by definition of K, and applying c_L we conclude that $c_L(K) \subseteq N_L(P^0)$. Now the fact that $c_L(K)$ is L-cr, together with another application of Lemma 2.9.20, puts $c_L(H)$ inside an R-Levi subgroup of P. So we can finally conclude that $\mathcal{P}(H)$ yields a semisimplification of H, and hence $\mathcal{P}(H)$ has all the desired properties. $\qquad\square$

Remark 8.8.3 If H is a non-trivial unipotent subgroup of G^0 then it is not clear in general whether the R-parabolic subgroup $P_{\mathrm{BT}}(H)$ from Theorem 2.9.23 must contain Γ. But we can apply the same trick as in the proof of Theorem 8.1.1. Set $K := \bigcap_{\sigma_P \in (\Delta_G)^H} N_G(P)$; we saw above that $\Gamma = N_G(K)$. We can choose a Borel subgroup B of G^0 such that $H \subseteq R_u(B)$. Then $K \subseteq N_G(B)$ and $N_G(B)$ normalises $R_u(B)$, so K normalises $K \cap R_u(B)$, which contains H. Now $K \cap R_u(B)$ is a normal unipotent subgroup of K. It follows from Proposition 2.3.2 that H is contained in the largest normal unipotent subgroup N of K. Clearly $\Gamma = N_G(K)$ normalises N. So the subgroup $P_{\mathrm{BT}}(N)$ contains Γ, and hence gives a Γ-stable centre for $(\Delta_G)^H$; moreover, $H \subseteq N \subseteq R_u(P_{\mathrm{BT}}(N))$.

We've already seen an example showing that $N_G(H)$ can be properly contained in Γ (Exercise 7.5). Here is another example with H non-G-cr.

Example 8.8.4 Suppose $p > 0$. Let $G = \mathrm{SL}_3$ acting on $X := \mathrm{SL}_3$ by conjugation. Let $T \subseteq B$ be the standard diagonal maximal torus and upper triangular Borel subgroup of G, and denote the simple roots by α and β as usual. Let $H = U_\alpha U_{\alpha+\beta}$, a unipotent subgroup of G. Let $\mathscr{S}$ be the set of all parabolic subgroups of G that contain H and let Γ be the stabiliser in G of $\mathscr{S}$. Then $\Gamma \supseteq N_G(H) = P_\beta$. It is easily seen that G does not stabilise $\mathscr{S}$, so $\Gamma = P_\beta$.

By Exercise 5.3, there is a finitely generated—and hence finite—subgroup H_1 of H such that $\mathscr{S}$ is the set of all parabolic subgroups of G that contain H_1. Choose a non-trivial finite subgroup F of U_α such that $H_2 := F U_{\alpha+\beta}$ contains H_1; then $\mathscr{S}$ is the set of all parabolic subgroups of G that contain H_2. We claim that $N_G(H_2)$ is properly contained in Γ. To see this, first note that

$$N_G(H_2) \subseteq N_G(H_2^0) = N_G(U_{\alpha+\beta}).$$

Now a direct calculation shows that $N_G(U_{\alpha+\beta}) = B$. With a little more effort one can show that

$$N_G(H_2) = SU_\alpha U_\beta U_{\alpha+\beta}, \tag{8.8.1}$$

where $S = \ker(\alpha) \subseteq T$. It follows from (8.8.1) that the Borel-Tits parabolic subgroup $P_{\mathrm{BT}}(H_2)$ is B. On the other hand, the parabolic subgroup $P_{\mathrm{opt}}(\mathscr{A}_n, \mathcal{S}_n, C_n)$ for $K = H_2$ constructed as in the proof of Theorem 8.1.1 is P_β (see Exercise 8.12).

8.9 Interpretation in Terms of the Strong Tits Centre Conjecture

In this section we discuss the relationship between our optimality results, the Tits Centre Theorem (Theorem 7.4.1) for the simplicial building Δ_G and the Strong Tits Centre Conjecture (Conjecture 7.8.3) for the vector building $V_G(\mathbb{Q})$. In the process we consider the question of whether the objects yielded by these constructions are unique.

First we consider Δ_G. Suppose G is connected. Let $\mathrm{Aut}_0(\Delta_G)$ be the group of automorphisms of Δ_G that arise from elements of G. It is clear from our formulation of Theorem 8.1.1 that if H is a non-G-cr subgroup of G then $\mathcal{P}(H)$ gives an $\mathrm{Aut}_0(\Delta_G)$-centre of $(\Delta_G)^H$. The Tits Centre Theorem tells us that $(\Delta_G)^H$ has a centre σ_P. It does not, however, guarantee that P yields a semisimplification of H, or even that P is a witness that H is not G-cr; recall that the latter condition is equivalent to saying that σ_P is unopposed in $(\Delta_G)^H$ (see Remark 7.4.5). The construction of $\mathcal{P}(H)$ via optimality yields a centre with extra properties.

Similar remarks apply to the vector building $V_G(\mathbb{Q})$ (without the assumption that G is connected). Let $\mathrm{Aut}_0(V_G(\mathbb{K}))$ be the group of automorphisms of $V_G(\mathbb{K})$ that come from the action of G. The point ζ_{opt} from the optimality construction in the proof of Theorem 8.1.1 is an $\mathrm{Aut}_0(V_G(\mathbb{K}))$-centre of $V_G(\mathbb{Q})^H$ (recall from Lemma 7.5.2 that H fixes $\eta \in V_G(\mathbb{Q})$ if and only if $H \subseteq P_\eta$). Because $H \subseteq P_{\zeta_{\mathrm{opt}}}$ but H is not contained in any R-Levi subgroup of $P_{\zeta_{\mathrm{opt}}}$, ζ_{opt} is an unopposed $\mathrm{Aut}_0(V_G(\mathbb{K}))$-centre of $V_G(\mathbb{Q})^H$.

These results suggest that one can use geometric invariant theory as a proxy for the Tits Centre Theorem 7.4.1/Strong Tits Centre Conjecture 7.8.3, or as a way to prove special cases. On the other hand, the Tits Centre Theorem can sometimes be used as a proxy when the optimality theory breaks down: for instance, over a non-algebraically closed field (see Sect. 11.6 and the proof of Theorem 11.7.4). These ideas are taken further in [21]. Here is one natural question which arises.

Open Problem 8.9.1 Let $C \subseteq V_G(\mathbb{Q})$ be a polyhedral cone of finite type. Is it the case that $C = N(\mu)$ for some stately function $\mu\colon V_G(\mathbb{Q}) \to \mathbb{Q}$?

The answer is yes by Proposition 8.6.11 if C is a destabilising locus. If the answer in general is yes then, given a polyhedral cone C of finite type, we can find an unopposed $\mathrm{Aut}_0(V_G(\mathbb{K}))$-centre of C by optimising the associated function f over C, using Theorem 8.5.5. This would prove a version of the Strong Tits Centre Conjecture for such cones.

We chose $\mathcal{P}(H)$ using a construction involving optimality. There may be more than one candidate for $\mathcal{P}(H)$ but this is not important for the applications of Theorem 8.1.1: any choice of $\mathcal{P}(H)$ satisfying the properties in the theorem will do. We do not know whether the subgroup $P_{\mathrm{opt}} = P_{\mathrm{opt}}(\mathcal{A}_n, \mathcal{S}_n, C_n, \|\cdot\|)$ that appears in our proof of Theorem 8.1.1 depends on the choice of n and $\|\cdot\|$ (cf. Examples 8.7.1 and 8.7.2).

Here is an example showing that $\mathcal{P}(H)$ need not be unique.

Example 8.9.2 Let G be simple of type B_2 and let $X = G$ with the conjugation action of G. Fix a Borel subgroup B of G and a maximal torus T of B; then B has roots $\alpha, \beta, \alpha + \beta, 2\alpha + \beta$, where α is short and β is long. Let $H = U_{\alpha+\beta}U_{2\alpha+\beta}$. Let $\mathcal{S}$ be the set of parabolic subgroups of G that contain H and set $\Gamma = N_G(\mathcal{S})$. Then $B \subseteq \Gamma$ because B normalises H.

We claim that $\Gamma = N_G(H) = B$. If $\Gamma \neq B$ then $\Gamma = P_\alpha$ or $\Gamma = P_\beta$ or $\Gamma = G$. Choose representatives n_α, n_β of the simple reflections s_α, s_β in the Weyl group of G, respectively. To establish the claim, it is enough to show that $n_\sigma, n_\beta \notin \Gamma$. Now $n_\beta B n_\beta^{-1}$ has roots $\alpha+\beta$, $-\beta, \alpha, 2\alpha + \beta$, so $n_\beta B n_\beta^{-1} \in \mathcal{S}$; but $n_\alpha(n_\beta B n_\beta^{-1})n_\alpha^{-1}$ has roots $\alpha + \beta$, $-2\alpha - \beta$, $-\alpha$, β, so $n_\alpha(n_\beta B n_\beta^{-1})n_\alpha^{-1} \notin \mathcal{S}$. Hence s_α does not stabilise $\mathcal{S}$. Likewise, $n_\alpha B n_\alpha^{-1}$ has roots $-\alpha, 2\alpha + \beta, \alpha + \beta, \beta$, so $n_\alpha B n_\alpha^{-1} \in \mathcal{S}$; but $n_\beta(n_\alpha B n_\alpha^{-1})n_\beta^{-1}$ has roots $-\alpha - \beta$, $2\alpha + \beta, \alpha, -\beta$, so $n_\beta(n_\alpha B n_\alpha^{-1})n_\beta^{-1} \notin \mathcal{S}$. Hence s_β does not stabilise $\mathcal{S}$. We deduce that $\Gamma = N_G(H) = B$, as claimed.

It follows that each of B, P_α and P_β satisfies the properties of $\mathcal{P}(H)$, so each of them gives an $\mathrm{Aut}_0(\Delta_G)$-centre for $(\Delta_G)^H$. It is clear that $P_{\mathrm{BT}}(H) = B$. On the other hand, $P_{\mathrm{opt}}(\mathcal{A}_n, \mathcal{S}_n, C_n)$ is P_α (see Exercise 8.15).

8.10 Historical Remarks and References

The existence of an optimal destabilising cocharacter was proved by Kempf [89], answering a question raised by Mumford and Fogarty (see the discussion in [130, Appendix to Ch. 2, part C]). A similar construction was discovered independently by Rousseau [150]. Shortly afterwards Hesselink developed some of these notions further [75].

Our presentation is based on that of Kempf, but incorporates some ideas of Hesselink. Kempf considers $\mathcal{S}$-instability, but only for a single point x in a G-variety X. Hesselink considers uniform $\mathcal{S}$-instability for an arbitrary subset $\mathcal{A}$ of X, but only[4] when $\mathcal{S} = \{y\}$ for some G-fixed $y \in X$. We have combined these two notions in Sect. 8.6, following the approach of [28, §4]. The proof of our main theorem Theorem 8.6.15 follows the same

---

[4] Kempf's first draft of [89] deals with the special case when X is a G-module and $\mathcal{S} = \{0\}$ (see [88]). Hesselink (private communication) explains that he learnt of this earlier work from a talk given by Kempf in 1977, and when he wrote his paper [75] he was not aware of the extension to more general $\mathcal{S}$ in the final version of [89].

lines as Kempf's proof of his [89, Thm. 4.2], but with the addition of uniform instability. Our notion of a stately function is essentially the numerical function associated to an admissible state, in the language of states and their numerical functions: see [89, §2]. In our language, Kempf takes two stately functions μ and ν, and optimises μ on $N(\nu)$, although he does not work explicitly with subsets of $V_G(\mathbb{K})$. The authors observed in [21] that only one of the two states needs to be admissible; the other need only satisfy a weaker condition we termed quasi-admissibility. We have captured this generalisation here by carrying out our optimisation over subsets of $V_G(\mathbb{K})$.

The notion of cone for $Y_G(\mathbb{K})$ rather than $V_G(\mathbb{K})$ was introduced in [21, Def. 2.4]; our cones correspond to "saturated cones" in the language of that paper.

The construction of $P_{\mathrm{opt}}(\mathcal{A}, \mathcal{S})$ in the context of uniform $\mathcal{S}$-instability for arbitrary $\mathcal{A}$ and $\mathcal{S}$ was carried out in [28, §4]. Theorem 8.6.15 is a generalisation of [28, Thm. 4.5] (for $k = \overline{k}$): the former is the special case $C = \mathscr{D}_{\mathcal{A}}(\mathbb{Q})$ and asserts only that $N_G(\mathcal{A})$ normalises $P_{\lambda_{\mathcal{A}}}$, not that it is contained in $P_{\lambda_{\mathcal{A}}}$. Theorem 8.1.1 is a slight strengthening of [28, Thm. 5.16].

Hesselink considers the dependence of P_{opt} on the choice of length function. He proves that under suitable hypotheses on the G-variety X, P_{opt} does not depend on the choice of $\| \cdot \|$ [75, (7.2) Thm.]. His result, however, requires $\mathcal{S}$ to consist of a single point, so we cannot apply it to the subgroup $P_{\mathrm{opt}}(\mathcal{A}_n, \mathcal{S}_n, C_n)$ constructed in the proof of Theorem 8.1.1.

The question of whether or not an optimal destabilising R-parabolic subgroup of G yields a centre for the corresponding convex cone is discussed in [21, §5].

Example 8.9.2 is a slight variation on [75, Ex. 8.5].

There is a version of Theorem 8.5.5 for $\mathbb{K} = \mathbb{R}$. One obtains an optimal ray $\mathcal{R}$ in C, but we do not get an analogue of λ_{opt} in general because $\mathcal{R}$ need not contain any elements of V_G. This variation is useful if one wishes to study the Strong Tits Centre Conjecture for $V_G(\mathbb{R})$ rather than $V_G(\mathbb{Q})$.

Kempf, Rousseau and Hesselink proved some optimality results over a non-algebraically closed field k; we discuss these in Sect. 11.6.

The idea of proving cases of the Strong Tits Centre Conjecture using geometric invariant theory is pursued in [21], where we proved the existence of a centre in some special cases within a framework which is similar to that in this chapter. Those ideas have since matured into the approach we present here. Rousseau used a version of the Bruhat-Tits fixed point theorem (see Remark 7.4.3) to prove his version of the optimality results of this chapter: see [150, Prop. 2.3, Thm. 3.4, Prop. 5.1]. The essential idea of Rousseau's approach is to find a closed convex subset of $V_G(\mathbb{R})$ which is contained in the destabilising locus, does not contain 0, and is stabilised by the same automorphisms as the destabilising locus. One can then apply the Bruhat-Tits theorem to find a centre for this closed subset, which in turn gives rise to an optimal destabilising cocharacter. As mentioned in Sect. 7.10, there are also "metric geometry" versions of these results in some very special cases, see [6] for example.

8.11 Exercises

Exercise 8.1 Let V be a rational G-module and let $v \in V$. Let T be a maximal torus of G and $\lambda \in Y_T$. Show that (8.2.4) holds for every $\lambda \in Y_T$.

Exercise 8.2 Let E be a finite-dimensional Euclidean space and let $v, w \in E$ such that $\|v + w\| = \|v\| = \|w\|$. Show that $w = av$ or $v = aw$ for some $a \geq 0$. Deduce that if $\|v\| = \|w\| = 1$ then $v = w$.

Exercise 8.3 Show that the condition in Definition 8.4.2 does not depend on the choice of maximal torus T_0.

Exercise 8.4 Show that the condition in Definition 8.5.1(ii) does not depend on the choice of maximal torus T_0.

Exercise 8.5 Let C be a polyhedral cone of finite type in $V_G(\mathbb{Q})$ and let $\mu : C \to \mathbb{K}$ be stately. Let $g \in G$. Show that $g \cdot C$ is a polyhedral cone of finite type and $g \cdot \mu$ is stately.

Exercise 8.6 Prove the following uniform version of Theorem 3.3.6. Let X be a G-variety and let $\varnothing \neq \mathcal{A} \subseteq X$. Let $\lambda \in Y_G$. Suppose there exists $g \in G$ such that $\lim_{a \to 0} \lambda(a) \cdot x = g \cdot x$ for all $x \in \mathcal{A}$. Show that there exists $u \in R_u(P_\lambda)$ such that $\lim_{a \to 0} \lambda(a) \cdot x = u \cdot x$ for all $x \in \mathcal{A}$.

[*Hint*: Let G act diagonally on X^n and consider $\lim_{a \to 0} \lambda(a) \cdot (x_1, \ldots, x_n)$ for $(x_1, \ldots, x_n) \in \mathcal{A}^n$.]

Exercise 8.7 Suppose G is connected. Let X be G with the conjugation action and let H be a connected reductive subgroup of G. Show that there does not exist a closed G-stable subset S of X such that H is properly uniformly S-stable.

Exercise 8.8 Let $\mathcal{A} \subseteq X$. Show that $\mathscr{D}_{\mathcal{A}}(\mathbb{Q})$ is $V_G(\mathbb{Q})$-cr if and only if $\mathcal{A}$ is not properly uniformly unstable.

Exercise 8.9 Let $r, t \in \mathbb{N}$ with $r \leq t$ and let $n_1, \ldots, n_t \in \mathbb{N}$. Define $f : k^* \to k^r$ by $f(a) = (f_1(a), \ldots, f_r(a))$, where $f_i(a) = a^{n_i}$ for $1 \leq i \leq t$ and $f_i(a) = 0$ for $t < i \leq r$. Show that $f^{-1}(0, \ldots, 0)$ has coordinate ring $k[Z]/(Z^n)$, where $n := \min_{1 \leq i \leq t} n_i$.

Exercise 8.10 Let X be a G-variety, let $x \in X$ such that $G \cdot x$ is not closed, and let S be the unique closed G-orbit contained in $\overline{G \cdot x}$. Set $\mathcal{A} = \{x\}$ and $C = \mathcal{D}_x(\mathbb{Q})$. Let K be a G-completely reducible subgroup of G_x. Show that $C_G(K)$ and $N_G(K)$ are optimal subgroups in the sense of Remark 8.6.20.

Exercise 8.11 With the notation as in Example 8.6.22, find $\lambda_{\mathrm{opt}}(\mathcal{A})$ for some other subsets $\mathcal{A}$ of U. Find an example where $P_{\mathrm{opt}}(\mathcal{A})$ is not minimal among the parabolic subgroups P such that $\mathcal{A} \subseteq R_u(P)$. Find an example where $P_{\mathrm{opt}}(\mathcal{A})$ is not maximal among the parabolic subgroups P such that $\mathcal{A} \subseteq R_u(P)$.

Exercise 8.12 Show that $P_{\mathrm{opt}}(\mathcal{A}_n, \mathcal{S}_n, C_n) = P_\beta$ in Example 8.8.4.
   [*Hint*: Adapt the arguments of Example 8.6.22/Exercise 8.11.]

Exercise 8.13 Suppose that G is connected, and let P be a parabolic subgroup of G. Show that P itself is the unique parabolic subgroup of G that satisfies the properties of $\mathcal{P}(P)$ from Theorem 8.1.1. Give an example which shows the same is not necessarily true if we take G non-connected and P an R-parabolic subgroup of G.

Exercise 8.14 Let $X = G$ with G acting by conjugation and let $\mathcal{S} = \{1\}$. Let $\mathcal{A} = R_u(P)$, where P is a parabolic subgroup of G^0. Let $C = \mathscr{D}_{\mathcal{A}}(\mathbb{Q})$. Show that $P_{\mathrm{opt}}(\mathcal{A}) = N_G(P)$.

Exercise 8.15 Show that $P_{\mathrm{opt}}(\mathcal{A}_n, \mathcal{S}_n, C_n) = P_\alpha$ in Example 8.9.2.
   [*Hint*: Use the method of Example 8.6.22/Exercise 8.11/Exercise 8.12. The unique minimal weight set is $\{\alpha + \beta, 2\alpha + \beta\}$. Write λ_{opt} as $r\alpha^\vee + s\beta^\vee$. One can construct a W-invariant inner product $(\cdot\,,\cdot)$ on $X_T(\mathbb{Q})$ as in [79, Appendix]. Identifying $\alpha^\vee$ with $\frac{2\alpha}{(\alpha,\alpha)}$ and $\beta^\vee$ with $\frac{2\beta}{(\beta,\beta)}$ allows us to port $(\cdot\,,\cdot)$ over to $Y_T(\mathbb{Q})$ and obtain a length function.]

Applications to G-Complete Reducibility

9

In this chapter we collect some immediate applications of the optimality theory of the previous chapter in the context of G-complete reducibility, including some significant generalisations of earlier results from this book. A theme of the chapter is that the optimality constructions allow proofs which are very short and uniform (free of any case checks), demonstrating the power of the geometric approach.

The chapter begins with our generalised version of the "Clifford Theory" result Corollary 7.4.4 (Theorem 9.1.1), which historically was the result that motivated much of the geometric approach; see Sect. 9.5. The proof is a completely straightforward application of the methods of Chap. 8. We give various companion results on G-complete reducibility for normal subgroups (Sect. 9.1), centralisers and normalisers (Sect. 9.2), complements to a normal subgroup (Sect. 9.3), and semisimplification (Sect. 9.4), extending and generalising many results from earlier in the book, especially Chaps. 4 and 5.

9.1 Normal Subgroups of G-Completely Reducible Subgroups

Here is a more general version of Corollary 7.4.4. Recall that this was proved above using the Tits Centre Theorem 7.4.1, which has a case-by-case proof. Further, the proof only works as given for connected G. Our proof of the following is completely free of case checks and works equally well in the non-connected case.

Theorem 9.1.1 *Let H be a subgroup of G with normal subgroup N. If H is G-completely reducible, then so is N. In particular, if H is G-completely reducible, then so is H^0.*

Proof Suppose N is not G-cr. Let $P = \mathcal{P}(N)$ be the R-parabolic subgroup from Theorem 8.1.1. Since H normalises N, H is contained in P. Since N is not contained

© The Author(s), under exclusive license to Springer Nature Switzerland AG 2026
M. Bate et al., *G-Complete Reducibility, Geometric Invariant Theory and Spherical Buildings*, Oberwolfach Seminars 57, https://doi.org/10.1007/978-3-032-08866-6_9

in any R-Levi subgroup of P, H cannot be contained in any R-Levi subgroup of P either. Hence H is not G-cr. $\qquad\square$

We can immediately generalise Corollary 5.4.12. Recall that a subgroup of G is called *regular* if it is normalised by a maximal torus of G.

Proposition 9.1.2 *Let H be a regular reductive subgroup of G. Then H is G-completely reducible.*

Proof Let T be a maximal torus of G normalising H. Then HT is G-cr by Corollary 5.4.12, and hence the normal subgroup H of HT is G-cr by Theorem 9.1.1. $\qquad\square$

Remark 9.1.3 Serre proves a converse to Theorem 9.1.1 in [158, Property 5] under the assumption that the index of N in H is finite and coprime to $\operatorname{char} k = p$. Examples show that this restriction cannot be removed. For instance, let U be a non-trivial finite unipotent subgroup of G^0. Then, thanks to Theorem 2.9.23, there exists an R-parabolic subgroup P of G such that $U \subseteq R_u(P)$. In particular, U is not G-cr, but clearly $U^0 = \{1\}$ is.

Our optimality results allow us to give a significant strengthening of Serre's result from the previous remark, allowing for the quotient H/N to be linearly reductive. Before giving the proof we collect some preliminary results. We begin with the following special case of Theorem 5.4.10(ii).

Corollary 9.1.4 *Suppose $N \subseteq H$ are subgroups of G, N is normal in H and N is G-completely reducible. Then H is G-completely reducible if and only if H is $N_G(N)$-completely reducible.*

Proof This follows from Theorem 5.4.10(ii), setting $K = N$ and $M = N_G(N)$. Note that since N is G-cr, $N_G(N)$ is reductive by Theorem 5.4.1. $\qquad\square$

Proposition 9.1.5 *Suppose $N \subseteq H$ are subgroups of G and N is normal in G. Then H is G-completely reducible if and only if H/N is G/N-completely reducible.*

Proof The forward implication follows from Lemma 4.2.7(ii)(a). For the reverse implication, let $\pi : G \to G/N$ be the canonical projection. Suppose P is an R-parabolic subgroup of G containing H. Then $\pi(P)$ is an R-parabolic subgroup of G/N containing H/N by Lemma 2.9.21(i). Since H/N is G/N-cr, there is a Levi subgroup M of $\pi(P)$ such that $H/N \subseteq M$. Then $H \subseteq \pi^{-1}(H/N) \subseteq \pi^{-1}(M)$, and $\pi^{-1}(M)$ is an R-Levi subgroup of $\pi^{-1}(\pi(P))$ by Remark 2.9.22. But $\pi^{-1}(\pi(P)) = P$ since $P \supseteq N$. This shows that H is G-cr. $\qquad\square$

We can now prove the most general result relating complete reducibility of H to complete reducibility of H/N inside a suitable quotient group, when N is normal in H.

Theorem 9.1.6 *Suppose that $N \subseteq H$ are subgroups of G with N normal in H.*

(i) If N is G-completely reducible, then H is G-completely reducible if and only if H/N is $N_G(N)/N$-completely reducible.

(ii) If H/N is linearly reductive, then H is G-completely reducible if and only if N is G-completely reducible.

Proof (i). By Corollary 9.1.4, H is G-cr if and only if H is $N_G(N)$-cr. By Proposition 9.1.5 applied to the inclusions $N \subseteq H \subseteq N_G(N)$, it follows that H is $N_G(N)$-cr if and only if H/N is $N_G(N)/N$-cr. This proves the result.

(ii). Suppose H/N is linearly reductive. If N is G-cr, then $N_G(N)$ is a reductive group by Theorem 5.4.1 and so is the quotient $N_G(N)/N$. Thus H/N is a linearly reductive subgroup of $N_G(N)/N$. By Lemma 4.2.1, H/N is $N_G(N)/N$-cr, and hence H is G-cr, by part (i). On the other hand, if H is G-cr, then N is G-cr, by Theorem 9.1.1.

$\square$

Remark 9.1.7   (i). Note that the finite linearly reductive groups are those whose orders are coprime to $p = \mathrm{char}k$, so we obtain Serre's result Remark 9.1.3 as a special case of part (ii).

(ii). It is easy to construct examples where H and N are both G-cr, even when H/N is not linearly reductive. For example, we could just take $G = M \times N$ to be the direct product of two reductive groups. Then part (i) of the corollary implies that a subgroup of the form $H = K \times N$ is G-cr if and only if K is M-cr, and in this case $H/N \cong K$.

(iii). For a naturally occurring example of the previous phenomenon where the index of N in H is finite, let N be a maximal torus in $G = \mathrm{GL}_m$ and let $H = N_G(N)$. Then N and H are both G-cr (e.g., by Corollary 5.4.12), but the index of N in H is $m!$, which is divisible by p as long as $m \geq p$.

9.2   Centralisers and Normalisers Revisited

We revisit the question of how complete reducibility interacts with centralisers and normalisers. With the optimality results at our disposal, we can improve some of our earlier results. For earlier context for the next two results, see Remarks 5.4.11(ii) and (iii).

Corollary 9.2.1 *Let H be a subgroup of G. Then H is G-completely reducible if and only if $N_G(H)$ is G-completely reducible.*

Proof The forward implication is in Remark 5.4.11(ii), and the reverse follows from Theorem 9.1.1. □

Corollary 9.2.2 *Let H be a subgroup of G. If H is G-completely reducible, then so is $C_G(H)$.*

Proof Since $C_G(H)$ is normal in $N_G(H)$, the result follows from Theorem 9.1.1 and Corollary 9.2.1. □

Remark 9.2.3 (i). Observe that in the case $G = \mathrm{GL}(V)$, Corollary 9.2.2 is just a consequence of Wedderburn's Theorem. For if H is a subgroup of $\mathrm{GL}(V)$, then $C_G(H)$ consists of the invertible elements of the endomorphism ring $\mathrm{End}_H(V)$. Then if H acts semisimply on V we see that $\mathrm{End}_H(V)$ is isomorphic to a direct product of matrix rings by Wedderburn's Theorem, and this direct product also acts semisimply on V, so $C_G(H)$ does.

(ii). The converse of Corollary 9.2.2 is not true. For an easy example, let G be connected reductive and not a torus, and let H be a Borel subgroup of G. Then $C_G(H) = Z(G)$ is G-cr, but H is not.

Next we obtain a generalisation of Theorem 5.4.4, which is the special case when S is a subtorus of G.

Corollary 9.2.4 *Let S be a linearly reductive group acting on the reductive group G by automorphisms and let $M = C_G(S)$. Then M is reductive, and for any subgroup H of M, H is M-completely reducible if and only if H is G-completely reducible.*

Proof Let $\widetilde{G} = S \ltimes G$. Then $\widetilde{G}$ is reductive, so $C_{\widetilde{G}}(S)$ is reductive by Proposition 2.4.9. Hence the normal subgroup $M = G \cap C_{\widetilde{G}}(S)$ is also reductive. Clearly $C_{\widetilde{G}}(S) = Z(S) \times M$. Denote this group by $\widetilde{M}$. Since G is normal in $\widetilde{G}$, and M is normal in $\widetilde{M}$, H is G-cr if and only if H is $\widetilde{G}$-cr, and H is M-cr if and only if H is $\widetilde{M}$-cr, by Proposition 4.2.9. We may therefore replace G with $\widetilde{G}$ and M with $\widetilde{M}$ to reduce to the case that S is a subgroup of G.

Let $\mathbf{h}$ be a generic tuple for H. Then H is G-cr (resp., M-cr) if and only if $G \cdot \mathbf{h}$ (resp., $M \cdot \mathbf{h}$) is closed, by Theorem 5.3.3. The result now follows by ascent and descent for closed orbits (Corollaries 8.6.25 and 3.3.7). □

Remark 9.2.5 If we allow S to be a non-smooth linearly reductive group scheme, then Corollary 9.2.4 can fail; see Exercise 9.4.

Corollary 9.2.4 provides situations in which both Questions 4.1.8 and 4.1.9 have positive answers. Here is a very basic example of how it can be applied.

Example 9.2.6 Suppose that $p \neq 2$. Let V be a finite-dimensional k-vector space, and let $f : V \times V \to k$ be a non-degenerate symmetric or alternating form on V. Let $G = GL(V)$. Then for each $g \in G$ there is a unique element $g^* \in G$ such that $f(gv, w) = f(v, g^*w)$ for all $v, w \in V$. The map $g \mapsto g^*$ defines an automorphism of G of order 2; let S be the corresponding group of order 2 acting on G. Then S is linearly reductive, since $p \neq 2$, so Corollary 9.2.4 applies. Let $M = C_G(S)$; so $M \cong O(V)$ when f is symmetric and $M \cong Sp(V)$ when f is alternating. We conclude that for any subgroup H of M, H is M-completely reducible if and only if it is G-completely reducible, that is, if and only if V is a semisimple H-module; for a direct proof of this fact, see Exercise 4.11. Note also that if it happens that $H \subseteq SO(V)$ then we get the same conclusion with $M = SO(V)$, e.g., by the above argument and an application of Proposition 4.2.9.

Note that Example 4.1.10 shows the equivalence from the previous example does *not* hold in general when $p = 2$; see also see Exercise 4.10. We return to these ideas in Sect. 10.4 below where we look at some specific examples and applications and discuss how such characteristic restrictions naturally occur.

9.3   Normal Subgroups and their Complements

Theorem 9.1.6(ii) above provides a useful criterion to ensure that G-complete reducibility of H and of N are equivalent. But there are many examples where H and N are G-completely reducible, but H/N is not linearly reductive. In this section we briefly consider what can be said when there is a complement (or partial complement) to the normal subgroup N of H; that is, we have a first look at the following question.

Question 9.3.1 Let H, N be subgroups of G with N normal in H. Let M be any subgroup of H such that $MN = H$. Is it true that H is G-completely reducible if and only if M and N are G-completely reducible?

If H is G-cr and M is also normal in H, then M and N are G-cr, by Theorem 9.1.1. When M is not assumed to be normal in H, it is easy to construct examples where H is G-cr but M is not: we can just take $G = H = N$ and M to be a non-G-cr subgroup of G. The next example shows that even when M is a complement to N in H it need not be G-cr; see also Example 9.3.5 below. (The problem in both these examples is that N can fail to normalise M.)

Example 9.3.2 Let $p = 2$, let H be the symmetric group S_3 embedded irreducibly in $G = GL_2$ and let N be the subgroup of H of order 3. Any subgroup M of H of order 2 is a complement to N, but such a subgroup, being a unipotent subgroup of G^0, cannot be G-completely reducible by Theorem 2.9.23.

The other implication of Question 9.3.1 is also problematic in general. Even in the best possible case when M and N are connected disjoint commuting subgroups of G, so that M is a complement to N and M is normal in H, the answer is no. It turns out that this is a phenomenon which only really occurs in low characteristic, so we delay further discussion of this until the next chapter (see Sect. 10.6). For now, we finish this section by giving two positive results and a further example, all of which follow quickly from what we have already seen.

Proposition 9.3.3 *Suppose N is a G-completely reducible subgroup of G and M is a subgroup of $C_G(N)$. Then M is $C_G(N)$-completely reducible if and only if MN is G-completely reducible.*

Proof Suppose MN is not G-cr and let P be a witness to the non-G-complete reducibility of MN. Then $N \subseteq P$ and N is G-cr, so there is $\lambda \in Y_{C_G(N)}$ with $P = P_\lambda$. Since MN is not contained in any R-Levi subgroup of P, we cannot have M contained in any R-Levi subgroup of $P_\lambda(C_G(N))$, and hence M is not $C_G(N)$-cr.

For the converse, suppose MN is G-cr, and Q is some R-parabolic subgroup of $C_G(N)$ containing M. Writing $Q = P_\lambda(C_G(N))$ for some $\lambda \in Y_{C_G(N)}$, we have $MN \subseteq P_\lambda$ and hence $MN \subseteq L_\mu$ for some μ with $P_\lambda = P_\mu$ (because MN is G-cr). Since μ centralises MN, it centralises N, and we have found $\mu \in Y_{C_G(N)}$ with $P_\lambda(C_G(N)) = P_\mu(C_G(N))$ and $M \subseteq L_\mu(C_G(N))$, so M is $C_G(N)$-cr. $\qquad\square$

Corollary 9.3.4 *Suppose N is G-completely reducible and M is a subgroup of $C_G(N)$. If M is $C_G(N)$-completely reducible, then M is G-completely reducible.*

Proof Under these hypotheses, MN is G-cr, by Proposition 9.3.3. But M is a normal subgroup of MN, so M is G-cr, by Theorem 9.1.1. $\qquad\square$

In Sect. 10.6 we show that the converse of Corollary 9.3.4 is not true in general, but it is true in many important special cases, notably those covered in Theorem 10.6.1 below.

Example 9.3.5 Recall Example 4.1.10. We have $p = 2$ and $m \geq 4$ is even. Let H be the maximal rank subgroup $\mathrm{Sp}_m \times \mathrm{Sp}_m$ of $G := \mathrm{Sp}_{2m}$, and let M be Sp_m embedded diagonally in H. Then H is G-cr but M is not. Now let $N = \{1\} \times \mathrm{Sp}_m$ and $M' = \mathrm{Sp}_m \times \{1\}$. Then N and M' are G-cr by Theorem 9.1.1, since they are normal in H. Note that M and M' are both complements to N in H, so we have two different complements, one being G-cr and the other not.

It is also easy to check in this situation that $H = N_G(N)^0$, and it follows from Proposition 4.2.12 and Lemma 4.2.7(i) that M is $N_G(N)$-cr. This shows that Corollary 9.3.4 is false if we replace $C_G(N)$ with $N_G(N)$.

9.4 Semisimplification and Normal Subgroups

Recall Definition 5.5.1 of the semisimplification of a subgroup H of G: given H, we say a pair (P, L) yields the semisimplification H' of H if P is an R-parabolic subgroup of G containing H, L is an R-Levi subgroup of P, and $H' = c_L(H)$ is G-cr. Recall also that any two semisimplifications of H are G-conjugate by Proposition 5.5.8. (By construction, if H is not already G-cr, the optimal destabilising R-parabolic subgroup $\mathcal{P}(H)$ of Theorem 8.1.1 gives a particularly natural choice for P.) Now suppose N is a normal subgroup of H. It is clear that $N' := c_L(N)$ is a normal subgroup of H', since the map $c_L : P \to L$ is a surjective homomorphism. But now we see that N' is also G-cr by Theorem 9.1.1. Thus (P, L) also yields a semisimplification of N, and we have shown:

Proposition 9.4.1 *Let H, N be subgroups of G with N normal in H. Then there exist semisimplifications N' of N and H' of H such that N' is normal in H'. In fact, for any pair (P, L) yielding a semisimplification H' of H, the group $N' := c_L(N)$ is a semisimplification of N which is normal in H'.*

Example 9.4.2 We noted above that if N is a normal subgroup of H and H is not G-cr, the optimal destabilising R-parabolic $\mathcal{P}(H)$ of Theorem 8.1.1 simultaneously semisimplifies H and N. One might hope that if N is not G-cr, then we could also take the optimal destabilising R-parabolic $\mathcal{P}(N)$ (or its normaliser) to semisimplify H and N; after all, one of the key properties from Theorem 8.1.1 implies that $\mathcal{P}(N)$ contains H, since $H \subseteq N_G(N)$. However, easy examples show that this will not work in general. Suppose M is any connected reductive group and P is a proper parabolic subgroup of M. Now set $G = M \times M$ and let $N = P \times 1$; it is easy to see that $\mathcal{P}(N) = P \times M$ using a slight variation on the argument of Exercise 8.13. Hence if we let K be *any* non-M-cr subgroup of M then the subgroup $H = P \times K$ of G contains N as a normal subgroup but is not semisimplified by $\mathcal{P}(N)$.

If $H \subseteq M \subseteq G$ with M reductive, then we can semisimplify H as a subgroup of G and as a subgroup of M. Since we have little control in general over the relationship between G-complete reducibility and M-complete reducibility, there is no reason to expect that these semisimplifications should be related. However, in one case we can say something positive.

Lemma 9.4.3 *Let H and M be subgroups of G with $H \subseteq M \subseteq G$ and M reductive. Suppose H is G-completely reducible but not M-completely reducible. Let H' be a* *semisimplification of H in M. Then H' is G-conjugate to H, and in particular is also a semisimplification of H in G.*

Proof Suppose $(P_\lambda(M), L_\lambda(M))$ is a pair yielding H', so that we can write $H' = c_\lambda(H)$ for some $\lambda \in Y_M$. Since H is G-cr, H' is G-conjugate to H by Corollary 5.3.4.   □

For examples of the set-up in the previous lemma, see Example 4.1.10 above and Example 10.4.5 below; note in particular that the non-M-cr subgroups in the second example are constructed as certain G-conjugates of an M-cr subgroup.

9.5   Historical Remarks and References

Theorem 9.1.1 was first proved for strongly reductive subgroups of G in [118, Thm. 2], and we deduced the version above in [18, Thm. 3.10] after proving that strong reductivity is equivalent to complete reducibility (see the historical remarks to Chap. 4, Sect. 4.4). The proof presented here is rather different, but the essential idea is the same: by using tuples, one can turn the problem into a geometric invariant theory question and then employ the machinery of optimal destabilising cocharacters. This result gave an affirmative answer to a question posed by Serre, [158, p. 24]. We repeat that historically [18, Thm. 3.10] precedes the proof of the Tits Centre Theorem 7.4.1. The success of these geometric methods in answering this question of Serre provided much of the motivation for the development of a lot of the tools and techniques outlined in this book; see particularly the sequence of papers [18], [26], [28], [17], [23] where there is extensive discussion of the "G-cr motivation" for more general geometric invariant theory results.

Most of the other results in this chapter can be found in [18] or [19]; we have presented several results in a slightly different way, and the systematic development of the previous chapters has allowed us to streamline and otherwise improve many of the proofs. Corollary 9.2.4 has been generalised in several ways. Here is one such result, which is [26, Cor. 5.8] (note that the hypotheses hold if S is linearly reductive): let S be an algebraic group acting on G by automorphisms such that S acts semisimply on $\mathfrak{g}$ and $C_{\mathfrak{g}}(S) = \mathrm{Lie}(C_G(S))$. Let $H \subseteq M$ be subgroups of G such that $M^0 = C_G(S)^0$ and S normalises H. Then M is reductive and H is G-completely reducible if and only if H is M-completely reducible. Proposition 9.4.1 was first observed in [23, Thm. 5.4], although in that paper we work over an arbitrary field as we do in Chap. 11 below.

9.6   Exercises

Exercise 9.1 For a subgroup H of G, let $W(H) = N_G(H)/H$.

(i) Show that if H is G-cr and $M = N_G(H)$, then $W(M)$ is finite.

(ii) Find an example of a reductive G and a reductive subgroup H of G such that $W(M)$ is infinite, where again $M = N_G(H)$.

Exercise 9.2 Let S be a linearly reductive subgroup of G and let H be a subgroup of $C_G(S)$. Show that H is G-cr if and only if HS is G-cr.

Exercise 9.3 Let S be a linearly reductive group acting on the reductive group G by automorphisms and let $M = C_G(S)$. Prove that if H is an M-cr subgroup of M then H is G-cr, without directly appealing to Corollary 8.6.25.

[*Hint*: Prove the contrapositive, using the properties of $\mathcal{P}(H)$.]

Exercise 9.4 If you know about group schemes, do the following: find an example of a non-smooth linearly reductive subgroup scheme S of some G and a (smooth) subgroup H of $C_G(S)$ such that H is $C_G(S)$-cr but not G-cr.

[*Hint*: Look at Exercise 4.10(iii).]

Exercise 9.5 Consider Liebeck's Example 4.1.10: so $p = 2$, $V = W \oplus W'$ is a direct sum of two isomorphic symplectic spaces W and W', $G = \mathrm{GL}(V)$, $M = \mathrm{Sp}(V)$, $K = \mathrm{Sp}(W) \times \mathrm{Sp}(W')$, and H is the diagonal embedding of $\mathrm{Sp}(W)$ in K. Recall that H is G-cr but not M-cr. Work out what the semisimplification of H in M is.

Large Versus Small Characteristic **10**

While some results for G-complete reducibility hold without any restrictions on the characteristic p of the underlying field, such as Lemma 4.1.6 and Theorem 9.1.1, for instance, others do require a bound on p. In this section we explore various results of the latter kind along with some counterexamples when p is small. As above, guiding questions for this chapter are Questions 4.1.8 and 4.1.9 on ascent and descent of complete reducibility. We make the following definition.

Definition 10.0.1 Let M be a reductive subgroup of G. We say the pair (G, M) has the *ascent property* if whenever H is an M-completely reducible subgroup of M, H is G-completely reducible. We say the pair (G, M) has the *descent property* if whenever H is a G-completely reducible subgroup of M, H is M-completely reducible.

A typical example of an M which satisfies both the ascent and descent property is an R-Levi subgroup of G (see Theorem 5.4.4). Basic examples of when the ascent property fails are when H is a non-G-cr subgroup of G; then we can take $H = M$. Examples of the failure of the descent property tend to be more subtle, but one which we use a lot in various contexts is Liebeck's Example 4.1.10. Another important example is introduced in the next section.

The chapter is laid out as follows. In Sect. 10.1 we introduce a key example which illustrates many of the issues which can arise in small characteristic. In Sects. 10.2 and 10.3 we give a summary of some of the early results of Serre concerning G-complete reducibility, as developed in his influential notes [158, 159]. In Sects. 10.4–10.6 we outline some of the work of the authors and others concerning ascent and descent of complete reducibility in various situations, including the important notion of a reductive pair due to Richardson. In Sect. 10.7 we briefly outline the basic ideas around Serre's notion of saturation for unipotent elements and subgroups of reductive groups.

© The Author(s), under exclusive license to Springer Nature Switzerland AG 2026

M. Bate et al., *G-Complete Reducibility, Geometric Invariant Theory and Spherical Buildings*, Oberwolfach Seminars 57, https://doi.org/10.1007/978-3-032-08866-6_10

10.1   An Instructive Example

In this section we introduce an example which serves to illustrate much of the pathological behaviour which can occur in small characteristic. This example was first described in [26, §7]; here we give the key definitions and properties but suppress most of the more involved (but straightforward) calculations, referring the reader to *loc. cit.* for the full details.

Let G be a simple algebraic group of type G_2 and suppose that $p = 2$. Fix a maximal torus T of G and a Borel subgroup B of G that contains T, and label the simple roots with respect to (B, T) as α and β, where α is short and β is long. The positive roots are α, β, $\alpha + \beta$, $2\alpha + \beta$, $3\alpha + \beta$ and $3\alpha + 2\beta$. For each root γ we fix an isomorphism κ_γ from $\mathbb{G}_a = (k, +)$ onto the root group U_γ in such a way that the equations in [77, 33.5] hold; these equations give explicit expressions for commutators $[\kappa_\gamma(a), \kappa_\delta(b)]$.

Given a root γ, we define $G_\gamma = \langle U_\gamma \cup U_{-\gamma} \rangle \cong \mathrm{SL}_2$ (the isomorphism with SL_2 follows since G is simply connected). Set $s_\gamma = \kappa_\gamma(1)\kappa_{-\gamma}(1)\kappa_\gamma(1) \in N_{G_\gamma}(T)$. Then $T_\gamma := T \cap G_\gamma$ is a maximal torus of G_γ and s_γ represents the non-trivial element of the Weyl group $W(G_\gamma, T_\gamma)$. Since $G_\gamma \cong \mathrm{SL}_2$, T_γ is the image of the coroot $\gamma^\vee$.

Let $M = G_\alpha G_{3\alpha+2\beta}$ (note that G_α and $G_{3\alpha+2\beta}$ commute with each other because $\alpha \pm (3\alpha + 2\beta)$ is not a root). Then $T \subseteq M$ and M is semisimple and G-ir. A quick calculation with the commutator relations shows that $C_G(M)$ is trivial and it is easy to check from this that $N_G(M) = M$. Further, if we let $0 \neq z \in \mathrm{Lie}(T_\alpha)$ be any nonzero element, we have $M = C_G(z)$.

Let $L = \langle G_\alpha \cup T \rangle$, a Levi subgroup of G, and let $P = \langle L \cup B \rangle$, the standard parabolic subgroup of G with L as its Levi subgroup. Then $R_u(P)$ is generated by the root groups corresponding to the positive roots with a nonzero contribution from β. If we choose $\lambda = \alpha^\vee + 2\beta^\vee \in Y_T$, then we can check that $P = P_\lambda$ and $L = L_\lambda$. Note that $L \subseteq M$, so L is also a Levi subgroup of M, and $P \cap M = P_\lambda(M)$ is the corresponding parabolic subgroup of M. Choose an element $t \in T_\alpha$ of order 3 and set $H := \langle t, s_\alpha \rangle \subseteq G_\alpha$. It is easily checked that H is not contained in any Borel subgroup of L, so H is L-ir, and thus H is G-cr and M-cr.

Define $u(a) = \kappa_\beta(a)\kappa_{3\alpha+\beta}(a) \in R_u(P)$ for $a \in k$, and for each such a let

$$H_a := u(a)Hu(a)^{-1} = \langle t, s_\alpha\kappa_{3\alpha+2\beta}(a^2) \rangle \subseteq M.$$

The second equality here uses the fact that $t \in T_\alpha$, so commutes with $u(a)$, and also the following calculation using the commutator relations:

$$u(a)s_\alpha u(a)^{-1} = \kappa_\beta(a)\kappa_{3\alpha+\beta}(a)\kappa_\alpha(1)\kappa_{-\alpha}(1)\kappa_\alpha(1)\kappa_\beta(a)\kappa_{3\alpha+\beta}(a)$$

$$= s_\alpha\kappa_{3\alpha+2\beta}(a^2). \tag{10.1.1}$$

Since $u(a) \in R_u(P)$ and $H \subseteq L$ we have $H_a \subseteq P$ for each $a \in k$. Note that $H_0 = H$ and we can see that $c_L(H_a) = c_L(u(a)Hu(a)^{-1}) = H$ for every $a \in k$. Moreover, $C_G(H) = G_{3\alpha+2\beta}$ and $N_G(H) = HG_{3\alpha+2\beta} \subseteq M$.

Lemma 10.1.1 *For any* $0 \neq a \in k$, $H_a \subseteq M$ *is G-completely reducible but not M-completely reducible.*

Proof Since H_a is G-conjugate to H and H is G-cr, it follows that H_a is G-cr. We have observed that $H_a \subseteq P_\lambda(M) = P \cap M$ and we know that $c_L(H_a) = H$. Since H is M-cr, to show that H_a is not M-cr it therefore suffices to show that H_a is *not* M-conjugate to H, by Corollary 5.3.4. But $u(a) \notin M$ and $N_G(H) \subseteq M$, so H and H_a are not M-conjugate and we're done. □

Remark 10.1.2 We see that the pair (G, M) does not have the descent property, giving a negative answer to Question 4.1.9. Note that in this case the subgroup $M = C_G(z)$ is the centraliser of a semisimple element of $\mathrm{Lie}(G)$, giving another example for the failure of Corollary 9.2.4 when the linearly reductive subgroup of G is replaced with a subalgebra of $\mathrm{Lie}(G)$ spanned by semisimple elements; see also Exercise 9.4.

We return to this example several times below.

10.2   *G*-Complete Reducibility and Reductivity

Any G-completely reducible subgroup H of G is reductive by Lemma 4.1.6. The converse is true in characteristic 0, but fails in positive characteristic. For example, any non-commutative connected reductive group H in any positive characteristic has non-semisimple representations, and hence we can find a connected reductive but non-G-cr image of H inside some $G = \mathrm{GL}_m$. Further, once we have such an example in GL_m, then we can find one in GL_r for all $r \geq m$ by embedding GL_m in GL_r (say as the "top left factor" of a Levi subgroup). Thus if we fix p and let G vary, there is no hope that reductivity of subgroups can be equivalent to G-complete reducibility.

However, there is an alternative point of view: since connected reductive groups are classified by root data, it makes sense to fix a Dynkin type and let the *prime* vary. In this scenario, we might hope that reductive subgroups are "generically" G-cr in the sense that if the prime p is large enough then reductivity and G-complete reducibility should be equivalent. One result along these lines which has a very clean statement is due to Serre, [159, Thm. 4.4]. Given a connected reductive group G, define an invariant $a(G)$ of G, as follows: for G simple, set $a(G) = \mathrm{rk}(G) + 1$, where $\mathrm{rk}(G)$ is the rank of G. For connected reductive G with simple factors $G_1, \ldots, G_r$, let $a(G) = \max\{1, a(G_1), \ldots, a(G_r)\}$. Clearly if M is a connected reductive subgroup of G then $a(M) \leq a(G)$.

Theorem 10.2.1 *Suppose G is connected reductive with* $p \geq a(G)$. *If H is a connected subgroup of G then H is reductive if and only if H is G-completely reducible.*

Discussion of Proof Theorem 10.2.1 is really a collection of results known for the different Dynkin types. First one easily reduces the question to the case that G is simple: e.g., by applying Remark 4.2.8. Then the result in type A follows from a result of Jantzen: if H is a connected reductive group and V is an H-module with $p \geq \dim(V)$, then V is semisimple, [81, Prop. 3.2]. This result was extended to other classical types by McNinch [124] (with the appropriate changes to the bound on p). For exceptional groups, the proof relies on classification results of Liebeck and Seitz [99]: for almost all primes, they provide a table of conjugacy classes of semisimple subgroups for each exceptional group, and from these tables one can read off the required result.   □

Remark 10.2.2   (i). We can extend the result to non-connected H under the hypothesis that the index $[H : H^0]$ is coprime to p, using Theorem 9.1.1 and Remark 9.1.3.

(ii). If $M \subseteq G$ are connected reductive groups, then $a(G) \geq a(M)$. Hence if $p \geq a(G)$ and H is a connected subgroup of M then H is M-cr if and only if it is reductive if and only if it is G-cr. Thus when $p \geq a(G)$, every such pair (G, M) has both the ascent and descent property for connected subgroups.

We finish this section by noting that the tables of Liebeck and Seitz have continued to be expanded upon and improved in the years since [99] was published. By now, we have almost complete information about conjugacy classes of connected reductive subgroups of exceptional groups in all characteristics, and about which ones of these are G-cr; see Sect. 12.1 for further discussion of this.

10.3   *G*-Complete Reducibility and Semisimple Modules

In this section, we assume G is connected. We saw Jantzen's result [81, Prop. 3.2] in the discussion of the proof of Theorem 10.2.1: if H is a connected reductive group and V is a module for H with $p \geq \dim(V)$ then V is semisimple, so (the image of) H is GL(V)-cr. We briefly look at some results of a similar nature relating G-complete reducibility of subgroups to the semisimplicity of G-modules.

Let V be a finite-dimensional G-module. Let T be a maximal torus of G and $\Phi = \Phi(G, T)$ the roots of G. Let $\Phi^+ \subset \Phi$ be a system of positive roots. We define

$$n(V) = \max \left\{ \sum_{\alpha \in \Phi^+} \langle \lambda, \alpha^\vee \rangle \right\},$$

where the maximum is taken over all T-weights λ of V. On [158, p. 20], Serre observes that if V is *non-degenerate* (i.e., if the connected kernel of the representation is a torus), then $n(V) \geq h(G) - 1$. The following is the main objective in [158]; see also [159, Thm. 5.4].

Theorem 10.3.1 *Let V be a G-module with $p > n(V)$. Let H be a subgroup of G.*

(i) If H is G-completely reducible, then V is a semisimple H-module. In particular, V is a semisimple G-module.

(ii) If V is non-degenerate and semisimple as an H-module, then H is G-completely reducible.

The proof of part (i) of this theorem uses the notion of saturation, which we discuss in Sect. 10.7 below. We note also that the proof of part (ii) uses Theorem 10.2.1 above, so again relies on the case checking which goes into that theorem.

For the case that $V = \mathfrak{g}$ is the adjoint module (which is non-degenerate), note that the weights of V are just the roots, and so $n(V)$ is just twice the sum of the coefficients of the highest root, which is $2(h(G) - 1) = 2h(G) - 2$, where $h(G)$ is the Coxeter number of G (see the equivalent characterisations of $h(G)$ in Sect. 2.4). We get the following special case of the theorem, which is [159, Cor. 5.5].

Corollary 10.3.2 *Let H be a subgroup of G. Suppose that $p > 2h - 2$. Then $\mathfrak{g}$ is a semisimple H-module if and only if H is G-completely reducible.*

Remark 10.3.3   (i). It is a result of Serre that when $G = \mathrm{GL}(V)$ the forward implication in Corollary 10.3.2 does not require any restrictions on p, see [156, Thm. 3.3].

(ii). For a uniform proof of the forward implication under an extra hypothesis on H see also Theorem 10.5.5. Applying Theorem 10.5.13 then shows that for the forward implication to hold it suffices to assume that p is *pretty good* for G.

(iii). It is shown in [100, Cor. 3] that if G is simple of exceptional type and H is simple of rank at least 2, then it suffices to require that $p > 7$ for the reverse implication of Corollary 10.3.2 to hold. However, Serre observes that in many cases the bound in Corollary 10.3.2 is sharp for the reverse implication, [159, Rem. 5.6].

Here is a simple example of the failure of the conclusion of Theorem 10.3.1(i) if $p \leq n(V)$.

Example 10.3.4 Let $p > 2$. The adjoint representation of $H := \mathrm{PGL}_p(k)$ on $\mathfrak{h}$ has a composition factor of dimension $m := p^2 - 2$. So we can regard H as an SL_m-irreducible subgroup of $G := \mathrm{SL}_m$. Now $\mathfrak{g}$ is simple as a G-module, as p is coprime to m. However, $\mathfrak{g}$ is not semisimple as an H-module: for the H-submodule $\mathfrak{h}$ of $\mathfrak{g}$ is not semisimple. Note that $n(\mathfrak{g}) = 2m - 2$.

Liebeck's Example 4.1.10 above provides an example where Theorem 10.3.1(ii) fails without the restriction on p: in that case we have a subgroup H of $G = \mathrm{Sp}_{2m}$ in characteristic 2 which acts semisimply on the natural module but is not G-cr.

10.4   Graph Automorphisms and Maximal Rank Subgroups

Recall Corollary 9.2.4: in the terminology introduced at the start of the chapter, if S is a linearly reductive subgroup acting on G and $M = C_G(S)$, then the pair (G, M) has the ascent and descent properties. Typical applications of this are when S is the group generated by a graph automorphism of G, or S is the group generated by a semisimple element of G.

An important instance of this idea has already been given in Example 9.2.6: the classical groups can be realised as centralisers of outer automorphisms of order 2, so when $p \neq 2$ Corollary 9.2.4 applies. We give one further example along the same lines.

Example 10.4.1 Suppose that $p > 3$. Let $G = \mathrm{SO}_8$ and let H be the subgroup of G of type G_2 that is the fixed point subgroup of G under the triality graph automorphism of G. Let K be a closed subgroup of H. Then K is H-completely reducible if and only if it is G-completely reducible. By our assumption on p the natural 8-dimensional module $V_G(8)$ of G splits as an H-module into a direct sum $V_G(8)|_H = V_H(7) \oplus k$ of the simple 7-dimensional H-module $V_H(7)$ and the trivial module. It follows from Corollary 9.2.4 and Example 9.2.6 that K is H-completely reducible if and only if $V_H(7)$ is a semisimple K-module. (We observe that this equivalence does also hold in characteristic 3: see [159, Ex. 3.2.2(c)].)

Suppose G is connected and $s \in G$ is a semisimple element (so that the subgroup generated by s is linearly reductive). Then s lies in some maximal torus T of G, so $T \subseteq C_G(s)$; that is, the centraliser is a maximal rank subgroup of G. In good characteristic, there is a converse to this statement. First recall that the closed subsystems of the root system corresponding to maximal semisimple rank subgroups of a simple group G are determined by means of the algorithm of Borel and de Siebenthal [34]; see also [37, Ex. 4, Ch. VI, §4]. In brief, one attaches an extra node corresponding to the lowest root to the Dynkin diagram for the root system of G, giving the *affine extension* of that diagram. Then one deletes a node from this extended diagram, obtaining the Dynkin diagram of a closed subsystem; the maximal rank subgroup for this closed subsystem is the one generated by the maximal torus and the corresponding root groups. Then one iterates this procedure with the (factors of the) resulting maximal rank subsystem to obtain all such maximal rank subsystem subgroups of G.

Now it follows from work of Carter-Deriziotis (see [44, Prop. 11] or [79, §2.15]) that when the characteristic is good for G, all such subgroups arise as the connected centralisers of semisimple elements of G. This allows us to prove the following:

Theorem 10.4.2 *Suppose that G is connected, and that p is good for G. Let M be a regular connected reductive subgroup of G. Then the pair (G, M) has both the ascent and descent properties.*

Proof Let T be a maximal torus that normalises M, and let H be a subgroup of M. Then M is normal in MT, so H is M-cr if and only if H is MT-cr, by Proposition 4.2.9. Thus we may assume that $T \subseteq M$.

As M is a regular reductive subgroup of G, it is G-cr, by Proposition 9.1.2. Thanks to Theorem 5.4.4 and Corollary 5.4.6, we may assume that M is G-ir. Then, since $C_G(M)^0 = Z(G)^0$ by Lemma 4.2.2(ii), we see that M has maximal semisimple rank. Finally, the result follows from Corollary 9.2.4, together with the algorithm of Borel and de Siebenthal and the work of Carter-Deriziotis described before the theorem (applied iteratively if necessary to obtain M). $\qquad\square$

Here is a basic example of how to apply this theorem.

Example 10.4.3 Let G be of type E_8 and suppose that $p \neq 5$. Let M be a maximal rank subgroup of G of type $A_4 A_4$; then M is the centraliser in G of an element of order 5. Hence a subgroup H of M is G-cr if and only if it is M-cr.

We give two examples to show that some condition on p is necessary for Theorem 10.4.2 to hold.

Example 10.4.4 Recall Example 4.1.10. In that example we have $p = 2$ and some even integer $m \geq 4$. Then the symplectic group $M = \mathrm{Sp}_{2m}$ contains a maximal rank subgroup $K := \mathrm{Sp}_m \times \mathrm{Sp}_m$, which has a subgroup $H \cong \mathrm{Sp}_m$ embedded diagonally. We saw that H is K-cr but not M-cr, so (M, K) does not have the ascent property, even though K is a regular reductive subgroup of M.

Example 10.4.5 The example in Sect. 10.1 has $p = 2$, G a simple group of type G_2 and M a maximal rank subgroup of G. We saw that M admits a family of subgroups $H_a \cong S_3$, where $a \in k^*$, such that each H_a is G-cr but not M-cr. Thus (G, M) does not have the descent property even though M is a regular reductive subgroup of G. Note also that $M = C_G(z)$ is the centraliser of a semisimple element of $\mathrm{Lie}(G)$ in this case; see Remark 10.1.2.

10.5 Separability and Reductive Pairs

Now we consider the interaction of subgroups of G with the Lie algebra $\mathrm{Lie}\,G = \mathfrak{g}$ of G. We wish to apply Richardson's tangent space argument (Theorem 3.8.1) to the setting of G-complete reducibility. In order to do so, we need two key definitions. First is the notion of a separable subgroup.

Definition 10.5.1 We say a subgroup H of G is *separable in G* if the Lie algebra centraliser

$$C_{\mathfrak{g}}(H) = \{X \in \mathfrak{g} \mid \mathrm{Ad}(h)(X) = X \text{ for all } h \in H\}$$

of H equals the Lie algebra of $C_G(H)$ (that is, if the scheme-theoretic centraliser of H in G is smooth). There is always an inclusion $\mathrm{Lie}(C_G(H)) \subseteq C_{\mathfrak{g}}(H)$, so if the latter properly contains the former, then we say that H is *non-separable in G*. Since all algebraic groups are smooth in characteristic 0 (see Remark 3.5.3), non-separable subgroups can occur only in positive characteristic.

Remark 10.5.2 The terminology in Definition 10.5.1 is motivated as follows. Suppose that $\mathbf{h} \in H^n$ is a generic tuple for H. Then by Proposition 3.5.2 the orbit map $\kappa_{\mathbf{h}} : G \to G \cdot \mathbf{h}$ is separable if and only if $\mathrm{Lie}(G_{\mathbf{h}}) = \mathfrak{g}_{\mathbf{h}}$. But $\mathfrak{g}_{\mathbf{h}} = C_{\mathfrak{g}}(H)$, by Lemma 5.2.5, so we see that this holds if and only if H is separable in G. That is, the separable subgroups of G are those for which the orbits of generic tuples are separable.

Example 10.5.3 Any subgroup H of $G = \mathrm{GL}_m$ is separable in G. For separability means precisely that the centralisers of H in GL_m and in $\mathrm{Lie}(\mathrm{GL}_m) = \mathrm{Mat}_m$ have the same dimension, and for GL_m this holds, because the centraliser of H in GL_m is the open subset of invertible elements of the centraliser of H in Mat_m.

Remark 10.5.4 There is a cohomological condition which guarantees separability: by [57, II, §5, 2.8], H is separable in G if $H^1(H, \mathfrak{g})$ is trivial. In particular, then, linearly reductive subgroups are separable, by Lemma 2.10.2. This result is also proved by Richardson in [143, Lem. 4.1].

We can now prove the following striking result; compare with Corollary 10.3.2 above.

Theorem 10.5.5 *Let H be a separable subgroup of G. If $\mathfrak{g}$ is semisimple as an H-module, then H is G-completely reducible.*

Proof Suppose, for a contradiction, that H is not G-cr, but that $\mathfrak{g}$ is H-semisimple. Let $\mathbf{h} \in H^n$ be a generic tuple for H. Then $G \cdot \mathbf{h}$ is not closed by Theorem 5.3.3, so we may find $\lambda \in Y_G$ such that $\mathbf{h}' := \lim_{a \to 0} \lambda(a) \cdot \mathbf{h} = c_\lambda(\mathbf{h}) \in G^n$ lies outside the G-orbit of $\mathbf{h}$. This means that $\dim(G \cdot \mathbf{h}') < \dim(G \cdot \mathbf{h})$, and thus $\dim(G_{\mathbf{h}'}) > \dim(G_{\mathbf{h}})$ by Lemma 2.7.6. Since H is separable in G, we can deduce that $\dim(\mathfrak{g}_{\mathbf{h}'}) > \dim(\mathfrak{g}_{\mathbf{h}})$.

Now consider the images of $\mathbf{h}$ and $\mathbf{h}'$ under the adjoint representation; say $\mathbf{x} = \mathrm{Ad}_G(\mathbf{h})$ and $\mathbf{x}' = \mathrm{Ad}_G(\mathbf{h}')$ inside $\mathrm{GL}(\mathfrak{g})^n$. Let also $\mu = \mathrm{Ad} \circ \lambda \in Y_{\mathrm{GL}(\mathfrak{g})}$. Then $\mathbf{x}' = c_\mu(\mathbf{x})$; also $\mathbf{x}$ is a generic tuple for the image M of H inside $\mathrm{GL}(\mathfrak{g})$. But now, since $\mathfrak{g}$ is H-semisimple, the subgroup M is $\mathrm{GL}(\mathfrak{g})$-cr, and hence $\mathbf{x}'$ is $\mathrm{GL}(\mathfrak{g})$-conjugate to $\mathbf{x}$ by Theorem 5.3.3 again. Since $\mathfrak{g}_{\mathbf{h}}$ (resp., $\mathfrak{g}_{\mathbf{h}'}$) is the set of fixed points in $\mathfrak{g}$ of the components of $\mathbf{x}$ (resp., $\mathbf{x}'$), we deduce that $\mathfrak{g}_\mathbf{h}$ and $\mathfrak{g}_{\mathbf{h}'}$ are $\mathrm{GL}(\mathfrak{g})$-conjugate and hence, in particular, they have the same dimension. This is a contradiction. $\square$

Open Problem 10.5.6 Does the conclusion of Theorem 10.5.5 hold without the separability assumption on the subgroup H?

Here is a special case of Problem 10.5.6 with a representation-theoretic flavour. Suppose $p > 0$, let $m = p^2 - 1$ and consider groups of type A_m. The adjoint modules for the simply connected form SL_m and the adjoint form PGL_m are indecomposable but not semisimple. There is, however, a so-called intermediate form G whose adjoint module splits as the sum of a simple $(m^2 - 2)$-dimensional module and a 1-dimensional trivial module. The first summand is isomorphic to $\mathfrak{psl}_m := \mathfrak{sl}_m/\mathfrak{z}(\mathfrak{sl}_m)$; note that $\mathfrak{sl}_m$ has a 1-dimensional centre because p divides m.

Open Problem 10.5.7 Does there exist a non-completely reducible representation $\rho \colon M \to \mathrm{GL}_m$ of a semisimple group M such that M acts semisimply on $\mathfrak{psl}_m$?

Note that if M acts semisimply on $\mathfrak{sl}_m$ then ρ is completely reducible by [26, Thm. 4.4]. The point is that having M act semisimply on $\mathfrak{psl}_m$ is *a priori* a weaker condition than having M act semisimply on $\mathfrak{sl}_m$. A positive answer to Problem 10.5.7 would yield a negative answer to Problem 10.5.6. To see this, let $\pi \colon \mathrm{SL}_m \to G$ be the canonical projection, and set $H = \pi(\rho(M))$ (note that $\rho(M) \subseteq \mathrm{SL}_m$ since M is semisimple). Then H is not G-cr by Lemma 4.2.7(b), since $\rho(M)$ is not SL_m-cr; but H acts semisimply on $\mathfrak{psl}_m$, so H acts semisimply on $\mathfrak{psl}_m \oplus k \cong \mathfrak{g}$.

Although there are some cases where separability of subgroups is guaranteed (as above, if $G = \mathrm{GL}_m$, for example, or if the subgroup is linearly reductive), in general the phenomenon of separability is closely tied together with the characteristic p of the field k and with the root datum of G. We give some examples to illustrate this.

Example 10.5.8 Let $p = 2$ and let $G = \mathrm{SL}_2$. Then $\mathrm{Lie}(C_G(G)) = \mathrm{Lie}(Z(G)) = \{0\}$, but $C_\mathfrak{g}(G) = Z(\mathfrak{g})$ is the one-dimensional centre of $\mathfrak{sl}_2$, so G is not separable in itself. A similar thing happens for SL_m in characteristic p whenever p divides m.

Remark 10.5.9 If G itself is non-separable in G, as in Example 10.5.8, then there exists a reductive group $\widehat{G}$ such that $\widehat{G}$ is generated by G and a central torus, and $\widehat{G}$ is separable in $\widehat{G}$; see [116, Thm. 4.5]. For example, when $G = \mathrm{SL}_p$ we can take $\widehat{G} = \mathrm{GL}_p$.

Example 10.5.10 Let $p = 2$, let $G = \mathrm{PGL}_2(k)$ and let T be a maximal torus of G. Then $N_G(T)$ is non-separable in G.

Example 10.5.11 For the example introduced in Sect. 10.1, p is 2 and G is a simple group of type G_2 with a Levi subgroup L. Then one can check that the subgroup H of that example is separable in L, but not separable in G: the key point is that the sum of root elements $e_\beta + e_{3\alpha+\beta}$ is fixed by H but does not lie in $\text{Lie}(C_G(H)) = \mathfrak{g}_{3\alpha+2\beta}$; see [26, Prop. 7.11].

Remark 10.5.12 The previous example shows that we can have $H \subseteq M \subseteq G$ with H separable in M but not in G. Conversely, by putting $H = M = \text{SL}_p$ inside $G = \text{GL}_p$ we see that H can be non-separable in M whilst being separable in G.

As the examples above might suggest, if we fix a group G of a given Dynkin type, then the existence of non-separable subgroups in G is a phenomenon which only occurs in a finite number of positive characteristics p. In fact, Herpel has given a complete characterisation of when non-separable subgroups can occur [73]: to do this he introduces the notion of a *pretty good prime*, which is somewhere in between being good and very good (i.e., very good $\implies$ pretty good $\implies$ good, but not vice versa). As an example, p is pretty good for GL_p, but not very good. Note, however, that if G is semisimple then p is pretty good for G if and only if p is very good for G, [73, Lem. 2.12(c)]. The following result is [73, Thm. 1.1 and Rem. 4.4]. We give a sketch of the proof in some special cases below once we have introduced the notion of a reductive pair: see Remark 10.5.25.

Theorem 10.5.13 *Let G be a connected reductive group. Then $p = 0$ or p is a pretty good prime for G if and only if every subgroup of G is separable in G.*

As an illustration of the machinery we have developed, we prove a variation on Theorem 10.2.1; the bound on p is worse but we do not rely on the detailed classification results of [99].

Proposition 10.5.14 *Let H be a connected reductive subgroup of G and suppose $p \geq \dim(G)$. Then H is G-completely reducible.*

Proof By the theorem of Jantzen cited in the proof of Theorem 10.2.1, H acts semisimply on $\mathfrak{g}$. Since p is very good for G, Theorem 10.5.13 implies that H is separable in G. The result now follows from Theorem 10.5.5. $\square$

We now introduce the notion of a reductive pair, due to Richardson [143].

Definition 10.5.15 We call (G, M) a *reductive pair* if M is a reductive subgroup of G and $\text{Lie}(G)$ decomposes as an M-module into a direct sum

$$\text{Lie}(G) = \text{Lie}(M) \oplus \mathfrak{n},$$

for some M-module $\mathfrak{n}$, where M acts via the restriction of the adjoint map $\text{Ad} : G \to \text{GL}(\text{Lie}(G))$.

Example 10.5.16 Some basic examples of reductive pairs are the pairs of the form $(\mathrm{GL}(V), G)$ where G is an orthogonal or symplectic group in characteristic not 2, and the pairs of the form $(\mathrm{GL}(\mathfrak{g}), G)$ where G is an adjoint type exceptional simple group in good characteristic; see Exercises 10.1 and 10.2. It is also clear from the definition that if (G, M) and (M, K) are reductive pairs, then so is (G, K).

Example 10.5.17 Let M be a reductive subgroup of G containing a maximal torus T of G, so that in particular, M is regular reductive. If the root system $\Phi(M, T)$ is a closed subsystem of $\Phi(G, T)$, then (G, M) is a reductive pair. The complement to $\mathrm{Lie}(M)$ in $\mathrm{Lie}(G)$ is the sum of the root spaces of $\mathrm{Lie}(G)$ corresponding to the roots in $\Phi(G, T)$ that lie outside $\Phi(M, T)$.

This is no longer valid if $\Phi(M, T)$ is not a closed subsystem. For instance, suppose $p = 2$ and let G be of type B_2. Then G has a regular reductive subgroup M of type A_1^2 generated by the short root subgroups of G, but $\mathrm{Lie}(M)$ does not admit an M-stable complement in $\mathrm{Lie}G$: see Exercise 10.3.

We can now give the two main results about reductive pairs. The first is due to Slodowy [164, Thm. 1].

Proposition 10.5.18 *Suppose that (G, M) is a reductive pair. Let $n \in \mathbb{N}$ and let $\mathbf{m} \in M^n$ be such that $G \cdot \mathbf{m}$ is separable. Then $G \cdot \mathbf{m} \cap M^n$ is a finite union of M-orbits. Moreover, each of these orbits is separable as an M-orbit and closed in $G \cdot \mathbf{m} \cap M^n$.*

Proof We wish to apply Theorem 3.8.1. Let $X = G^n$ and $Y = G \cdot \mathbf{m} \cap M^n$. Then Y is an M-stable locally closed subset of X. We want to verify the hypotheses of Theorem 3.8.1, so let $\mathbf{y} = (y_1, \ldots, y_n) \in Y$ be any point. Since $G \cdot \mathbf{y} = G \cdot \mathbf{m}$, we have $Y \subseteq G \cdot \mathbf{y}$, so

$$T_{\mathbf{y}}(G \cdot \mathbf{y}) \cap T_{\mathbf{y}}(Y) = T_{\mathbf{y}}(Y).$$

Thus we need to show that $T_{\mathbf{y}}(M \cdot \mathbf{y}) = T_{\mathbf{y}}(Y)$. The inclusion left-to-right is obvious.

For the reverse inclusion, note first that since $\mathbf{y} \in G \cdot \mathbf{m}$, the G-orbit of $\mathbf{y}$ is separable. Let $\kappa_{\mathbf{y}} : G \to G \cdot \mathbf{y}$ denote the orbit map, and let $\mathbf{1} = (1, \ldots, 1) \in G^n$. We wish to identify $T_{\mathbf{y}}(X)$ with $\mathfrak{g}^n = T_{\mathbf{1}}(X)$, by "translating the origin". Explicitly, if we apply the isomorphism of varieties $\phi : X \to X$ given by $(g_1, \ldots, g_n) \mapsto (g_1 y_1^{-1}, \ldots, g_n y_n^{-1})$, then $\phi(\mathbf{y}) = \mathbf{1}$ and $\psi := \phi \circ \kappa_{\mathbf{y}} : G \to X$ is given by

$$\psi(g) = (g y_1 g^{-1} y_1^{-1}, \ldots, g y_n g^{-1} y_n^{-1}).$$

Now applying Exercise 2.1 in each coordinate we see that for each $a \in \mathfrak{g}$ we get

$$(d\psi)_1(a) = ((1 - \mathrm{Ad}(y_1))(a), \ldots, (1 - \mathrm{Ad}(y_n))(a)) \in \mathfrak{g}^n. \tag{10.5.1}$$

Let $\mathfrak{n}$ be an M-stable complement to $\mathfrak{m}$ in $\mathfrak{g}$. Equation (10.5.1) shows that $(d\psi)_1(\mathfrak{m}) \subseteq \mathfrak{m}^n$ and $(d\psi)_1(\mathfrak{n}) \subseteq \mathfrak{n}^n$.

Separability implies that the derivative $(d\kappa_\mathbf{y})_1 : \mathfrak{g} \to T_\mathbf{y}(G \cdot y)$ is surjective with kernel equal to $\mathrm{Lie}(G_\mathbf{y})$, by Proposition 3.5.2, and hence the kernel of $(d\psi)_1$ is the same since ϕ is an isomorphism. Let $b \in T_\mathbf{y}(Y)$, and let $a \in \mathfrak{g}$ be such that $(d\kappa_\mathbf{y})_1(a) = b$. Write $a = a_1 + a_2$ with $a_1 \in \mathfrak{m}$ and $a_2 \in \mathfrak{n}$. It is clear that ϕ maps $Y = G \cdot \mathbf{m} \cap M^n$ into M^n, so since $(d\kappa_\mathbf{y})_1(a) = b \in T_\mathbf{y}(Y)$, we have $(d\psi)_1(a) = (d\psi)_1(a_1) + (d\psi)_1(a_2) \in \mathfrak{m}^n$. Hence $(d\psi)_1(a_2) = 0$, and so $(d\kappa_\mathbf{y})_1(a_2) = 0$, which is to say that $b = (d\kappa_\mathbf{y})_1(a_1)$ lies in the image of $\mathfrak{m}$. We have shown that

$$T_\mathbf{y}(M \cdot \mathbf{y}) \subseteq T_\mathbf{y}(Y) \subseteq (d\kappa_\mathbf{y})_1(\mathfrak{m}).$$

But it is obvious that $(d\kappa_\mathbf{y})_1(\mathfrak{m}) \subseteq T_\mathbf{y}(M \cdot \mathbf{y})$ since the restriction of the orbit map $\kappa_\mathbf{y}$ to M is the orbit map $M \to M \cdot \mathbf{y}$. Thus we obtain the equality required, and the hypotheses of Theorem 3.8.1 hold for X and Y.

We have shown that the derivative of the orbit map $M \to M \cdot \mathbf{y}$ is surjective for all choices of $\mathbf{y} \in Y$, and hence all of these orbit maps are separable. Theorem 3.8.1 also implies that there are finitely many M-orbits in Y, and each one is a union of irreducible components of Y. In particular, then, each M-orbit is closed in Y and there are finitely many of them. $\qquad\square$

Remark 10.5.19 Suppose $p = 0$, M is a reductive subgroup of G and $\mathbf{m} \in M^n$. Then (G, M) is automatically a reductive pair since M is linearly reductive, and $G \cdot \mathbf{m}$ is separable (see Sect. 3.5). Hence the hypotheses of Proposition 10.5.18 hold. We used this observation at the end of Sect. 6.7 in the proof of Theorem 6.1.1.

Remark 10.5.20 Guralnick [69, Thm. 1.2] has shown that when (G, M) is *any* pair of reductive groups and $m \in M$, then the intersection of the G-conjugacy class of m with M is a finite union of M-conjugacy classes—that is, the previous result holds when $n = 1$ without assuming that (G, M) is a reductive pair or that the orbit map $G \to G \cdot m$ is separable. However, we can use the example from Sect. 10.1 to illustrate that this fails as soon as $n > 1$ when the orbit map is not separable, even in the presence of a reductive pair.

Recall in Sect. 10.1 we had $p = 2$, $G = G_2$ and $M = G_\alpha G_{3\alpha+2\beta}$ a maximal rank subgroup of G. Let $(m_1, m_2) = (s_\alpha, t\kappa_{3\alpha+2\beta}(1)) \in M^2$ and for each $a \in k$ set

$$(m_1(a), m_2(a)) := u(a) \cdot (m_1, m_2),$$

where we recall that $u(a) = \kappa_\beta(a)\kappa_{3\alpha+\beta}(a)$. Note that (10.1.1) together with the fact that $u(a)$ commutes with t and $\kappa_{3\alpha+2\beta}(1)$ implies that each $(m_1(a), m_2(a))$ belongs to M^2. It is shown in [26, Ex. 7.15] that the pairs $(m_1(a), m_2(a))$ are pairwise non-conjugate under the action of M on M^2, and hence we see that the intersection $G \cdot (m_1, m_2) \cap M^2$ contains an infinite number of M-orbits. The argument to show that the tuples are pairwise non-conjugate is very similar to the proof of Lemma 10.1.1: by calculating the stabiliser of the tuple (m_1, m_2) and showing that it is contained in M, we can use the fact that the conjugating elements $u(a)$ do not belong to M. Note finally that this example when $n = 2$ immediately gives an example for all $n \geq 2$ by using the tuple $(m_1, m_2, 1, \ldots, 1) \in M^n$.

Theorem 10.5.21 *Suppose that (G, M) is a reductive pair. Let H be a subgroup of M such that H is a separable subgroup of G. Then H is separable in M. Moreover, if H is G-completely reducible, then it is also M-completely reducible.*

Proof We apply Proposition 10.5.18 to a generic tuple $\mathbf{h} \in H^n$ for H. Since H is a separable subgroup of G, the orbit $G \cdot \mathbf{h}$ is separable, by Remark 10.5.2. Now the proposition tells us that each of the M-orbits in $G \cdot \mathbf{h} \cap M^n$ is separable, so in particular $M \cdot \mathbf{h}$ is. We deduce that H is separable in M by Remark 10.5.2 again. In addition, the proposition implies that $M \cdot \mathbf{h}$ is closed in $G \cdot \mathbf{h} \cap M^n$. Hence, if H is G-cr—so that $G \cdot \mathbf{h}$ is closed in G^n by Theorem 5.3.3—then $M \cdot \mathbf{h}$ is closed in M^n, so H is M-cr by Theorem 5.3.3. □

The following examples illustrate the failure of the conclusion of Theorem 10.5.21 if one of the hypotheses is not met.

Example 10.5.22   (i). We again get an example from Sect. 10.1. Here the pair (G, M) is a reductive pair by Example 10.5.17, and yet the subgroups H_a for $a \in k^*$ are G-cr but not M-cr by Lemma 10.1.1. The problem is that none of the subgroups H_a is separable in G, so this key hypothesis of Theorem 10.5.21 does not hold. Interestingly, the subgroup $H = H_0$ itself *is* M-cr.

(ii). Consider Liebeck's Example 4.1.10 with $H \cong \mathrm{Sp}_m \subseteq M = \mathrm{Sp}_{2m} \subseteq G = \mathrm{GL}_{2m}$ in characteristic 2. Then H is G-cr but not M-cr, and H is separable in G, hence (G, M) cannot be a reductive pair. One can also check that H is not separable in M: see Exercise 10.4.

Example 10.5.23 The following example is due to Litterick, Stewart and Thomas; see [105, Ex. 2.40, Ex. 2.48] where the subgroups discussed below are described in more detail. Let $p = 2$ and let G be simple of type F_4. Then G contains two distinct conjugacy classes of subgroups isomorphic to G_2, each of which is G-cr. For a representative of one of the classes we can take H_1 to be a subgroup of type G_2 inside a Levi subgroup of F_4 of type B_3 (this sequence of subgroups $G_2 \subseteq B_3 \subseteq F_4$ exists in all characteristics), and because $p = 2$ for a representative of the second class we can take H_2 to be a subgroup of type G_2 inside a C_3 Levi subgroup of F_4. Now G has a maximal rank subgroup M of type B_4 containing H_1 and a maximal rank subgroup M' of type C_4 containing H_2; these maximal rank subgroups are swapped by the special isogeny of G which switches long and short roots in characteristic 2.

The subgroup H_1 is not M-cr and H_2 is not M'-cr, despite H_1 and H_2 being G-cr. Further, (G, M) is a reductive pair, so H_1 is not separable in G. On the other hand, (G, M') is not a reductive pair, and one can show by direct calculation that H_2 *is* separable in G. So we are seeing the failure of Theorem 10.5.21 under weaker hypotheses in two different ways. The groups H_1 and H_2 are connected, which is also rare in these examples.

Example 10.5.3 shows that the separability hypothesis is automatically satisfied for the case $G = \mathrm{GL}(V)$. We obtain an immediate consequence of Theorem 10.5.21; part (ii) of the following result is in the spirit of Serre's Theorem 10.3.1(ii) above.

Corollary 10.5.24 *Suppose* $(\mathrm{GL}(V), M)$ *is a reductive pair, and H is a subgroup of M. Then:*

(i) H is separable in M.
(ii) If V is a semisimple H-module, then H is M-completely reducible.

We deduce from Corollary 10.5.24(i) that not every M can be embedded in some $\mathrm{GL}(V)$ in such a way that $(\mathrm{GL}(V), M)$ is a reductive pair: this applies, for instance, to $M = \mathrm{SL}_2$ when $p = 2$, because $H = M$ is not a separable subgroup of M (Example 10.5.8). This is a small characteristic phenomenon.

Remark 10.5.25 We finish this section by giving an indication of how the proof of the forward direction of Theorem 10.5.13 goes in the case when G is simple and p is very good. In type A_m, the hypothesis that p is very good means that the pair $(\mathrm{GL}_m, \mathrm{SL}_m)$ is a reductive pair, since the scalar matrices are a complement in $\mathfrak{gl}_m$ to the traceless matrices. In type B_m, the pair $(\mathrm{GL}_{2m+1}, \mathrm{SO}_{2m+1})$ is a reductive pair; similarly for type C_m we have $(\mathrm{GL}_{2m}, \mathrm{Sp}_{2m})$, and for type D_m we get $(\mathrm{GL}_{2m}, \mathrm{SO}_{2m})$. For exceptional types, we can take G to be the adjoint group of that type and then $(\mathrm{GL}(\mathfrak{g}), G)$ is a reductive pair. Since every subgroup of a general linear group is separable, we conclude the same for the simple groups of each type listed, by Corollary 10.5.24.

10.6   Commuting G-Completely Reducible Subgroups

We return to the following problem, which we considered briefly in Sect. 9.3: if A and B are commuting subgroups of G, how is G-complete reducibility of A and B related to G-complete reducibility of the product AB? We give some examples below to show that even when G and A and B are connected, we can have A and B both G-cr whilst AB is not. However, this is a low characteristic phenomenon, as our next result demonstrates:

Theorem 10.6.1 *Suppose that G is connected and that p is good for G or p > 3. Let A and B be commuting connected subgroups of G. Then AB is G-completely reducible if and only if both A and B are G-completely reducible.*

Remark 10.6.2 (i). Note that the forward implication holds by Theorem 9.1.1 without any connectedness hypotheses or any assumption on p.

(ii). One consequence of Theorem 10.6.1 is that the converse of Corollary 9.3.4 is true for connected groups if p is good or $p > 3$; this follows from Proposition 9.3.3. Thus if A is a connected G-cr subgroup of G and p is good for G or $p > 3$, then the pair $(G, C_G(A))$ has both the ascent and descent properties for connected subgroups.

We now give a brief overview of the proof of the reverse implication in Theorem 10.6.1. The first step is to reduce to the case that G is simple, and to show that the result is insensitive to isogeny, so we can prove it for a chosen representative of each isogeny class of simple groups. By Proposition 4.2.9 and Corollary 9.2.4, the result for SL(V), Sp(V) and SO(V) follows from the result for GL(V), which is the following; the argument indicated is due to Rudolf Tange. Note again that no connectedness hypothesis is needed for A and B in this case.

Lemma 10.6.3 *Let V be a finite-dimensional vector space over k. Suppose A and B are commuting subgroups of* GL(V). *Then V is semisimple for the product AB if and only if V is semisimple for A and B.*

Proof We need to prove the reverse implication, since the forward implication is just a consequence of Clifford's Theorem. The subgroups A and B generate subalgebras C and D of End(V) which act semisimply on V, and hence C and D are semisimple algebras [93, Ch. XVII, Prop. 4.7]. Then the tensor product $C \otimes_k D$ is also semisimple, by [93, Ch. XVII, Thm. 6.4], and hence its image $E \subseteq$ End(V) under the map $c \otimes d \mapsto cd$ is also semisimple by [93, Ch. XVII, Prop. 2.2 and §4], and hence E acts semisimply on V. But E is just the subalgebra generated by AB, so AB acts semisimply on V. □

Having established the result for classical types, we then need to deal with exceptional types. For this, we can invoke the classification results of Liebeck and Seitz [99] described in the discussion of the proof of Theorem 10.2.1. For each Dynkin type we have a table listing all the semisimple subgroups of a given simple G, together with information about centralisers. These tables make the result automatic if $p > 7$, because all connected reductive subgroups are G-cr. For smaller primes, a more careful analysis of the possible cases which can arise allows us to whittle down to the stated condition p is good or $p > 3$. The key is Proposition 9.3.3: given a subgroup A, the possible B we can choose all lie inside $C_G(A)$, and the tables of [99] allow us to check that in all cases B is $C_G(A)$-cr. It is perhaps of interest that the case $G = E_8$, $p = 5$ doesn't throw up any issues, even though 5 is a bad prime for type E_8: this is essentially because 5 is a good prime for all the interesting centralisers $C_G(A)$ that can arise.

This concludes our discussion of the proof of Theorem 10.6.1. We finish the section by collecting some examples. Our first one shows that, even in good characteristic, Theorem 10.6.1 fails for disconnected groups. Note that we cannot have an example of this kind inside a connected group: for a non-trivial unipotent subgroup A can never be G-cr if G is connected by Theorem 2.9.23.

Example 10.6.4 ([19, Ex. 5.1]) Suppose $p = 2$ and $m \geq 4$ is even. Define $\phi \in \mathrm{Aut}(\mathrm{GL}_{2m})$ by $\phi(g) = J(g^t)^{-1}J^{-1}$, where g^t denotes the matrix transpose of g, $J = \begin{pmatrix} 0 & I_m \\ I_m & 0 \end{pmatrix}$ and I_m is the $m \times m$ identity matrix. Set $A = \langle \phi \rangle$ and $G = A \ltimes \mathrm{GL}_{2m}$. Let $B = \mathrm{Sp}_m$ embedded diagonally in the maximal rank subgroup $\mathrm{Sp}_m \times \mathrm{Sp}_m$ of Sp_{2m}, and consider the canonical embedding of Sp_{2m} in G^0. We can identify Sp_{2m} with $C_G(A)^0$. By Example 4.1.10, B is G^0-cr but not Sp_{2m}-cr.

Observe that Sp_{2m} is G^0-ir. Thus if $\lambda \in Y_{G^0} = Y_G$ with $A\mathrm{Sp}_{2m} \subseteq P_\lambda$, then λ belongs to $Y_{Z(G^0)}$ and $R_u(P_\lambda) = \{1\}$, so $A\mathrm{Sp}_{2m} \subseteq P_\lambda = L_\lambda$. This shows that $A\mathrm{Sp}_{2m}$ is G-cr. Thus the normal subgroup A of $A\mathrm{Sp}_{2m}$ is G-cr (Theorem 9.1.1). By Proposition 4.2.12, B is G-cr. However, B is not Sp_{2m}-cr, so Proposition 4.2.12 implies that B is not $A\mathrm{Sp}_{2m}$-cr. Note that $A\mathrm{Sp}_{2m} = C_G(A)$, so B is not $C_G(A)$-cr.

Thus we have commuting subgroups A and B which are G-cr, but such that B is not $C_G(A)$-cr. Hence, by Proposition 9.3.3, AB is not G-cr. In particular, even though $p = 2$ is good for $G^0 = \mathrm{GL}_{2m}$, Theorem 10.6.1 fails for these subgroups of the disconnected group G.

We now present an extension of Liebeck's Example 4.1.10 above which shows the failure of Theorem 10.6.1 for connected groups in bad characteristic. This example was also outlined to the authors by Liebeck.

Example 10.6.5 Suppose $p = 2$. We show that there exist connected commuting subgroups A, B of Sp_8 such that A, B are Sp_8-cr but AB is not.

Let $A = B = \mathrm{SL}_2$ and let V_A, V_B be the natural modules for A, B respectively. Choose symplectic forms $(\cdot, \cdot)_A$, $(\cdot, \cdot)_B$ for V_A, V_B respectively. Then $\mathrm{SL}(V_A) = \mathrm{Sp}(V_A)$ and $\mathrm{SL}(V_B) = \mathrm{Sp}(V_B)$. Set $W := V_A \otimes V_B$ with the symplectic form $(\cdot, \cdot)_W$ given by

$$(u_1 \otimes u_2, v_1 \otimes v_2)_W := (u_1, v_1)_A (u_2, v_2)_B.$$

Then A, B and $A \times B$ act on W in the obvious way, and these actions preserve $(\cdot, \cdot)_W$. Below we shall be interested in A-stable subspaces of W. For $v \in V_B$, set

$$V_A \otimes v := \{u \otimes v \mid u \in V_A\},$$

a subspace of W. Since W is the A-module direct sum of the irreducible A-modules $V_A \otimes v_1$ and $V_A \otimes v_2$, where v_1 and v_2 are any two linearly independent vectors in V_B, we see that A acts completely reducibly on W and the proper non-trivial A-stable subspaces of W are precisely the subspaces of the form $V_A \otimes v$ for some $v \in V_B$. In particular, the A-stable subspaces of W have dimension 0, 2 or 4.

There exists an A-module isomorphism $\phi_A : V_A \to V_A^*$ corresponding to the symplectic form on V_A. Define $\phi_B : V_B \to V_B^*$ analogously and identify W^* with $V_A^* \otimes V_B^*$ via the $(A \times B)$-module isomorphism $\psi := \phi_A \otimes \phi_B$; this is precisely the isomorphism corresponding to the symplectic form on W, and gives rise to a symplectic form $(\cdot , \cdot)_{W^*}$ on W^* given by

$$(\psi(w_1), \psi(w_2))_{W^*} := (w_1, w_2)_W.$$

Consider the $(A \times B)$-module $U := W \oplus W^*$ endowed with the direct sum symplectic form, which we denote by $(\cdot , \cdot)_U$. The $(A \times B)$-action preserves the symplectic structure, so we can regard A, B and $A \times B$ as subgroups of $\mathrm{Sp}(U)$. If M is an A-stable subspace of U then we have a short exact sequence of A-modules

$$\{0\} \to M \cap W \to M \to \pi_2(M) \to \{0\},$$

where $\pi_2 : U \to W^*$ is the canonical projection. Since the second and fourth term have even dimension, M must have even dimension. In particular, if M is isotropic then M has dimension 0, 2 or 4. We claim that if M is any 4-dimensional A-stable isotropic subspace of U, then there exists an A-stable isotropic complementary subspace. To establish this, we observe that such a subspace M must either be of the form $V_A \otimes v \oplus V_A^* \otimes f$ for some $0 \neq v \in V_A$ and $0 \neq f \in V_A^*$, or of the form $M_\theta := \mathrm{graph}(\theta)$, where $\theta : W \to W^*$ is an A-module isomorphism (note that W and W^* are not isotropic). Any subspace of the first type is isotropic, so we can take a complement to be $V_A \otimes v' \oplus V_A^* \otimes f'$, where v and v' (resp., f and f') are linearly independent in V_A (resp., V_A^*). Given M of the form M_θ, choose any $0 \neq v \in V_B$. The A-stable subspace $\theta(V_A \otimes v)$ of W^* is of the form $V_A^* \otimes f$ for some $f \in V_B^*$. Choose $g \in V_B^*$ such that f and g are linearly independent; then M_θ and $V_A \otimes v \oplus V_A^* \otimes g$ intersect trivially, and the latter subspace is isotropic. This proves the claim.

We now prove that A is $\mathrm{Sp}(U)$-cr; the analogous result for B follows by symmetry. The parabolic subgroups of $\mathrm{Sp}(U)$ are precisely the stabilisers of flags

$$\mathcal{F} = (M_1 \subset M_2 \subset \cdots \subset M_r)$$

of isotropic subspaces of U. Moreover, a parabolic subgroup P is opposite to the stabiliser of $\mathcal{F}$ if and only if P is the stabiliser of a flag $(N_1 \subset N_2 \subset \cdots \subset N_r)$ of isotropic subspaces such that $U = M_i^\perp \oplus N_i$ for each i. Let $\mathcal{F}$ be a flag of A-stable isotropic subspaces of U. There are only three possible types of flag to check. If $\mathcal{F}$ has the form

(M) with $\dim M = 4$, then we are done, by the previous claim (note that $M = M^{\perp}$). Now suppose that $\mathcal{F}$ has the form $(M_1 \subset M_2)$, where $\dim M_1 = 2$ and $\dim M_2 = 4$. By the previous claim, there exists a 4-dimensional A-stable isotropic subspace N_2 of U such that $U = M_2^{\perp} \oplus N_2$. Since A acts completely reducibly on U, there exists an A-stable complement N_1 to $M_1^{\perp} \cap N_2$ in N_2. Then N_1 is an isotropic A-stable complement to $M_1^{\perp}$ and $N_1 \subset N_2$, as required. Finally, suppose that $\mathcal{F}$ has the form (M) with $\dim M = 2$. It is easy to show, by listing the possibilities for M as in the previous claim, that M is contained in an A-stable 4-dimensional isotropic subspace of U, and we can use the argument of the second case to prove that $M^{\perp}$ has an A-stable isotropic complement in U. This completes the proof that A is $\mathrm{Sp}(U)$-cr.

We finish by proving that the isotropic $(A \times B)$-stable subspace $M_{\psi} = \{(w, \psi(w)) \mid w \in W\}$ does not admit an isotropic $(A \times B)$-stable complement, which proves that $A \times B$ is not $\mathrm{Sp}(U)$-cr. Suppose that N is such a complement. It follows from the discussion above on A-stable subspaces of U that N must be of the form M_{θ} for some $(A \times B)$-module isomorphism $\theta : W \to W^{*}$. Since $A \times B$ acts irreducibly on W and W^{*}, we must have $\theta = a\psi$ for some $a \in k$, by Schur's Lemma. For $u, v \in W$, we have

$$((u, \theta(u)), (v, \theta(v)))_U = ((u, a\psi(u)), (v, a\psi(v)))_U$$

$$= (u, v)_W + a^2(\psi(u), \psi(v))_{W^*}$$

$$= (1 + a^2)(u, v)_W.$$

As N is isotropic, this expression is identically 0, so we must have $a = 1$. But then $N = M_{\psi}$, a contradiction.

10.7 Saturation and G-Complete Reducibility

In this part we briefly recall Serre's notion of saturation and its relevance for G-complete reducibility from [158] and [159, §5]. We assume that G is connected for the rest of this section. The key idea is that we can associate to each unipotent element $u \in G$ a one-parameter subgroup (i.e., a copy of the additive group $\mathbb{G}_a$) containing u. This is straightforward to do when $\mathrm{char}(k) = 0$, so we concentrate on the case where $p = \mathrm{char}(k) > 0$. The most basic situation is when $G = \mathrm{GL}(V)$ for some finite-dimensional vector space V and $u \in G$ has order p. Then we can let $\varepsilon = u - 1 \in \mathrm{End}(V)$, we have $\varepsilon^p = 0$, and this allows us to write

$$u^t := (1 + \varepsilon)^t = 1 + t\varepsilon + \binom{t}{2}\varepsilon^2 + \cdots + \binom{t}{p-1}\varepsilon^{p-1}, \tag{10.7.1}$$

for each $t \in \mathbb{G}_a$ [159, §5.1]. Then $\{u^t \mid t \in \mathbb{G}_a\}$ is a closed connected subgroup of $\mathrm{GL}(V)$ isomorphic to $\mathbb{G}_a$, and $u = u^1$ is contained in this subgroup.

To generalise this idea to arbitrary G, some extra hypotheses are needed which allow the construction of *exponential* and *logarithm* maps between the set of nilpotent elements of the Lie algebra and the set of unipotent elements of G. There is further discussion of this in Sect. 10.8 below; for brevity we stick to Serre's original development and make the additional assumption that the derived group $[G, G]$ is simply connected.

Let $\mathcal{U}$ be the subvariety of G consisting of all unipotent elements, and let $\mathcal{N}$ be the subvariety of $\mathrm{Lie}(G)$ consisting of all nilpotent elements. When $p \geq h(G)$, every unipotent element of G has order p (see [180, Order Formula 0.4]). Fix a maximal torus T of G and a Borel subgroup B of G containing T, and let $U = R_u(B)$. When $p \geq h(G)$, because the nilpotency class of $\mathrm{Lie}(U)$ is at most $h(G)$, we can (and do) view $\mathrm{Lie}(U)$ as an algebraic group with multiplication given by the Baker-Campbell-Hausdorff formula (cf. [37, Ch. II §6]). The following result is [158, Thm. 3]; for a detailed proof, see [5, §6].

Theorem 10.7.1 *Let $p \geq h(G)$. There is a unique isomorphism of varieties* $\log : \mathcal{U} \to \mathcal{N}$ *such that the following hold:*

  (i) $(\log \circ \sigma)(u) = ((d\sigma)_1 \circ \log)(u)$ *for any* $\sigma \in \mathrm{Aut}(G)$ *and any* $u \in \mathcal{U}$;

  (ii) the restriction of $\log$ *to U defines an isomorphism of algebraic groups* $U \to \mathrm{Lie}(U)$ *whose tangent map is the identity on* $\mathrm{Lie}(U)$;

(iii) $\log(x_\alpha(t)) = tX_\alpha$ *for any* $\alpha \in \Phi$ *and any* $t \in \mathbb{G}_a$.

Now let $\exp : \mathcal{N} \to \mathcal{U}$ be the inverse morphism to $\log$. We then define

$$u^t := \exp(t \log u),$$

for any $u \in \mathcal{U}$ and any $t \in \mathbb{G}_a$. Then $\{u^t \mid t \in \mathbb{G}_a\} \cong \mathbb{G}_a$ is a closed connected subgroup of G containing u. Following Serre [158, Lecture 3] we make the following definition:

Definition 10.7.2 Let $p \geq h(G)$. A subgroup H of G is *saturated (in G)* if $u^t \in H$ for any unipotent element u of H and any $t \in \mathbb{G}_a$. For a subgroup H, its *saturated closure* H^{sat} is the smallest saturated subgroup of G containing H.

Theorem 10.7.1 implies that the saturation process is well-behaved in the presence of automorphisms of G. We collect together some other good properties of saturation.

Remark 10.7.3   (i). Basic examples of saturated subgroups of G are parabolic subgroups, Levi subgroups, and more generally, arbitrary centralisers of subgroups of $\mathrm{Aut}(G)$ [158, Lecture 3]. We deduce that H^{sat} is contained in precisely the same parabolic and Levi subgroups as H, and so H is G-cr if and only if H^{sat} is G-cr.

(ii). Connected regular reductive subgroups of G are saturated: see [10, Rem. 5.9]. In particular, the simple factors of G are saturated (and in fact the $\log$ and $\exp$ maps for G restrict to the corresponding maps for the simple factors; see [10, Lem. 5.10]).

(iii). A connected reductive subgroup need not be saturated in general. For instance, the adjoint representation of SL_p in $G := GL(\text{Lie}(SL_p))$ is not saturated in G in characteristic p [158, p. 18]; see [10, Ex. 5.17] for details.

(iv). Suppose $F_q : G \to G$ is a standard q-power Frobenius endomorphism of G. Then $F_q(u^t) = F_q(u)^{t^q}$ for all $u \in \mathcal{U}, t \in \mathbb{G}_a$ (see [10, Prop. 6.1]). One deduces from this that if H is an F_q-stable subgroup of G, then so is H^{sat}. These facts are used in [10, §6] to investigate the interaction of saturation with finite groups of Lie type; see also [134] for the case of GL_m.

(v). Saturation is compatible with semisimplification, in the following sense: if H is a subgroup of G and H' is a semisimplification of H yielded by the pair (P, L), then $(H')^{\text{sat}}$ is a semisimplification of H^{sat}, also yielded by the pair (P, L): see [10, Cor. 5.2].

One of the crucial properties of a saturated subgroup H is that the index $[H : H^0]$ is coprime to p: see [158, Property 3]. (To prove this, one observes that any element of order p in the quotient H/H^0 must be the image of a unipotent element of H; but by saturation, any such element is contained in some connected subgroup of H, and hence lies in H^0.) Now Remark 9.1.3 (or Theorem 9.1.6) implies that for a saturated subgroup H, H is G-cr if and only if H^0 is G-cr. Serre is able to deduce the following [158, Thm. 8]:

Theorem 10.7.4 *Suppose $p \geq h(G)$ and H is a subgroup of G. The following are equivalent:*

(i) H is G-completely reducible;
(ii) $(H^{\text{sat}})^0$ is reductive.

Remark 10.7.5 If M is a saturated connected reductive subgroup of G with $[M, M]$ simply connected and $p \geq \max\{h(M), h(G)\}$, then saturation in M coincides with saturation in G (see [158, Property 2, Thm. 4]). Using this result, we see that under this hypothesis on p the pair (G, M) has both the ascent and descent property: for if $H \subseteq M$, then H is G-cr if and only if $(H^{\text{sat}})^0$ is reductive if and only if H is M-cr, by two applications of Theorem 10.7.4.

As we noted earlier, saturation is also a key ingredient in Serre's proof of Theorem 10.3.1: see [158, Thm. 6].

10.8   Historical Remarks and References

Sections 10.2, 10.3 and 10.7 are heavily based on the notes [158] arising from Serre's "Moursund Lectures" in Oregon in 1998. These lectures and the accompanying notes were instrumental in publicising the notion of G-complete reducibility to a wide section of the algebraic groups community. Sections 10.1, 10.4, 10.5, 10.6 are based primarily on works of the authors, particularly parts of [18], [19] and [26]. Uchiyama has found several similar examples to that given in Sect. 10.1 for G simple of type E_6, E_7 and E_8 in characteristic 2 [186], [189]. Theorem 10.5.5 is [18, Thm. 3.46]; some variations on this theorem are given in [26, §4].

Reductive pairs were introduced by Richardson [143], as part of his beautiful uniform proof of the finiteness of the number of unipotent conjugacy classes for reductive groups in good characteristic. The main idea follows the scheme of the sketch proof of Theorem 10.5.13 at the end of Sect. 10.5: after reducing to the simple case and replacing G with an isogenous group if necessary, one takes an embedding of G in some $GL(V)$ such that $(GL(V), G)$ is a reductive pair, and then applies the tangent space argument Theorem 3.8.1 and its Corollary 3.8.2 to deduce the result for G from the result for $GL(V)$ (which follows from the Jordan Normal Form). For small values of p, there need not exist any such reductive pair $(GL(V), G)$ (see the paragraph following Corollary 10.5.24). Nonetheless Lusztig proved without any restriction on the characteristic, using some very deep mathematics, that G has only finitely many unipotent conjugacy classes [113]. As we mentioned in Remark 10.5.20, Guralnick extended this result to non-connected reductive groups [69, Thm. 3.3]; he then used this to prove that if M is a reductive subgroup of G and $m \in M$, then the intersection of the G-conjugacy class of m with M splits up as a finite union of M-conjugacy classes [69, Thm. 1.2]. Our example in the remark shows that this fails for orbits of n-tuples for $n > 1$.

The examples of reductive pairs in Example 10.5.16 can be found in [164, I.3], where Slodowy also discusses various other constructions giving reductive pairs and proves Proposition 10.5.18. Our application to G-complete reducibility (Theorem 10.5.21) is [18, Thm. 3.6]. Serre provided us with many further examples of reductive pairs in [18, Rem. 3.34]. See also [10, §3] for further discussion and extensions.

Above we described saturation for G with simply connected derived subgroup under the hypothesis that $p \geq h(G)$. This hypothesis ensures that every unipotent element of G has order p. In [154], Seitz effectively settled the saturation problem, for he was able to achieve saturation for unipotent elements of order p in any simple group G under the weaker hypothesis that p is good for G. This was achieved by Seitz in connection with his notion of *good* A_1 subgroups of G, [154, Thm. 1.3]. In [165, Prop. 4.1], Sobaje gives a simplified proof. We observe that the isomorphism of varieties $\log : \mathcal{U} \to \mathcal{N}$ from Theorem 10.7.1 is a so-called *Springer map*. For a discussion of the connection between Springer maps and saturation, see [10], [11, §4], [166] and the references therein.

10.9 Exercises

Exercise 10.1 Prove that (GL_{2m}, Sp_{2m}) is a reductive pair when the characteristic is not 2. Do the same for (GL_m, SO_m).

Exercise 10.2 The *trace form* on $\mathrm{Lie}(\mathrm{GL}_m) = \mathrm{Mat}_m$ is given by

$$(X, Y) := \mathrm{tr}(XY),$$

where $\mathrm{tr} : \mathrm{Mat}_m \to k$ denotes the usual matrix trace.

  (i) If this is unfamiliar to you, prove that the trace form gives a non-degenerate GL_m-invariant bilinear form on Mat_m.

 (ii) Now suppose G is a simple algebraic group of adjoint type, so that $\mathrm{Ad} : G \to \mathrm{GL}(\mathfrak{g})$ has trivial kernel. Viewing G as a subgroup of $\mathrm{GL}(\mathfrak{g})$ via Ad, show that the restriction of the trace form to $\mathfrak{g}$ is the Killing form on $\mathfrak{g}$.

(iii) Deduce that if the Killing form on $\mathfrak{g}$ is non-degenerate, then $(\mathrm{GL}(\mathfrak{g}), G)$ is a reductive pair.

This argument shows in particular that $(\mathrm{GL}(\mathfrak{g}), G)$ is a reductive pair when G is an adjoint simple group of exceptional type in good characteristic, as claimed in Example 10.5.16 (the characteristic restriction ensures the Killing form is non-degenerate).

Exercise 10.3 Verify the final claim of Example 10.5.17: if $p = 2$, $G = B_2$ and M is the regular subgroup of type A_1^2 generated by the short root subgroups, then (G, M) is *not* a reductive pair even though M is regular.

    [*Hint*: Since there is a maximal torus $T \subseteq M$, any M-stable complement to $\mathfrak{m}$ in $\mathfrak{g}$ has to be a sum of root spaces.]

Exercise 10.4 Let $H \cong \mathrm{Sp}_m$ embedded diagonally in the subgroup $K \cong \mathrm{Sp}_m \times \mathrm{Sp}_m \subseteq M = \mathrm{Sp}_{2m}$ in characteristic 2, as in Liebeck's Example 4.1.10. Verify the claim in Example 10.5.22(ii) that H is not separable in M.

    [*Hint*: Look at Exercise 4.10(iii).]

Exercise 10.5 In the discussion of the proof of Theorem 10.2.1 we saw Jantzen's result: if H is a connected reductive group and V is an H-module with $p \geq \dim(V)$, then V is semisimple for H. Combine this with Corollary 10.5.24 to prove the following: if G is a connected reductive group and $p \geq (\dim(V))^2$ for some non-degenerate G-module V, then a connected subgroup H of G is G-cr if and only if it is reductive.

Exercise 10.6 Suppose $p = 2$ and $G = \mathrm{SL}_2$. Then (10.7.1) has a much simplified form: $\varepsilon = u - 1$ satisfies $\varepsilon^2 = 1$ and so $u^t = 1 + t\varepsilon$ for all unipotent u and $t \in \mathbb{G}_a$.

  (i) Show that the saturation of the group generated by $u = \begin{pmatrix} 1 & 1 \\ 0 & 1 \end{pmatrix}$ is the usual upper triangular root group in G.

(ii) What is the saturation of the group generated by $w = \begin{pmatrix} 0 & 1 \\ 1 & 0 \end{pmatrix}$?

(iii) What is the saturation of $N_G(T)$, where T is the diagonal maximal torus in G?

Exercise 10.7 Suppose G is connected reductive, the derived group $[G, G]$ is simply connected, and $p \geq h(G)$, so that the results on saturation from Sect. 10.7 hold.

(i) Deduce from Theorem 10.7.1 that for any automorphism $\sigma : G \to G$, $u \in \mathcal{U}$ and $t \in \mathbb{G}_a$ we have $\sigma(u^t) = \sigma(u)^t$.

(ii) Show that for every $\lambda \in Y_G$ and $u \in \mathcal{U}$ we have

$$(\lambda(a)u\lambda(a)^{-1})^t = \lambda(a)u^t\lambda(a)^{-1} \tag{10.9.1}$$

for all $a \in k^*$, $t \in \mathbb{G}_a$. Use this to show directly that parabolic subgroups of G are saturated, as claimed in Remark 10.7.3(i).

G-Complete Reducibility over an Arbitrary Field 11

In this chapter, we drop the assumption that the field k is algebraically closed, so we let k be an arbitrary field, with algebraic closure $\overline{k}$. We continue to denote the characteristic of k by $p \geq 0$. Recall that an algebraic field extension k_1/k is *separable* if every element of k_1 has a separable minimal polynomial over k (no repeated roots in any extension field of k); we denote the separable closure of k by k_s. Recall that k is *perfect* if $p = 0$ or if the map $x \mapsto x^p$ is surjective in the case that $p > 0$. If k is perfect, then every algebraic extension of k is separable and hence, in particular, the separable closure is the algebraic closure; that is, $\overline{k} = k_s$.

There are various approaches to studying linear algebraic groups defined over non-algebraically closed fields. We choose to follow a more traditional approach, as in [32], since this fits with the majority of the literature on G-complete reducibility and also makes certain notions easier to write down. Over the next few sections we collect together the key definitions and results, roughly following the scheme of the earlier chapters in the book. That is, in Sect. 11.1 we deal with some generalities about affine varieties and algebraic groups before specialising to reductive groups in Sect. 11.2 (cf. Chaps. 2 and 3). In Sects. 11.3 and 11.4 we introduce the notion of G-complete reducibility and related geometric notions (cf. Chaps. 4, 5 and 6), before interpreting these ideas in the language of buildings in Sect. 11.5 (cf. Chap. 7). After that, in Sect. 11.6 we briefly discuss how the optimality ideas from Chap. 8 carry over to this new setting. Once this preliminary material has been established, we finish the current chapter by collecting in Sects. 11.7 and 11.8 a selection of highlights from the theory illustrating in particular how the notion of G-complete reducibility behaves (or doesn't!) in the presence of algebraic extensions and Steinberg endomorphisms. For further references and discussion, see Sect. 11.9.

© The Author(s), under exclusive license to Springer Nature Switzerland AG 2026

M. Bate et al., *G-Complete Reducibility, Geometric Invariant Theory and Spherical Buildings*, Oberwolfach Seminars 57, https://doi.org/10.1007/978-3-032-08866-6_11

11.1   Linear Algebraic k-Groups

We keep our standing assumption that "variety" means "affine variety". A variety X over $\overline{k}$ is said to be *k-defined* if there is a k-structure $k[X]$ on the coordinate ring $\overline{k}[X]$. That is, if when we view $\overline{k}[X]$ as a k-algebra via the inclusion of k in $\overline{k}$, there is a k-subalgebra $k[X]$ such that $\overline{k}[X] = k[X] \otimes_k \overline{k}$. We also say that *X is a k-variety* or *X has a k-structure*. A *k-morphism* (or *k-defined morphism*) between k-varieties Y and X is a morphism of varieties $\phi : Y \to X$ for which the comorphism ϕ^* induces a k-algebra homomorphism $k[X] \to k[Y]$. If Y is a closed subvariety of X then we say that Y is *k-defined* if there is a (necessarily unique) k-structure on Y such that the inclusion map $i : Y \to X$ is k-defined.

A *linear algebraic k-group H* (or just *algebraic k-group*) is a linear algebraic group H over $\overline{k}$ such that H is k-defined and the multiplication and inverse maps are also k-defined. We say "H is a k-group" or "H is k-defined" or "H has a k-structure". A *k-homomorphism* between k-groups is a k-defined homomorphism of algebraic groups. A *k-subgroup* (or *k-defined subgroup*) K of a k-group H is a closed subgroup K which is k-defined as a subvariety.

Given a k-defined variety X and any algebraic extension[1] k_1/k, we can also view X as a k_1-defined variety. Formally, we form the *base change* X_{k_1} of X to k_1 via the k_1-structure $k_1[X_{k_1}] := k[X] \otimes_k k_1$ on $\overline{k}[X]$. We only ever need to deal here with algebraic extensions, so we will always have $\overline{k_1} = \overline{k}$ and hence $X_{k_1}(\overline{k_1}) = X(\overline{k})$. Thus, following our convention that a variety is identified with its $\overline{k}$-points, we see that we can think of X and X_{k_1} as the same variety, just defined over different subfields of $\overline{k}$. Recall that the $\overline{k}$-points of X are in 1-1 correspondence with the $\overline{k}$-algebra homomorphisms $\overline{k}[X] \to \overline{k}$. If X is k-defined, then every k-algebra homomorphism $k[X] \to k$ induces a $\overline{k}$-algebra homomorphism $\overline{k}[X] \to \overline{k}$ after tensoring with $\overline{k}$, and in this way we can identify a subset $X(k) \subseteq X$ of k-points. Similarly, for any intermediate extension field $k \subseteq k_1 \subseteq \overline{k}$ we get a subset of k_1-points $X(k_1) = X_{k_1}(k_1)$. In this situation, if H is a k-group and k_1/k is any algebraic extension then $H(k_1)$ identifies as a subgroup of H.

We use the following quite a few times in what follows, so we single it out for referencing purposes; it is [127, Cor. 1.17].

Lemma 11.1.1 *Suppose X is a k-defined variety. Then $X(k_s)$ is dense in X. In particular, if H is a k-group then $H(k_s)$ is dense in H.*

Note that the previous result implies that if H is a k-group and H_1 and H_2 are two k-defined subgroups of H, then $H_1(k_s) = H_2(k_s)$ if and only if $H_1 = H_2$. Given a k-variety X there is an action of the absolute Galois group $\Gamma = \mathrm{Gal}(k_s/k)$ on $X(k_s)$, and there is also an action on k_s-defined morphisms between k-varieties; see [32, AG 14.3]. The Γ-action on $X(k_s)$ extends to an action on all of $X(\overline{k})$. One sees in *loc. cit.* that a

---

[1] or, more generally, a homomorphism of fields $k_1 \to k$.

k_s-defined morphism $\phi : X \to Y$ is k-defined if and only if the restriction $X(k_s) \to Y(k_s)$ is Γ-equivariant. Similarly, if Y is a closed subvariety of a k-variety X, then Y is k-defined if and only if Y is k_s-defined and $Y(k_s)$ is Γ-stable [32, AG 14.4].

Example 11.1.2 Let $k = \mathbb{C}$ and $H = \mathrm{SL}_2$ with coordinate ring

$$\mathbb{C}[H] = \frac{\mathbb{C}[X_{11}, X_{12}, X_{21}, X_{22}]}{(X_{11}X_{22} - X_{12}X_{21} - 1)},$$

where the X_{ij} are the usual coordinate functions on 2×2 matrices. Then there are two distinct $\mathbb{R}$-structures on H. The first is the "obvious" one, letting $\mathbb{R}[H]$ be the $\mathbb{R}$-subalgebra generated by the X_{ij}. This gives $H(\mathbb{R}) = \mathrm{SL}_2(\mathbb{R})$. The second is

$$\mathbb{R}[H] = \frac{\mathbb{R}[A, B, C, D]}{(A^2 + B^2 + C^2 + D^2 - 1)},$$

where we identify $\mathbb{R}[H]$ as an $\mathbb{R}$-subalgebra of $\mathbb{C}[H]$ via $A + iB \mapsto X_{11}$, $A - iB \mapsto X_{22}$, $C + iD \mapsto X_{12}$, $-C + iD \mapsto X_{21}$. For this $\mathbb{R}$-structure, we have $H(\mathbb{R}) = \mathrm{SU}(2)$; see Exercise 11.2.

A k-defined torus T is *split* if T is k-isomorphic to a direct product of copies of the multiplicative group over k. Any k-defined torus T contains a unique maximal split subtorus; the *k-rank of* T is the dimension of this maximal split subtorus. For a k-defined torus T there is some finite Galois extension k_1/k such that T_{k_1} is split (i.e., T splits over k_1). In particular, any k-defined torus splits over the separable closure k_s. If H is a k-group, then H contains a maximal torus (i.e., a maximal torus over $\overline{k}$) which is k-defined; we call such a subtorus a *maximal k-torus of H*. We also have *maximal split tori*; these are the k-defined subtori in H which are maximal amongst split k-subtori of H. Note that the maximal k-tori of H need not be $H(k)$-conjugate, but any two maximal split tori of H are $H(k)$-conjugate; see [47, Prop. A.2.10, Thm. C.2.3]. The *split rank of H* is the k-rank of a maximal split torus in H.

We write $Y_{H,k}$ for the set of *k-cocharacters of H* (i.e., k-homomorphisms from the multiplicative group over k into H). The image of any k-cocharacter of H is obviously k-split, and a maximal split torus T of H is generated by the images of k-cocharacters evaluating in T. If T is k-split, then every cocharacter of T is k-defined, since splitness implies precisely that $Y_{T,k} \cong \mathbb{Z}^r$, where r is the rank of T.

Suppose now that M is a k-group and H is a subgroup of M. In this situation, we define $H(k) = H \cap M(k)$, which makes sense even if H is not itself k-defined as a subgroup of M. It also makes sense to talk about a k-torus of H—we mean a subtorus of H which is k-defined as a subtorus of M—and to ask whether such a subtorus is split or not. Any two maximal split tori of H are $H(k)$-conjugate: to see this, consider the subgroup of H generated by all such tori, which is k-defined. Note finally that even if H is k-defined, it is not necessarily the case that $C_M(H)$ is k-defined. More generally, if X is a G-variety over

k and $x \in X(k)$ then the stabiliser G_x need not be k-defined. However, if H is linearly reductive (and in particular if H is a k-torus) then $C_M(H)$ is k-defined.

11.2   Reductive k-Groups and R-Parabolic Subgroups

There is a structure theory for connected reductive k-groups which mirrors that in the algebraically closed case, with some modifications. We begin by collecting together some of the details of that theory. Extending the convention in force for the rest of the book, G always denotes a reductive k-group; if we need G to be connected we will specify so. All of the material on connected reductive k-groups below can be found in [32, Ch. V], for example.

11.2.1 Connected Reductive k-Groups

We say that G is *isotropic* if it has nonzero split rank (contains a non-trivial split torus), and *anisotropic* otherwise. We call the k-defined parabolic subgroups of G *k-parabolic subgroups*, and their k-defined Levi subgroups are *k-Levi subgroups*. There are proper k-parabolic subgroups in G if and only if G contains a non-central k-split torus. Any k-parabolic subgroup P contains a k-Levi subgroup, and $R_u(P)(k)$ acts simply transitively on the set of k-Levi subgroups of P. The minimal k-parabolic subgroups in G are all $G(k)$-conjugate. The intersection of any two k-parabolic subgroups of G contains the centraliser of some maximal split torus of G and hence, in particular, contains a maximal split torus of G.

Fix a maximal split torus T of G and let P be a minimal k-parabolic subgroup of G containing T. Then the subgroup $L = C_G(T)$ is a Levi k-subgroup of P. The *relative Weyl group* of G is the finite group $W_k = N_G(T)/C_G(T) \cong N_G(T)(k)/C_G(T)(k)$. The *relative root system* $\Phi_k = \Phi_k(G, T)$ of G is the set of T-weights on the Lie algebra $\mathfrak{g}$. It does form a root system, but is possibly non-reduced. The choice of P determines a base Π_k of the relative root system.

For every $\lambda \in Y_{G,k}$, the corresponding parabolic and Levi subgroups $P_\lambda \supseteq L_\lambda$ are k-defined. Conversely, every pair (P, L) consisting of a k-parabolic subgroup P and a k-Levi subgroup L of P has the form (P_λ, L_λ) for some $\lambda \in Y_{G,k}$.

Example 11.2.1   Note that it is not true in general that the centraliser $C_G(S)$ of a k-defined torus S in a connected reductive k-group G is a Levi subgroup of a k-parabolic subgroup (this is in contrast to the algebraically closed case, where every $C_G(S)$ is a Levi subgroup of some parabolic subgroup; see Corollary 2.9.15). For example, we could take the real form of $G = \mathrm{SL}_2$ for $k = \mathbb{C}$ with $G(\mathbb{R}) = \mathrm{SU}(2)$ from Example 11.1.2 above. There are no proper $\mathbb{R}$-defined parabolic subgroups in G at all, but there is an $\mathbb{R}$-defined maximal torus T such that $T(\mathbb{R}) \cong S^1$ is the usual realisation of the circle group inside $\mathrm{SU}(2)$.

If one (hence all, by $G(k)$-conjugacy) of the maximal k-tori in G is split, then we call the group G itself *split*. When G is split it contains a Borel subgroup defined over k, and the relative root system and combinatorics of k-parabolic and k-Levi subgroups of G is essentially the same as it is over $\overline{k}$. Since any k-torus splits over a finite separable extension of k, we can always split a connected reductive group G by passing to a suitable finite separable extension; further, if $k = k_s$, then every reductive k-group is split. In this case the description of R-parabolic subgroups and R-Levi subgroups from Sect. 2.9 carries over: e.g., P_λ is generated by a maximal k-torus T such that $\lambda \in Y_{T,k}$ and the root groups U_α for $\alpha \in \Phi(G, T)$ such that $\langle \lambda, \alpha \rangle \geq 0$. In fact, we can make this description work for non-split G as well. First we pass to k_s and form P_λ as above, regarding λ as an element of Y_{G,k_s}. Then we observe that the resulting subgroup is k-defined, because the set of roots α such that $\langle \lambda, \alpha \rangle \geq 0$ is Galois-stable.

11.2.2 Non-Connected Reductive k-Groups

There is one subtlety in the theory of non-connected reductive k-groups which merits some careful attention (although it does not in the end affect the theory of G-complete reducibility, see Proposition 11.3.4 below). We noted in the previous subsection that for a connected reductive k-group G all the k-parabolic subgroups have the form P_λ for $\lambda \in Y_{G,k}$. This is *not* the case in general for R-parabolic subgroups when G is not connected.

Example 11.2.2   For a straightforward example of the phenomenon above, consider the $\mathbb{R}$-algebra $A = \mathbb{R}[X, Y \mid X^2 + Y^2 = 1]$. Then $B := A \otimes_{\mathbb{R}} \mathbb{C} \cong \mathbb{C}[Z, Z^{-1}]$, the algebra of complex Laurent polynomials in Z, where the isomorphism over $\mathbb{C}$ comes from $X + iY \mapsto Z$, $X - iY \mapsto Z^{-1}$. Since B is the coordinate algebra of the multiplicative group $T := \mathbb{G}_m$ over $\mathbb{C}$, we see that A provides an $\mathbb{R}$-structure on T. Note also that since there is no $\mathbb{R}$-algebra isomorphism of A with $\mathbb{R}[Z, Z^{-1}]$, T is not split over $\mathbb{R}$ when endowed with this $\mathbb{R}$-structure; hence, in particular $Y_{T,\mathbb{R}} = \{0\}$.

We can let the finite group F of order 2 act on T by inversion, giving a non-connected reductive group $G = FT$. For any non-trivial cocharacter $\lambda \in Y_T = Y_G$ we have $P_\lambda = L_\lambda = T$, so this is an $\mathbb{R}$-defined R-parabolic subgroup of G. But there is no $\mu \in Y_{T,\mathbb{R}}$ with $T = P_\mu$, because $Y_{T,\mathbb{R}}$ is trivial.

As above, let $\Gamma = \mathrm{Gal}(k_s/k)$ denote the absolute Galois group of k and recall that there is an action of Γ on the k_s-points $G(k_s)$ and on k_s-defined morphisms. This gives rise to an action on k_s-defined subgroups and on Y_{G,k_s}. We denote both of these actions by a dot, so for $\gamma \in \Gamma$ we have $H \mapsto \gamma \cdot H$ for a k_s-defined subgroup H of G, and $\lambda \mapsto \gamma \cdot \lambda$ for $\lambda \in Y_{G,k_s}$. It is not hard to prove that this set-up is compatible with the formation of R-parabolic and R-Levi subgroups in the following sense.

Lemma 11.2.3 *Suppose $\lambda \in Y_{G,k_s}$ and $\gamma \in \Gamma$. Then $P_{\gamma \cdot \lambda} = \gamma \cdot P_\lambda$ and $L_{\gamma \cdot \lambda} = \gamma \cdot L_\lambda$.*

To get around the sort of problem in Example 11.2.2, we use part (ii) of the following. We also use this result to record some basic properties of R-parabolic and R-Levi subgroups in this setting.

Lemma 11.2.4 *Suppose P is a k-defined R-parabolic subgroup of G.*

(i) *There is $\lambda \in Y_{G,k}$ such that $P \subseteq P_\lambda$ and $P^0 = P_\lambda^0$.*
(ii) *Given λ as in part (i), $L_\lambda \cap P$ is a k-defined R-Levi subgroup of P. In particular, P has k-defined R-Levi subgroups.*
(iii) *For each k-defined maximal torus T of P, there is a unique k-defined R-Levi subgroup of P containing T.*
(iv) *$R_u(P)$ is k-defined, and given any k-defined R-Levi subgroup L of P, the isomorphism $L \ltimes R_u(P) \to P$ is k-defined. In particular, every $g \in P(k)$ has a unique expression $g = lu$ with $u \in R_u(P)(k)$ and $l \in L(k)$.*
(v) *Any two k-defined R-Levi subgroups of P are conjugate by a unique element of $R_u(P)(k)$.*

Proof (i). Let T be a k-defined maximal torus of P. There is a finite Galois extension k_1/k such that T splits over k_1, and then we can find $\mu \in Y_{T,k_1}$ such that $P = P_\mu$. Let $\lambda = \sum \gamma \cdot \mu$ be the sum of the Galois conjugates of μ. Then λ is k-defined and it is reasonably straightforward to check that $P_\mu \subseteq P_\lambda$ and $P_\lambda^0 = P_\mu^0$.

(ii). Let L be the unique R-Levi subgroup of P containing T, which exists by Lemma 2.9.10(ii). Since L is G-cr (over $\bar{k}$), there is an R-Levi subgroup of P_λ containing L, and this must be L_λ since L_λ is the unique R-Levi subgroup of P_λ containing T, again by Lemma 2.9.10(ii). The subgroup $L_\lambda \cap P = L_\lambda(P)$ is the centraliser in the k-defined group P of the image of λ, so is k-defined, and it contains L. But P is the semidirect product of L and $R_u(P) = R_u(P_\lambda)$, so we must in fact have $L = L_\lambda \cap P$.

(iii). The existence of a k-defined R-Levi subgroup containing T follows from (ii), and uniqueness from Lemma 2.9.10(ii) again.

(iv). Since $R_u(P) = R_u(P^0)$ the first statement follows from the connected case. The second and third statements follow quickly since the multiplication map $G \times G \to G$ is k-defined.

(v). Let L and M be two k-defined R-Levi subgroups of P, and let $u \in R_u(P)(k) = R_u(P^0)(k)$. Then $uLu^{-1} = M$ if and only if $uL^0u^{-1} = M^0$, by (iii), so the result follows from the connected case.

$\square$

11.3   *G*-Complete Reducibility over *k*

Definition 11.3.1 Suppose H is a subgroup of G. We say that H is *G-completely reducible over k* (*G-cr over k*) if whenever H is contained in a k-defined R-parabolic subgroup P of G, there is a k-defined R-Levi subgroup L of P with $H \subseteq L$.

Remark 11.3.2 Note that we do not require H to be k-defined in this definition. We also extend the definition to cover arbitrary abstract subgroups H of G.

Example 11.3.3 Suppose U is a k-vector space and let $V := U \otimes_k \bar{k}$. Then $G = \mathrm{GL}(V)$ or $\mathrm{SL}(V)$ gets a k-structure from the k-structure on V in such a way that we can identify $G(k)$ with $\mathrm{GL}(U)$ or $\mathrm{SL}(U)$. We say a subspace Y of V is *k-defined* if Y has the form $Y = X \otimes_k \bar{k}$ for a subspace X of U. Then the k-parabolic subgroups of G are the stabilisers of flags of k-defined subspaces of V.

Now it is not hard to see that an abstract subgroup H of G is G-completely reducible over k if and only if for every H-stable k-defined subspace Y of V there is an H-stable k-defined subspace Y' of V with $V = Y \oplus Y'$. If $H \subseteq G(k)$, this is the case if and only if H acts semisimply on U (viewed as a module for H over k).

Recall from Example 11.2.2 above that it is not necessarily true that every k-defined R-parabolic subgroup of G has the form P_λ for some $\lambda \in Y_{G,k}$. However, this does not cause a problem in the theory of G-complete reducibility over k, as we now show.

Proposition 11.3.4 *Suppose H is a subgroup of G. Then H is G-completely reducible over k if and only if for every $\lambda \in Y_{G,k}$ such that $H \subseteq P_\lambda$, there exists $\mu \in Y_{G,k}$ such that $P_\lambda = P_\mu$ and $H \subseteq L_\mu$.*

Proof The forward implication is obvious. For the other direction, given any k-defined R-parabolic subgroup P of G containing H, let $\lambda \in Y_{G,k}$ be as given by Lemma 11.2.4(i). By hypothesis, we can replace λ with $\mu \in Y_{G,k}$ such that $P \subseteq P_\lambda = P_\mu$ and $H \subseteq L_\mu$. But then $H \subseteq L_\mu \cap P$, which is an R-Levi subgroup of P, by Lemma 11.2.4(ii). $\square$

We now have two very natural questions:

Question 11.3.5 (Ascent/Descent Through Field Extensions)   If k'/k is an algebraic extension, what is the relationship between G-complete reducibility over k and over k'?

Question 11.3.6 (Rational Versus Geometric G-complete Reducibility)   How many results from the algebraically closed situation can be ported over to this rational situation?

Providing a satisfactory answer to the first of these questions will take up most of the rest of this chapter. In situations where the answer to Question 11.3.5 is understood, we can then try to address the second question, for example by moving backwards and forwards between k and $\bar{k}$. Using the optimality theory of Chap. 8, one can show that there is no obstruction to this at all if the field k is perfect; that is, in this situation we have the best possible answer to Question 11.3.6, see Remark 11.7.13(ii). However, if the field k is not perfect, we immediately run into problems, as the next examples show.

Example 11.3.7 Suppose k is an imperfect field of characteristic 2. Then the squaring map $x \mapsto x^2$ is not surjective on k, so we may find an element $a \in k \setminus k^2$, and form a purely inseparable extension $k_1 = k(t)$, where $t^2 = a$. Consider the algebraic group G whose $\bar{k}$-points are given by

$$G = \left\{ \begin{pmatrix} x & ay \\ y & x \end{pmatrix} \;\middle|\; x, y \in \bar{k}, x^2 + ay^2 \neq 0 \right\}.$$

This group is clearly a k-subgroup of the group GL_2. The natural module k^2 is irreducible for $G(k)$, since $G(k)$ has a single orbit on the nonzero vectors in k^2: for each $x, y \in k$ not both 0 we have

$$\begin{pmatrix} x & ay \\ y & x \end{pmatrix} \begin{bmatrix} 1 \\ 0 \end{bmatrix} = \begin{bmatrix} x \\ y \end{bmatrix}. \tag{11.3.1}$$

However, the group G is triangularisable over k_1: it has a simultaneous eigenvector $\begin{bmatrix} t \\ 1 \end{bmatrix}$, and one can see that $G \cong \mathbb{G}_m \times \mathbb{G}_a$ over k_1 (and hence this also holds over $\bar{k}$). Note that this eigenvector is *not* in the $G(\bar{k})$-orbit of $\begin{bmatrix} 1 \\ 0 \end{bmatrix}$, since if we set $x = t$ and $y = 1$ in (11.3.1) we obtain a matrix which is not invertible! It is now easy to see that G is GL_2-cr over k but not GL_2-cr over k_1 (or GL_2-cr).

A similar example can be given in any positive characteristic p: these are all so-called *Weil restrictions* of the multiplicative group across purely inseparable field extensions (given a finite extension of fields k_1/k, the Weil restriction functor $\mathrm{R}_{k_1/k}$ turns k_1-groups into k-groups; for full details, see [47, §A.5]). In the case that $k_1 = k(a)$ is a purely inseparable extension of degree p generated by the element a, we can take G to be the group whose $\bar{k}$-points are the units of the ring $A := \bar{k} \otimes_k k_1$; since A has a $\bar{k}$-basis $1 \otimes 1, 1 \otimes a, \ldots, 1 \otimes a^{p-1}$, the elements of G can be written as $p \times p$ matrices with entries in $\bar{k}$, and hence we get a subgroup of GL_p. It turns out that this subgroup is GL_p-ir over k, but not GL_p-cr. See [18, Ex. 5.11], which was originally communicated to the authors by G. McNinch, and Exercise 11.4.

Example 11.3.8 We can refine the example introduced in Sect. 10.1 to give an example of the failure of the descent of G-complete reducibility to k. Suppose that $p = 2$ and k is an imperfect field. Let $G = G_2$ be split over k so that we can choose a maximal torus T of G which is split over k and a k-defined Borel subgroup B of G containing T. All the root homomorphisms κ_γ are then defined over k and we can also ensure that the subgroup $H = \langle s_\alpha, t \rangle \subseteq G(k)$.

Since k is imperfect, we can find $a \in \overline{k} \setminus k$ such that $a^2 \in k$. Then it is shown in [26, Ex. 7.22] that H_a is not G-cr over k; since H_a *is* G-cr, this shows that G-complete reducibility cannot be descended through arbitrary algebraic field extensions. The proof that H_a is not G-cr over k is very similar to the idea in the proof of Lemma 10.1.1: if H_a were G-cr over k then it would be $G(k)$-conjugate to H, but knowledge of the normaliser of H and the fact that $u(a) \notin G(k)$ (since $a \notin k$) allows one to show that H_a and H are not conjugate by an element of $G(k)$.

We refer to [7, Thm. 1.3] for a similar example when $p = 2$ and H is a connected subgroup of a connected simple group G of type F_4. Note that in [7, Thm. 1.3], the authors also provide an example of a connected subgroup H' of G of type F_4 which is G-cr over k but not G-cr, as in Example 11.3.7.

All known examples of the failure of the descent of G-complete reducibility occur in characteristic 2, so we pose the following:

Open Problem 11.3.9 Does there exist a reductive group G defined over a field k of odd characteristic and a k-subgroup H of G such that H is G-cr but not G-cr over k?

11.4 Cocharacter-Closed Orbits and G-Complete Reducibility over k

We wish to provide a "rational analogue" of the geometric approach to G-complete reducibility, where we can study properties of subgroups of a k-group G via properties of orbits in the action of G on G^n for suitable n. In order to set this up, it is convenient to expand our remit to cover the action of k-groups on k-varieties more generally. To this end, we call a variety X a *k-defined G-variety* if X is a G-variety which is k-defined and the action of G on X is also k-defined. One can show that the construction of the quotient variety $X /\!\!/ G$ from Sect. 3.1 makes sense for arbitrary k (see [8, §2]) but for the most part we concentrate on orbits and their properties.

Suppose until further notice that X is a k-defined G-variety. Recall Definition 3.3.1; we extend these ideas to this new setting.

Definition 11.4.1 With the notation as above, we make the following definitions:

(i) Let $\mathcal{O}$ be a $G(k)$-orbit in X, and let $x \in X$. We say $\mathcal{O}$ is *accessible from x over k* if there exists $\lambda \in Y_{G,k}$ such that $\lim_{a \to 0} \lambda(a) \cdot x$ exists and lies in $\mathcal{O}$.

(ii) Given a subset Z of X, we say that Z is *cocharacter-closed over k* if for every $x \in Z$, all the points accessible from x over k also lie in Z.

(iii) Given a subset Z of X, we define the *cocharacter-closure of Z over k*, denoted $\overline{Z}^c$, to be the smallest subset of X such that $Z \subseteq \overline{Z}^c$ and $\overline{Z}^c$ is cocharacter-closed over k. (This makes sense because the intersection of cocharacter-closed subsets is clearly cocharacter-closed.)

Note that we do not insist the orbit O is k-defined, or is the orbit of a k-point, and nor do we insist that $x \in X(k)$. When it is clear from the context that we are working over a fixed field k we sometimes omit the "over k" from the above terms.

Example 11.4.2 If $X = G^m$, $\lambda \in Y_{G,k}$ and $\mathbf{g} \in P_\lambda^m$ then $G(k) \cdot c_\lambda(\mathbf{g})$ is accessible from $\mathbf{g}$ over k.

Remark 11.4.3 Note that the cocharacter-closure of an orbit need not be Zariski closed, even if we are working over $\overline{k}$. For example, in the situation of Example 3.3.5, if G is a simple group over $k = \overline{k}$, $u \in G$ is a regular unipotent element of G and $v \in G$ is a distinguished non-regular unipotent element, then $v \in \overline{G \cdot u}$ but $G \cdot v$ is *not* accessible from any point of $\overline{G \cdot u} \setminus G \cdot v$, by the argument given in that example, so $\overline{G \cdot u} \setminus G \cdot v$ is cocharacter-closed; hence $\overline{G \cdot u}^c$ is properly contained in the Zariski closure $\overline{G \cdot u}$. See also Exercise 11.3.

The relevance to the theory of G-complete reducibility is the following result. Note that in this theorem we do not require the subgroup H to be k-defined, and nor do we require the generic tuple to be a k-point; indeed, there may be situations where there are *no* generic tuples which are k-points, but we can still apply the theorem. See also Lemma 11.4.14.

Theorem 11.4.4 ([17, Thm. 9.3]) *Let H be a subgroup of G and let $\mathbf{h} \in H^n$ be a generic tuple of H. Then H is G-completely reducible over k if and only if $G(k) \cdot \mathbf{h}$ is cocharacter-closed over k.*

Recall that the Hilbert-Mumford Theorem 3.3.2 says that over $\overline{k}$ an orbit is closed if and only if it is cocharacter-closed, so the above result is a direct generalisation of Theorem 5.3.3 to arbitrary fields. We give the proof of this theorem at the end of this section, once we have collected some general machinery. Since the proofs of these technical results are quite involved, we only give sketches of the main ideas here. First of all, we note that there is a version over k of the "Levi descent" result, Corollary 2.10.10.

Proposition 11.4.5 *Let $x \in X$, suppose S is a k-defined torus of G_x, set $L = C_G(S)$ and suppose $\lambda \in Y_{L,k}$ is such that $x' = \lim_{a \to 0} \lambda(a) \cdot x$ exists. If x' is $R_u(P_\lambda)(k)$-conjugate to x, then x' is $R_u(P_\lambda(L))(k)$-conjugate to x.*

Sketch Proof Let $u \in R_u(P)(k)$ be such that $x' = u \cdot x$. Then $u^{-1} \cdot \lambda$ fixes x by Lemma 3.2.8, so evaluates in G_x. Let H be the k-subgroup of G generated by S and the image of $u^{-1} \cdot \lambda$. Since $S \subseteq P$ and $u^{-1} \cdot \lambda$ evaluates in P, we have $H \subseteq P$. Choose a maximal k-torus S' of H with $S \subseteq S'$ and let $h \in H(k)$ such that $\mu := (hu^{-1}) \cdot \lambda \in Y_{S',k}$. Then μ fixes x, $P_\lambda = P_\mu$, and λ and μ both lie in $Y_{L,k}$. Thus the R-Levi subgroups $L_\lambda(L)$ and $L_\mu(L)$ are conjugate by some $u_0 \in R_u(P_\lambda(L))(k)$ (Lemma 11.2.4(v)). Now one can show that $x' = u_0 \cdot x$. For full details, see [17, Prop. 2.11]. $\qquad\square$

Our next result gives an analogous descent result, but this time rather than descending to a Levi subgroup, we wish to descend through a separable field extension.

Theorem 11.4.6 *Let $x \in X$ and suppose $G_x(k_s)$ is Γ-stable. Let $\lambda \in Y_{G,k}$ be such that $x' = \lim_{a \to 0} \lambda(a) \cdot x$ exists. If x' is $R_u(P_\lambda)(k_s)$-conjugate to x, then x' is $R_u(P_\lambda)(k)$-conjugate to x.*

Sketch Proof One can quickly reduce to the case that G is connected, since $R_u(P_\lambda) \subseteq G^0$. Let $u \in R_u(P_\lambda)(k_s)$ such that $x' = u \cdot x$; then by Lemma 3.2.8, $\mu = u^{-1} \cdot \lambda \in Y_{G,k_s}$ fixes x, so the image of μ lies in G_x. Now generate a subgroup H of G with all the Galois conjugates of (the k_s-points of) the image of μ. The hypothesis that $G_x(k_s)$ is Γ-stable allows us to conclude that $H \subseteq G_x$. Since H is (essentially by construction) k-defined, it has a k-defined maximal torus S. We can now find a k-defined Levi subgroup of P containing S, which is conjugate to L_λ by some $u_0 \in R_u(P_\lambda)(k)$. After some further work, one can show that in fact $x' = u_0 \cdot x$. For full details see [28, Thm. 3.1]. $\qquad\square$

Remark 11.4.7 Recall that our earlier Levi descent result was proved in Sect. 2.10 using 1-cohomology arguments (see Proposition 2.10.8). It would be nice to have such a "cohomological proof" which works over k, but this seems to be difficult to implement in practice. However, a proof of Theorem 11.4.6 using Galois cohomology should be possible, as follows. First note that the unipotent radical of any k-defined R-parabolic subgroup of G is *split* (it admits a composition series over k with each factor isomorphic to the additive group); see [47, Prop. 2.1.20]. For a split unipotent k-group U the Galois cohomology $H^1(\Gamma, U)$ is trivial, where Γ is the absolute Galois group, see [125, Prop. 30]. The vanishing of this cohomology set should then allow us to descend $R_u(P_\lambda)(k_s)$-conjugacy to $R_u(P_\lambda)(k)$-conjugacy. We leave it to the interested reader to check the details.

Note that neither of the previous results requires that x is a k-point of X, but the hypothesis in Theorem 11.4.6 does hold in particular when $x \in X(k)$. Here is an example which shows that the result can fail without some condition on the stabiliser.

Example 11.4.8 Suppose k is a field which is not separably closed, let $G = \mathrm{SL}_2$ over k, and let G act on itself by conjugation. Choose any $y \in k_s \setminus k$ and $t \in k \setminus \{-1, 0, 1\}$. Let

$$u = \begin{pmatrix} 1 & y \\ 0 & 1 \end{pmatrix} \text{ and } x' = \begin{pmatrix} t & 0 \\ 0 & t^{-1} \end{pmatrix}.$$

Let λ be the diagonal cocharacter given by $\lambda(a) = \begin{pmatrix} a & 0 \\ 0 & a^{-1} \end{pmatrix}$ and let $x = ux'u^{-1}$. Then the limit $\lim_{a \to 0} \lambda(a) \cdot x = x'$ is $R_u(P_\lambda)(k_s)$-conjugate to x, but x' is *not* $R_u(P_\lambda)(k)$-conjugate to x.

Our final technical result is used three times below; it is in fact the key result which allows much of the development of the theory of cocharacter-closed orbits over k.

Proposition 11.4.9 *Let $x \in X$, let S be a maximal k-split torus of G_x, and let $L = C_G(S)$. Suppose there exists $\lambda \in Y_{G,k}$ such that $\lim_{a \to 0} \lambda(a) \cdot x$ exists but is not $R_u(P_\lambda)(k)$-conjugate to x. Then there exists $\tau \in Y_{L,k}$ such that $\lim_{a \to 0} \tau(a) \cdot x$ exists and lies outside $G(k) \cdot x$.*

Sketch Proof The key is to argue that without loss one can "line up" S and λ so that they commute with each other. More precisely, since S is split we can first find a cocharacter $\mu \in Y_{S,k}$ such that $L = L_\mu$ and μ fixes the same points of X as S. Then, after replacing S, λ and μ with suitable conjugates if necessary we may get to the situation where λ and μ commute. Then τ is constructed as the cocharacter $\lambda + n\mu$ for suitably large n. Since λ is assumed to take x outside its $R_u(P_\lambda)(k)$-orbit, no $R_u(P_\lambda)(k)$-conjugate of λ fixes x, by Lemma 3.2.8, and hence $\mathrm{Im}(\tau)$ and S generate a split torus S' of rank larger than the rank of S. This torus S' stabilises the limit point $x' := \lim_{a \to 0} \tau(a) \cdot x$, so x' cannot be $G(k)$-conjugate to x, by the maximality of S. For full details, see [17, Prop. 4.1] $\square$

We can now give a rational version of the Hilbert-Mumford Theorem 3.3.2. This is a central result in the study of rational G-actions on varieties (just as its analogue over $\bar{k}$ is central in the study of orbits in the geometric setting).

Theorem 11.4.10 (Rational Hilbert-Mumford Theorem) *Suppose $x \in X$.*

  (i) *There is a unique cocharacter-closed $G(k)$-orbit in X which is accessible from x over k.*
 (ii) *The orbit from part (i) is the unique cocharacter-closed $G(k)$-orbit in $\overline{G(k) \cdot x}^c$.*
(iii) *If $x' \in X$ and $G(k) \cdot x' \subseteq \overline{G(k) \cdot x}^c$, then the unique cocharacter-closed $G(k)$-orbits in $\overline{G(k) \cdot x}^c$ and $\overline{G(k) \cdot x'}^c$ are equal.*

Sketch Proof To construct the cocharacter-closed orbit in (i), we use Proposition 11.4.9 repeatedly. First choose a maximal k-split torus S of G_x. If $G(k) \cdot x$ is not cocharacter-closed then, using Proposition 11.4.9, we can find $\tau \in Y_{G,k}$ such that τ commutes with S

and $x' = \lim_{a \to 0} \tau(a) \cdot x$ exists and lies outside $G(k) \cdot x$. If $G(k) \cdot x'$ is cocharacter-closed, then we're done. If not, then note that $G_{x'}$ contains the split torus S' generated by S and the image of τ—this torus has rank strictly larger than the rank of S—and repeat the argument with x' and S' in place of x and S. Since the rank of a split torus in G is bounded above, this process terminates.

Once the existence of the orbit in (i) is established, one needs to show that it is actually accessible from x (one can take a single limit rather than having to take limits repeatedly), and to establish the remaining claims. For the details, see [17, Prop. 4.1, Thm. 4.3]. $\square$

This result shows that we might expect the cocharacter-closed orbits to play the same role over k as the closed orbits do over $\bar{k}$ (indeed, it is part of the Hilbert-Mumford Theorem over $\bar{k}$ that cocharacter-closed orbits are closed). This expectation is reinforced by the following rational version of Theorem 3.3.6 for cocharacter-closed orbits.

Lemma 11.4.11 *Suppose $x \in X$ is such that $G(k) \cdot x$ is cocharacter-closed. Then for every $\lambda \in Y_{G,k}$ such that $x' = \lim_{a \to 0} \lambda(a) \cdot x$ exists, x' is $R_u(P_\lambda)(k)$-conjugate to x.*

Sketch Proof Suppose for a contradiction that x is not $R_u(P_\lambda)(k)$-conjugate to x, and let S be a maximal split torus of G_x. Then we can use Proposition 11.4.9 again to construct a cocharacter τ commuting with S such that the limit $\lim_{a \to 0} \tau(a) \cdot x$ exists and is not $G(k)$-conjugate to x. But this contradicts the fact that $G(k) \cdot x$ is cocharacter-closed. See [17, Cor. 5.1]. $\square$

With this result, we can now prove Theorem 11.4.4:

Proof of Theorem 11.4.4 Let H be a subgroup of G with generic tuple $\mathbf{h} \in H^n$. Suppose first that H is G-cr over k. Then given any $\lambda \in Y_{G,k}$ such that $\mathbf{h}' := \lim_{a \to 0} \lambda(a) \cdot \mathbf{h}$ exists, we have $H \subseteq P_\lambda$ and so we can find $u \in R_u(P_\lambda)(k)$ with $H \subseteq L_{u \cdot \lambda}$. But then $u \cdot \lambda$ fixes $\mathbf{h}$, which is equivalent to $\mathbf{h}' = u^{-1} \cdot \mathbf{h}$ by Lemma 3.2.8, which implies that $\mathbf{h}' \in G(k) \cdot \mathbf{h}$. This proves that $G(k) \cdot \mathbf{h}$ is cocharacter-closed.

Suppose conversely that $G(k) \cdot \mathbf{h}$ is cocharacter-closed, and let $\lambda \in Y_{G,K}$ be such that $H \subseteq P_\lambda$. Then $\mathbf{h}' = \lim_{a \to 0} \lambda(a) \cdot \mathbf{h}$ exists, and we can apply Lemma 11.4.11 to find some $u \in R_u(P_\lambda)(k)$ with $\mathbf{h}' = u^{-1} \cdot \mathbf{h}$. Now another application of Lemma 3.2.8 implies that $u \cdot \lambda$ fixes $\mathbf{h}$ and hence $H \subseteq L_{u \cdot \lambda}$, a k-defined R-Levi subgroup of P_λ. It follows that H is G-cr over k. $\square$

This result provides the link over k between G-completely reducible subgroups and cocharacter-closed orbits in G^n, which is our rational analogue of Theorem 5.3.3. Note that Lemma 11.4.11 gives an extension of Theorem 3.3.6, but only for points with cocharacter-closed orbits, whereas over $\bar{k}$ the theorem applies for arbitrary points. It is not known whether the result holds in general without the assumption of cocharacter-closedness. We formulate this as an open problem.

Open Problem 11.4.12 Does the full analogue of Theorem 3.3.6 hold over k? That is, suppose $x \in X$ and $\lambda \in Y_{G,k}$ are such that $x' = \lim_{a \to 0} \lambda(a) \cdot x$ exists. If x' is $G(k)$-conjugate to x is it true that x' is $R_u(P_\lambda)(k)$-conjugate to x?

The fact that we do not have a full answer to Open Problem 11.4.12 is closely related to the fact that at the time of writing we also do not have an analogue over arbitrary fields of the optimality results from Chap. 8; see the next section. One way around these problems is to impose extra hypotheses of k-definedness on the stabiliser of a point. We give a result along these lines, showing that we can descend Theorem 3.3.6 to k as long as we are looking at a point with k-defined stabiliser.

Theorem 11.4.13 *Suppose $x \in X$ is such that G_x^0 is k-defined. Then for any $\lambda \in Y_{G,k}$ such that $x' = \lim_{a \to 0} \lambda(a) \cdot x$ exists and is G-conjugate to x, in fact x' is $R_u(P_\lambda)(k)$-conjugate to x.*

Sketch Proof By applying Lemma 3.2.8 (over $\bar{k}$) we can find some $u \in R_u(P_\lambda)$ such that $u \cdot \lambda$ fixes x. This means that $u \cdot \lambda$ evaluates in G_x, and we can consider the intersection $G_x^0 \cap P_\lambda = G_x^0 \cap P_{u \cdot \lambda} = P_{u \cdot \lambda}(G_x^0)$. The hypothesis that G_x^0 is k-defined allows us to deduce that this intersection is also k-defined, and hence contains some k-defined maximal torus S. By conjugating suitably with an element of $R_u(P_\lambda)(k)$ we can replace λ with a conjugate which in fact evaluates in S, and then show that this λ fixes x, which implies that the original x' is $R_u(P_\lambda)(k)$-conjugate to x as claimed. For details see [17, Thm. 7.1]. $\qquad\square$

We finish this section by resolving another potential issue for the study of G-complete reducibility over k: even if we restrict attention to k-subgroups of G, there is no guarantee that we can find a generic tuple $\mathbf{h} \in H(k)^n$ for H. Although Theorem 11.4.4 does not demand that $\mathbf{h}$ is a k-point, for some applications it is useful to have this extra condition (e.g., see Corollary 11.7.12).

Suppose H is a k-subgroup of G. Since the k_s-points of H are dense in H (Lemma 11.1.1), we can at least find a generic tuple $\mathbf{h} \in H(k_s)^n$ for some n. Letting the absolute Galois group Γ act on the k_s-points as usual, by making the tuple longer if necessary we may arrange for Γ to permute the entries of $\mathbf{h}$. Now consider the quotient $X = G^n /\!/ S_n$, where S_n acts on tuples by permuting the entries in the obvious way. It follows from [8, §2.2] that X is k-defined. Let $x \in X$ be the image of $\mathbf{h}$; then $x \in X(k)$ since it is a k_s-point which is Γ-fixed by construction. Let k_1/k be an algebraic field extension.

Lemma 11.4.14 *With notation as just set up, we have $G_x^0 = C_G(H)^0$ and H is G-completely reducible over k_1 if and only if $G(k_1) \cdot x$ is cocharacter-closed over k_1.*

For details of the proof see [17, Lem. 6.3, Rem. 9.4], where one can also find versions which work for arbitrary k-defined G-varieties.

11.5 *G*-Complete Reducibility over a Field and Buildings

In this section we indicate how to extend the ideas from Chap. 7 to the setting of this chapter. Recall that to a (possibly non-connected) reductive group G we attach a simplicial building $\Delta = \Delta_G = \Delta_{G^0}$, which has a simplex σ_P for each parabolic subgroup of G^0 and a partial order given by $\sigma_P \leq \sigma_Q$ if and only if $P \supseteq Q$. We can do the same over k: given a reductive k-group G, let $\Delta_k = \Delta_{G,k} = \Delta_{G^0,k}$ denote the subset of Δ consisting of the σ_P where P is a k-parabolic subgroup of G^0. One can check that Δ_k is still a spherical building; its apartments now correspond to the maximal k-split tori of G (hence of G^0), and each one is isomorphic to the Coxeter complex attached to the relative Weyl group W_k of G^0. Note that the second building axiom—that any two simplices lie in a common apartment—corresponds in this setting to the fact that the intersection of any two k-parabolic subgroups of G^0 contains a maximal k-split torus of G^0.

All the notions introduced in Sects. 7.2 and 7.3 go through in this setting with suitable changes being made. In particular, the conjugation action of $G(k)$ on G^0 induces an action of $G(k)$ by simplicial building automorphisms on Δ_k. Serre's criterion from Sect. 7.2.6 still applies: a subset $\Sigma \subseteq \Delta_k$ is a convex subcomplex if and only if whenever P, Q, R are k-parabolic subgroups of G^0 with $\sigma_P, \sigma_Q \in \Sigma$ and $P \cap Q \subseteq R$, then $\sigma_R \in \Sigma$. Note, however, that the inclusion $P \cap Q \subseteq R$ must be an inclusion of algebraic groups, not just k-points. We also note that a subcomplex of Δ_k need not be a subcomplex of Δ, since not every overparabolic of a k-parabolic subgroup of G^0 need be k-defined.

Two simplices in Δ_k are opposite in Δ_k if and only if the intersection of the corresponding k-parabolic subgroups is a common k-Levi subgroup of each one, so we can define what it means for a convex subcomplex of Δ_k to be Δ_k-*completely reducible* (or Δ_k-*cr*). If H is a subgroup of G, we can form the subset

$$\Delta_k^H = \{\sigma_P \in \Delta_k \mid H \subseteq N_G(P)\}.$$

If H is a subgroup of G^0 then this is a convex subcomplex of Δ_k. For a general subgroup of G, Lemma 7.9.5(i) goes through without any problems, so H is G-cr over k if and only if every simplex in Δ_k^H has an opposite in Δ_k^H.

Remark 11.5.1 (i). If H is an arbitrary subgroup of G then there can be a difference between H normalising a parabolic subgroup P of G^0 and $H(k)$ normalising P, so the set Δ_k^H is not necessarily a "fixed point subcomplex". Note, however, that if A is a subgroup of $G(k)$, so that A acts on Δ_k, then the subcomplex Δ_k^A is the fixed point subcomplex for A.

 (ii). The proof of the Tits Centre Theorem 7.4.1 is valid for all spherical buildings, so the theorem applies in particular to the buildings of the form $\Delta_{G,k}$.

(iii). We also immediately obtain a version of Corollary 7.4.4, with an almost identical proof. Suppose A is a subgroup of $G^0(k)$, and suppose A_1 is a normal subgroup of A. Then if A is G^0-cr over k, so is A_1. However, we can do substantially better than this once we have worked out how to deal with separable field extensions in the context of G-complete reducibility; see Theorem 11.7.7 below.

The constructions of Sect. 7.5 leading to the definition of the vector building and the geometric spherical building go through over k with the appropriate changes made. Specifically, for $\mathbb{K} = \mathbb{R}$ or $\mathbb{Q}$ and each maximal split torus T of G, one can form the $\mathbb{K}$-vector space $Y_{T,k}(\mathbb{K}) = Y_{T,k} \otimes_{\mathbb{Z}} \mathbb{K}$, and then we can patch together these spaces over all the maximal split tori of G using a relation analogous to (7.5.3):

$$(T, \lambda) \sim (T', \lambda') \iff \text{ there exists } g \in L_\lambda^0(k) \text{ with } (T', \lambda') = (gTg^{-1}, g \cdot \lambda).$$

This allows us to form the set $Y_{G,k}(\mathbb{K}) = \left(\bigsqcup_T Y_{T,k}(\mathbb{K}) \right) / \sim$, where the union is over the maximal split tori of G. Further, since two maximal split tori of L_λ^0 are conjugate by an element of L_λ^0 if and only if they are conjugate by an element of $L_\lambda^0(k)$, we see that we may view $Y_{G,k}(\mathbb{K})$ as a subset of $Y_G(\mathbb{K})$. Now we can form the vector building $V_{G,k}(\mathbb{K})$ of G over k by using the obvious modification of the relation (7.5.4). We can show that the restriction of the map $\varphi_G : Y_G(\mathbb{K}) \to V_G(\mathbb{K})$ to $Y_{G,k}(\mathbb{K})$ is the natural projection $Y_{G,k}(\mathbb{K}) \to V_{G,k}(\mathbb{K})$, so we continue to use the notation φ_G for this restriction, and we view $V_{G,k}(\mathbb{K})$ as a subset of the vector building $V_G(\mathbb{K})$. The vector building $V_{G,k}(\mathbb{K})$ has as apartments the subsets $V_{T,k}(\mathbb{K})$ for the maximal split tori T of G. The *geometric spherical building* $\Delta_{G,k}(\mathbb{K})$ *of G over k* is the image of $V_{G,k}(\mathbb{K}) \setminus \{0\}$ inside $\Delta_G(\mathbb{K})$.

Given an element $\lambda \in Y_{G,k}(\mathbb{K})$ the R-parabolic subgroup P_λ is k-defined, and so is the R-Levi subgroup L_λ of P_λ. Similarly for an element $\zeta \in V_{G,k}(\mathbb{K})$ the parabolic subgroup P_ζ is k-defined. However, the converse is not true: that is, if $\zeta \in V_G(\mathbb{K})$ is such that P_ζ is k-defined, then ζ need not be an element of $V_{G,k}(\mathbb{K})$. To see this latter point, just take any situation in which there is a k-defined P which does not have the form P_λ for any $\lambda \in Y_{G,k}$. However, note that if $\lambda \in Y_{G,k}(\mathbb{K})$ and $u \in R_u(P_\lambda)$ is *any* element, then $\varphi_G(\lambda) = \varphi_G(u \cdot \lambda) \in V_{G,k}(\mathbb{K})$, so it is possible for $\varphi_G(\lambda)$ to lie in $V_{G,k}(\mathbb{K})$ even if λ is not in $Y_{G,k}(\mathbb{K})$.

For a subset $C \subseteq V_{G,k}(\mathbb{K})$ and a maximal k-split torus T of G, we let $C_T = C \cap V_{T,k}(\mathbb{K})$. Then we can extend the notions of what it means for C to be a cone, to be polyhedral, and to be of finite type in $V_{G,k}(\mathbb{K})$ (for finite type we consider only translation by elements of $G(k)$, which makes sense since all the maximal split tori are $G(k)$-conjugate).

The notion of a linear map makes sense for buildings of reductive groups over a field, and in this situation we have new examples of such maps.

Example 11.5.2 When k'/k is a field extension and $G' = G_{k'}$ is the base change of G to k', then the inclusion $V_{G,k}(\mathbb{R}) \to V_{G,k'}(\mathbb{R})$ is linear; in particular, the inclusion of $V_{G,k}(\mathbb{K})$ in $V_G(\mathbb{K})$ is linear.

More generally, if $\alpha : k \to k'$ is a field homomorphism and G is a reductive k-group, then we can form the base change $^\alpha G$ of G along α, which is a reductive k'-group. We get an induced linear map $\kappa_\alpha : V_{G,k}(\mathbb{R}) \to V_{^\alpha G,k'}(\mathbb{R})$; on cocharacters, this is the map $\lambda \mapsto {}^\alpha \lambda$ sending k-defined cocharacters of G to k'-defined cocharacters of $^\alpha G$.

We now record some facts used below about the interaction of buildings with Galois extensions. Suppose k/k_0 is a Galois extension with Galois group Γ, and suppose G is a reductive k_0-group. Then there is an action of Γ on the set of k-parabolic subgroups of the base change $(G^0)_k = (G_k)^0$ which induces an action of Γ on the simplicial building $\Delta_{G,k}$, such that the set of fixed points identifies with Δ_{G,k_0}.

This extends in a very natural way to the vector building setting, as a particular case of Example 11.5.2. For each element of Γ, which is a map $\gamma : k \to k$ fixing k_0, the base change $^\gamma(G_k)$ is a k-group which is naturally isomorphic as a k-group to G_k, because $^\gamma(G_k)$ and G_k have a common k_0-structure (i.e., that given by G as a k_0-group). Thus γ induces a linear *automorphism* of $V_{G,k}(\mathbb{K})$. It can be shown that

$$P_{\gamma \cdot \zeta} = \gamma \cdot P_\zeta \text{ for all } \gamma \in \Gamma \text{ and all } \zeta \in V_{G,k}(\mathbb{K}). \tag{11.5.1}$$

In this way, Γ acts on $V_{G,k}(\mathbb{K})$ and the vector building $V_{G,k_0}(\mathbb{K})$ can be identified with the subset of Γ-fixed elements in $V_{G,k}(\mathbb{K})$.

Example 11.5.3 Recall Example 7.5.4 above where we considered the linear automorphism κ_p of $V_{G,k}(\mathbb{K})$ induced by a standard p-power Frobenius morphism, where $G = \mathrm{GL}_m$ over $k = \overline{\mathbb{F}_p}$. As previewed in that example, there is also the field automorphism $\alpha_p : k \to k$ given by raising to the power p, and base change along this map induces a linear map $V_{G,k}(\mathbb{K}) \to V_{^{\alpha_p}G,k}(\mathbb{K})$. Since $^{\alpha_p}G$ is naturally isomorphic to G (via the $\mathbb{F}_p$-structure on G), this is an example of the set-up just described above, and we obtain a linear automorphism κ of $V_{G,k}(\mathbb{K})$. It is reasonably straightforward to see that κ induces the same permutation of the apartments as κ_p, with the difference between the two maps being that κ_p also scales everything by a factor of p.

A key result which allows us to apply some of the optimality results from Chap. 8 is as follows: identifying $V_{G,k}(\mathbb{K})$ as a subset of $V_G(\mathbb{K})$, then $V_{G,k}(\mathbb{K})$ is a polyhedral cone of finite type inside $V_G(\mathbb{K})$. To prove this result, the first thing we need is an extension of the definition of the cone $C(P)$ for a parabolic subgroup P of G from Lemma 8.4.4.

Definition 11.5.4 For a k-parabolic subgroup P of G^0, we let $C(P, k) = \{\zeta \in V_{G,k}(\mathbb{K}) \mid P_\zeta \supseteq P\}$.

Remark 11.5.5 (i). By Lemma 11.2.4(i), if P is a k-parabolic subgroup of G^0 and Q is any k-defined R-parabolic subgroup of G with $Q^0 = P$, then there is some $\lambda \in Y_{G,k}$ with $P_\lambda \supseteq Q$ and $P_\lambda^0 = Q^0 = P$. In this situation we have $\varphi_G(\lambda) \in C(P, k)$.

(ii). Given a k-parabolic subgroup P of G^0, there is a difference in general between $C(P)$ and $C(P, k)$. The latter set contains only those ζ for which $P_\zeta \supseteq P$ and P_ζ is k-defined, which could be a smaller set than $C(P)$.

Lemma 11.5.6 *Let P be a k-parabolic subgroup of G^0. Then $C(P, k)$ is a polyhedral cone of finite type in both $V_{G,k}(\mathbb{K})$ and $V_G(\mathbb{K})$.*

Proof It is clear that the first assertion follows from the second. One can also prove that $C(P, k)$ is a polyhedral cone of finite type in $V_{G,k}(\mathbb{K})$ using the argument from Lemma 8.4.4 with appropriate change: one just needs to work with maximal k-split tori rather than maximal tori, to use the relative root system, and to restrict attention to conjugating elements from $G(k)$.

To see that $C(P, k)$ is a polyhedral cone of finite type in $V_G(\mathbb{K})$ follows a similar line of proof but it takes a bit more work, so we give the details; the main issue is that we are viewing $C(P, k)$ as a subset of $V_G(\mathbb{K})$, so we need to be able to deal with conjugation by elements of G which are not k-points. Since G^0 has finite index in G, there is no harm in replacing G with G^0 when we are checking the finite type condition. Hence we may assume without loss that $G = G^0$ is connected. Note that $C(P, k)$ is a convex cone since it is closed under positive scalar multiplication and contains all geodesics between points (if P_ζ, $P_\eta \supseteq P$, then also $P_{t\zeta + (1-t)\zeta} \supseteq P$ for all $t \in [0, 1]$).

Now we claim that for every k-parabolic subgroup Q of G, the subset $C(Q, k)$ is a polyhedral cone in $V_T(\mathbb{K})$ for every maximal torus T of G contained in Q. Certainly for such a pair $T \subseteq Q$, we have $C(Q, k) \subseteq V_T(\mathbb{K})$. If T_1, T_2 are maximal tori of Q then $T_2 = qT_1q^{-1}$ for some $q \in Q$; conjugation by q gives a linear map from $V_{T_1}(\mathbb{K})$ to $V_{T_2}(\mathbb{K})$, and we see that $C(Q, k)$ is a polyhedral cone in $V_{T_1}(\mathbb{K})$ if and only if it is a polyhedral cone in $V_{T_2}(\mathbb{K})$. So it is enough to show that $C(Q, k)$ is a polyhedral cone in $V_{T_0}(\mathbb{K})$, where T_0 is a k-defined maximal torus of Q containing a maximal split torus S_0 of G. We have $C(Q, k) = C(Q, k)_{T_0} \subseteq V_{S_0,k}(\mathbb{K})$ by definition, and since $C(Q, k)$ is a polyhedral cone in the subspace $V_{S_0,k}(\mathbb{K})$ by the first paragraph of the proof, it is still a polyhedral cone when viewed as a subset of $V_{T_0}(\mathbb{K})$. But now, returning attention to our fixed k-parabolic subgroup P, for any maximal torus T of G we see that $C(P, k)_T$ is the union of certain $C(Q, k)$ for k-parabolic subgroups Q of G such that $P \subseteq Q$ and $T \subseteq Q$. Since there are only finitely many of these, we see that $C(P, k)_T$ is a cone which is the finite union of polyhedral cones in $V_T(\mathbb{K})$, and hence is itself polyhedral by Lemma 2.2.2.

To show that $C(P, k)$ has finite type, we need to fix a maximal torus T_0 of G and consider all the cones $g^{-1} \cdot C(P, k)_{gT_0g^{-1}}$ as g runs over the elements of G. Again, by Exercise 8.3, we are free to choose T_0, so let us pick $T_0 \subseteq P$ to be k-defined and containing some maximal split torus S_0 of G. Given $g \in G$, suppose Q is a k-parabolic subgroup of G containing gT_0g^{-1}, so that $C(Q, k) \subseteq V_{gT_0g^{-1}}(\mathbb{K})$. Then Q contains some maximal split torus S of G which is $G(k)$-conjugate to S_0, say $hSh^{-1} = S_0$ for $h \in G(k)$. It follows from [32, Prop. 20.6] that any k-parabolic subgroup of G containing S also contains $C_G(S)$

(since that is true for a minimal k-parabolic subgroup containing S). Therefore, since $T :=$ $h^{-1}T_0 h$ is a k-defined maximal torus of G containing S, we have $T \subseteq C_G(S) \subseteq Q$. We deduce that T is Q-conjugate to $gT_0 g^{-1}$, say $qTq^{-1} = gT_0 g^{-1}$ for $q \in Q$. Note that since $h \in G(k)$ we have $h \cdot C(Q, k) = C(hQh^{-1}, k)$, and since $q \in Q$ we can deduce that $hq^{-1} \cdot C(Q, k) = h \cdot C(Q, k) = C(hQh^{-1}, k) \subseteq V_{T_0}(\mathbb{K})$. Since $qh^{-1}T_0 hq^{-1} = gT_0 g^{-1}$, we see that $g^{-1} \cdot C(Q, k)$ is some $N_G(T_0)$-conjugate of $C(hQh^{-1}, k)$. Since there are only finitely many $C(hQh^{-1}, k)$ contained in $V_{T_0}(\mathbb{K})$, and only finitely many $N_G(T_0)$-conjugates of such subcones, we see that there are only finitely many possible $g^{-1} \cdot C(Q, k) \subseteq V_{T_0}(\mathbb{K})$ as Q runs over all the k-parabolic subgroups of G. Now each $g^{-1} \cdot C(P, k)_{gT_0 g^{-1}}$ is a union of such subsets; this is enough to conclude that $C(P, k)$ has finite type. $\qquad\square$

We can now prove our result.

Theorem 11.5.7 $V_{G,k}(\mathbb{K})$ *is a polyhedral cone of finite type in* $V_G(\mathbb{K})$.

Proof Let $C = V_{G,k}(\mathbb{K})$. It is clear that C is a convex cone. For each maximal torus T of G, C_T is a union of cones $C(P, k)$, where P runs over the k-parabolic subgroups of G containing T. Each $C(P, k)$ is polyhedral inside $V_T(\mathbb{K})$, by Lemma 11.5.6, and there are only finitely many P containing T, so C_T is a polyhedral cone by Lemma 2.2.2. Thus C is a polyhedral cone in $V_G(\mathbb{K})$. The fact that C has finite type also follows from Lemma 11.5.6: if we fix a maximal torus T_0 of G, then given $g \in G$, the subset $g^{-1} \cdot C_{gT_0 g^{-1}}$ is a finite union of subsets of the form $g^{-1} \cdot C(P, k)_{gT_0 g^{-1}}$ and there are only finitely many possibilities, since each $C(P, k)$ has finite type. $\qquad\square$

Remark 11.5.8 It immediately follows that if Σ is a convex subcomplex of $\Delta_{G,k}$ then C_Σ is a polyhedral cone of finite type in both $V_{G,k}(\mathbb{K})$ and $V_G(\mathbb{K})$, by the analogous proof to Corollary 8.4.5 (the more general statement there also holds over k).

11.6 Optimality over an Arbitrary Field

We would like to extend the constructions from Chap. 8 to the setting of k-defined G-varieties. We keep our assumption from Sect. 11.4 that X is a k-defined G-variety.

Open Problem 11.6.1 Formulate and prove a rational version of the Optimal Hilbert-Mumford Theorem 8.6.7.

To be precise: given such an X and some $x \in X$ such that $G(k) \cdot x$ is not cocharacter-closed over k, we want to construct some optimal destabilising $\lambda \in Y_{G,k}$ (or a class of such cocharacters) as in Theorem 8.6.23, with the property that $G_x(k) \subseteq P_\lambda$. (One can ask for a rational version of Theorem 8.6.15 as well, but for simplicity we stick with the simpler case.)

An obvious approach is to apply the methods of Sect. 8.6, but there is a fundamental obstacle. Suppose $G(k) \cdot x$ is not cocharacter-closed over k but $G \cdot x$ is closed (as happens in Example 11.3.8, for example). For any closed G-stable subset S of X, if $\{x\}$ is uniformly S-unstable then $\overline{G \cdot x}$ meets S, so $x \in S$. Hence $\{x\}$ is not properly uniformly S-unstable, and Theorem 8.6.7 does not apply.

At the time of writing, we do not know how to get around this problem. We do obtain a partial rational version of Theorem 8.6.15 under a stronger hypothesis.

Theorem 11.6.2 *Let S be a closed G-stable subset of X and let $\mathcal{A} \subseteq X$. Let $C \subseteq \mathcal{D}_{\mathcal{A}}(\mathbb{Q})$ be a polyhedral cone of finite type in $V_G(\mathbb{K})$. Suppose there exists $\lambda \in Y_{G,k}$ such that λ destabilises $\mathcal{A}$ into S and $\varphi_G(\lambda) \in C$. Suppose also that $\mathcal{A} \nsubseteq S$. Then there exists $\lambda_{\mathcal{A},k} \in Y_{G,k}$ such that $\lambda_{\mathcal{A},k}$ destabilises $\mathcal{A}$ into S, $N_{G(k)}(\mathcal{A}) \cap N_{G(k)}(C) \subseteq P_{\lambda_{\mathcal{A},k}}$ and $\varphi_G(\lambda_{\mathcal{A},k}) \in C$.*

Proof Let $C' = C \cap V_{G,k}(\mathbb{Q})$. Since $V_{G,k}(\mathbb{Q})$ is a polyhedral cone of finite type by Theorem 11.5.7, this is a polyhedral cone of finite type (it is clear that the intersection of polyhedral cones is polyhedral, and that it has finite type follows from Lemma 8.4.3). The result now follows from Theorem 8.6.15, taking $\lambda_{\mathcal{A},k}$ to be $\lambda_{\mathrm{opt}}(\mathcal{A}, S, C')$. $\square$

Remark 11.6.3   (i). In fact the proof of Theorem 11.6.2 goes through without the assumption that the G-variety X is k-defined.

(ii). Hesselink proved Theorem 11.6.2 in the special case when S consists of a single point and $C = \mathcal{D}_{\mathcal{A}}(\mathbb{Q}) \cap V_{G,k}(\mathbb{Q})$ (see [75, Thm. 5.2]).

As one would expect, Theorem 11.6.2 yields a rational version of Theorem 8.6.23 as well. Suppose that $x \in X$ and there exists $\lambda \in Y_{G,k}$ such that $x' := \lim_{a \to 0} \lambda(a) \cdot x$ exists and does not belong to $G \cdot x$. Set $\mathcal{A} = \{x\}$, $S = \overline{G \cdot x} \setminus G \cdot x$ and $C = V_{G,k}(\mathbb{Q}) \cap \mathcal{D}_x(\mathbb{Q})$. Note that S is closed by Lemma 2.7.4(ii).[2] We obtain an optimal destabilising cocharacter $\lambda_{\mathcal{A},k}$ as in Theorem 11.6.2. Clearly $G_x(k)$ fixes x and stabilises C, so $G_x(k)$ is contained in $P_{\lambda_{\mathcal{A},k}}$.

Under some extra hypotheses we obtain a variation on Theorem 11.6.2 which allows us to work with a separably closed field. Recall that Γ denotes the Galois group $\mathrm{Gal}(k_s/k)$, and this acts on $V_{G,k_s}(\mathbb{K})$; see the discussion after Example 11.5.2. In order to state and prove the theorem, we need the notion of a "length function defined over k". This is discussed in [89, §4]; we reproduce some of those ideas in a slightly more general context

---

[2] We do not take S to be the unique closed orbit contained in $\overline{G \cdot x}$, as we did in the proof of Theorem 8.6.23, because we don't know that this orbit is accessible from x via a k-defined cocharacter.

(at various points, Kempf assumes his field is perfect so that he can ascend to and descend from $\overline{k}$, but that is not actually necessary). Given G defined over k, when we construct the length function on G as in Sect. 2.5 we do so by picking a maximal torus T_0 and an arbitrary positive-definite $\mathbb{Z}$-valued bilinear form on Y_{T_0} and then averaging to make it W-invariant. We may choose T_0 to be a k-defined, and then since T splits over a *finite* Galois extension of k, we may further average the form so that it is Γ-invariant. This gives rise to a length function on $Y_G(\mathbb{K})$ with the additional property that $\|\gamma \cdot \lambda\| = \|\lambda\|$ for all $\lambda \in Y_{G,k_s}(\mathbb{K})$ and $\gamma \in \Gamma$. We call such a length function a *k-defined length function*.

Theorem 11.6.4 *Let S be a closed G-stable k-defined subset of X and let $\mathcal{A} \subseteq X$. Let $C \subseteq \mathcal{D}_{\mathcal{A}}(\mathbb{Q})$ be a polyhedral cone of finite type. Assume that $\mathcal{A}$ and C are Γ-stable, that $\|\cdot\|$ is k-defined and that $\mathcal{A} \nsubseteq S$. Suppose there exists $\lambda \in Y_{G,k_s}$ such that λ destabilises $\mathcal{A}$ into S and $\varphi_G(\lambda) \in C$. Then there exists $\lambda_{\mathcal{A},k} \in Y_{G,k}$ such that $\lambda_{\mathcal{A},k}$ destabilises $\mathcal{A}$ into S, $N_{G(k_s)}(\mathcal{A}) \cap N_{G(k_s)}(C) \subseteq P_{\lambda_{\mathcal{A},k}}$ and $\varphi_G(\lambda_{\mathcal{A},k}) \in C$.*

Proof The hypotheses of Theorem 11.6.2 are satisfied for the field k_s. Let $\lambda_{\mathcal{A},k_s} \in Y_{G,k_s}$ be the cocharacter from Theorem 11.6.2; then $N_{G(k_s)}(\mathcal{A}) \cap N_{G(k_s)}(C) \subseteq P_{\lambda_{\mathcal{A},k_s}}$ and $\varphi_G(\lambda_{\mathcal{A},k_s}) \in C$. We claim that $\zeta_{\mathcal{A},k_s}$ is fixed by Γ.

The main thing to show is that the function $f_{\mathcal{A}}$ from Notation 8.6.13 is Γ-invariant: that is, $f_{\mathcal{A}}(\gamma \cdot \zeta) = f_{\mathcal{A}}(\zeta)$ for all $\gamma \in \Gamma$ and all $\zeta \in \mathcal{D}_{\mathcal{A}}(\mathbb{Q})$. Since the length function is k-defined, this amounts to showing that $\mu_{\mathcal{A}}(\gamma \cdot \zeta) = \mu_{\mathcal{A}}(\zeta)$ for every $\zeta \in \mathcal{D}_{\mathcal{A}}(\mathbb{Q})$. We are assuming S to be k-defined, and it is straightforward to check that the proof of Lemma 3.2.1 goes through over k: that is, we can find a *k-defined G-* and Γ-equivariant map $\psi : X \to V$ to a *k-defined G*-module V such that the scheme-theoretic preimage $\psi^{-1}(0)$ is S. Now given a k_s-defined (hence k_s-split) maximal torus T of G, the action of the Galois group Γ on cocharacters and characters behaves well in that

$$\langle \gamma \cdot \lambda, \gamma \cdot \chi \rangle = \langle \lambda, \chi \rangle, \tag{11.6.1}$$

for all $\lambda \in Y_{T,k_s}$, $\gamma \in \Gamma$ and $\chi \in \Phi_T(V)$. Therefore, if $\lambda \in Y_{T,k_s}$ and $v \in V_{\lambda,\geq 0}(k_s)$ we see that $\gamma \cdot v \in V_{\gamma \cdot \lambda,\geq 0}(k_s)$ (note that $\gamma \cdot \lambda \in Y_{\gamma \cdot T,k_s}$, and if χ is a T-weight in the support of v then $\gamma \cdot \chi$ is a $(\gamma \cdot T)$-weight in the support of $\gamma \cdot v$). Since the subset $\mathcal{A}$ is Γ-stable and our map ψ is Γ-equivariant, the set of all χ appearing in the support of elements of $\psi(\mathcal{A})$ is Γ-stable, and (11.6.1) gives the required result: the value of $\mu_{\mathcal{A}}$ is Γ-invariant, because taking the minimal value of the pairing of λ with all χ appearing is the same as taking the minimal value of the pairing of $\gamma \cdot \lambda$ with all $\gamma \cdot \chi$ appearing.

Given the previous paragraph, it follows from the uniqueness in the optimality construction that $\zeta := \zeta_{\mathcal{A},k_s}$ is fixed by Γ, as claimed. It follows from (11.5.1) that P_ζ is fixed by Γ, so P_ζ is k-defined. We can choose a k-defined maximal torus T of P_ζ and a cocharacter $\tau \in Y_T$ such that $\varphi_G(\tau) = \zeta$. Since T is split over k_s, τ is automatically k_s-defined. For any $\gamma \in \Gamma$ we have

$$\varphi_G(\gamma \cdot \tau) = \gamma \cdot \varphi_G(\tau) = \gamma \cdot \zeta = \zeta = \varphi_G(\tau),$$

so $\gamma \cdot \tau = \tau$ (since φ_G gives a bijection from $Y_T(\mathbb{K})$ to $V_T(\mathbb{K})$). Hence τ is k-defined and so $\lambda_{\mathcal{A},k} := \tau$ has the desired properties. $\square$

Corollary 11.6.5 *Suppose the hypotheses of Theorem 11.6.4 hold and suppose also that k is perfect. Then we can choose the optimal destabilising cocharacter $\lambda_{\mathcal{A}} = \lambda_{\mathrm{opt}}(\mathcal{A}, \mathcal{S}, C)$ from Theorem 8.6.7 to be k-defined.*

Recall that the optimal destabilising cocharacter is not unique: given one such cocharacter τ, we can replace it with $u \cdot \tau$ for any $u \in R_u(P_\tau)$. The corollary says there is a choice that is k-defined.

Proof Since λ uniformly destabilises $\mathcal{A}$ into $\mathcal{S}$ and $\mathcal{A} \not\subseteq \mathcal{S}$, $\mathcal{A}$ is properly uniformly $\mathcal{S}$-unstable, so the hypotheses of Theorem 8.6.7 are satisfied. Hence we obtain an optimal destabilising cocharacter $\lambda_{\mathcal{A}} = \lambda_{\mathrm{opt}}(\mathcal{A}, \mathcal{S}, C)$ from Theorem 8.6.7. Now the cocharacter $\lambda_{\mathcal{A},k_s}$ from the proof of Theorem 11.6.4 is defined to be $\lambda_{\mathrm{opt}}(\mathcal{A}, \mathcal{S}, C')$, where $C' = C \cap V_{G,k_s}(\mathbb{Q})$. But k is perfect, so $k_s = \overline{k}$, so $C' = C$. Hence we can take $\lambda_{\mathcal{A}}$ and $\lambda_{\mathcal{A},k_s}$ to be the same. Now the argument of Theorem 11.6.4 tells us that there is a k-defined cocharacter $\lambda_{\mathcal{A},k}$ in $R_u(P_{\lambda_{\mathcal{A},k_s}}) \cdot \lambda_{\mathcal{A},k_s}$. Hence we can choose both $\lambda_{\mathcal{A}}$ and $\lambda_{\mathcal{A},k_s}$ to be $\lambda_{\mathcal{A},k}$. $\square$

Remark 11.6.6 Here is a special case of Corollary 11.6.5, which was proved by Kempf (see [89, Thm. 4.2]): if k is perfect, $\| \cdot \|$ is k-defined, $x \in X(k)$ and $G \cdot x$ is not closed then the optimal destabilising cocharacter $\lambda_{\{x\}} = \lambda_{\mathrm{opt}}(\{x\}, O, \mathscr{D}_x(\mathbb{Q}))$ from Theorem 8.6.7 is k-defined. (Here O is the unique closed orbit contained in $\overline{G \cdot x}$.) This follows from Corollary 11.6.5, taking $\mathcal{A} = \{x\}$, $\mathcal{S} = O$ and $C = \mathscr{D}_x(\mathbb{Q})$.

11.7   Ascent and Descent Results

We first look at answers to Question 11.3.5. We continue to assume that X is a k-defined G-variety. For separable extensions we have the best possible descent result.

Theorem 11.7.1 (Galois Descent) *Let $x \in X$. Let k_1/k be a separable algebraic extension. Suppose $G_x(k_s)$ is Γ-stable. If $G(k_1) \cdot x$ is cocharacter-closed over k_1, then $G(k) \cdot x$ is cocharacter-closed over k.*

In particular, if a k-subgroup H of G is G-completely reducible over k_1 then it is G-completely reducible over k.

Proof Suppose $\lambda \in Y_{G,k}$ is such that $x' = \lim_{a \to 0} \lambda(a) \cdot x$ exists. Then since $\lambda \in Y_{G,k_1}$ also, by Lemma 11.4.11 there is some $u \in R_u(P_\lambda)(k_1)$ with $x' = u \cdot x$. Since k_1/k

is separable, we have $R_u(P_\lambda)(k_1) \subseteq R_u(P_\lambda)(k_s)$. But then $x' \in R_u(P_\lambda)(k) \cdot x$ by Theorem 11.4.6, and hence $x' \in G(k) \cdot x$. We see that $G(k) \cdot x$ is cocharacter-closed over k, as required.

Now suppose H is a k-subgroup of G. Pick a generic tuple $\mathbf{h} \in H^n$ for H. Then $G_{\mathbf{h}} = C_G(H)$ by Lemma 5.2.5(i), and $C_G(H)$ is Γ-stable because H is k-defined. The statement about G-complete reducibility now follows from Theorem 11.4.4. $\quad\square$

Remark 11.7.2 Note that this result implies in particular that if k is a perfect field then for every $x \in X$ such that $G \cdot x$ is closed and G_x is Γ-stable, $G(k) \cdot x$ is cocharacter-closed over k; this follows since $\overline{k}/k$ is separable when k is perfect.

In terms of G-complete reducibility, we deduce that if k is perfect then any k-subgroup of G which is G-cr is also G-cr over k. Recall that this is *not* true in general without the hypothesis that k is perfect: see Example 11.3.8.

Remark 11.7.3 Let k, G, u, x and λ be as in Example 11.4.8. Then $x' := \lim_{a \to 0} \lambda(a) \cdot x$ is not $R_u(P_\lambda)(k)$-conjugate to x, so $G(k) \cdot x$ is not cocharacter-closed over k. On the other hand, it is not hard to see that $G(k_s) \cdot x$ is cocharacter-closed over k_s (we leave this to the reader to check). This shows that Theorem 11.7.1 can fail without the hypothesis that $G_x(k_s)$ is Γ-stable.

There are various versions of the corresponding Galois ascent result, but none which apply in such complete generality. In terms of G-complete reducibility, the cleanest statement we can obtain is for connected G.

Theorem 11.7.4 (Galois Descent and Ascent for Connected G) *Suppose k_1/k is a separable algebraic field extension, and G is connected. A k-subgroup H of G is G-completely reducible over k_1 if and only if it is G-completely reducible over k.*

Proof By Theorem 11.7.1, we only need to deal with the reverse implication, and for that we may assume that $k_1 = k_s$ (for we can descend from k_s to any other separable extension using Theorem 11.7.1). Let Γ be the absolute Galois group. The hypothesis that G is connected allows us to use the buildings of G over k and k_s. As explained in Sect. 11.5, Γ can be made to act on Δ_{G,k_s} by building automorphisms in such a way that $\Delta_{G,k}$ can be identified as the set of Γ-fixed simplices. Now, using the same reduction as at the start of the proof of Corollary 7.4.4, we may assume that H is contained in no proper Levi k-subgroup of G (since a Levi k-subgroup is also a Levi k_s-subgroup). Since H is assumed to be k-defined, the subcomplex Δ^H_{G,k_s} is Γ-stable. If H is not G-cr over k_s, then this subcomplex is not Δ_{G,k_s}-cr, so Theorem 7.4.1 provides a non-empty Γ-fixed simplex in Δ^H_{G,k_s}. Because it is Γ-fixed, this simplex corresponds to a k-parabolic subgroup of G containing H. Since we have reduced to the case that H is contained in no proper k-Levi subgroup of G, we see that H cannot be G-cr over k. This proves the result, by contrapositive. $\quad\square$

Remark 11.7.5 The proof of Theorem 11.7.4 uses the Tits Centre Theorem 7.4.1 in an essential way, which requires that we have a subcomplex in the building to work with. Although we believe the result of the theorem should still be true when G is not connected, examples like Example 7.9.2 show that it is not completely straightforward to apply the building-theoretic framework to the non-connected setting. Another possible route to this result (both in the connected and non-connected setting) would be an analogue of the optimality results in Chap. 8: that is, any way to overcome the problems outlined in Sect. 11.6 would likely lead to a quick and easy proof of Theorem 11.7.4.

Open Problem 11.7.6 Prove Theorem 11.7.4 without the assumption that G is connected.

We can now, as promised, give an extension of the Clifford Theory result Theorem 9.1.1 to arbitrary fields, at least in the case that G is connected.

Theorem 11.7.7 *Suppose G is connected, H is a k-defined subgroup of G, and N is a k-defined normal subgroup of H. Then if H is G-completely reducible over k, so is N.*

Proof By Theorem 11.7.4, it is enough to prove the result under the additional hypothesis that $k = k_s$ is separably closed. In this case, $H(k)$ is dense in H by Lemma 11.1.1, so H is G-cr over k if and only if $H(k)$ is, and similarly for N and $N(k)$. Now we can apply Remark 11.5.1(iii) to get the result. □

In view of Corollary 9.2.2 it is natural to ask the following question.

Open Problem 11.7.8 Suppose H is G-completely reducible over k. Must $C_G(H)$ be G-completely reducible over k?

We cannot answer Problem 11.7.8 in general. One technical obstacle is that $C_G(H)$ need not be k-defined, as we pointed out in Sect. 11.1. For an example of this, let $p = 2$ and let H be the subgroup of the simple group G of type G_2 defined in [187, §4.1]. Then H is k-defined and is G-cr over k, but it follows from the description of $C_G(H)$ given in [187, Rem. 4.6] that $C_G(H)$ is not k-defined. Here is a partial result, which is a slight variation on a theorem of Uchiyama [188, Thm. 1.5].

Proposition 11.7.9 *Let H be a subgroup of G such that H is G-completely reducible over k, and suppose $C_G(H)$ is reductive. Then $HC_G(H)$ is G-completely reducible over k. Moreover, if G is connected and H and $C_G(H)$ are k-defined then $C_G(H)$ is G-completely reducible over k.*

Proof Let $\mu \in Y_{G,k}$ such that $HC_G(H) \subseteq P_\mu$. Since H is G-cr over k, there exists $\lambda \in Y_{G,k}$ such that $P_\mu = P_\lambda$ and $H \subseteq L_\lambda$. Then $\lambda \in Y_{C_G(H)}$ and $C_G(H) \subseteq P_\lambda$. Since $C_G(H)$ is reductive by hypothesis, we have $C_G(H) \subseteq L_\lambda$ by Lemma 2.9.9. Hence

$HCG_G(H) \subseteq L_\lambda$. It follows from Proposition 11.3.4 that $HC_G(H)$ is G-cr over k. This proves the first assertion. The second follows from Theorem 11.7.7. $\qquad\square$

Obtaining results about the ascent and descent of cocharacter-closedness (and hence G-complete reducibility) through arbitrary algebraic field extensions necessitates imposing extra hypotheses. Here is one such result which imposes a fairly natural condition on the stabiliser.

Theorem 11.7.10 *Suppose $x \in X$ is such that G_x^0 is k-defined.*

(i) For any algebraic extension k_1/k, if $G(k_1) \cdot x$ is cocharacter-closed over k_1, then $G(k) \cdot x$ is cocharacter-closed over k.

(ii) If $x \in X(k)$ and k_1/k is separable, then $G(k_1) \cdot x$ is cocharacter-closed over k_1 if and only if $G(k) \cdot x$ is cocharacter-closed over k.

Sketch Proof (i). Suppose that $\lambda \in Y_{G,k}$ is such that $x' = \lim_{a \to 0} \lambda(a) \cdot x$ exists. Then cocharacter-closedness of $G(k_1) \cdot x$ implies in particular that x' is G-conjugate to x. Now apply Theorem 11.4.13 to deduce that x' is $R_u(P_\lambda)(k)$-conjugate (and hence $G(k)$-conjugate) to x.

(ii). We prove the contrapositive. We may assume that $k_1 = k_s$, by part (i). Suppose that $G(k_s) \cdot x$ is not cocharacter-closed over k_s. There exists $\lambda \in Y_{G,k_s}$ such that $x' = \lim_{a \to 0} \lambda(a) \cdot x$ exists but x' is not $G(k_s)$-conjugate to x. It follows from Theorem 11.4.13 that x' lies outside the geometric orbit $G \cdot x$. We can assume if we wish that $\| \cdot \|$ is k-defined. The result now follows from Theorem 11.6.4, taking $\mathcal{A} = \{x\}$, S to be $\overline{G \cdot x'}$ and $C = \mathcal{D}_x(\mathbb{Q})$.

$\qquad\square$

Remark 11.7.11   (i). Part (i) of the theorem allows us to remove the separability hypothesis from Theorem 11.7.1 at the cost of imposing a stronger hypothesis on G_x.

(ii). The above sketch gives an alternative quicker proof of some of the results in [17]. See [17, Thm. 5.7, Cor. 7.2] for example.

(iii). The hypothesis that the stabiliser is k-defined holds in particular if $x \in X(k)$ and k is perfect, or if $x \in X(k)$ and the orbit map $G \to G \cdot x$ is separable: see [168, Prop. 12.1.2]. In the latter case, the point is that separability of the orbit map implies that the scheme-theoretic stabiliser $\mathcal{G}_x$ is smooth (see Remark 3.5.3). Since $\mathcal{G}_x$ is k-defined, it follows that G_x is k-defined as well.

We finish this part of our discussion by applying the previous result to the study of G-complete reducibility.

Corollary 11.7.12 *Suppose H is a k-subgroup of G, and suppose that at least one of the following holds:*

(i) $C_G(H)^0$ *is k-defined;*
(ii) *k is perfect;*
(iii) *H is a separable subgroup of G;*
(iv) *p is a pretty good prime for* G^0.

For any algebraic extension k_1/k, if H is G-completely reducible over k_1 then H is G-completely reducible over k. If k_1/k is separable, then H is G-completely reducible over k_1 if and only if H is G-completely reducible over k.

Proof Let X and $x \in X(k)$ be as in Lemma 11.4.14. Then H is G-cr over k (resp., H is G-cr over k_1) if and only if $G(k) \cdot x$ is cocharacter-closed over k (resp., $G(k_1) \cdot x$ is cocharacter-closed over k_1). If (i) holds then G_x^0 is k-defined by construction. If (ii) holds then G_x^0 is k-defined by Remark 11.7.11(iii). If (iii) holds, then the separability of H implies that the scheme-theoretic centraliser of H in G is smooth and hence k-defined (cf. Remark 11.7.11(iii)), and this implies that G_x^0 is k-defined. If (iv) holds, then all subgroups of G are separable in G, and hence (iii) holds automatically. Hence, in all cases we can apply Theorem 11.7.10. $\square$

Remark 11.7.13 (i). Note that hypothesis (iii) in the corollary holds if H is a k-defined linearly reductive subgroup of G—see Remark 10.5.4. It then follows from the corollary and from Lemma 4.2.1 that H is G-cr over k.

(ii). Part (ii) of Corollary 11.7.12 shows that for a perfect field k, a k-subgroup of G is G-cr over k if and only if it is G-cr (over $\overline{k}$). Thus for perfect fields we have the best possible answer to Questions 11.3.5 and 11.3.6.

We now turn to Levi ascent and descent results.

Theorem 11.7.14 *Suppose $x \in X$ and $S \subseteq G_x$ is a k-defined torus. Let $L = C_G(S)$.*

(i) If $G(k) \cdot x$ is cocharacter-closed over k then $L(k) \cdot x$ is cocharacter-closed over k.
(ii) If S is k-split, then $G(k) \cdot x$ is cocharacter-closed over k if and only if $L(k) \cdot x$ is cocharacter-closed over k.

In particular, for H a subgroup of L, if H is G-completely reducible over k then H is L-completely reducible over k, and if, moreover, S is k-split, then H is G-completely reducible over k if and only if H is L-completely reducible over k.

Proof (i). If $G(k) \cdot x$ is cocharacter-closed, then for any $\lambda \in Y_{L,k}$ such that $x' = \lim_{a \to 0} \lambda(a) \cdot x$ exists, x' is $R_u(P_\lambda)(k)$-conjugate to x by Lemma 11.4.11. But then Proposition 11.4.5 implies that x' is $R_u(P_\lambda(L))(k)$-conjugate to x, so $x' \in L(k) \cdot x$, as required.

(ii). This follows from another application of Proposition 11.4.9: if $G(k) \cdot x$ is not cocharacter-closed, then we can find a $\tau \in Y_{L,k}$ such that $x' := \lim_{a \to 0} \tau(a) \cdot x$ is not $G(k)$-conjugate to x. But then x' is not $L(k)$-conjugate to x either, so we're done.

Now the statements about G-complete reducibility follow from Theorem 11.4.4. $\qquad\square$

Combining this theorem with our Galois ascent and descent results we deduce the following in case G is connected.

Corollary 11.7.15 *Suppose G is connected, suppose S is a k-torus in G and let $L = C_G(S)$. If H is a k-subgroup of L then H is G-completely reducible over k if and only if H is L-completely reducible over k.*

Proof We may find a finite separable extension k_1/k such that S is split over k_1. Then H is G-cr over k if and only if H is G-cr over k_1 by Theorem 11.7.4, and this holds if and only if H is L-cr over k_1, by Theorem 11.7.14. Now a final application of Theorem 11.7.4 (this time to L) gives the required result. $\qquad\square$

Recall from Example 11.2.1 that not all subgroups of the form $L = C_G(S)$ arise as R-Levi subgroups of k-defined R-parabolic subgroups. However, if we restrict attention to those subgroups L that do arise in this way, we can get a general Levi ascent and descent result for non-connected groups too. Furthermore, we don't need the subgroup H to be k-defined.

Corollary 11.7.16 *Suppose L is a k-defined R-Levi subgroup of some k-defined R-parabolic subgroup of G. If H is a subgroup of L then H is G-completely reducible over k if and only if H is L-completely reducible over k.*

Proof Let P be a k-defined R-parabolic subgroup of G containing L as an R-Levi subgroup. It follows from Lemma 11.2.4(i) and (ii) that we can find some $\lambda \in Y_{G,k}$ such that $P^0 = P_\lambda^0$ and $L = L_\lambda \cap P$. Since L therefore has finite index in L_λ, the analogue over k of Proposition 4.2.12 says that H is L-cr over k if and only if H is L_λ-cr over k. So we may replace L with L_λ. But now L is the centraliser of a split torus (the image of λ), so Theorem 11.7.14(ii) gives the result. $\qquad\square$

Remark 11.7.17 (i). In case G is connected and $H \subseteq G(k)$, Corollary 11.7.16 was first proved by Serre [159, Prop. 3.2] using the building-theoretic argument outlined in Remark 7.3.10(i).

(ii). We in fact believe that the equivalence in Theorem 11.7.14(ii) (and hence in particular the ascent and descent of G-complete reducibility) should hold without the splitness hypothesis on S. However, as yet, there is no proof of this; it would follow if we had a version of the optimality construction valid over arbitrary fields.

11.8 Complete Reducibility and Steinberg Endomorphisms

The starting point for our discussion here is the following, which is a special case of Corollary 11.7.12(ii); see also Remark 11.7.13(ii). Let q be a power of p.

Theorem 11.8.1 *Suppose that $H \subseteq G$ and both G and H are defined over $\mathbb{F}_q$. Then H is G-completely reducible if and only if it is G-completely reducible over $\mathbb{F}_q$.*

Now let G be connected reductive and suppose $\sigma : G \to G$ is a *Steinberg endomorphism* of G, i.e., a surjective endomorphism of G that fixes only finitely many points; see Steinberg [173] for a detailed discussion (and for this terminology, see [66, Def. 1.15.1b]). The set of all Steinberg endomorphisms of G is a subset of all isogenies $G \to G$ (see [173, 7.1(a)]) which encompasses in particular all (generalised) Frobenius endomorphisms, i.e., endomorphisms of G some power of which is a Frobenius endomorphism F_q corresponding to some $\mathbb{F}_q$-rational structure on G. Recall from Example 7.5.3(iv) that such an endomorphism gives rise to an automorphism of the vector building of G (and of the geometric and simplicial spherical buildings).

There is a relatively straightforward extension of the notion of G-complete reducibility in this setting. Let σ be a Steinberg endomorphism of G and let H be a subgroup of G. Following [74], we say that H is *σ-completely reducible* (or *σ-cr* for short), provided that whenever H lies in a σ-stable parabolic subgroup P of G, it lies in a σ-stable Levi subgroup of P. In the special case that σ is a standard Frobenius morphism F_q of G, then a subgroup H of G is defined over $\mathbb{F}_q$ if and only if it is F_q-stable and so H is G-completely reducible over $\mathbb{F}_q$ if and only if it is F_q-completely reducible: it is in this sense that we get an extension of the usual notion of G-complete reducibility over a field. We have the following generalization of Theorem 11.8.1 to arbitrary Steinberg endomorphisms of G (the special case of Theorem 11.8.2 when $\sigma = F_q$ gives Theorem 11.8.1).

Theorem 11.8.2 ([74, Thm. 1.4]) *Let σ be a Steinberg endomorphism of the connected reductive group G. Let H be a σ-stable subgroup of G. Then H is G-completely reducible if and only if H is σ-completely reducible.*

Remark 11.8.3 Theorem 11.8.2 is false without the σ-stability condition on H. For instance, a maximal torus T of G is always G-cr, cf. [18, Lem. 2.6]. But it may happen that T is contained in a σ-stable Borel subgroup of G, without being itself σ-stable. Then T clearly fails to be σ-cr. In the other direction, G may contain a maximal parabolic subgroup

P of G that is not σ-stable. The only σ-stable parabolic subgroup of G containing P is G itself. Then P is σ-cr for trivial reasons, whereas a proper parabolic subgroup of G is not G-cr.

Since the proof of Theorem 11.8.2 ties in with the building-theoretic version of G-complete reducibility from Sect. 7.3 and the Tits Centre Theorem 7.4.1, we give most of the details. The key is the following:

Lemma 11.8.4 *Let σ be a Steinberg endomorphism of G. Let H be a σ-stable subgroup of G. If H is not G-completely reducible, then there exists a proper σ-stable parabolic subgroup of G containing H.*

Proof Recall that σ induces a simplicial automorphism κ_σ of Δ_G. Let H be a σ-stable subgroup of G which is not G-cr; then the fixed point subcomplex $(\Delta_G)^H$ is κ_σ-stable and is not Δ_G-cr (Remark 7.3.3). Hence, by the Tits Centre Theorem 7.4.1, there is a non-empty κ_σ-fixed simplex σ_P of $(\Delta_G)^H$, which corresponds to a proper σ-stable parabolic subgroup of G containing H.   □

We can now prove the theorem:

Proof of Theorem 11.8.2 First suppose that H is σ-cr. If H is not contained in any proper σ-stable parabolic subgroup of G, then it is G-cr by Lemma 11.8.4, so we can assume that there is a proper σ-stable parabolic subgroup P of G containing H. Choose P minimal with these properties; then since H is σ-cr it is contained in some σ-stable Levi subgroup L of P. But now if Q is any σ-stable parabolic subgroup of L containing H, then $QR_u(P)$ is a proper parabolic subgroup of P by Lemma 2.9.10(iv), it contains H, and is σ-stable since Q is and $R_u(P)$ obviously is too. This implies that $QR_u(P) = P$, by minimality, and hence $Q = L$. Thus H is contained in no proper σ-stable parabolic subgroup of L, and hence no proper parabolic subgroup of L at all, by Lemma 11.8.4 again. So H is L-ir, and hence G-cr by Theorem 5.4.4.

For the converse, suppose H is G-cr and suppose that P is a σ-stable parabolic subgroup of G containing H. Since H is G-cr, there is some Levi subgroup L of P that contains H; this means that the set

$$\Lambda = \{M \mid M \text{ is a Levi subgroup of } P,\, H \subseteq M\}$$

is non-empty. It is clearly also σ-stable. We show that it contains a σ-stable element, by a variation of a standard argument.

Let $U = R_u(P)$, and note that every element of Λ is of the form uLu^{-1} for some $u \in R_u(P)$, and in this case we get $u^{-1}Hu \subseteq L$. But for any $h \in H$ we can write $u^{-1}hu = u^{-1}(huh^{-1})h \in UH$, and hence $u^{-1}Hu \subseteq UH$. Since $H \subseteq L$ and $L \cap U = \{1\}$, we have $UH \cap L = H$, so we see that u in fact normalises H. In fact, u centralises H, since $[N_U(H), H] \subseteq H \cap U = \{1\}$. So the group $C_U(H)$ acts transitively on Λ.

Now write $L = L_\lambda$ for some $\lambda \in Y_G$, and note that λ commutes with H, so $\lambda \in C_G(H)$. Thus for any $u \in C_U(H)$ and $a \in k^*$ we have $\lambda(a)u\lambda(a)^{-1} \in C_U(H)$. Since $\lim_{a \to 0} \lambda(a)u\lambda(a)^{-1} = 1$, we see that $u \in C_U(H)^0$; i.e., $C_U(H)$ is connected. But now we can apply the Lang-Steinberg theorem (see [173, Thm. 10.1]) to conclude that Λ contains an element fixed by σ. That is, we have found a σ-stable Levi subgroup of P containing H. $\qquad\qquad\qquad\qquad\qquad\qquad\qquad\qquad\qquad\qquad\qquad\qquad\qquad\qquad\square$

11.9   Historical Remarks and References

Whilst there is already some material on rationality questions and G-complete reducibility in the authors' original paper [18, §5], the paper [28] is really the first place in which this material is systematically developed. It is here that the notion of a cocharacter-closed orbit is introduced, together with many of the important tools used in this chapter (and, indeed, the rest of the book). The other significant source for this chapter is the paper [17], where the rational Hilbert-Mumford Theorem 11.4.10 was proved along with some ascent and descent results. See [20] for Theorem 11.7.4. For the material on G-complete reducibility and buildings over a field we have relied heavily again on Serre's Bourbaki Seminar [159]. See also the references on buildings given in Sect. 7.10. The most up-to-date source for our approach to vector buildings over an arbitrary field is [25], where we develop the material for arbitrary smooth connected groups, not just reductive ones (note that in [25] we use the terminology "vector edifice", since we are discussing more general objects).

All three of Kempf [89], Rousseau [150], and Hesselink [75] provide versions of their optimality constructions which work over a non-algebraically closed field. Our approach in Chap. 8 has allowed us to beef up these results in Sect. 11.6; our Theorem 11.6.4 is new. For a further discussion of the Strong Centre Conjecture and applications to geometric invariant theory in the context of edifices, see [25, §7].

We cited some of the papers of Uchiyama in the main body of the chapter. He has a series of papers including [187], [188], [191], and [7] (the last with Bannuscher and Litterick) which provide examples of subgroups which are G-cr over k but not G-cr over $\bar{k}$, or vice versa. Most of these examples arise as variations on the construction of the example in Sect. 10.1 above, using the idea of Example 11.3.8 to leverage the imperfectness of the field k.

We note that Lemma 11.8.4 is a generalization of (a special case of) [98, Prop. 2.2 and Rem. 2.4]. The proof of Lemma 11.8.4 given in [74, Prop. 2.1] consists of a reduction to the case when H is finite, covered in [98, Prop. 2.2 and Rem. 2.4].

We refer to [68, Thm. 1.3(iii)] for a building-theoretic interpretation of σ-complete reducibility.

11.10 Exercises

Exercise 11.1 Verify the claim in Example 11.2.2, that the $\mathbb{R}$-algebras $\mathbb{R}[Z, Z^{-1}]$ and $\dfrac{\mathbb{R}[X, Y]}{X^2 + Y^2 - 1}$ become isomorphic as $\mathbb{C}$-algebras when we tensor them with $\mathbb{C}$.

Exercise 11.2 Fill in the details of Example 11.1.2. That is, show that the given $\mathbb{R}$-algebras do give $\mathbb{R}$-structures on the coordinate ring $\mathbb{C}[SL_2]$, and show that the two groups of $\mathbb{R}$-points arising are $SL_2(\mathbb{R})$ and $SU(2)$ (the 2×2 unitary matrices of determinant 1), respectively.

Exercise 11.3 Let $G = \mathbb{G}_m$ be the multiplicative group over $\mathbb{C}$ with $\mathbb{R}$-structure given by the algebra of real Laurent polynomials $\mathbb{R}[Z, Z^{-1}]$ (i.e., the usual split structure of $\mathbb{G}_m$ over $\mathbb{R}$). Let G act on $X = \mathbb{A}^1$ by squares: that is, $t \cdot a := t^2 a$ for all $t \in G$ and $a \in X$.

 (i) Show that there are two orbits of G in X, but three orbits of $G(\mathbb{R})$ in $X(\mathbb{R})$.
(ii) For each $G(\mathbb{R})$-orbit O in part (i) find the cocharacter-closure $\overline{O}^c$ and compare it to $\overline{O} \cap X(\mathbb{R})$, where $\overline{O}$ denotes the Zariski closure of O inside all of X.

Exercise 11.4 Fill in the details of Example 11.3.7. That is, taking k'/k to be a purely inseparable extension of degree p, construct a subgroup H of GL_p such that H is GL_p-cr over k but not GL_p-cr over k' (or over $\overline{k}$).

Exercise 11.5 Let $p = 2$ and let $G = GL_2$. Find a k-defined subgroup of H of G such that H is G-cr over k and $C_G(H)$ is k-defined but not reductive.

Variations, Applications and Future Directions **12**

In this chapter we briefly describe some topics which build on ideas from earlier in the book. Most of them are to do with G-complete reducibility one way or another, but we also look at some wider applications of geometric invariant theory methods to the study of subgroups of G. Unless stated otherwise, we keep our convention that k is algebraically closed and G is a reductive group over k.

12.1 Subgroup Structure of Reductive Groups

As mentioned in the introduction, one motivation for studying G-complete reducibility is its application to the subgroup structure of reductive groups. We discuss this in more detail and give a brief overview of some work in the subject over the last few decades; for a far more thorough survey, see [105]. We focus on the following problem.

Open Problem 12.1.1 Suppose G is connected. Find all the connected reductive subgroups of G, up to conjugacy.

Note that if $\mathrm{char}(k) = 0$ then there is a bijective correspondence between connected subgroups of G and Lie subalgebras of $\mathfrak{g}$ [77, Ch. V], so one can reduce this to a study of the Lie algebra, and the classification essentially follows from work of Dynkin [61], [62]. In positive characteristic, not only does this correspondence break down, but there are extra complications; for one thing, reductive subgroups of G need not be G-cr.

For the rest of this section we assume that $p > 0$. There are various reductions to make the subgroup classification problem more tractable. By the Borel-Tits Theorem 2.9.23, a maximal connected subgroup of G is either a maximal parabolic subgroup, or it is

© The Author(s), under exclusive license to Springer Nature Switzerland AG 2026
M. Bate et al., *G-Complete Reducibility, Geometric Invariant Theory and Spherical Buildings*, Oberwolfach Seminars 57, https://doi.org/10.1007/978-3-032-08866-6_12

G-irreducible. The positive-dimensional maximal subgroups[1] in simple algebraic groups were classified by Seitz [152], [153] and Liebeck-Seitz [101] (with a correction in [52]). Now, if p is sufficiently large relative to, say, the rank of G then one would expect things to be more straightforward: for instance, a connected reductive subgroup H is G-cr if $p \geq a(G)$ by Theorem 10.2.1, hence if H is not maximal then it is contained in some proper Levi subgroup L of G. This suggests proceeding by induction on $\dim(G)$. However, we have seen that there are examples of connected reductive non-G-cr subgroups in every positive characteristic, so this expectation needs to be tempered, especially when the characteristic is small relative to the rank of G (and note that such non-G-cr examples can be maximal amongst connected *reductive* subgroups: for instance, when $p = 2$ the image of SL_2 under the adjoint representation is clearly a maximal connected reductive subgroup of SL_3 and is not SL_3-cr (Exercise 4.3)). Note also that problems of ascent and descent of complete reducibility also come into play when trying to employ this inductive approach. Nevertheless, it turns out from the work of Liebeck and Seitz discussed above that for a simple group G there are only finitely many conjugacy classes of maximal positive-dimensional reductive subgroups. (This is not true for arbitrary semisimple groups; see Exercise 12.1 for an example.) To find maximal connected reductive subgroups in, say, SL_m, we end up finding all m-dimensional irreducible representations of all connected reductive groups in that given dimension. This is easier in principle than finding all the m-dimensional representations of connected reductive groups—simple modules for connected reductive groups are classified by their highest weight—but this problem is far from being solved in any real sense (there are no dimension formulae for simple modules in general, let alone information about decomposition numbers, etc.). The procedure for the other classical groups is similar but harder, especially when $p = 2$. For exceptional simple groups, however, the maximal subgroup problem is tractable and the possibilities can be investigated case-by-case.

We briefly mention some further reductions for Problem 12.1.1 which suggest ways to narrow the search further. First, tools like Lemma 4.2.7 give hope that the problem for general reductive G does reduce to the case that G is simple (but see Exercise 12.1 below for an indication of one potential issue). Second, suppose H is a connected reductive but not semisimple subgroup of G; then H contains a non-trivial central torus S, and so we may replace G with $L = C_G(S)$. Then we may quotient by the centre of L and reduce to the case that H is semisimple. Further, if H has simple components $H_1, \ldots, H_r$ and we assume that H is G-cr, then H_1 is G-cr by Theorem 9.1.1. We also get $H_2 \cdots H_r \subseteq C_G(H_1)^0$, which is reductive by Theorem 5.4.1, and $H_2 \cdots H_r$ is $C_G(H_1)^0$-cr, by Proposition 9.3.3. This suggests a systematic reduction in many instances to the case that the subgroup H is simple.

---

[1] This in fact encompasses all maximal subgroups when $p > 0$: for by Proposition 6.1.4 no finite subgroup is maximal.

A final preliminary fact to note is that, unlike for maximal subgroups, there are infinitely many conjugacy classes of connected reductive subgroups in almost every simple G (except when G has type A_1; see Exercise 12.7). However, the program to classify them for G exceptional has seen huge advances over the past few decades. It was begun by Liebeck and Seitz [99], [101] and continued by many others, among them Lawther, Litterick, Stewart, Testerman, and Thomas: e.g., see [95], [104], [105], [106], [107], [175], [176], [179], [180]. We now describe some aspects of their approach.

Consider a finite set S of connected reductive groups (e.g., the set of simple exceptional groups). Let $\widehat{S}$ be the set obtained from S by adding in all the Levi subgroups of all the elements of S, i.e., one representative for each isomorphism class; then $\widehat{S}$ is still finite. Let $G \in S$. Let H be a connected reductive subgroup of G. By an inductive argument, we can assume that we have found all the connected reductive subgroups of every Levi subgroup L of G up to L-conjugacy. So (assuming that we can control the fusion[2] of L-conjugacy classes of subgroups in G), we may assume without loss that H is not contained in any proper Levi subgroup of G.

Here the theory of G-complete reducibility comes into play. Either H is G-completely reducible or it is not. If H is G-cr then H is G-ir, since it is not contained in any proper Levi subgroup of G. So we have reduced the problem to finding the connected G-ir subgroups of G, and such a subgroup is automatically reductive, by Lemma 4.1.6. Whilst this still requires considerable effort, it is easier than the general problem of finding all connected reductive subgroups of G.

So suppose H is not G-cr. Choose a subgroup P of G which is minimal amongst the parabolic subgroups of G that contain H, and fix a Levi subgroup L of P; note that P and hence L are proper since we are assuming H is not G-cr. Then $c_L(H)$ is an L-ir subgroup of L by Corollary 5.4.7. By assumption, we already know the L-ir subgroups of L. Hence it is enough to fix P and L and find all the connected reductive subgroups H of P such that $c_L(H) = M$, where M is a fixed L-ir subgroup of L. In practice, it is convenient to consider instead homomorphisms ρ from a fixed connected reductive group H to P such that $c_L \circ \rho = \sigma$, where $\sigma : H \to L$ is a fixed homomorphism with L-ir image. Describing these comes down to calculating the non-abelian 1-cohomology $H^1(H, R_u(P))$, by Lemma 2.10.4.

Non-abelian 1-cohomology is hard to compute, but one can reduce a certain amount of the complexity to the abelian case. The essential idea is that a short exact sequence $1 \to N \to U \to A \to 1$ of unipotent groups with an H-action induces a long exact sequence of cohomology sets—see [105, §1.4.3] or [177] for an explanation—and in principle one can use these long exact sequences to reconstruct $H^1(H, U)$ from cohomology sets for the simpler objects N and A. So, inductively, one can try to find $H^1(H, U)$ by piecing together abelian 1-cohomologies of the form $H^1(H, A)$, where the A are H-modules which are

---

[2] One says that two subgroups of L *fuse* if they become conjugate when regarded as subgroups of G.

subquotients of U (the actual process is more complicated than this: the piecing together involves understanding non-abelian 2-cohomology as well as non-abelian 1-cohomology).

The program described above has been almost completely carried out when S is the family of simple exceptional groups over a fixed algebraically closed field. Let $G \in S$. The G-ir subgroups were determined up to conjugacy by Thomas in [182] and the G-cr subgroups were determined up to conjugacy by Litterick and Thomas [107] (all of this builds on the aforementioned work of Lawther, Liebeck, Seitz, Stewart and Testerman). The simple non-G-cr subgroups of G were determined up to conjugacy by Litterick and Thomas for p good in [106]. The connected reductive non-G-cr subgroups with simple components of rank at least 3 were determined up to conjugacy by Litterick and Thomas for p bad in [108]. Work is in progress to determine the connected reductive non-G-cr subgroups with a simple component of rank less than 3. This will complete the classification program for S the set of simple exceptional groups.

Here is one interesting observation following from this work, which complements earlier results in this book. Recall our rigidity result that G has only countably many conjugacy classes of G-cr subgroups, even when k is uncountable; see Theorem 6.1.2. However, when k is uncountable, $G = F_4$ and either $p = 2$ or $p = 3$, then G admits uncountably many pairwise non-conjugate connected reductive subgroups [105, Thm. 2.42]. Lond found similar examples when $G = B_4$ and $p = 2$, see [109, §6.5]; here the conjugacy classes are parametrised by some positive-dimensional variety over k, so if k is uncountable then we get uncountably many conjugacy classes of such subgroups.

We end this summary by stressing that the geometric techniques presented in this book do not play a large part in the classification of the connected reductive subgroups of the simple exceptional groups. The work cited above primarily uses the classical (i.e., non-geometric) theory of reductive groups, along with deep and thorough analyses of the structure of the simple exceptional groups, representation theory and non-abelian cohomology. We view the geometric invariant theory approach as a complement to these classical methods; one which can also provide shorter conceptual proofs in some cases. We have seen many situations in this book where geometric invariant theory yields useful information about subgroups of a reductive group: for instance, if H is a G-cr subgroup of G then we obtain other G-cr subgroups for free, such as $N_G(H)$ and $C_G(H)$, and one does not need a classification to deduce this. Here is another example: a connected reductive overgroup of a regular unipotent element is G-ir. This result is discussed in Sect. 12.2. Note that this result was proved in [181] using case-by-case methods, but the geometric techniques provide a very quick proof, see Theorem 12.2.3.

For a situation where the classical methods and the geometric methods work together, see [19]: as we described in Sect. 10.6 above, this work uses tables from [99] along with geometric invariant theory techniques. (Note that now there are far more comprehensive tables available than those in [99], one can improve the bounds given in Theorem 10.6.1 in many cases; at the time of writing, no counterexamples have been found with $p = 3$. See [105, Rem. 1.12].)

12.2 Reductive Overgroups of Regular and Distinguished Unipotent Elements

In this section we give our short proof from [24] that a connected reductive subgroup of G containing a regular unipotent element is G-irreducible. This was a result first proved in [181, Thm. 1.2], utilising much of the work discussed in Sect. 12.1 describing the subgroup structure of reductive algebraic groups. Here is a brief history: for simple G, Saxl and Seitz determined the maximal positive-dimensional subgroups containing a regular unipotent element of G in [151], building on work of Suprunenko [178]. Subsequently, these classifications have been extended and refined, for example by Testerman-Zalesski [181], Guralnick-Malle [70] and Craven [51].

We begin our proof by recalling a result of Steinberg.

Lemma 12.2.1 *Let H be an algebraic group and let $\sigma : H \to H$ be any surjective homomorphism. Then σ stabilises a Borel subgroup of H. In particular, for every $x \in H$ there is a Borel subgroup of H normalised by x.*

Proof The first statement is [173, Thm. 7.2]. The second follows by applying this to the endomorphism of H given by conjugation by x. $\qquad\square$

Recall that if G is connected then $g \in G$ is *regular* if $\dim(C_G(g))$ is minimal, and likewise for elements of $\mathrm{Lie}(G)$. We need two properties of regular unipotent and regular nilpotent elements for connected reductive groups.

Lemma 12.2.2 *Assume G is connected, and let $u \in G$ be unipotent. Then:*

 (i) u is regular if and only if u is contained in a unique Borel subgroup B of G;
(ii) if u is regular and P is a parabolic subgroup of G with $u \in R_u(P)$, then $P = B$.

Similarly, any regular nilpotent element $e \in \mathrm{Lie}(G)$ is contained in a unique Borel subalgebra $\mathrm{Lie}(B)$, and if P is a parabolic subgroup of G such that $e \in \mathrm{Lie}(R_u(P))$, then $P = B$.

Proof Part (i) is [174, §3.7, Thm. 1], and the analogue for the Lie algebra is [83, Cor. 6.8]. If P is a parabolic subgroup containing u, then P contains B, and with respect to a suitable choice of maximal torus T of B, we may write $u = \prod_{\alpha \in \Phi^+} x_\alpha$, where each $x_\alpha \in U_\alpha$ and $x_\alpha \neq 1$ for each simple root α, cf. [174, §3.7, Thm. 1]. Since u has a non-trivial contribution from each simple root group, u can only lie in $R_u(P)$ if $P = B$. The analogous argument works equally well for e, which has a standard form involving a non-trivial contribution from each root space $\mathrm{Lie}(U_\alpha)$ relative to any simple root α, cf. [83, 6.7(1)]. $\qquad\square$

We can now prove the main result of this section; this proof is identical to that given in
[24, Thm. 3.2].

Theorem 12.2.3 *Let $H \subseteq G$ be connected reductive groups. If H contains a regular
unipotent element of G, or* $\mathrm{Lie}(H)$ *contains a regular nilpotent element of* $\mathrm{Lie}(G)$*, then H
is G-irreducible.*

Proof Suppose $u \in H$ is a regular unipotent element of G. Let B be a Borel subgroup of
H containing u, let S be a maximal torus of B, and write $B = P_\lambda(H)$ for some $\lambda \in Y_S$.
Then u belongs to $R_u(B)$, so $\lim_{t \to 0} \lambda(t)u\lambda(t)^{-1} = 1$, so $u \in R_u(P_\lambda(H)) \subseteq R_u(P_\lambda)$. It
follows from Lemma 12.2.2(ii) that P_λ is the unique Borel subgroup of G containing u.

Now let $B^- = P_{-\lambda}(H)$ be the opposite Borel subgroup of H with respect to the
maximal torus S of H. The Borel subgroups B and B^- of H are conjugate, say by $x \in H$.
Let $v = xux^{-1} \in P_{-\lambda}(H) \subseteq H$. Since v is H-conjugate to u, v is also a regular unipotent
element of G belonging to H. The argument of the first paragraph shows that $P_{-\lambda}$ is the
unique Borel subgroup of G containing v.

Now suppose P is a parabolic subgroup of G containing H. Then P contains u and v,
and hence must contain Borel subgroups normalised by u and v, by Lemma 12.2.1. But
a Borel subgroup of P is a Borel subgroup of G, so uniqueness forces P to contain the
opposite Borel subgroups P_λ and $P_{-\lambda}$ of G. This implies that $P = G$, so H is G-ir, as
required.

The proof in the case that $\mathrm{Lie}(H)$ contains a regular nilpotent element of $\mathrm{Lie}(G)$ is
essentially the same— given a parabolic subgroup P of G containing H, $\mathrm{Lie}(P)$ must
contain a pair of opposite Borel subalgebras of $\mathrm{Lie}(G)$, and therefore $\mathrm{Lie}(P) = \mathrm{Lie}(G)$,
which means that $P = G$.   □

Remark 12.2.4   (i). For a more general version of Theorem 12.2.3 for possibly non-
connected H and possibly non-connected G, generalising both [181, Thm. 1.2] and
[114, Thm. 1], we refer to [24, Thm. 1.1].

(ii). The part of Theorem 12.2.3 for regular unipotent elements of G was initially proved
in [181] using case-by-case methods depending on each Dynkin type of the simple
factors of G.

There is a nice analogue of Theorem 12.2.3 for finite groups of Lie type. Recall the
material on Steinberg endomorphisms for connected reductive G from Sect. 11.8. Note
that if $\sigma : G \to G$ is a Steinberg endomorphism and H is a connected reductive and
σ-stable subgroup of G, then the restriction of σ to H is a Steinberg endomorphism of H.
Then we have the following counterpart of Theorem 12.2.3 for finite groups of Lie type.

Theorem 12.2.5 ([24, Thm. 1.3]) *Let $H \subseteq G$ be connected reductive groups and suppose
σ is a Steinberg endomorphism of G such that H is σ-stable. Suppose that H contains a
regular unipotent element of G. Then H_σ is G-irreducible.*

Proof It follows from the proof of Theorem 12.2.3 that if H contains a regular unipotent element of G, then that element is regular unipotent in H; hence the regular unipotent elements of H are the regular unipotent elements of G contained in H, since these elements form a single H-conjugacy class in H. It follows from [169, III.1.19] applied to H that we may find a regular unipotent element u of G lying in H_σ. Since u is fixed by σ, the unique Borel subgroup B of H containing u is σ-stable. By [173, Cor. 10.10], there is a σ-stable maximal torus S in B, and the opposite Borel subgroup B^- to B in H with respect to S is also σ-stable. Again thanks to [173, Cor. 10.10], B and B^- are conjugate by an element $x \in H_\sigma$. Thus $v = xux^{-1}$ is a regular unipotent element of G which belongs to B^- and H_σ. The rest of the proof of Theorem 12.2.3 now goes through for H_σ. $\square$

We finish this section with some comments about a natural generalisation of the preceding result, where we consider overgroups of *distinguished* unipotent elements. Recall that a unipotent element u of G is distinguished if every torus in the centraliser $C_G(u)$ is central in G. Regular unipotent elements in G are distinguished [169, III 1.14] (or [45, Prop. 5.1.5]). The converse is true in type A, since a distinguished unipotent element must clearly consist of a single Jordan block. In [90, Thm. 6.5], Korhonen proves the following:

Theorem 12.2.6 *Suppose G is simple and p is good for G. Let H be a reductive subgroup of G. Suppose H^0 contains a distinguished unipotent element of G of order p. Then H is G-irreducible.*

One can easily extend this theorem to arbitrary connected reductive G by reducing to the simple case: see [11, Rem. 6.3]. Korhonen's proof of Theorem 12.2.6 depends on checks for the various possible Dynkin types for simple G. E.g., for G simple of exceptional type, Korhonen's argument relies on case-by-case investigations from [106], where all connected reductive non-G-cr subgroups are classified in the exceptional type groups in good characteristic. For classical G, Korhonen requires a classification of SL_2-representations on which a non-trivial unipotent element of SL_2 acts with at most one Jordan block of size p. In [11, Thm. 1.2], we present a short and uniform proof of Theorem 12.2.6 under an extra hypothesis, using the good A_1 subgroups discussed in Sect. 10.8. We also formulate a counterpart of Korhonen's theorem for overgroups of u which are finite groups of Lie type in [11, Thm. 1.6], and we discuss versions of the result without the restriction on the order of u and for distinguished nilpotent elements of $\mathrm{Lie}(G)$. Note that if one is willing to put up with a worse bound on p, then the result follows very quickly from results we have already seen in the book: for example, if $p \geq a(G)$, where $a(G)$ is as in Theorem 10.2.1, then a connected reductive subgroup H of G is automatically G-cr. If H also contains a distinguished unipotent element u of G, then H cannot be contained in any proper Levi subgroup of G (since that would imply a non-central torus of G centralising H and hence u). Hence H is G-ir.

12.3 G-Complete Reducibility of Subalgebras of Lie(G)

The notion of G-complete reducibility for subalgebras of $\mathfrak{g} = \mathrm{Lie}(G)$ was first defined by McNinch [126] for algebraically closed k and was developed further in [27, §3]; the non-algebraically closed case was first studied in [16, §5]. The approach via geometric invariant theory stems from work of Richardson [147], as in the subgroup case. The theory closely parallels the theory for subgroups developed in the earlier chapters. Some arguments are actually easier in the subalgebra case, as we need not worry about non-connected groups. There are, however, some extra problems for subalgebras, which can be traced back to the failure of certain normalisers to be smooth. Throughout this section we assume that G is connected.

Generalising Example 2.7.1(ii) and in analogy with Example 2.7.2, we consider the simultaneous diagonal action of G on the n-fold Cartesian product of $\mathfrak{g}$ via the adjoint action:

$$g \cdot (x_1, \ldots, x_m) := (\mathrm{Ad}(g)(x_1), \ldots, \mathrm{Ad}(g)(x_m))$$

for $g \in G$ and $(x_1, \ldots, x_m) \in \mathfrak{g}^m$.

Definition 12.3.1 ([16, Def. 5.3]) A subalgebra $\mathfrak{h}$ of $\mathfrak{g}$ is G-*completely reducible over k* (G-cr over k) if for any k-parabolic subgroup P of G such that $\mathfrak{h} \subseteq \mathrm{Lie}(P)$, there is a k-Levi subgroup L of P such that $\mathfrak{h} \subseteq \mathrm{Lie}(L)$. As in the subgroup case, we say that $\mathfrak{h}$ is G-*completely reducible* if it is G-completely reducible over $\overline{k}$.

For $k = \overline{k}$, this notion is due to McNinch; see [126] and also [28, §5.3].

Remark 12.3.2 As in the case of G-complete reducibility for subgroups, the notion of G-complete reducibility for subalgebras generalises the concept of semisimplicity from representation theory. For, if $\mathfrak{h}$ is a subalgebra of $\mathfrak{gl}(V)$ or $\mathfrak{sl}(V)$, then $\mathfrak{h}$ is $\mathrm{GL}(V)$-cr or $\mathrm{SL}(V)$-cr if and only if V is a semisimple $\mathfrak{h}$-module: see Exercise 12.3.

We now give the link to geometric invariant theory. For this, we need the following definition.

Definition 12.3.3 Let $\mathfrak{h}$ be a Lie algebra. We call $\mathbf{x} = (x_1, \ldots, x_m) \in \mathfrak{h}^m$, for some $m \in \mathbb{N}$, a *generating tuple for* $\mathfrak{h}$ if $\{x_1, \ldots, x_m\}$ is a generating set for $\mathfrak{h}$ as a Lie algebra.

The next theorem is [16, Thm. 5.4]; see also [126, Thm. 1(i)] for the case $k = \overline{k}$.

Theorem 12.3.4 *Let $\mathfrak{h}$ be a subalgebra of $\mathfrak{g}$. Let $\mathbf{x} \in \mathfrak{g}^m$ be a generating tuple for $\mathfrak{h}$. Then $\mathfrak{h}$ is G-completely reducible over k if and only if the $G(k)$-orbit of $\mathbf{x}$ is cocharacter-closed in $\mathfrak{g}^m$ over k.*

We also have an interpretation of complete reducibility in building-theoretic terms. The argument given below forms part of the proof of [126, Lem. 4]; it provides the key link between G-complete reducibility and the building Δ of G.

Proposition 12.3.5 *Suppose $\mathfrak{h}$ is a subalgebra of $\mathfrak{g}$ and let $\Delta = \Delta_{G,k}$ denote the building of G over k. Then $\Delta^{\mathfrak{h}} := \{\sigma_P \mid \mathfrak{h} \subseteq \mathrm{Lie}(P)\}$ is a convex subcomplex of Δ, and it is Δ-completely reducible if and only if $\mathfrak{h}$ is G-completely reducible over k.*

Sketch Proof For the first assertion, the key observation is that for the intersection of any two connected subgroups P and Q of G that are normalised by T, we have $\mathrm{Lie}(P \cap Q) = \mathrm{Lie}(P) \cap \mathrm{Lie}(Q)$ [32, Cor. 13.21]. Now the result follows from Serre's criterion [159, Prop. 3.1] for recognising a convex subcomplex of Δ: a collection Σ of simplices is a convex subcomplex if whenever P, Q, R are k-parabolic subgroups of G with $\sigma_P, \sigma_Q \in \Sigma$ and $P \cap Q \subseteq R$, then $\sigma_R \in \Sigma$. See Sects. 7.2.6 and 11.5.

To see that $\mathfrak{h}$ is G-cr over k if and only if $\Delta^{\mathfrak{h}}$ is Δ-cr, the key again is the smoothness of the intersection of two k-parabolic subgroups. This implies that for two k-parabolic subgroups P and Q of G, we can detect whether or not they are opposite (and hence correspond to opposite simplices of Δ) on the level of their Lie algebras. $\square$

With the above results in hand, one can prove analogues for subalgebras of many of the results from this book for subgroups, including the rationality results in Chap. 11. Rather than listing all such results, we concentrate on where there are significant differences in the theory. We first give a positive result in one direction, before showing why the converse is not true. This next result is [126, Thm. 1(ii)] when k is algebraically closed. See also [28, Thm. 5.27, Ex. 5.29], but note that the proofs given there (which use the technology of optimal destabilising R-parabolic subgroups) do not go through over arbitrary k. This is why we use the Tits Centre Theorem 7.4.1 instead.

Theorem 12.3.6 ([12, Thm. 3.17]) *Suppose H is a k-defined subgroup of G. If H^0 is G-completely reducible over k, then so is $\mathrm{Lie}(H)$.*

Proof It is no loss to assume that H is connected since $\mathrm{Lie}(H) = \mathrm{Lie}(H^0)$. Let $\mathfrak{h} = \mathrm{Lie}(H)$. Using Theorem 11.7.4 and its analogue for subalgebras, which follows in a very similar manner, it is enough to prove the result over the separable closure k_s of k. This implies that we may assume that $H(k)$ is dense in H by Lemma 11.1.1. Now suppose L is a minimal Levi k-subgroup of G containing H. Since $L = L_\lambda = C_G(\mathrm{Im}(\lambda))$ for some $\lambda \in Y_{C_G(H),k}$, Corollary 11.7.15 and its analogue for subalgebras imply that H and $\mathfrak{h}$ are G-cr over k if and only if they are L-cr over k, so we may replace G with L. Then H is G-cr over k but is not contained in any proper Levi k-subgroup of G, so H is G-ir over k.

We now proceed with the proof of the theorem by contradiction. Suppose that $\mathfrak{h}$ is not G-cr over k. The subgroup $H(k)$ acts on the building Δ by simplicial automorphisms, and the subcomplex $\Delta^{\mathfrak{h}}$ is stabilised by the action of $H(k)$ since H stabilises its own

Lie algebra. We are assuming that $\mathfrak{h}$ is not G-cr over k, so this subcomplex is not Δ-cr, by Proposition 12.3.5, and hence there is a non-empty $H(k)$-fixed simplex σ in $\Delta^{\mathfrak{h}}$ by Theorem 7.4.1. This simplex has the form σ_P for some proper k-parabolic subgroup P of G, and the fact that σ_P is $H(k)$-fixed translates into the fact that P is normalised by $H(k)$. Since parabolic subgroups are self-normalising, this shows that $H(k) \subseteq P$, which in turn gives $H \subseteq P$ because $H(k)$ is dense in H. This contradicts the conclusion of the first paragraph, that H is G-ir over k, so we are done. $\qquad\square$

The converse to the above result is false, even when the field is algebraically closed. The following example is due to the second author; it appears in [126, §1].

Example 12.3.7 Suppose $p > 0$. Let H be a non-trivial connected semisimple group and let $\varrho_i : H \to \mathrm{SL}(V_i)$ be representations of H for $i = 1, 2$ with ϱ_1 semisimple and ϱ_2 not semisimple. Let $\varrho : H \to G := \mathrm{SL}(V_1 \oplus V_2)$ be the representation given by $h \mapsto \varrho_1(h) \oplus \varrho_2(F(h))$, where $F_p : H \to H$ is the Frobenius endomorphism of H. Let J be the image of H under ϱ in G. Since $V_1 \oplus V_2$ is not semisimple as a J-module, J is not G-cr (cf. Example 4.1.4). However, $\mathrm{Lie}(J) = \mathrm{Im}\, d\varrho_1 \oplus 0 \subseteq \mathfrak{g}$ *is* G-cr, showing that the converse of Theorem 12.3.6 fails.

Remark 12.3.8 If k is perfect, then Theorem 12.3.6 also holds with the hypothesis that H (instead of H^0) is G-cr. For, as has already been observed, if k is perfect then H is G-cr over k if and only if H is G-cr (Corollary 11.7.12), and the same holds for $\mathrm{Lie}(H)$. Now over $\overline{k}$, if H is G-cr then H^0 is G-cr too, thanks to Theorem 9.1.1, and the hypotheses of the theorem hold.

Let H be a reductive subgroup of G and let $n \in \mathbb{N}$. The inclusion of $\mathfrak{h}^n$ in $\mathfrak{g}^n$ gives rise to a map ψ from $\mathfrak{h}^n /\!\!/ H$ to $\mathfrak{g}^n /\!\!/ G$. It is natural to ask whether the analogue of the Finiteness Theorem 6.2.1 holds: that is, whether ψ is finite. It turns out that the answer is no in general: see [12, Ex. 4.2]. On the other hand, the answer is yes if we impose a mild restriction on p (see [12, Prop. 4.4]). For in this case one can show that the nullcone $\mathcal{N}(\mathfrak{h}^n)$ is equal to $\mathcal{N}(\mathfrak{g}^n) \cap \mathfrak{h}^n$ and apply an argument similar to the proof of Theorem 12.8.1(a).

We finish this section with an example which shows that the obvious analogue of our Clifford Theory result Theorem 9.1.1 is not true in the Lie algebra setting.

Example 12.3.9 Suppose k is algebraically closed of characteristic $p = 2$ and let $G = \mathrm{PGL}_2$. Let $\mathfrak{h}$ be the abelian subalgebra of $\mathfrak{g}$ spanned by $x_1 := \begin{pmatrix} 0 & 1 \\ 0 & 0 \end{pmatrix}$ and $x_2 := \begin{pmatrix} 0 & 0 \\ 1 & 0 \end{pmatrix}$. It is easy to check that $\mathfrak{h}$ is not contained in $\mathrm{Lie}(B)$ for any Borel subgroup B of G, so $\mathfrak{h}$ is G-cr.

Now let $\mathfrak{m}$ be the subalgebra of $\mathfrak{g}$ spanned by x_1. Then $\mathfrak{m}$ is clearly not G-cr, since it is contained in the Lie algebra of the upper triangular Borel subgroup, but not in the Lie algebra of any maximal torus ($\mathfrak{m}$ consists of nilpotent elements!). This shows that it is not necessarily the case that an ideal of a G-cr subalgebra is G-cr.

Note that in this case $C_G(\mathfrak{m})$ is (the image of) the group of upper unitriangular matrices, but $C_{\mathfrak{g}}(\mathfrak{m}) = \mathfrak{h}$. Hence $C_G(\mathfrak{m})$ is not smooth. A similar calculation shows that $C_G(\mathfrak{h})$ is not smooth either. It is this failure of smoothness that contributes to the existence of such examples. For much more on this, see [12], where this example appears as part of Example 2.9.

12.4   Relative G-Complete Reducibility

In this section we discuss the paper [27], where we gave an extension of some of the results from Chap. 5. Here is the basic idea: let H be a reductive subgroup of G and consider the action of H on G^n by simultaneous conjugation. Then we wish to characterise the closed H-orbits in G^n is a similar way to how Theorem 5.3.3 characterises the closed G-orbits. This characterisation turns out to be a fairly natural generalisation of the notion of G-complete reducibility, which we dub "relative complete reducibility".

Definition 12.4.1 Let K be a subgroup of G and let H be any reductive subgroup of G. We say that K is *relatively G-completely reducible with respect to H* if for every $\lambda \in Y_H$ such that K is contained in P_λ, there exists $\mu \in Y_H$ such that $P_\lambda = P_\mu$ and $K \subseteq L_\mu$. We sometimes use the shorthand *relatively G-cr with respect to H*.

Remark 12.4.2   (i). When $H = G$, Definition 12.4.1 coincides with the usual definition of G-complete reducibility, Definition 4.1.1. Since the definition is given in terms of cocharacters of H, a subgroup of G is relatively G-completely reducible with respect to H if and only if it is relatively G-completely reducible with respect to H^0, so we can assume that the subgroup H is connected if we wish.

 (ii). If $K \subseteq H$, then K is relatively G-cr with respect to H if and only if K is H-cr (this follows from Lemma 12.4.3(ii) below).

(iii). In characteristic zero a subgroup of G is G-completely reducible if and only if it is reductive, see Remark 4.1.7(i). We don't know of any simple characterisation of relative G-complete reducibility in this case (but see item (iv)).

 (iv). If we take $H \subseteq Z(G)$, then all subgroups of G are relatively G-cr with respect to H. In particular, it is possible for non-reductive (even unipotent) subgroups to be relatively G-cr with respect to a subgroup H.

 (v). Let $H \subseteq M \subseteq G$ be any chain of subgroups such that H is G-cr but not M-cr (see Example 4.1.10, for example). Then obviously H is not relatively G-cr with respect to M, so we see that G-complete reducibility need not imply relative complete reducibility with respect to a subgroup.

The following technical lemma helps with the proof of the main result of this section.

Lemma 12.4.3 *Let H be a reductive subgroup of G.*

(i) Let $\lambda, \mu \in Y_H$ such that $P_\lambda = P_\mu$ and let u be the element of $R_u(P_\lambda(H))$ such that $u L_\lambda(H) u^{-1} = L_\mu(H)$. Then $u L_\lambda u^{-1} = L_\mu$.

(ii) Let K be a subgroup of G. Then K is relatively G-completely reducible with respect to H if and only if for every $\lambda \in Y_H$ such that $K \subseteq P_\lambda$ there exists $u \in R_u(P_\lambda(H))$ such that $K \subseteq L_{u \cdot \lambda}$.

Proof For (i) set $\mu' = u \cdot \lambda \in Y_H$, and note that $P_{\mu'} = P_\lambda$. Then $L_\mu(H) = L_{\mu'}(H)$ so μ and μ' commute, and hence L_μ and $L_{\mu'}$ contain a common maximal torus, so they are equal by Lemma 2.9.10. Part (ii) is now clear from part (i) and Definition 12.4.1. $\square$

Before giving the main result on relative complete reducibility, we remark that the formalism of generic tuples is well adapted to this situation. Let H be a reductive subgroup of G, let K be a subgroup of G, and let $\mathbf{k} \in K^n$ be a generic tuple for K. Then it follows from Lemma 5.2.5(ii) that K is relatively G-cr with respect to H if and only if $\mathcal{G}(\mathbf{k})$ is. The following result, which is an analogue of Theorem 5.3.3, solves the problem of characterising the closed H-orbits in G^n. It was originally given in [27, Thm. 3.5(iii)]; we leave the proof as Exercise 12.4 since it is a relatively straightforward adaptation of the proof of Theorem 5.3.3, using Lemma 12.4.3 in a couple of places.

Theorem 12.4.4 *Let H be a reductive subgroup of G. Let K be a subgroup of G and let $\mathbf{k} \in K^n$ be a generic tuple for K. Then $H \cdot \mathbf{k}$ is closed in G^n if and only if K is relatively G-completely reducible with respect to H.*

We refer to [4] for further results on relative complete reducibility and to [68, Thm. 1.3(ii)] where it is shown that this notion can be captured in a building-theoretic manner in the sense of Sect. 7.3.

12.5 G-Complete Reducibility Versus G^0-Complete Reducibility

For most of the book we have not needed to worry too much about whether or not our reductive group G is connected, and on the whole this has helped the development of the theory: even if one is interested mainly in connected reductive groups, one must often consider non-connected groups. For instance, natural subgroups of a connected group, such as normalisers and centralisers, are often non-connected. However, in applications it is of interest to try to relate G-complete reducibility of subgroups of G to G^0-complete reducibility of appropriate subgroups of G^0. First recall that if $H \subseteq G^0$ then H is G-cr if and only if H is G^0-cr (Remark 4.2.10), so this question is only of interest when $H \cap G^0 \neq H$. Moreover, if $p = 0$ then H is G-cr if and only if H^0 is G^0-cr if and only if H^0 is reductive, so without loss we assume $p > 0$.

In [15, Thm. 5.3], we describe an algorithm which, starting with the pair (H, G), where H is a subgroup of G, reduces the question of whether H is G-completely reducible in a finite number of steps to questions of complete reducibility in various connected groups. Rather than describing the full algorithm here, we now describe an important special case.

Let H be a subgroup of G, and say that H *acts on G^0 by outer automorphisms* if for each $1 \neq h \in H$, conjugation by h gives a non-inner automorphism of G^0. In other words, we may identify H with a subgroup of the group of (algebraic) outer automorphisms of G^0. Now suppose also that G^0 is simple; then H is cyclic except for possibly when G^0 is of type D_4. In this situation, if H is also of order coprime to p then any generator for H is semisimple and hence H is G-cr by Theorem 5.3.3, as semisimple conjugacy classes are closed. On the other hand, if H is cyclic and unipotent (that is, H is a p-group) then H can be G-cr or non-G-cr; recall that we observed in Remark 4.1.7(ii) that a p-group can still be G-cr in the non-connected setting.

The following result is [15, Cor. 4.5]. It gives a criterion for G-complete reducibility of a subgroup H of G in terms of the centraliser of H in G^0 (i.e., the fixed points of the automorphisms in H), and is a key ingredient in the algorithm described in [15, §5]. In case H is cyclic, the theorem also follows from a result due to Guralnick and Malle [71, Thm. 2.3] and [15, Cor. 3.2]. The forward implication follows from Theorem 5.4.1, but the reverse takes some work. The hardest case to deal with is when G is of type D_4 and $p = 3$, because then there are two distinct classes of non-inner automorphisms of order 3, one of which has a fixed-point subgroup which is non-reductive.

Theorem 12.5.1 *Suppose G^0 is simple and H acts on G^0 by outer automorphisms. Then H is G-completely reducible if and only if $C_{G^0}(H)$ is reductive.*

We finish this section with an example which indicates the sort of problem that can arise in this sort of analysis; the example is [15, Ex. 6.2].

Example 12.5.2 Let Γ be a finite group acting transitively on a finite set I. Let $i_0 \in I$. Let $\rho : \Gamma \to M$ be a homomorphism to a simple group M such that $\rho(C_\Gamma(i_0))$ is not M-cr. We set

$$G = \Gamma \ltimes \prod_{i \in I} M,$$

where Γ acts on the product by permuting the indices. Clearly, $G^0 = \prod_i M$. Let $d : M \to G^0$ be the diagonal embedding. As Γ commutes with the image of d, we may form the subgroup

$$H = \{\gamma d(\rho(\gamma)) \mid \gamma \in \Gamma\}$$

of G, a finite subgroup of G. The algorithm described in [15, Thm. 5.3] is then used to show that H is not G-cr.

To see this example concretely, take $\Gamma = \mathrm{PGL}_2(q)$ for q a sufficiently large power of p, $I = \mathrm{PGL}_2(q)/B(q)$ where B is a Borel subgroup of PGL_2, and consider I as a transitive Γ-set by left translation. Let $M = \mathrm{PGL}_2$, and let $\rho : \Gamma \hookrightarrow M$ be the canonical embedding. If we take $i_0 = B(q)$, then $\rho(C_\Gamma(i_0)) = B(q)$ is not M-cr, for q large enough, as B is not M-cr. In this example, we have $H \cap G^0 = 1$ and $C_{G^0}(H) = 1$ (as $C_M(B(q)) = 1$ for q sufficiently large). In particular, this shows that H may fail to be G-cr even if $C_{G^0}(H)$ and $H \cap G^0$ are both G^0-cr.

12.6 Semisimplification Over Non-Algebraically Closed Fields

Recall the notion of a semisimplification of a subgroup of G from Sect. 5.5 and its connection with G-complete reducibility. In this section we indicate how to make that material go through over a non-algebraically closed field; this material is based on [23] (note also that there are analogues of all of this for subalgebras of the Lie algebra, but we do not go into that; the interested reader is referred to [12]). Let k be a field (not necessarily algebraically closed) and let G be a reductive k-group.

Definition 12.6.1 Let H be a subgroup of G. A subgroup H' of G is said to be a k-*semisimplification of H (for G)* if there exist an R-parabolic k-subgroup P of G and an R-Levi k-subgroup L of P such that $H \subseteq P$, $H' = c_L(H)$ and H' is G-completely reducible over k (or equivalently by Corollary 11.7.16, L-completely reducible over k). We say *the pair (P, L) yields H'*.

Remark 12.6.2 (a). In analogy with the case $k = \bar{k}$, a subgroup H which is G-cr over k is a k-semisimplification of itself.

(b). Since all the k-defined R-Levi subgroups of a given k-defined R-parabolic subgroup P are conjugate by $R_u(P)(k)$ (Lemma 11.2.4(v)), if (P, L) yields a k-semisimplification of H then so does (P, M) for any other k-defined R-Levi subgroup of P; hence it makes sense to simply say that P yields a k-semisimplification of H.

Remark 12.6.3 Let $\mathbf{h} = (h_1, \ldots, h_m) \in H^m$ be a generic tuple for H. Then for any $\lambda \in Y_{G,k}$ with $H \subseteq P_\lambda$, $c_\lambda(\mathbf{h})$ is a generic tuple for $c_\lambda(H)$ by Lemma 5.2.5(iii). Hence by Theorem 11.4.4, $c_\lambda(H)$ is a k-semisimplification of H if and only if $G(k) \cdot c_\lambda(\mathbf{h})$ is cocharacter-closed over k. It follows from Theorem 11.4.10 that H admits at least one k-semisimplification: for we can choose $\lambda \in Y_{G,k}$ such that $G(k) \cdot c_\lambda(\mathbf{h})$ is cocharacter-closed over k, so $c_\lambda(H)$ is a k-semisimplification of H, yielded by (P_λ, L_λ).

Lemma 12.6.4 *Suppose that H' is a k-semisimplification of H. Then there is $\lambda \in Y_{G,k}$ such that H' is yielded by the pair (P_λ, L_λ).*

Proof Suppose H' is yielded by the pair (P, L). By Lemma 11.2.4(i) and (ii), we can find some $\lambda \in Y_{G,k}$ such that $P \subseteq P_\lambda$ and $L = L_\lambda \cap P$. But then $H' = c_L(H) = c_\lambda(H)$. $\square$

We can now state the main result from [23, Thm. 4.5], which is the analogue over k of Proposition 5.5.8.

Theorem 12.6.5 *Let H be a subgroup of G. Then any two k-semisimplifications of H are $G(k)$-conjugate.*

Proof Let H_1, H_2 be k-semisimplifications of H. By Lemma 12.6.4, there exist $\lambda_1, \lambda_2 \in Y_{G,k}$ such that $(P_{\lambda_1}, L_{\lambda_1})$ realises H_1 and $(P_{\lambda_2}, L_{\lambda_2})$ realises H_2. Let $\mathbf{h} \in H^m$ be a generic tuple for H. Then $c_{\lambda_i}(\mathbf{h})$ is a generic tuple for H_i for $i = 1, 2$, and each orbit $G(k) \cdot c_{\lambda_i}(\mathbf{h})$ is cocharacter-closed over k and accessible from $\mathbf{h}$ over k (Example 11.4.2). It follows from the uniqueness result in Theorem 11.4.10 that the closed subset $C_{\mathbf{h}} := \{g \in G \mid g \cdot c_{\lambda_1}(\mathbf{h}) = c_{\lambda_2}(\mathbf{h})\}$ contains a k-point.

Pick $g \in C_{\mathbf{h}}$. If $H_2 = gH_1g^{-1}$ then we are done. Otherwise there exists $h \in H$ such that $gc_{\lambda_1}(h)g^{-1} \notin H_2$ or $g^{-1}c_{\lambda_2}(h)g \notin H_1$. Without loss assume the former. We can repeat the above argument, replacing $\mathbf{h}$ with the generic tuple $\mathbf{h}' := (\mathbf{h}, h) \in H^{m+1}$; note that $C_{\mathbf{h}'}$ is properly contained in $C_{\mathbf{h}}$. The result now follows by a descending chain condition argument. $\square$

Let H be a subgroup of G and let P be minimal among the k-defined R-parabolic subgroups that contain H. Then as in Corollary 5.4.7, it is easy to see that $c_L(H)$ is L-ir over k. Conversely, if P is an R-parabolic k-subgroup with R-Levi k-subgroup L such that $P \supseteq H$ and $c_L(H)$ is L-ir over k then a similar argument shows that P is minimal among the R-parabolic k-subgroups containing H. We see that $c_L(H)$ is a k-semisimplification of H (the proof of the corollary goes through over k with the appropriate changes, since the same is true of the key result Lemma 2.9.10)

Example 12.6.6 Suppose $\mathrm{char}(k) = 0$. It follows from Remark 11.7.13(ii) that a k-subgroup of G is G-cr if and only if it is G-cr over k. Now let H be a k-subgroup of G and let P be an R-parabolic subgroup of G with R-Levi subgroup L such that $P \supseteq H$. Then the above and Remark 4.1.7(i) imply that $c_L(H)$ is a k-semisimplification of H if and only if $R_u(H) \subseteq R_u(P)$.

Remark 12.6.7 Given a reductive k-group G and a subgroup H of G, it makes sense to consider the semisimplification (over $\bar{k}$) of H as well as the k-semisimplification. Since it is possible that H is G-cr over k but not G-cr, or vice versa (see Examples 11.3.7 and 11.3.8), there is no direct relation between the notions of k-semisimplification and semisimplification.

Using canonical parabolic subgroups of G that are attached to a given subgroup H of G, either by means of the optimality formalism from Chap. 8 and Sect. 11.6, or via the Tits Centre Theorem 7.4.1, one can derive the following analogue over k of Proposition 9.4.1, giving a link between the k-semisimplification of H and a normal subgroup N of H.

Theorem 12.6.8 *Let H be a k-subgroup of G and let N be a normal k-subgroup of H. Suppose at least one of the following holds:*

(i) k is perfect.
(ii) G is connected.

Then there is an R-parabolic subgroup P of G such that P yields both a k-semisimplification of H and a k-semisimplification of N. In particular, there exist k-semisimplifications H' and N' of H and N such that N' is normal in H'.

Sketch Proof If (i) holds then a subgroup is G-cr over k if and only if it is G-cr (over $\bar{k}$), by Remark 11.7.13(ii), so the result follows from Proposition 9.4.1. If (ii) holds then we can use an argument involving the building $\Delta_{G,k}$: if H is not G-cr over k then we can find an appropriate parabolic subgroup P with $\sigma_P \in (\Delta_{G,k})^H$ using the Tits Centre Theorem 7.4.1.

For full details, see [23, Thm. 4.5]. □

12.7   Külshammer's Question

In this section we study the following question. Let H_1 and H_2 be subgroups of G having a large overlap. Must H_1 and H_2 be equal? For instance, suppose H_1 and H_2 are connected and reductive, and H_1 and H_2 contain a common Borel subgroup. We can use geometric ideas to prove that $H_1 = H_2$, as follows. The homogeneous space $X := G/H_2$ is affine by [145, Thm. A] (see also Proposition 3.1.15). We can let H_1 act on X by left multiplication (see Example 3.1.3). Let x be the coset $H_2 \in G/H_2$. Then $(H_1)_x = H_1 \cap H_2$. But $H_1 \cap H_2$ contains a Borel subgroup of H_1, so $(H_1)_x = H_1$, thanks to Lemma 3.1.14. Hence $H_1 \subseteq H_2$. We have $H_2 \subseteq H_1$ by symmetry, so $H_1 = H_2$.

We consider an application of geometric invariant theory to the situation when H_1 and H_2 have a maximal unipotent subgroup in common. We follow the presentation of [110]; Theorem 12.7.2 below is a special case of [110, Thm. 1.1].

Lemma 12.7.1 *Let X be a G-variety and let $x \in X$. Let $\lambda \in Y_G$ such that $x' := \lim_{a \to 0} \lambda(a) \cdot x$ exists and $G \cdot x'$ is closed. Let U be a subgroup of P_λ which is normalised by $\lambda(k^*)$. Then $\dim(U_{x'}) \geq \dim(U_x)$.*

Proof Let $S_\lambda = \lambda(k^*)U$ and set $C := S_\lambda \cdot x$. Let $D = \overline{C}$. Then $x' \in D$. Now $\dim(U_y) = \dim(U_x)$ for all $y \in C$ since $\lambda(k^*)$ normalises U. As C is dense in D, we have $\dim(U_{x'}) \geq \dim(U_x)$ by Lemma 2.7.4(i). $\qquad\qquad\square$

Theorem 12.7.2 *Let H_1 and H_2 be connected semisimple subgroups of G. Suppose there is a subgroup U of $H_1 \cap H_2$ such that U is a maximal unipotent subgroup of both H_1 and H_2. Then H_1 and H_2 are $N_G(U)$-conjugate.*

Proof Let $X := G/H_2$ with G and H_1 acting by left multiplication. Let x be the coset $H_2 \in G/H_2$. Note that $(H_1)_x = H_1 \cap H_2 \supseteq U$. If $H_1 \cdot x$ is closed then take λ to be $0 \in Y_{H_1}$. If $H_1 \cdot x$ is not closed then take $\lambda \in Y_{H_1}$ to be an optimal destabilising cocharacter for the H_1-action as in Theorem 8.6.23. In both cases, $x' := \lim_{a \to 0} \lambda(a) \cdot x$ exists, $H_1 \cdot x'$ is closed and $(H_1)_x \subseteq P_\lambda$; in particular, $U \subseteq P_\lambda$. Let $H = (H_1)_{x'}$; then H is reductive by Lemma 3.1.15(ii).

Since U is a maximal unipotent subgroup of H_1, U contains $R_u(P_\lambda(H_1))$, so $U = U'R_u(P_\lambda(H_1))$ for some maximal unipotent subgroup U' of $L_\lambda(H_1)$. Hence $\lambda(k^*)$ normalises U. Moreover, U is connected since H_1 is. We have $\dim(U_{x'}) \geq \dim(U_x) = \dim(U)$ by Lemma 12.7.1, so $U \subseteq H$. It is not hard to deduce that $H = H^0 = H_1$; we leave this as an exercise (Exercise 12.2). There exists $g \in G$ such that $x' = gH_2 = g \cdot x$, so $(H_1)_{x'} \subseteq G_{x'} = gG_x g^{-1} = gH_2 g^{-1}$. Hence

$$H_1 \subseteq gH_2 g^{-1}. \tag{12.7.1}$$

By symmetry, $H_2 \subseteq \widetilde{g} H_1 \widetilde{g}^{-1}$ for some $\widetilde{g} \in G$. Hence $H_1 = gH_2 g^{-1}$. $\qquad\square$

It is natural to ask what happens in Theorem 12.7.2 if we relax the requirement that H_1 and H_2 are connected and semisimple. We know from Proposition 2.3.1 that maximal unipotent subgroups exist and are unique up to conjugacy. Our first observation is that the conclusion of Theorem 12.7.2 fails if we allow H_1 and H_2 to be connected reductive subgroups: just take H_1 and H_2 to be two non-conjugate tori, for example.

A convenient way to formulate the problem is as follows.

Question 12.7.3 ([110, Question 1.3]) Let H be an algebraic group and let U be a maximal unipotent subgroup of H. Fix a homomorphism $\sigma \colon U \to G$. Are there only finitely many conjugacy classes of homomorphism $\rho \colon H \to G$ such that $\rho|_U = \sigma$?

The answer is yes if H is connected and semisimple [110, Thm. 1.2]. Note that this does not follow directly from Theorem 12.7.2: the point is that we can have many different homomorphisms $\rho \colon H \to G$ with the same image. If H is finite, however, then the set of homomorphisms $\rho \colon H \to G$ is finite, so the settings of Theorem 12.7.2 and Question 12.7.3 are closely related in this case.

Question 12.7.3 was, indeed, first posed by Külshammer for H finite: see [92, §2] and also [164, I.5]. In this case a maximal unipotent subgroup is the trivial group when $p = 0$ and a Sylow p-subgroup of H when $p > 0$. Külshammer showed that the answer is yes when $G = \mathrm{GL}_m$ for some $m \in \mathbb{N}$: see [92, §2] or [164, I.5]. He asked whether the answer is yes for every finite group H and every connected reductive group G. Cram-Martin gave an example to show that the answer can be no if we allow G to be non-reductive (see [50]). Slodowy proved that the answer is yes for connected reductive G if there is an embedding of G in some GL_m such that (GL_m, G) is a reductive pair; this is the case under some mild assumptions on p (cf. Remark 10.5.25). We sketch the argument, since it illustrates several of the ideas discussed in this book.

Pick generators $\gamma_1, \ldots, \gamma_n$ for H. We have a commutative diagram of inclusions

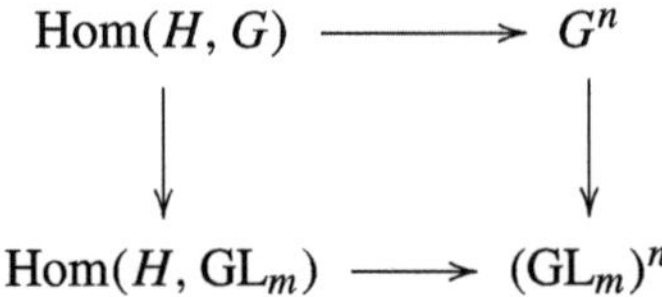

(see Sect. 6.5). Let $\rho \in \mathrm{Hom}(H, G)$. We may regard ρ as an element of $(\mathrm{GL}_m)^n$. The orbit $\mathrm{GL}_m \cdot (\rho(\gamma_1), \ldots, \rho(\gamma_n))$ is separable by Remark 10.5.2, as every subgroup of GL_m is separable (Example 10.5.3); hence $\mathrm{GL}_m \cdot \rho$ is separable. It follows from Proposition 10.5.18 that

$$\mathrm{GL}_m \cdot \rho \cap \mathrm{Hom}(H, G) \text{ is a finite union of } G\text{-conjugacy classes.} \tag{12.7.2}$$

We know the answer to Question 12.7.3 is yes for GL_m, and it is now straightforward to deduce that the answer to Question 12.7.3 is yes for G, using (12.7.2).

The proof presented above is essentially the same as Slodowy's proof (see [164, II.4]), but Slodowy works directly with the representation varieties $\mathrm{Hom}(\Gamma, G)$ and $\mathrm{Hom}(\Gamma, \mathrm{GL}_m)$ rather than with the spaces of tuples. He proves an analogue of Proposition 10.5.18 by applying Richardson's tangent space argument Theorem 3.8.1; he shows that the hypotheses of Theorem 3.8.1 hold using the cohomological interpretation of tangent spaces to representation varieties discussed in Sect. 6.10. He takes H to be finite but in fact his argument applies to $\mathrm{Hom}(\Gamma, G)$ and $\mathrm{Hom}(\Gamma, \mathrm{GL}_m)$ for any finitely generated group Γ; Proposition 10.5.18 is the special case when $\Gamma = \Gamma_n$.

The authors gave a counterexample to Question 12.7.3 for $G = G_2$ when $p = 2$ and $H = D_{2q} \times C_2$, where D_{2q} is the dihedral group of order $2q$ for $q > 3$ odd: see [22, Thm. 1.2]. This example is in fact a very minor perturbation of part of the G_2 example from Sect. 10.1. Uchiyama found similar examples for G of type D_4 [190, Thm. 1.9] and E_6 [189, Thm. 1.14]. The authors also showed that the conclusion of Proposition 10.5.18 can fail if the hypotheses are not met (see [22, Rem. 2.1(i)]); see Remark 10.5.20, [186, Thm. 1.12] and [189, Thm. 1.11] for further examples.

So far all known counterexamples to Question 12.7.3 are for $p = 2$.

Open Problem 12.7.4 Are there any counterexamples to Question 12.7.3 with $p > 2$?

The counterexample of [22, Thm. 1.2] sheds light on some interesting behaviour of non-abelian 1-cohomology, as we now explain. Observe first that if in Question 12.7.3 we replace the words "Are there only finitely many conjugacy classes of homomorphism $\rho : H \to G$" with "Are there only countably many conjugacy classes of homomorphism $\rho : H \to G$ with G-cr image" then the answer is yes for every H and every G. This follows from Theorem 6.1.2 and Exercise 6.6. Now suppose H is finite. We now get something stronger: we get a positive answer to Question 12.7.3 for H if we replace "Are there only countably many conjugacy classes of homomorphism $\rho : H \to G$ with G-cr image" with "Are there only finitely many conjugacy classes of homomorphism $\rho : H \to G$ with G-cr image", using Theorem 6.2.2 instead of Theorem 6.1.2. Hence if we want to find a counterexample to Question 12.7.3 for finite H then we need to work with homomorphisms $\rho : H \to G$ with non-G-cr image.

Translating Question 12.7.3 into the language of 1-cohomology leads to the following question. Suppose an algebraic group H acts on a unipotent group V. Restriction of functions gives a map $i : Z^1(H, V) \to Z^1(U, V)$, and this descends to give a map $j : H^1(H, V) \to H^1(U, V)$.

Question 12.7.5 Are the fibres of j finite?

The answer to Question 12.7.5 is yes if V is a vector space and H acts linearly on V [110, Lem. 6.3]. It turns out, however, that the answer is no in general: the $H = D_{2q}$ counterexample to Question 12.7.3 discussed above also yields a counterexample to Question 12.7.5. See [22, §3] and [110, §5] for further discussion.

12.8 Fixed Points and Quotient Maps

We have seen several results which hold for a reductive subgroup H of G if $p = 0$ but not when $p > 0$. For instance, H is G-cr in characteristic 0, so $C_G(H)$ is reductive by Theorem 5.4.1, but this is no longer true if $p > 0$ (see Example 5.4.3). On the other hand, if H is G-cr then $C_G(H)$ is reductive. This suggests that in certain contexts, one should replace reductive subgroups of G in characteristic zero with G-cr subgroups of G in positive characteristic.

We give another example of this phenomenon, taken from [13]. Let X be a G-variety and let H be a reductive subgroup of G. Then $N_G(H)$ stabilises X^H, the set of H-fixed points in X. The inclusion of X^H in X is $N_G(H)$-equivariant, so if $N_G(H)$ is reductive then it gives rise to a map $\psi : X^H /\!/ N_G(H) \to X /\!/ G$.

Theorem 12.8.1 *Let H be a reductive subgroup of G.*

(a) Suppose H is G-completely reducible. Then for any G-variety X, the induced map $\psi\colon X^H /\!/ N_G(H) \to X /\!/ G$ is finite.

(b) Suppose H is not G-completely reducible and $N_G(H)$ is reductive. Then there exists a G-variety X such that the map $\psi\colon X^H /\!/ N_G(H) \to X /\!/ G$ is not finite.

Suppose $p = 0$. In this case part (b) holds vacuously, since reductive subgroups of G are G-cr, and part (a) was proved by Luna using étale slices [112]. Different techniques are needed in positive characteristic. The original proof of (a) given in [13] is rather elaborate. We give a quicker proof here using Proposition 3.4.3.

Let M be a G-cr subgroup of G. Let V be a G-module and let W be an M-stable subspace of V. The inclusion $i\colon W \to V$ gives rise to a map $\psi\colon W /\!/ M \to V /\!/ G$, which corresponds to a k-algebra homomorphism from $k[V]^G$ to $k[W]^M$.

Proposition 12.8.2 *Suppose that for all $w \in W$, if w is 0-unstable for G then w is 0-unstable for M. Then the map ψ is finite.*

Proof The hypothesis implies that $\mathscr{N}_M(W) = \mathscr{N}_G(V) \cap W$. Hence $\mathscr{N}_M(W)$ is the set of zeroes of $\{f|_W \mid f \in k[V]_+^G\}$ (see Sect. 3.4). So there exist $f_1, \ldots, f_t \in k[V]_+^G$ for some $t \in \mathbb{N}$ such that the images h_i of f_i in $k[W]_+^M$ for $1 \le i \le t$ have $\mathscr{N}_M(W)$ as their set of zeroes. It follow from Proposition 3.4.3 that $k[W]^M$ is generated by $h_1, \ldots, h_t$ as a $k[V]^G$-module. Hence ψ is finite, as required. $\qquad\square$

Proof of Theorem 12.8.1(a) We can embed X G-equivariantly in a G-module V. Let $M = N_G(H)$, which is G-cr by Corollary 9.2.1. The maps $X /\!/ G \to V /\!/ G$ and $X^H /\!/ M \to V^H /\!/ M$ are finite by Lemma 3.6.3, so without loss using Lemma 2.1.2 we can replace X with V. Let $v \in V^H$ and suppose v is 0-unstable with respect to G. Then $0 \in \overline{G \cdot v}$. By Theorem 8.6.23 there exists $\lambda \in Y_G$ such that λ destabilises v to 0 and $G_v \subseteq P_\lambda$; in particular, $H \subseteq P_\lambda$. But H is G-cr, so there exists $u \in R_u(P_\lambda)$ such that $u \cdot \lambda$ centralises H. Then $u \cdot \lambda \in Y_{C_G(H)^0} \subseteq Y_M$ and $u \cdot \lambda$ destabilises v to 0. This shows that v is 0-unstable for M as well. We deduce from Proposition 12.8.2 (taking $W = V^H$) that Ψ is finite. $\qquad\square$

Remark 12.8.3   (i). The proof in fact shows that the map $X^H /\!/ C_G(H)^0 \to X /\!/ G$ is finite. But this is not as much of a strengthening as it first appears: for $N_G(H)$ is a finite extension of $HC_G(H)^0$, and $X^H /\!/ HC_G(H)^0 \cong X^H /\!/ C_G(H)^0$ as H acts trivially on X^H.

(ii). The proof of Theorem 12.8.1(b) is achieved by a fairly direct construction using ideas we have also seen in the book; see [13, Lem. 4.3].

12.9 A Model-Theoretic Perspective

Some of the results in this book can be expressed in model-theoretic language, and this point of view can be useful. We touch very briefly on the relevant ideas. For an introduction to model theory and its applications to algebraic geometry, see Chapters 1–3 of Marker's book [115].

We work in the language of rings [115, §1.1]. A *formula* is a well-formed expression made using the ring operations $+$ and $\cdot$, the constants 0 and 1, the equality symbol $=$, variables $X, Y \ldots$, the quantifiers $\exists$ (there exists) and $\forall$ (for all), and the symbols $\wedge$ (and), $\vee$ (or) and $\neg$ (not). For instance,

$$(\exists Y)\, Y^2 = X \tag{12.9.1}$$

and

$$(\forall X)(\exists Y)\, Y^2 = X^2 \wedge \neg(Y = X) \tag{12.9.2}$$

are formulas. For a more precise definition, see [115, Def. 1.1.5]. A variable in a formula is *bound* if it is quantified over and *free* otherwise. For instance, in (12.9.1) the X is free and the Y is bound, while both variables are bound in (12.9.2). A formula with no free variables is a *sentence*: so (12.9.2) is a sentence but (12.9.1) is not. In any given ring a sentence is either true or false: for instance, (12.9.2) is false in any field of characteristic 2, and true in any other field. We write $\psi(\mathbf{X})$ to represent a formula with free variables $X_1, \ldots, X_m$.

A *theory* is a set of sentences. For instance, the theory ACF is the set of all sentences that are true for every algebraically closed field. If p is either 0 or a fixed prime then the theory ACF_p is the set of all sentences that are true for every algebraically closed field of characteristic p. The theory ACF_p is *complete* [115, Def. 2.2.1]: this means that for any given sentence ψ, either ψ is true for every algebraically closed field of characteristic p or ψ is false for every algebraically closed field of characteristic p. On the other hand, ACF is not complete: for instance, the sentence $1 + 1 = 0$ is true in any algebraically closed field of characteristic 2, but false for $\mathbb{C}$.

One application of these ideas to algebraic geometry is the Lefschetz Principle (see [115, Cor. 2.2.10]): this says that if ψ is a sentence then ψ is true for algebraically closed fields of characteristic 0 if and only if for all but finitely many primes p, ψ is true for algebraically closed fields of characteristic p. For examples in the context of algebraic groups and Lie algebras, see [121] and [123].

A key concept in model theory is the notion of a definable set. Let $\psi(\mathbf{X}, \mathbf{Y})$ be a formula with free variables $X_1, \ldots, X_m$ and $Y_1, \ldots, Y_n$ (we allow $n = 0$). Let R be a ring and let $\mathbf{a} = (a_1, \ldots, a_n) \in R$. We define

$$D(\psi, \mathbf{a})_R = \{(x_1, \ldots, x_m) \in R^m \mid \psi(x_1, \ldots, x_m, a_1, \ldots, a_n) \text{ holds}\}.$$

We call a subset of the form $D(\psi, \mathbf{a})_R$ a *definable subset* of R^m. If $a_1, \ldots, a_n$ belong to some subset A of R then we say that $D(\psi, \mathbf{a})_R$ is *A-definable*.

From now on we take R to be our algebraically closed field k. Here is an important example. Let $f \in k[X]$: say, $f(X) = a_n X^n + \cdots + a_1 X + a_0$. Then the set of zeroes $Z(f) \subseteq k$ is definable. To see this, observe that $Z(f)$ is the definable set $D(\psi, \mathbf{a})_k$, where $\mathbf{a} = (a_0, \ldots, a_n)$ and $\psi(X, \mathbf{Y})$ is the formula

$$Y_n X^n + \cdots + Y_1 X + Y_0 = 0.$$

By a similar argument, if f is a polynomial over k in m variables then $Z(f)$ is a definable subset of k^m. Pushing this further, one can show that any (Zariski-)closed subset of k^m is definable: if $V \subseteq k^m$ is the set of zeroes of some polynomials $f_1, \ldots, f_r$ then there exist a formula $\psi(\mathbf{X}, \mathbf{Y})$ and a tuple $\mathbf{a}$ of elements of k (obtained from the coefficients of the f_i) such that $V = D(\psi, \mathbf{a})_k$. Moreover, this construction is compatible with base change: if k'/k is an algebraically closed field extension then $V_{k'} = D(\psi, \mathbf{a})_{k'}$.

We can consider formulas where the variables represent elements of G. To do this, we regard G as the closed subset of k^t given by the set of zeroes of some polynomials $h_1, \ldots, h_s$ in t variables, for some $t \in \mathbb{N}$. So we interpret, for example, $(\forall g \in G)$ as

$$(\forall Z_1) \cdots (\forall Z_t)\, h_1(Z_1, \ldots, Z_t) = 0 \wedge \cdots \wedge h_s(Z_1, \ldots, Z_t) = 0.$$

For further details see [121, §2].

The definable subsets of k^m are completely characterised by the following result, which follows from the fact that ACF_p has *quantifier elimination* [115, Def. 3.1.1].

Theorem 12.9.1 ([115, Cor. 3.2.8]) *Let $m \in \mathbb{N}$. Let $D \subseteq k^m$. Then D is definable if and only if D is constructible.*

We use Theorem 12.9.1 in the proof of Lemma 12.10.6. Definable sets are insensitive to algebraically closed field extensions in the following sense.

Proposition 12.9.2 *Let $\psi(\mathbf{X}, \mathbf{Y})$ be a formula and let $\mathbf{a}$ be a tuple of elements of k. Let k' be an algebraically closed field extension of k. Then $D(\psi, \mathbf{a})_k$ is non-empty if and only if $D(\psi, \mathbf{a})_{k'}$ is non-empty.*

Proof Let $\phi(\mathbf{Y})$ be the formula $(\exists X_1) \cdots (\exists X_m)\, \psi(\mathbf{X}, \mathbf{Y})$. Since ACF_p has quantifier elimination, it follows from [115, Prop. 3.1.14] that ACF_p is *model complete* [115, Def. 3.1.13]. Hence the inclusion of k in k' is an *elementary embedding* [115, Def. 2.3.1]. So $\phi(\mathbf{a})$ holds in k if and only if $\phi(\mathbf{a})$ holds in k', and the result follows. $\qquad\square$

For example, let $m \in \mathbb{N}$. It can be shown that there exist a formula $\psi(\mathbf{X}, \mathbf{Y})$ and a tuple $\mathbf{a}$ of elements of k such that for any algebraically closed field extension k'/k we have

$$D(\psi, \mathbf{a})_{k'} = \{\mathbf{g} \in (G_{k'})^m \mid \mathcal{G}(\mathbf{g}) \text{ is } G_{k'}\text{-cr}\}.$$

Much of model theory is concerned with definable sets and their properties. There are, however, model-theoretic entities that can change their behaviour under field extensions. We have encountered similar phenomena already: in Sect. 6.4 we saw that if $k = \overline{\mathbb{F}_p}$, $\dim(G) > 0$ and $m \geq \kappa(G) + 1$ then $(G^m)_{\text{gen}}$ is empty, but $(G_{k'})^m_{\text{gen}}$ is dense in $(G_{k'})^m$ for any solid field extension k'/k. We deduce the following result from Proposition 12.9.2: there do not exist a formula $\psi(\mathbf{X}, \mathbf{Y})$ and a tuple $\mathbf{a}$ of elements of k such that for every algebraically closed field extension k' of k, $D(\psi, \mathbf{a})_{k'}$ is the set of generating tuples in $(G_{k'})^m$. It would be interesting to know whether one can describe the behaviour of generating tuples and rigidity of G-cr subgroups in model-theoretic terms. We do not pursue this further here.

12.10   Generic Stabilisers for G-Actions

Suppose we are given a family $\mathscr{S}$ of subgroups of G parametrised by a variety. The theme of this section is that such families must satisfy constraints that need not hold for an arbitrary set of subgroups. In this section we discuss the following example: we take $\mathscr{S}$ to be the family $\{G_x \mid x \in X\}$, where X is a G-variety. The results below are taken from [119]. Our main result Theorem 12.10.2 below is a slight variation on [119, Thm. 1.4]; we give a shorter and easier proof.

We consider the following question: what can we say about a generic stabiliser (that is, the stabiliser of a suitably general point $x \in X$)? Richardson proved the following result [144].

Theorem 12.10.1 *Suppose $p = 0$. Let X be a smooth G-variety. Then there exist a subgroup H of G and a non-empty open subset U of X such that the subgroups G_x for $x \in U$ are all conjugate to H.*

We call H a *principal stabiliser* for X. Clearly if X is irreducible then a principal stabiliser, if it exists, must be unique up to conjugacy.

Theorem 12.10.1 can fail if $p > 0$. In [119, Ex. 8.3] an example is given of an irreducible SL_2-variety X such that there is no principal stabiliser; the SL_2-orbits on X are all closed and every stabiliser subgroup is a finite unipotent subgroup. We can, however, obtain an analogue of Theorem 12.10.1 in positive characteristic using the notion of semisimplification from Sect. 5.5.

Theorem 12.10.2 ([119, Thm. 1.4]) *Let X be a G-variety. Then there is a non-empty open subset U of X such that the conjugacy classes $\mathcal{D}(G_x)$ for $x \in U$ are all equal.*

(Recall that if H is a subgroup of G then $\mathcal{D}(H)$ denotes the conjugacy class of semisimplifications of H—see Definition 5.5.11.) We sketch a proof of this result; for a complete proof, see [119].

We start by proving a kind of rigidity result.

Proposition 12.10.3 *Suppose k is uncountable and let X be a variety over k. Let Λ be a countable set and let $\{Y_i \mid i \in \Lambda\}$ be a family of closed subsets of X such that $\bigcup_{i \in \Lambda} Y_i = X$. Then there is a finite subset F of Λ such that $\bigcup_{i \in F} Y_i = X$.*

Proof Clearly the result holds if X is finite, so assume X is infinite. Choose non-constant $f \in k[X]$ and set $X_a = f^{-1}(a)$ for $a \in k$. By noetherian induction we can assume the result holds for any proper closed subset of X, so for each $a \in k$ there is a finite subset F_a of Λ such that $\bigcup_{i \in F_a} Y_i \supseteq X_a$. Since k is uncountable and Λ is countable, there is a finite subset F of Λ and an infinite subset A of k such that $F_a = F$ for all $a \in A$.

We claim that $X = \bigcup_{i \in F} Y_i$. Since the Y_i are closed, it is enough to show that $\bigcup_{i \in F} Y_i$ is dense in X. So let $U \subseteq X$ be non-empty and open. Then $f(U)$ is a cofinite subset of k, so $f(U) \cap A \neq \varnothing$. Hence U meets $f^{-1}(A)$, which is contained in $\bigcup_{i \in F} Y_i$, and we are done. $\square$

Remark 12.10.4 Proposition 12.10.3 still holds if we allow the Y_i to be constructible rather than closed: see Exercise 12.11.

Remark 12.10.5 It is not enough in Proposition 12.10.3 just to assume that k is transcendental over the prime field. For instance, let $X = k^2$, where k has transcendence degree 1 over the prime field; then k is countable. For any $(x, y) \in k^2$, x and y are algebraically dependent, so there is a nonzero polynomial $f_{x,y}$ in two variables such that $f_{x,y}(x, y) = 0$. So X is covered by the sets $f_{x,y}(u, v) = 0$, but is not covered by any finite union of these sets. This example shows that Lemma 2.4 of [119] is incorrect. This does not affect the other results in [119]: for these rely on [119, Cor. 2.5], which one can derive from Proposition 12.10.3 above, using Exercise 12.11.

For H a G-cr subgroup of G we define

$$B_H(X) = \{x \in X \mid \mathcal{D}(G_x) = G \cdot H\}.$$

Recall the partial order $\preceq$ from Exercise 6.7.

Lemma 12.10.6 *Suppose k is solid. Let X be a G-variety and let H be a G-completely reducible subgroup of G. Then $B_H(X)$ is constructible.*

Proof We can choose $r \in \mathbb{N}$ such that every subgroup of G admits a generic r-tuple (Remark 5.2.4(iii)). Let $n = \kappa(H) + 3 + r$. We can choose $\mathbf{h}_0 \in H^n$ such that $\mathbf{h}_0$ is a generating tuple for H, by Corollary 6.4.7. Let

$$S_1 = \{x \in X \mid \exists \mathbf{g} \in (G_x)^n \ \pi_{G^n}(\mathbf{h}_0) = \pi_{G^n}(\mathbf{g})\}$$

and let

$$S_2 = \{x \in X \mid \forall \mathbf{g} \in (G_x)^n \ \exists \mathbf{h} \in H^n \ \pi_{G^n}(\mathbf{g}) = \pi_{G^n}(\mathbf{h})\}.$$

It follows from Theorem 12.9.1 that S_1 and S_2 are constructible. To complete the proof, it is enough to show that

$$B_H(X) = S_1 \cap S_2.$$

Suppose $x \in B_H(X)$. Then $G \cdot H \preceq \mathcal{D}(G_x)$, so by Exercise 6.7(v) there exists $\mathbf{g} \in (G_x)^n$ such that $\pi_{G^n}(\mathbf{h}_0) = \pi_{G^n}(\mathbf{g})$. This means that $x \in S_1$. Likewise, $\mathcal{D}(G_x) \preceq G \cdot H$, and it follows from Exercise 6.7(v) that $x \in S_2$. Hence $x \in S_1 \cap S_2$.

Conversely, suppose $x \in S_1 \cap S_2$. Let H' be a semisimplification of G_x, yielded by P_λ for some $\lambda \in Y_G$. Since $x \in S_1$, there exists $\mathbf{g} \in (G_x)^n$ such that $\pi_{G^n}(\mathbf{g}) = \pi_{G^n}(\mathbf{h}_0)$. Then $c_\lambda(\mathbf{g}) \in (H')^n$ and $\pi_{G^n}(c_\lambda(\mathbf{g})) = \pi_{G^n}(\mathbf{h}_0)$, so it follows from Exercise 6.7(v) that $G \cdot H \preceq G \cdot H' = \mathcal{D}(G_x)$; in particular, $\dim(H) \leq \dim(H')$. The inclusion $\mathcal{D}(G_x) \preceq G \cdot H$ takes a little more work to establish, as we do not have an a priori bound on $\kappa(H')$. It is enough to show that $\kappa(H') \leq \kappa(H)$: for then there exists $\mathbf{g}_0 \in (G_x)^n$ such that $c_\lambda(\mathbf{g}_0)$ is a generating tuple for H', and the hypothesis that $x \in S_2$ implies that $\mathcal{D}(G_x) \preceq G \cdot H$, by Exercise 6.7(v).

So suppose $\kappa(H') > \kappa(H)$. Pick $\mathbf{g} = (g_1, \ldots, g_n) \in (G_x)^n$ such that: (i) $c_\lambda(g_1), \ldots, c_\lambda(g_{\kappa(H)+1})$ belong to distinct connected components of H'; (ii) $c_\lambda(g_{\kappa(H)+2})$ and $c_\lambda(g_{\kappa(H)+3})$ generate $(H')^0$; and (iii) $(g_{\kappa(H)+3+1}, \ldots, g_{\kappa(H)+3+r})$ is a generic tuple for G_x. Set $H'' := c_\lambda(\mathcal{G}(\mathbf{g}))$. Condition (iii) implies that $(c_\lambda(g_{\kappa(H)+3+1}), \ldots, c_\lambda(g_{\kappa(H)+3+r}))$ is a generic tuple for H' (Lemma 5.2.5(iii)). So $c_\lambda(\mathbf{g})$ is a generic tuple for H', and hence for H''. It follows that H'' is G-cr, since H' is. Condition (ii) implies that $\dim(H'') = \dim(H')$.

Since $x \in S_2$, there exists $\mathbf{h} \in H^n$ such that $\pi_{G^n}(c_\lambda(\mathbf{g})) = \pi_{G^n}(\mathbf{g}) = \pi_{G^n}(\mathbf{h})$. Now $c_\lambda(\mathbf{g})$ is a generating tuple for H'', so $G \cdot H'' \preceq G \cdot H$ by Exercise 6.7(v). Hence either (a) $\dim(H') = \dim(H'') \leq \dim(H)$, or (b) $\dim(H') = \dim(H'') = \dim(H)$ and $\kappa(H'') \leq \kappa(H)$. We proved above that $\dim(H) \leq \dim(H')$, so (b) must hold. But H'' has at least $\kappa(H) + 1$ distinct connected components by condition (iii), so we get a contradiction. This completes the proof. $\qquad\square$

Proof of Theorem 12.10.2 By the definition of the subsets $B_H(X)$, it is enough to show that $B_H(X)$ has non-empty interior in X for some G-cr subgroup H of G. Fix an uncountable algebraically closed field extension E of k. Base change gives us an action of G_E on X_E. If $x \in X_E$ then $M := \mathcal{D}((G_E)_x)$ is a G_E-cr subgroup of G_E, and we have $x \in B_M(X_E)$. Hence the subsets $B_M(X_E)$ cover X_E as M ranges over the G_E-cr subgroups of G. By Theorem 6.1.2 the set $\{B_M(X_E) \mid M \text{ is } G_E\text{-cr}\}$ is countable, so Lemma 12.10.6 and Exercise 12.11 imply that X_E has a covering by finitely of the $B_M(X_E)$: say, $X \subseteq B_{M_1}(X_E) \cup \cdots \cup B_{M_t}(X_E)$, where the M_i are G_E-cr subgroups of G.

By Theorem 6.1.1 we can take each M_i to be of the form $(H_i)_E$ for some G-cr subgroup H_i of G. Let $x \in X$. Then $\mathcal{D}((G_E)_x) = \mathcal{D}((G_x)_E)$ is G_E-conjugate to $(H_i)_E$ for some i, so $\mathcal{D}(G_x)$ is G-conjugate to H_i. Hence $x \in B_{H_i}(X)$. This shows that $X = \bigcup_{i=1}^{t} B_{H_i}(X)$. Since the $B_{H_i}(X)$ are constructible subsets of X, one of them must have non-empty interior in X. This completes the proof. $\qquad\square$

Remark 12.10.7 Theorem 12.10.2 is stated for quasi-projective varieties in [119, Thm. 1.4]. The proofs above can be modified to deal with this more general case: for instance, the argument for Lemma 12.10.6 shows that for any open affine subset U of X and any G-cr subgroup H of G, the set $\{x \in U \mid \mathcal{D}(G_x) = G \cdot H\}$ is constructible. We leave the details to the reader.

Given a G-variety X, set

$$X_{\mathrm{red}} = \{x \in X \mid G \cdot x \text{ has maximal dimension and } G_x \text{ is reductive}\}.$$

It can be shown that X_{red} is open [119, Thm. 1.1]. If $p = 0$ then $\mathcal{D}(G_x) = G \cdot G_x$ for any $x \in X_{\mathrm{red}}$ because reductive subgroups are G-cr; it follows from Theorem 12.10.2 in this case that there is a principal stabiliser if $X_{\mathrm{red}} \neq \varnothing$. We then recover the conclusion of Theorem 12.10.1 without the assumption that X is smooth. On the other hand, it can happen that X_{red} is empty. For instance, let V be the natural module for $G := \mathrm{GL}(V)$: then the nonzero elements of V form a single open orbit, and if $v \neq 0$ then G_v is a maximal parabolic subgroup of G.

12.11 Exercises

Exercise 12.1 Let $k = \overline{\mathbb{F}_p}$ and suppose $G = \mathrm{SL}_2 \times \mathrm{SL}_2$. For each power q of p, let $M_q \cong \mathrm{SL}_2$ be the subgroup of G embedded via the map $g \mapsto (g, F_q(g))$, where $F_q : \mathrm{SL}_2 \to \mathrm{SL}_2$ is the standard q-power Frobenius map.

(i) Show that M_q is maximal in G for each q.
(ii) Show that if q_1 and q_2 are different powers of p, then M_{q_1} and M_{q_2} are not G-conjugate, so G has infinitely many conjugacy classes of maximal subgroups.

Exercise 12.2 Suppose G is semisimple. Let U be a maximal unipotent subgroup of G and let H be a reductive subgroup of G such that $H \supseteq U$. Show that $H = G$.

[*Hint*: There is a Borel subgroup B of H containing U; write $B = P_\lambda(H)$ for some $\lambda \in Y_H$ and think about P_λ and $P_{-\lambda}$ as subgroups of G.]

Exercise 12.3 Fill in the details of Remark 12.3.2.

Exercise 12.4 Prove Theorem 12.4.4 by adapting the proof of Theorem 5.3.3. You should also find the characterisation in Lemma 12.4.3(ii) helpful.

Exercise 12.5 Let V be a finite-dimensional k-vector space and set $G = \mathrm{GL}(V)$. Recall that if K is a subgroup of G, then K is G-cr if and only if V is a semisimple module for K. This exercise gives an analogous interpretation for relative G-complete reducibility with respect to a smaller general linear group H inside G. Prove the following statement.

Let U be a subspace of V, and pick a direct complement $\tilde{U}$ to U. Let $H = \mathrm{GL}(U) \subseteq G$ embedded in the obvious way. Let K be a subgroup of $G = \mathrm{GL}(V)$. Then K is relatively G-completely reducible with respect to $H = \mathrm{GL}(U)$ if and only if the following two conditions hold:

(i) every K-submodule of V contained in U has a K-complement in V containing $\tilde{U}$;
(ii) every K-submodule of V containing $\tilde{U}$ has a K-complement in V contained in U.

Exercise 12.6 Let k be arbitrary (i.e., not necessarily algebraically closed), and let U be a finite-dimensional k-vector space. Let $H \subseteq \mathrm{GL}(U)$ be a subgroup, so that H acts in a natural way on U. Suppose

$$0 = U_0 \subset U_1 \subset U_2 \subset \cdots \subset U_r = U \qquad (12.11.1)$$

is a Jordan-Hölder sequence for U as an H-module over k: that is, each U_i is an H-module and each quotient U_i / U_{i-1} is simple.

Let $V = V_r = U \otimes_k \bar{k}$, and let $G = \mathrm{GL}(V)$ have the corresponding k-structure as in Example 11.3.3, so that $G(k) = \mathrm{GL}(U)$. Show that if we let P be the parabolic subgroup of G arising from (the base change of) the flag (12.11.1), then P gives a semisimplification of H over k.

Exercise 12.7 Suppose G is simple of rank ≥ 2. Let T be a maximal torus of G. Show that T contains infinitely many pairwise non-G-conjugate one-dimensional subtori, and hence that G contains infinitely many conjugacy classes of G-cr subgroups.

[*Hint*: Start by showing that if S_1 and S_2 are two one-dimensional subtori of G, then they are G-conjugate if and only if they are $N_G(T)$-conjugate.]

Exercise 12.8 Show that the answer to Question 12.7.3 is yes when $p = 0$ and H is finite.

Exercise 12.9 In this question we provide a counterexample to the conclusion of Theorem 12.8.1 when the hypothesis of complete reducibility is not met. Let $p = 2$, let $G = \mathrm{SL}_3$ and let $H \subseteq G$ be the image of the adjoint representation of SL_2. Then H is not G-cr, see Exercise 4.3.

Let X be the natural module for G. Show that $X /\!/ G$ is a single point but $X^H /\!/ N_G(H)$ is infinite, and hence the map in Theorem 12.8.1 is not finite in this case.

Exercise 12.10 Let $p > 0$ and let k be solid. Suppose $G = k^*$. Show that G_{gen} is not a definable subset of G.

Exercise 12.11 Let k be an uncountable field and let $t \in \mathbb{N}$. Let X be a constructible subset of k^t. Let $\{Y_i \mid i \in \Lambda\}$ be a family of constructible subsets of X such that $\bigcup_{i \in \Lambda} Y_i = X$.

(a) Use Proposition 12.10.3 to show there is a finite subset F of Λ such that $\bigcup_{i \in F} Y_i = X$.
(b) Show there exists $i \in F$ such that Y_i has non-empty interior in X.

Exercise 12.12 Let $H = \mathrm{SL}_2$ and let X be the closed subvariety of unipotent elements of H; we allow H to act on X by conjugation. Define

$$C = \{(x, h) \in X \times H \mid hxh^{-1} = x\}.$$

Find the irreducible components of C.

(Note: We may regard C as a parametrisation of the family of stabiliser subgroups $\{H_x \mid x \in X\}$, where H acts on X by conjugation. This example shows that even when X is irreducible and H and all the H_x are connected, we may still obtain a reducible variety. See [119, §7] for further discussion.)

Here we provide some hints and solutions and further references for the exercises.

Solution 2.1. Note that we can factorise β as

$$m \mapsto (m, m) \mapsto (m, m^{-1}) \mapsto (m, hm^{-1}h^{-1}) \mapsto mhm^{-1}h^{-1}.$$

The first map is the diagonal embedding of G into $G \times G$, the second is inversion on the second factor, the third is $\mathrm{Inn}(h)$ on the second factor, and the fourth is multiplication. Using the facts given in Sect. 2.3, we deduce that $(d\beta)_1$ is the composition

$$X \mapsto (X, X) \mapsto (X, -X) \mapsto (X, -\mathrm{Ad}(h)(X)) \mapsto X - \mathrm{Ad}(h)(X),$$

giving the claimed result. Now note that $\alpha = R_h \circ \beta$, which gives the claimed result for α.

Solution 2.2. Since H is connected, $H/R_u(H)$ is connected and reductive, so $H/R_u(H)$ has no non-trivial normal unipotent subgroups. Hence $R_u(H)$ must be the largest normal unipotent subgroup of H.

Solution 2.3. For the first part, let $1 \to N \to H \xrightarrow{\pi} M \to 1$ be an extension of reductive groups M and N. Then $\pi(R_u(H))$ is a connected normal unipotent subgroup of M, so $\pi(R_u(H)) = 1$ as M is reductive. But then $R_u(H)$ is a connected normal unipotent subgroup of N, so $R_u(H) = 1$ as N is reductive. This shows that H is reductive.

For the second part, suppose H is reductive, and let N be a normal subgroup of H. Then $N \cap H^0$ is a normal subgroup of the connected reductive group H^0, so $N \cap H^0$ is normal in H^0. Hence $N \cap H^0$ is reductive. It follows that $N^0 = (N \cap H^0)^0$ is reductive, so N is reductive.

© The Author(s), under exclusive license to Springer Nature Switzerland AG 2026
M. Bate et al., *G-Complete Reducibility, Geometric Invariant Theory and Spherical Buildings*, Oberwolfach Seminars 57, https://doi.org/10.1007/978-3-032-08866-6_13

Solution 2.4. (i). It is easy to write ϕ_A as the combination of maps which are clearly homomorphisms; the key is that raising to an integer power is a homomorphism of the multiplicative group $\mathbb{G}_m$. So ϕ_A is a homomorphism. Then one can see that for two matrices $A, B \in \mathrm{SL}_m(\mathbb{Z})$ we have $\phi_A \circ \phi_B = \phi_{AB}$, so these homomorphisms are all invertible since the matrices are.

(ii). We give the idea for $m = 2$. A symmetric $\mathrm{Aut}(G)$-invariant bilinear form on Y_T would give a symmetric $\mathrm{SL}_2(\mathbb{Z})$-invariant bilinear form on the module $\mathbb{Z}^2$. Suppose $(\cdot, \cdot)$ is such a form. Let e and f be the standard $\mathbb{Z}$-basis vectors of $\mathbb{Z}^2$. Set $m := (e, e)$ and $n := (e, f)$. The matrix $A = \begin{pmatrix} 1 & a \\ 0 & 1 \end{pmatrix} \in \mathrm{SL}_2(\mathbb{Z})$ satisfies $Ae = e$ and $Af = ae + f$, so we get

$$n = (e, f) = (Ae, Af) = (e, ae + f) = am + n.$$

Letting a vary, we conclude that $m = 0$.

Now let $v \in \mathbb{Z}^2$. There exist $A \in \mathrm{SL}_2(\mathbb{Z})$ and $a \in \mathbb{Z}$ such that $aA(e) = v$. We have

$$(v, v) = (aA(e), aA(e)) = a^2(A(e), A(e)) = a^2(e, e) = 0.$$

So for any $v_1, v_2 \in \mathbb{Z}^2$ we have

$$0 = (v_1 + v_2, v_1 + v_2) = (v_1, v_1) + (v_1, v_2) + (v_2, v_1) + (v_2, v_2) = (v_1, v_2) + (v_2, v_1).$$

Hence $(\cdot, \cdot)$ is antisymmetric. But this forces the form to be zero, as it is assumed to be symmetric.

Solution 2.5. Note that $f_i = \pi_i \circ (f_1 \times f_2)$ for $i = 1, 2$, where $\pi_i : X_1 \times X_2 \to X_i$ is the projection onto the ith factor. Now apply part (i) and (ii) of Lemma 2.8.2.

Solution 2.6. If $\widehat{f} : k \to X$ is the morphism giving the limit, then $\widehat{f}^{-1}(\overline{H \cdot x})$ is a closed subset of k (because $\widehat{f}$ is continuous) containing all nonzero points of k (because $\widehat{f}(a) = \lambda(a) \cdot x \in H \cdot x$ for all nonzero a). But this means the preimage also contains 0, and hence $\widehat{f}(0) = x' \in \overline{H \cdot x}$.

Solution 2.7. Note that $\chi \in \Phi_T(V)$ if and only if there exists $0 \neq v \in V$ with $t \cdot v = \chi(t)v$ for all $t \in T$. But this holds if and only if

$$(gtg^{-1}) \cdot (g \cdot v) = g \cdot (\chi(t)v) = \chi(t)(g \cdot v) = (g \cdot \chi)(gtg^{-1})(g \cdot v),$$

so if and only if $g \cdot \chi \in \Phi_{gTg^{-1}}(V)$. This also shows that $g \cdot V_\chi = V_{g \cdot \chi}$, which finishes part (i). Part (ii) follows from the H-invariance of the angle bracket pairing: that is, from (2.5.1).

Solution 2.8. We know already there is at least one. If Q_1 and Q_2 are opposites to P with respect to L then $(Q_1)^0$ and $(Q_2)^0$ are opposite to P^0 with respect to L^0, as is observed in the main text. Hence $(Q_1)^0 = (Q_2)^0$, by uniqueness of opposites in the connected case. To finish, note that Q_2 meets exactly the same connected components of G as Q_1 does because those are precisely the components met by the common R-Levi subgroup L; so if C is any connected component of G such that $L \cap C \neq \varnothing$ then there exists $g \in L \cap C$ such that $Q_2 \cap C = g(Q_2)^0 = g(Q_1)^0 = Q_1 \cap C$.

Solution 2.9 and 2.10. Both of these are just calculations using the definitions.

Solution 3.1. This is a standard sort of argument in this categorical setting. We can apply the categorical quotient condition for (Y, ϕ) to the map $\phi' : X \to Y'$ to deduce the existence of a unique map $\psi : Y \to Y'$ with $\phi' = \psi \circ \phi$. On the other hand, applying the categorical quotient condition for (Y', ϕ') to the map $\phi : X \to Y$ we get a unique map $\psi' : Y' \to Y$ with $\phi = \psi' \circ \phi'$. Putting these together, we deduce that $\phi = (\psi' \circ \psi) \circ \phi$. Since we also have $\phi = \mathrm{id}_Y \circ \phi$ we deduce from uniqueness that $\psi' \circ \psi = \mathrm{id}_Y$. Similarly, $\psi \circ \psi' = \mathrm{id}_{Y'}$, so ψ is an isomorphism and $\psi' = \psi^{-1}$.

Solution 3.2. (i). We use Proposition 3.1.5 repeatedly. If C is a connected component of X then $\pi_X(C)$ is connected. Note also that $\pi_X(G \cdot C) = \pi_X(C)$, so $\pi_X(C)$ is also closed. The connected components of $G \cdot C$ are all of the form $g \cdot C$ as g ranges over the elements of G, so if C' is another connected component of X such that $G \cdot C \cap G \cdot C'$ is non-empty, a whole component of X lies in the intersection, and hence $G \cdot C = G \cdot C'$. This shows that the connected components of $X /\!/ G$ are the images of the distinct G-orbits of connected components of X. Since $\pi_X(G \cdot C) = \pi_X(C)$, we have $G \cdot C \subseteq \pi_X^{-1}(\pi_X(C))$. Conversely, if C' is a connected component of X such that $G \cdot C \neq G \cdot C'$, then the argument above implies that $G \cdot C \cap G \cdot C' = \varnothing$. Hence the closed G-orbits in $G \cdot C$ are all distinct from those in $G \cdot C'$. This gives $\pi_X^{-1}(\pi_X(C)) = G \cdot C$.

(ii). The proof is similar to that of (i). It is clear that for an irreducible component C of X we have $\pi_X(C) = \pi_X(G \cdot C)$ and so $G \cdot C \subseteq \pi_X^{-1}(\pi_X(C))$. Now we are assuming X is visible, so X_s is dense in X by definition. This means that for every irreducible component C of X there is a non-empty open subset of $G \cdot C$ consisting of points x such that: $G \cdot x$ is closed; $\pi_X^{-1}(\pi(X)) = G \cdot x$; x does not lie in $G \cdot C'$ for any irreducible component C' of X with $G \cdot C' \neq G \cdot C$. This is enough to see that $\pi_X^{-1}(\pi_X(C)) = G \cdot C$.

(iii). The following works. Let $X_1 = \{(x_1, 0, 0) \mid x_1 \in k\}$, $X_2 = \{(0, x_2, 0) \mid x_2 \in k\}$, and $X_1 = \{(0, 0, x_3) \mid x_3 \in k\}$. Let $X = X_1 \cup X_2 \cup X_3$ and let k^* act on X by $a \cdot (x_1, x_2, x_3) = (ax_1, ax_2, x_3)$. Note that $\pi(X) = \pi(X_3)$ is irreducible and of dimension 1, but $\pi_X(X_1)$ and $\pi_X(X_2)$ consist of a single point.

Solution 3.3. We may identify $k[X_1 \times X_2]$ with $k[X_1] \otimes_k k[X_2]$, and it is easy to see that $(k[X_1] \otimes_k k[X_2])^{G_1 \times G_2} = k[X_1]^{G_1} \otimes_k k[X_2]^{G_2}$. Part (i) follows. Part (ii) is the special case of part (i) with $G_1 = 1$, $G_2 = G$, $X_1 = Y$ and $X_2 = X$. For part (iii), let G act on $M \times M$ by $g \cdot (m, m') = (m, m'g^{-1})$. The map $M \times M \to M /\!/ G$ given by $(m, m') \mapsto \pi_M(mm')$ is G-invariant, so we have a map from $(M \times M) /\!/ G \to M /\!/ G$. But $(M \times M) /\!/ G \cong M \times M /\!/ G$ by (ii), so we get a map from $M \times M /\!/ G = M \times M / G$ to $M /\!/ G = M / G$, and it is simple to check that this gives an action of M on M / G.

Solution 3.4. We proved in Lemma 3.2.7 that $\lim_{a \to 0} \lambda(a) \cdot (u \cdot v) = \lim_{a \to 0} \lambda(a) \cdot v$. This implies that $\lim_{a \to 0} \lambda(a) \cdot (u \cdot v - v) = 0$, and thus that $u \cdot v - v \in V_{\lambda, > 0}$, by Example 2.8.4.

Solution 3.5. Since $\lambda \in Y_H$, we have $x' \in \overline{H \cdot x}$, by Exercise 2.6. But if x' is not G-conjugate to x it cannot be H-conjugate to x either, and hence $H \cdot x$ is not closed.

Solution 3.6. (i). This is well known; it follows from the Jordan Normal Form for matrices.

(ii). Each orbit is represented by a matrix X in standard Jordan Normal Form, and for such a matrix any diagonal cocharacter $\lambda(a) = \mathrm{diag}(a^{r_1}, \ldots, a^{r_n})$ with $r_1 > r_2 > \cdots > r_n$ will do the trick (see Example 2.8.5). But then for any $g \in G$ the cocharacter $g \cdot \lambda$ will work for gXg^{-1}.

(iii). Try $X' = \begin{pmatrix} 0 & 1 & 1 & 0 \\ 0 & 0 & 0 & 0 \\ 0 & 0 & 0 & 1 \\ 0 & 0 & 0 & 0 \end{pmatrix}$. Then X' has kernel of dimension 2, hence 2 Jordan blocks. But $(X')^2 \neq 0$, so X' has a block of size greater than 2. Thus X' has partition $[3, 1]$ and must be conjugate to X. Now choose $\lambda(a) = \mathrm{diag}(a, a, 1, 1)$. In the limit of $\lambda(a) \cdot X'$ as $a \to 0$, the 1 in the $(1, 3)$-position of X' dies, so the limit is Y.

(iv). Note X and Y lie in different orbits, and the previous part shows that the orbit of Y is in the boundary of the orbit of X. So the orbit of Y has strictly lower dimension, by Remark 2.7.5. Mantra: you can't increase orbit dimension by taking limits.

(v). It is well known that in this set-up, $Y \in \overline{G \cdot X}$ if and only if $\pi_X \geq \pi_Y$ in the dominance order, see [83], for example. Hence we see that if $\lim_{a \to 0} \lambda(a) \cdot X \in G \cdot Y$ then we must have $\pi_X \geq \pi_Y$, which gives one direction.

For the other direction, we need to show that if $\pi_1 \geq \pi_2$ are two partitions, then we can find suitable representatives X and Y of the corresponding orbits and a cocharacter λ which takes X to Y in the limit. This is a fairly involved exercise in linear algebra. We complete the example when $n = 4$. The general case is more complicated, but the ideas are all here. The partitions of 4 are $[4]$, $[3, 1]$, $[2, 2]$, $[2, 1, 1]$, $[1, 1, 1, 1]$.

- Part (ii) gives the result in all cases when $\pi_2 = [1, 1, 1, 1]$.
- When $\pi_1 = [4]$, then in all cases we can take X to be the standard Jordan block of size 4 and λ to be a diagonal cocharacter; for example,

$$
X = \begin{pmatrix} 0 & 1 & 0 & 0 \\ 0 & 0 & 1 & 0 \\ 0 & 0 & 0 & 1 \\ 0 & 0 & 0 & 0 \end{pmatrix}, \; Y = \begin{pmatrix} 0 & 1 & 0 & 0 \\ 0 & 0 & 0 & 0 \\ 0 & 0 & 0 & 0 \\ 0 & 0 & 0 & 0 \end{pmatrix}, \; \lambda(a) = \begin{pmatrix} a^2 & 0 & 0 & 0 \\ 0 & a^2 & 0 & 0 \\ 0 & 0 & a & 0 \\ 0 & 0 & 0 & 1 \end{pmatrix}
$$

will have $\lim_{a \to 0} \lambda(a) \cdot X = Y$, thus getting from the orbit labelled $[4]$ to the orbit labelled $[2, 1, 1]$.
- The more difficult case $\pi_1 = [3, 1]$ and $\pi_2 = [2, 2]$ is part (iii).
- For the cases $\pi_1 = [3, 1]$ or $[2, 2]$ and $\pi_2 = [2, 1, 1]$ we can use the standard Jordan form for X and Y and the cocharacter λ given above.

Solution 3.7. It is clear that any homogeneous element of positive degree kills 0, and so any such element which is also G-invariant kills the nullcone. So $\mathscr{N}(V)$ is contained in the vanishing set of $k[V]_+^G$. Conversely, if v is any element outside the nullcone, then $\pi_G(v) \neq \pi_G(0) \in V /\!\!/ G$, so there is a G-invariant function $f \in k[V]^G$ such that $f(v) \neq f(0)$. Since $f - f(0)$ is also G-invariant, we may assume that $f(0) = 0$, so that f has positive degree. Since the action of G preserves degree, a G-invariant function is a sum of G-invariant homogeneous functions, so we may assume that f is homogeneous of positive degree. Since $f(v) \neq f(0) = 0$, this shows that v lies outside the vanishing set of $k[V]_+^G$, and we're done.

Solution 4.1. The key fact we use repeatedly is that submodules of completely reducible modules are completely reducible, which is obvious: if $W \subseteq U \subseteq V$ are submodules then for any complement W' to W in V the submodule $U \cap W'$ is a complement to W in U. Now in the situation of the exercise, suppose $U_1 \subset \cdots \subset U_r$ is an H-stable flag (i.e., a flag of H-submodules). Then complete reducibility means that there is a complement W_1 to U_1 in V. Now $U_2 \cap W_1$ is a submodule of W_1, and hence we may find a submodule W_2 of W_1 complementing $U_2 \cap W_1$. We claim that W_2 also complements U_2 in V. To see this, note that $U_2 \cap W_2 = (U_2 \cap W_2) \cap W_1$ since $W_2 \subseteq W_1$, and hence $U_2 \cap W_2 = 0$. Now

$$
V = U_1 + W_1 \subseteq U_2 + W_1 = U_2 + ((U_2 \cap W_1) + W_2) \subseteq U_2 + W_2
$$

and we're done.

We can continue in this way to build a flag $W_1 \supset W_2 \supset \dots \supset W_r$ of H-submodules with $V = U_i \oplus W_i$ for each i. (A similar argument using quotients instead of intersections also works.)

Solution 4.2. For (i), there is an obvious vector space complement to W given by the span of the standard basis vectors not in W; if U is another complement, then the change of basis vector which fixes the basis of W and maps a basis for U onto the remaining standard basis vectors should do the job. For (ii), take $H = \langle h \rangle$ where h belongs to a root subgroup of G, and take the conjugating element g to belong to the Weyl group.

Solution 4.3. If $p = 2$ then the centre $\mathfrak{z}$ of $\mathfrak{g}$ is a non-trivial G-stable subspace, but it has no G-stable complement. To see this concretely, note that if we take T to be the standard diagonal torus in G, then any complement to $\mathfrak{z}$ in $\mathfrak{g}$ would have to be spanned by T-weight vectors. The quotient $\mathfrak{g}/\mathfrak{z}$ is a copy of the simple G-module of highest weight 2 (which is the first Frobenius twist of the natural module since we are in characteristic 2) so a G-stable complement to $\mathfrak{z}$ in $\mathfrak{g}$ would have to be spanned by the two matrices $\begin{pmatrix} 0 & 1 \\ 0 & 0 \end{pmatrix}$ and $\begin{pmatrix} 0 & 0 \\ 1 & 0 \end{pmatrix}$. But the subspace spanned by these two is not G-stable. Finally, since $\mathfrak{z}$ and $\mathfrak{g}/\mathfrak{z}$ are both simple as G-modules, we see that $\mathfrak{g}$ is indecomposable.

Solution 4.4. Just look at pairs of upper unitriangular matrices; see [13, Ex. 3.2].

Solution 4.5. Either V is already simple, or there is a proper simple submodule S_1 with a complement T_1. Either T_1 is simple, in which case we're done, or it has a proper simple submodule S_2. Let W be a complement in V to S_2. Then $T_2 = W \cap T_1$ is a complement to S_2 in T_1 (because $S_2 \subset T_1$). Hence $V = S_1 \oplus S_2 \oplus T_2$. Continue in this way.

Solution 4.6. (i). Just look at the set of diagonal matrices satisfying the condition to lie in G; it has the correct rank to be maximal.

(ii). When $m = 2$, for example, there are three standard upper block triangular parabolics—the upper triangular Borel subgroup, and two others corresponding to the two simple roots. They are not too hard to realise with cocharacters: for example a cocharacter $\lambda(a) = \mathrm{diag}(a^2, a, a^{-1}, a^{-2})$ will evaluate in Sp_4 and will give rise to the upper triangular Borel. The corresponding Levi subgroups are the maximal torus $T \cong \mathbb{G}_m \times \mathbb{G}_m$ itself, an $\mathrm{SL}_2 \times \mathbb{G}_m$ with the SL_2 "in the middle", and a GL_2 diagonally embedded with a transpose inverse twist on the second factor. All other parabolics coming from diagonal cocharacters will contain T and hence will be $N_G(T)$-conjugate to a standard upper block triangular one.

(iii). Away from characteristic 2 a non-degenerate bilinear form on a space of dimension m, say, can be represented with respect to a suitable basis by an antidiagonal matrix with 1s on the antidiagonal. Then the same sort of results as in parts (i) and (ii) hold. In characteristic 2 one has to be more careful. You need to start with a quadratic form and the associated bilinear form may no longer be non-degenerate: in odd dimensions you will find a vector v in the radical of the bilinear form. See for example Exercise 5.6.

Solution 4.7. (i). We may fix a Borel $B \subseteq Q$ and maximal torus $T \subseteq B$, and then Q is a standard parabolic—$Q = P_I$ for some subset I of the simple roots. Then $Q = \bigcap_{\alpha \in I} P_\alpha$ exhibits Q as the intersection of the maximal parabolics containing Q.

(ii). Call the intersection $\widetilde{H}$. Suppose $H \subseteq Q$ for some parabolic Q. Then Q is the intersection of the maximal parabolics containing Q, by (i), and each of these lies in $\mathcal{M}(H)$, so $\widetilde{H} \subseteq Q$. The converse is obvious. Hence H and $\widetilde{H}$ are contained in the same parabolics of G. Since H is contained in a Levi subgroup of Q if and only if H is contained in some opposite parabolic to Q, this is enough to get the result.

Solution 4.8. The reverse implication is obvious. The forward implication can be proved as follows: following the hint, suppose P is a maximal parabolic of G containing H (if none exists, then we're done!), and let L be a Levi of P containing H. Let Q' be a maximal parabolic of L containing H (again, we're done if none exists). Then $Q' R_u(P)$ is a proper subparabolic of P (Lemma 2.9.12), and hence (by the previous exercises) there is another maximal parabolic Q of G such that $Q' R_u(P) \subseteq P \cap Q$. In particular, $H \subseteq Q$. Now H lies in a Levi subgroup M of Q. We may now choose a $\lambda \in Y_L$ such that $P = P_\lambda$ and $L = L_\lambda$. Note that since λ is central in L, λ evaluates in Q', and hence in Q. Choose a $\mu \in Y_M$ such that $Q = P_\mu$ and $M = L_\mu$. Then λ and μ are cocharacters of $C_Q(H)$, and so we may conjugate μ by something in $C_Q(H)$ so that the conjugate μ' commutes with λ and still centralises H. Since the conjugating element comes from Q, we don't change Q either. Now $P \cap Q$ is proper in P, and hence $P_{\mu'}(L)$ is proper in L. But it also contains Q', so $P_{\mu'}(L) = Q'$, and we see that H lies in the Levi subgroup $L_{\mu'}(L)$ of Q'. This shows that $(\star)$ passes down to L, so we can proceed by induction on $\dim(G)$.

For a building-theoretic proof of this result (for arbitrary convex subcomplexes), see [159, Thm. 2.2].

Solution 4.9. Since U is a submodule, so is its complement $U^\perp$. Since U is simple, this means that $U \cap U^\perp$ is either trivial or all of U. In the first case, we have $V = U \oplus U^\perp$, which gives the desired result. In the second case, U is isotropic, and hence has a complement since H is G-cr (an opposite parabolic subgroup in G to the stabiliser of U is the stabiliser of some isotropic subspace W with $U \oplus W^\perp = V$).

Solution 4.10. Let $(\cdot, \cdot)_W$ be the form on W, so that the form on V is defined by

$$(u + \phi(v), w + \phi(x)) := (u, w)_W + (v, x)_W$$

for each $u, v, w, x \in W$. Following the hint, it is clear that H acts irreducibly on W and W', and hence $V = W \oplus W'$ is a decomposition of V into simple H-modules. Hence V is semisimple, and H is GL(V)-cr. Suppose that U is an m-dimensional isotropic H-submodule. Then the projection of U to W (via the direct sum $V = W \oplus W'$) is either 0 or all of W, because W is simple, and similarly for W'. Thus we have $U = W$ or $U = W'$ or $U = \{w + a\phi(w)\}$ for some $a \in k$ (why? The projection maps are H-module isomorphisms, so must correspond to multiplication by some scalar, by Schur's lemma, and we can scale everything so that we only see the scalar in the second factor). Now we see that $(w_1 + a\phi(w_1), w_2 + a\phi(w_2)) = (1 + a^2)(w_1, w_2)_W$ for all $w_1, w_2 \in W$, so this space is isotropic if and only if $a = 1$, as required. Since H only stabilises one m-dimensional totally isotropic subspace, it is contained in the corresponding parabolic subgroup of G but no opposite of this parabolic, hence H is not G-cr. This answers (i) and (ii). For the original version of this argument, see [18, Ex. 3.45].

For (iii), recall that the Lie algebra of $M = \mathrm{Sp}(V)$ is the set of linear endomorphisms $\psi : V \to V$ such that $(\psi(u), v) + (u, \psi(v)) = 0$ for all $u, v \in V$. Given z as in the question and writing $u = u_1 + \phi(u_2)$ and $v = v_1 + \phi(v_2)$ with $u_1, u_2, v_1, v_2 \in W$, we see that

$$(z(u), v) + (u, z(v)) = (au_1, v_1)_W + (bu_2, v_2)_W + (u_1, av_1)_W + (u_2, bv_2)_W$$

$$= 2a(u_1, v_1)_W + 2b(u_2, v_2)_W = 0,$$

since we are in characteristic 2. Hence z is an element of Lie(M). Now because of the decomposition $V = W \oplus W'$, any endomorphism g of V can be represented as a matrix $\begin{pmatrix} g_1 & g_2 \\ g_3 & g_4 \end{pmatrix}$, where $g_1 : W \to W$, $g_2 : W \to W'$, $g_3 : W' \to W$ and $g_4 : W' \to W'$. For g to centralise z, we need g_2 and g_3 to be 0 because we've chosen $a \neq b$. For g to also preserve the form on V we then need $g_1 \in \mathrm{Sp}(W)$ and $g_2 \in \mathrm{Sp}(W')$, so $g \in K$. Conversely, any element of K does centralise z. This shows that $M_z = K$, as required.

Solution 4.11. (i). If U is a submodule, then so is $U^\perp$. Then GL(V)-complete reducibility implies that W exists. If W has a proper submodule, then the orthogonal complement of that submodule would intersect U in a proper submodule (by counting dimensions); but this is impossible, so W is simple. For all $u \in U$ there exists some $v \in V$ with $f(u, v) \neq 0$. But then $v = w + u'$ with $w \in W$ and $u' \in U^\perp$, so $(u, w) \neq 0$. With a bit more work, we can see that the form is therefore non-degenerate on $U \oplus W$.

   (ii). An argument we've seen before: since W is simple, $W \cap W^\perp = 0$ or all of W. The latter case is ruled out if W is assumed not to be isotropic, so we're in the former case, and then dimension tells us that $X = W \oplus W^\perp$. Since U and W are complementary submodules, we must have $W^\perp \cong U$ (via the projection from X to $W^\perp$).

(iii). It is pretty easy to see that U' is a submodule. Since U is isotropic and W and $W^\perp$ are orthogonal, given any $w_1 + w_1'$, $w_2 + w_2' \in U$ with $w_1, w_2 \in W$ and $w_1', w_2' \in W^\perp$, we have

$$f(w_1 + w_1', w_2 + w_2') = f(w_1, w_2) + f(w_1', w_2') = f(w_1 - w_1', w_2 - w_2') = 0.$$

Hence U' is isotropic. Now, if $w_1 + w_1' \in U$ and $w_1 - w_1' \in U$, then $2w_1 \in U \cap W$, so $w_1 = 0$ ($p \neq 2!!$). But then $w_1' \in U$, which implies $w_1' = 0$ also. Hence $U \cap U' = 0$. Now counting dimensions shows that $X = U \oplus U'$.

We've now shown that for every simple isotropic submodule U of V there is an isotropic submodule W such that $V = U^\perp \oplus W$, which in turn implies that $V = U \oplus W^\perp$. It's not hard to soup this up first to a general isotropic subspace, and then to flags of such subspaces.

Solution 5.1. In any such example it will always be the case that $\mathbf{h}' \neq c_\lambda(\mathbf{h})$ for any λ such that $P = P_\lambda$, so the result does not apply. (Equivalently, we can never have $gHg^{-1} = c_\lambda(H)$ in such a situation.)

Solution 5.2. Given $t \in \mathbb{N}$, if $\mathbf{h} \in (H_1)^t$ then define $C_{\mathbf{h}} = \{g \in G \mid g \cdot \mathbf{h} \in (H_2)^t\}$, which is non-empty since H_1 and H_2 are conjugate. If there is $g \in G$ with $g \cdot \mathbf{h}_1 \in (H_2)^n$ but $gH_1g^{-1} \neq H_2$ then there exists $h_{n+1} \in H_1$ such that $gh_{n+1}g^{-1} \notin H_2$. This implies that for the tuple $\mathbf{h}' = (h_1, \ldots, h_n, h_{n+1})$ we have $C_{\mathbf{h}} \supsetneq C_{\mathbf{h}'}$. Now we can use the descending chain condition on the closed subsets $C_{\mathbf{h}}$ of G to see that this process must terminate after a finite number of steps.

Solution 5.3. See [18, Lem. 2.10]. The idea is to find a large enough finitely generated subgroup inductively, using an argument very similar to the previous solution.

Solution 5.4. Suppose H is G-ir. Then Theorem 5.3.3 shows that the orbit of $\mathbf{h}$ is closed, and Lemma 4.2.2 shows that $C_G(H)^0 = G_{\mathbf{h}}^0 = Z(G)^0$, and hence the stabiliser of $\mathbf{h}$ has minimal possible dimension. Thus $\mathbf{h}$ is a stable point.

On the other hand, if $\mathbf{h}$ is a stable point, then H is G-cr since $G \cdot \mathbf{h}$ is closed. But $G_{\mathbf{h}}^0 = C_G(H)^0 = Z^0$ is central in G, so H is contained in no proper R-Levi subgroup of G, and hence in no proper parabolic subgroup of G because it is G-cr. Hence H is G-ir.

The second part of the exercise follows from Lemma 3.7.3 and the comments at the start of Sect. 3.7.

Solution 5.5. (i). As long as p does not divide m, the given representation is semisimple: $V = k^m$ decomposes as the sum of the one-dimensional representation X consisting of vectors whose components are all equal, and the $(m-1)$-dimensional representation Y consisting of vectors whose components sum to 0. In these cases, therefore, H is a semisimplification of itself. However, if $p \mid m$, then $X \subseteq Y$, and the module V is indecomposable but not simple. The module X is the trivial module, the module Y/X is an $(m-2)$-dimensional simple module, and the module V/Y is another copy of the trivial module. Therefore if we let P be the parabolic in GL_m which stabilises the flag $X \subset Y$, then P will semisimplify H to a block diagonal group with two trivial 1×1 blocks and an $(m-2) \times (m-2)$ block for the $(m-2)$-dimensional simple module.

(ii). Again, the solution depends on the characteristic p, as we saw in Exercise 4.3. When $p \neq 2$ the module is simple and there is nothing to do: H is G-irreducible and is the unique semisimplification of itself. When $p = 2$ the centre of the Lie algebra gives a one-dimensional H-stable subspace, and the quotient by this subspace is the Frobenius twist of the natural module for SL_2. Therefore a semisimplification of H is a block diagonal subgroup with a 1×1 trivial group, and a 2×2 block.

Solution 5.6. (i). Note that $(v, v) = q(v + v) + q(v) + q(v) = q(0) + 2q(v) = 0$, because $p = 2$. An alternating form can only be non-degenerate on an even-dimensional space, so there must be a nonzero radical. It is not hard to see that the radical is just one-dimensional, giving the existence of the vector v as claimed.

(ii). The parabolic P is the stabiliser of a one-dimensional space, so it has a Levi subgroup with block sizes 1 and $2m$. The corresponding Levi subgroup in GL_{2m+1} is $\mathrm{GL}_{2m} \times k^*$, but when we intersect this with SL_{2m+1}, we see that the element of k^* is determined by whatever (the determinant of) the $2m \times 2m$ matrix is, so $L \cong \mathrm{GL}_{2m}$ as claimed.

(iii). We have $H \subseteq P$ since the elements of H fix the form and hence, in particular, they stabilise the radical. Since we have chosen L so that the corresponding $2m$-dimensional space is non-degenerate, and H fixes the form, it is clear that $c_L(H) \subseteq \mathrm{Sp}_{2m}$. But $H \cap R_u(P) = 1$ as H is connected reductive, so $c_L(H) = \mathrm{Sp}_{2m}$ for dimension reasons.

(iv). If H were G-cr, then H would be conjugate to $c_L(H)$, but H and $c_L(H)$ are not even isomorphic. To see that $c_L(H)$ is the semisimplification of H, note that Sp_{2m} is GL_{2m}-ir, hence $c_L(H)$ is L-cr, hence G-cr by Theorem 5.4.4.

This example shows that the semisimplification of a reductive subgroup need not be isomorphic to the original subgroup. It is based on the constructions in [29, §2.1], where it is explained that although the restriction of c_L to H is a bijective morphism, it has a non-trivial scheme-theoretic kernel, which explains how we can have $H \ncong c_L(H)$.

Solution 5.7. We have $x_1 = \begin{bmatrix} 0 \\ 1 \end{bmatrix}$ and $x_2 = \begin{bmatrix} 1 \\ 0 \end{bmatrix}$, and these vectors are conjugate by $\begin{pmatrix} 0 & 1 \\ 1 & 0 \end{pmatrix} \in G$. The conclusions of Lemma 5.5.5 cannot possibly hold in this case since $P_{\lambda_1} = P_{\lambda_2} = T$ and neither has any non-trivial unipotent elements at all.

Solution 5.8. Pick a generic tuple $\mathbf{h} \in H^n$ for H. Then $\overline{G \cdot c_\lambda(\mathbf{h})}$ contains a unique closed G-orbit, and this is also the unique closed G-orbit contained in $\overline{G \cdot \mathbf{h}}$ since $\overline{G \cdot c_\lambda(\mathbf{h})} \subseteq \overline{G \cdot \mathbf{h}}$. Now apply Remark 5.5.4.

Solution 5.9. Given A, the set $C(A)$ of all matrices with the same characteristic polynomial as A is a closed and GL_m-invariant subset of Mat_m. Further, by considering Jordan Normal Form, there is a unique orbit O of semisimple elements in $C(A)$, represented by a diagonal matrix with this characteristic polynomial. It is easy to check (again using Jordan Normal Form) that the centraliser of such a semisimple element is maximal amongst the elements in $C(A)$, and hence by Lemma 2.7.6 that O has minimal dimension amongst the GL_m-orbits in $C(A)$—thus O is the unique closed orbit in $C(A)$. But for any A in standard Jordan Normal Form, letting $\lambda(a) = \mathrm{diag}(a^{r_1}, \ldots, a^{r_m})$ for any sequence of integers $r_1 > r_2 > \cdots > r_m$ we obtain a cocharacter λ with $\lim_{a \to 0} \lambda(a) A \lambda(a)^{-1} = A_s$, where A_s is the semisimple part of A. Hence the orbit is closed if and only if $A = A_s$.

Solution 5.10. Fix a maximal torus T of G. Write $n = p^r a$ where a is coprime to p. If $g \in G$ and $g^n = 1$ then there exists $h \in T \cap \overline{G \cdot g}$ such that $h^a = 1$.

Solution 6.1. (i). There are infinitely many primes, so the set P of primes is dense in k^*. Hence P^m is dense in $(k^*)^m$. We have $A = P^m \setminus C$, where C is the closed set of tuples $(x_1, \ldots, x_m) \in (k^*)^m$ such that the x_i are not all pairwise distinct. It follows that A is dense.

(ii). Similar to (i); just choose an infinite set of pairwise coprime polynomials from $\mathbb{F}_p[x]$.

Solution 6.2. (i). Take $a \in k^*$ such that a has infinite order. Then $(a, 0)$ and $(0, 1)$ topologically generate $k^* \ltimes k$.

(ii). Any finite subset of k generates a finite subgroup, since every non-trivial element of k has order p, so k is not topologically finitely generated. A similar argument works for $k^* \times k$.

Solution 6.3. Suppose $\beta_1, \ldots, \beta_m$ is another generating set. We can write $\beta_j = w_j(\gamma_1, \ldots, \gamma_n)$ for some word w_j. Define $f : G^n \to G^m$ by

$$f(h_1, \ldots, h_m) = (w_1(h_1, \ldots, h_n), \ldots, w_m(h_1, \ldots, h_n)).$$

Then f is a morphism and it is easy to check that f maps the copy of $\mathrm{Hom}(\Gamma, H)$ inside G^n to the copy of $\mathrm{Hom}(\Gamma, H)$ inside G^m. To see that f gives an isomorphism between these two copies of $\mathrm{Hom}(\Gamma, H)$, repeat the construction the other way around.

Solution 6.4. Following the hint, observe that we can identify H^n with $\mathrm{Hom}(\Gamma_n, H)$, as described in Sect. 6.5. We see that for $\phi \in \mathrm{Aut}(\Gamma_n)$ and $\rho \in \mathrm{Hom}(\Gamma_n, H)$, the homomorphism $\phi \cdot \rho$ is given by $(\phi \cdot \rho)(\gamma) = \rho(\phi^{-1}(\gamma))$. It is easy to check that this gives an action of $\mathrm{Aut}(\Gamma_n)$ on $\mathrm{Hom}(\Gamma_n, H)$.

Solution 6.5. We may identify $C(\Gamma, G)$ with $C(\Gamma, G_0)_k$. Since $C(\Gamma, G)$ is finite, $\pi_\Gamma(\rho)$ is a k_0-point. Hence there exists $\sigma \in \mathrm{Hom}(\Gamma, G_0)$ such that $\pi_\Gamma(\sigma) = \pi_\Gamma(\rho)$. We can assume without loss that $G_0 \cdot \sigma$ is closed. It follows that $G \cdot \sigma$ is closed by Lemma 3.9.1(ii). But $G \cdot \rho$ is also closed by Theorem 5.3.3. Since $\pi_\gamma(\sigma) = \pi_\Gamma(\rho)$, we conclude that σ and ρ are G-conjugate.

Solution 6.6. If $M = f(H)$ for some f then M is reductive, so there is no harm in taking M to be G. When $p = 0$, choose an embedding of G in some GL_m. By Remark 6.6.2 there are only countably many GL_m-conjugacy classes of homomorphisms from H to GL_m. Any two homomorphisms with image G differ by an element of $N_{\mathrm{GL}_m}(G)$; but $N_{\mathrm{GL}_m}(G)$ is a finite extension of $GC_{\mathrm{GL}_m}(G)$ by Proposition 2.4.8, so the result follows.

Now suppose $p > 0$. Choose an ascending chain of finite subgroups $M_1 \subseteq M_2 \subseteq \cdots$ of H such that $\bigcup_{n=1}^\infty M_n$ is dense in H, as in Proposition 2.6.3. If $f \colon H \to G$ is an epimorphism then $f|_{M_n}$ is G-ir for sufficiently large n (why?). Now $C(M_n, G)$ is finite for all n by Theorem 6.2.2. Show that f is determined up to conjugacy by the images of $f|_{M_n}$ in $C(M_n, G)$ for all n; the argument is very similar to the one in the proof of Lemma 6.6.3.

Solution 6.7. Part (i) is clear because if H is a semisimplification of M then gHg^{-1} is a semisimplification of gMg^{-1} for any $g \in G$. If $M = K$ in the definition then $G \cdot H = \mathcal{D}(K) = G \cdot K$. On the other hand, if M is a proper subgroup of K then for any $H' \in \mathcal{D}(H)$, H' is isomorphic to a quotient of M, so either (a) $\dim(H') < \dim(K)$, or (b) $\dim(H') = \dim(K)$ and $\kappa(H') < \kappa(K)$. The antisymmetry of $\preceq$ and the descending chain condition for $\preceq$ follow easily. This proves (ii) and (iv).

Now suppose $H \preceq K \preceq K'$. There exist subgroups M of K and M' of K' such that $\mathcal{D}(M) = G \cdot H$ and $\mathcal{D}(M') = G \cdot K$. We can choose $\lambda \in Y_G$ and $g \in G$ such that $c_\lambda(M') = gKg^{-1}$. There is a subgroup M_1' of M' such that $c_\lambda(M_1') = gMg^{-1}$. We have

$$\mathcal{D}(M_1') = \mathcal{D}(c_\lambda(M_1')) = \mathcal{D}(gMg^{-1}) = \mathcal{D}(M) = G \cdot H,$$

where the first equality is from Exercise 5.8. Hence $H \preceq K'$. This proves (iii).

Now consider part (v). If (a) holds then there exist $g \in G$, $\lambda \in Y_G$ and a subgroup M of K such that $c_\lambda(M) = gHg^{-1}$. So, given $n \in \mathbb{N}$ and $\mathbf{h} \in H^n$, there exists $\mathbf{k} \in M^n$ such that $c_\lambda(\mathbf{k}) = g \cdot \mathbf{h}$. We have

$$\pi_G(\mathbf{k}) = \pi_G(c_\lambda(\mathbf{k})) = \pi_G(g \cdot \mathbf{h}) = \pi_G(\mathbf{h}).$$

This proves (b). The implication (b) $\implies$ (c) follows from Corollary 6.4.7.

Suppose (c) holds. Let $\mathbf{h} \in H_{\mathrm{gen}}^n$ and $\mathbf{g} \in K^n$ such that $\pi_G(\mathbf{g}) = \pi_G(\mathbf{h})$. Now $G \cdot \mathbf{h}$ is closed because H is G-cr, so there exist $\lambda \in Y_G$ and $g \in G$ such that $c_\lambda(\mathbf{g}) = g \cdot \mathbf{h}$. Set $M = G(\mathbf{g})$. Then $c_\lambda(M) = gHg^{-1}$, so $H \preceq K$. This proves (a).

For part (vi), suppose that $H \preceq K$. There exists a subgroup M of K such that $\mathcal{D}(M) \in G \cdot H$. So there exist $\lambda \in Y_G$ and $g \in G$ such that $M \subseteq P_\lambda$ and $c_\lambda(M) = gHg^{-1}$. Then $M_E \subseteq K_E$, $M_E \subseteq (P_\lambda)_E = P_{\lambda_E}$ and

$$c_{\lambda_E}(M_E) = (c_\lambda)_E(M_E) = c_\lambda(M)_E = (gHg^{-1})_E = gH_Eg^{-1}.$$

Hence $H_E \preceq K_E$. Conversely, suppose $H_E \preceq K_E$. We use part (v). Let $n \in \mathbb{N}$, let $\mathbf{h} \in H^n$ and set $y = \pi_{G^n}(\mathbf{h})$. Then there exists $\mathbf{g} \in (K_E)^n$ such that $\pi_{(G_E)^n}(\mathbf{g}) = y$, so $(\pi_{(G_E)^n})^{-1}(y) \cap (K_E)^n$ is non-empty. But $(\pi_{(G_E)^n})^{-1}(y) \cap (K_E)^n$ is k-defined, so it has a k-point $\mathbf{g}'$, and we have $\pi_{G^n}(\mathbf{g}') = y$. It follows that $H \preceq K$.

Note that the definition of $\preceq$ in [119, Def. 5.1] applied to G-cr subgroups agrees with our definition by this exercise and by the results in [119, §5].

Solution 7.1. (i). The Cayley graph consists of a single hexagon, with the elements of S_3 in a cycle $1 \sim (1, 2) \sim (1, 2, 3) \sim (1, 3) \sim (1, 3, 2) \sim (2, 3) \sim 1$.

(ii). The reflection representation has two unit basis vectors $e_1 = e_{(1,2)}$ and $e_2 = e_{(2,3)}$ at an angle $2\pi/3$ to each other. The element $(1, 2)$ acts by negating e_1 and sending e_2 to $e_1 + e_2$, so it is the reflection in the line orthogonal to e_1. Similarly, $(2, 3)$ acts as the reflection in the line orthogonal to e_2. The other reflection $(1, 3) = (1, 2)(2, 3)(1, 2) = (2, 3)(1, 2)(2, 3)$ is therefore the reflection that sends

$$e_1 \to -e_1 \to -e_1 - e_2 \to -e_2 \qquad \text{and} \qquad e_2 \to e_1 + e_2 \to e_1 \to -e_1,$$

which is the reflection in the hyperplane orthogonal to $e_1 + e_2$.

(iii). The whole group $W = W_S$ itself is parabolic. It has a single coset and corresponds to the empty simplex. The trivial group $\{1\} = W_\varnothing$ is parabolic, and it has 6 cosets corresponding to the elements of W; these are the maximal elements of the complex. In between there are the parabolic subgroups $W_{(1,2)} = \{1, (1, 2)\}$, which has three cosets

$$\{1, (1, 2)\}, \{(2, 3), (1, 2, 3)\}, \{(1, 3), (1, 3, 2)\},$$

and the parabolic subgroup $W_{(2,3)} = \{1, (2, 3)\}$ which has three cosets

$$\{1, (2, 3)\}, \{(1, 2), (1, 3, 2)\}, \{(1, 3), (1, 2, 3)\}.$$

These are the minimal elements of the complex (the vertices). Thus we see that the Coxeter complex in this case has 6 vertices and 6 edges and is again a hexagon, but it is labelled as the dual to the Cayley graph from (i) (the edges are labelled by the elements of W and the vertices by one of the parabolics $W_{(1,2)}$ or $W_{(2,3)}$).

(iv). For A_3 the Cayley graph is (the 1-skeleton of) a permutahedron, which is a regular solid with square and hexagonal faces—the 24 elements of S_4 are the vertices and each edge is labelled by one of the three generators. The Coxeter complex is the dual solid: its faces are 24 triangles labelled by the elements of S_4; its edges are labelled by one of the parabolic subgroups $W_{(1,2)}$, $W_{(2,3)}$ or $W_{(3,4)}$; its vertices are labelled by one of the parabolics $W_{\{(1,2),(2,3)\}} \cong S_3$, $W_{\{(2,3),(3,4)\}} \cong S_3$ or $W_{\{(1,2),(3,4)\}} \cong S_2 \times S_2$; at each vertex, correspondingly, meet 6 or 4 faces depending on the order of the parabolic. The reflection representation is the simple 3-dimensional real representation of S_4 (e.g., acting as symmetries of a tetrahedron), and the hyperplanes of the reflections in this representation partition the unit sphere into precisely the same complex as you get by drawing the Coxeter complex on the surface of the sphere. See [1, Ch. 1] for much more on Coxeter complexes.

Solution 7.2. That any two flags lie in a common apartment amounts to suitably choosing a basis compatible with both of the flags. To do this, we just need to treat intersections of subspaces and their orthogonal complements carefully. It's a fairly nasty exercise in linear algebra.

To show that for any two apartments there is an isomorphism between them fixing their intersection, we can proceed as follows. Suppose we're in the symplectic case, so we have an alternating form and an apartment A corresponds to a choice of basis consisting of $\{e_i, f_i \mid 1 \leq i \leq n\}$ with $(e_i, e_j) = 0 = (f_i, f_j)$ and $(e_i, f_j) = \delta_{ij}$ for each i, j. Then another apartment A' corresponds to another such basis $\{e'_i, f'_i \mid 1 \leq i \leq n\}$. Let S be the set of isotropic subspaces formed from the A-basis, and T the set from the A'-basis. Then there is a finite list of subspaces $U_1, \ldots, U_r$ such that $U_i \in S \cap T$ for each i and U_i is minimal subject to this condition. This means that the U_i have pairwise trivial intersections, and each U_i has a basis consisting of some subset of the A-basis and also a basis consisting of some subset of the A'-basis. Now these subsets of each basis can be matched up to produce the isomorphism we are after. The orthogonal case is similar. The Coxeter group obtained in this case is the one of type $B_n \cong C_n$ or D_n depending on the type of form and the dimension. See also [1, Ex. 4.24].

Solution 7.3. This really is a direct translation of Exercise 4.8, since the maximal parabolics in G correspond to the vertices in Δ_G.

Solution 7.4. (i). This is obvious for $\lambda, \mu \in Y_T$ and $a, b \in \mathbb{Z}$, by Lemma 3.2.5. This is enough to deduce the result in general when $\mathbb{K} = \mathbb{Q}$, using Lemma 7.5.1. Then the result follows for $\mathbb{K} = \mathbb{R}$ because we can approximate a real virtual cocharacter arbitrarily closely by a rational one.

(ii). The direction ($\star$) implies ($\dagger$) follows from (i). For the reverse implication, we can choose a maximal torus T of $P \cap Q$ and work inside the apartment corresponding to T. We have the set of roots $\Phi_1 = \Phi(P, T) \subseteq \Phi(G, T)$, and we define

$$C_P := \{\zeta \in V_T(\mathbb{K}) \mid P_\zeta \supseteq P\} = \{\zeta \in V_T(\mathbb{K}) \mid \langle \zeta, \alpha \rangle \geq 0 \text{ for all } \alpha \in \Phi_1\}.$$

Do the same for Q, giving another subset $\Phi_2 = \Phi(Q, T)$ of roots and $C_Q \subseteq V_T(\mathbb{K})$. Since Σ is a subcomplex by assumption, we see that $C_P \cup C_Q \subseteq C_\Sigma$, and then ($\dagger$) implies that the convex hull of $C_P \cup C_Q$ inside $V_T(\mathbb{K})$ also lies in C_Σ. Now given $\zeta \in V_T(\mathbb{K})$, we have $P_\zeta \supseteq P \cap Q$ if and only if $\langle \zeta, \alpha \rangle \geq 0$ whenever $\alpha \in \Phi_1 \cap \Phi_2$; but this is just a condition which implies that ζ lies in the convex hull of C_P and C_Q, so we deduce ($\star$).

Solution 7.5. Following the hint, let $H \subseteq G$ be any example of a connected reductive G and a G-irreducible subgroup H of G such that H is not normal in G. Then Δ^H is empty, so its stabiliser is all of G; but H is not normal in G so $N_G(H)$ is a proper subgroup of G.

Solution 7.6. It is clear that for a parabolic subgroup P of G we have $H \subseteq P$ if and only if $\widetilde{H} \subseteq P$, and hence $\Sigma = \Delta^{\widetilde{H}}$. Certainly $N_G(\widetilde{H}) \subseteq N_G(\Delta^{\widetilde{H}})$, so we get $N_G(\widetilde{H}) \subseteq N_G(\Sigma)$. For the other inclusion, suppose $g \in N_G(\Sigma)$. Then g acts by conjugation on the set of parabolic subgroups P with P containing H, and hence g normalises $\widetilde{H}$.

Solution 7.7. (i). The elements of $m(V)$ are the simple H-submodules, so the first equivalence follows from Exercise 4.5 and the fact that a complement to a simple H-submodule U must be a maximal H-submodule W (for any submodule W' properly containing W would intersect U non-trivially, and hence contain all of U). The elements of $M(V)$ are the submodules with simple quotients, so a dual argument will work.

(ii). Given such U and W, the submodule $U \cap W$ is either 0 or all of U because U is simple. In the latter case we have $U \subseteq W$, as required, so we need to rule out $U \cap W = 0$. But in this case, $U \oplus W$ is a submodule properly containing W, so must be all of V, and then U and W are complementary, which contradicts $U \in m_0(V)$ and $W \in M_0(V)$.

(iii). By (i) and (ii), U_0 and W_0 are proper submodules, one contained in the other. Hence the flag $U_0 \subseteq W_0$ is a H-stable flag. Any automorphism of Δ permutes the vertices, and preserves opposition, and hence one which stabilises Δ^H must permute the set $m_0(V) \cup M_0(V)$. Actually, there are two flavours of automorphism, one preserves type (dimension) and one reverses type (dimension), so either the sets $m_0(V)$ and $M_0(V)$ are individually stabilised by an automorphism, or they are swapped. In any case, all the automorphisms must stabilise the flag $(U_0 \subseteq W_0) \in \Delta^H$, as required, and this flag has no complement by construction. See [128].

Solution 7.8. Up to a choice of orientation, the picture should have: a point for the origin $(0, 0)$; the ray for λ corresponding to the positive x-axis; the ray for μ corresponding to the positive y-axis; the subset of points giving the parabolic G corresponding to the line $y = x$, for B^+ the region $x > y$, and for B^- the region $x < y$; the subspace $Y_S(\mathbb{R})$ corresponding to the line $y = -x$.

Solution 7.9. For each maximal torus T of P_ζ we have a set of roots $\{\alpha_1, \ldots, \alpha_r\} \subseteq \Phi(G, T)$ for the unipotent radical $R_u(P_\zeta)$. The set

$$U_T = \{\eta \mid \langle \eta, \alpha_i \rangle > 0 \text{ for } 1 \leq i \leq r\}$$

is obviously an open subset of $V_T(\mathbb{K})$ containing ζ. Further, for each $\eta \in U_T$ we have $T \subseteq P_\eta$ and $R_u(P_\eta) \supseteq R_u(P_\zeta)$ by construction, so $P_\eta \subseteq P_\zeta$ for all $\eta \in U_T$. Now for any other maximal torus T' of P there exists some $g \in P_\zeta$ with $T' = gTg^{-1}$ and it is easy to see that $U'_{T'} = g \cdot U_T$ because $g \cdot \zeta = \zeta$ in this case. So the set

$$U = \{\eta \mid \eta \in U_T \text{ for some } T \subseteq P_\zeta\}$$

is an open subset of $V_G(\mathbb{K})$ with the required property.

   (Note: With a little more effort, one can show the result holds for non-connected G as well.)

Solution 8.1. This follows easily from the definitions, using (7.5.1).

Solution 8.2. This is very easy, and is often proved as part of the triangle inequality.

Solution 8.3. This follows from a straightforward calculation, using the fact that the maximal tori in G are all conjugate.

Solution 8.4. Again, this is a calculation using the definitions and the fact that all the maximal tori are conjugate. It is a bit more involved than the one in Exercise 8.3.

Solution 8.5. Just untangle the definitions.

Solution 8.6. Let $\mathbf{x} = (x_1, \ldots, x_n) \in \mathcal{A}^n$. Set

$$U_{\mathbf{x}} = \left\{ u \in R_u(P_\lambda) \,\middle|\, \lim_{a \to 0} \lambda(a) \cdot \mathbf{x} = u \cdot \mathbf{x} \right\}.$$

Now $U_{\mathbf{x}} \neq \varnothing$ by Theorem 3.3.6 and it is clear that if $\mathbf{x}' = (x_1, \ldots, x_n, x_{n+1})$ for some $x_{n+1} \in \mathcal{A}$ then $U_{\mathbf{x}'} \subseteq U_{\mathbf{x}}$. The result now follows from a descending chain condition argument.

Solution 8.7. Pick a maximal torus T of H. Then $\mathcal{S}$ must contain $\overline{H \cdot T}$ by a similar argument to the one in Example 8.6.6, so $H \subseteq \mathcal{S}$.

Solution 8.8. The proof of Theorem 7.7.1(ii) carries over, using Exercise 8.6 in place of Theorem 3.3.6.

Solution 8.9. Note $f^{-1}(0, \ldots, 0) = \bigcap_{1 \le i \le t} f_i^{-1}(0)$, and the ideal of $k[Z]$ corresponding to $f_i^{-1}(0)$ is (Z^{n_i}).

Solution 8.10. K is contained in P_{opt}, so K is contained in an R-Levi subgroup L of P_{opt}. We can choose λ_{opt} in such a way that $L = L_{\lambda_{\mathrm{opt}}}$; then λ_{opt} evaluates in $C_G(K)$. Hence $C_G(K)$ is optimal. It is clear from the definition that any reductive overgroup of an optimal subgroup is also an optimal subgroup, so $N_G(K)$ is optimal as well.

Solution 8.11. Case 1 of Example 8.6.22 already yields an example where $P_{\mathrm{opt}}(\mathcal{A}) = B$ is non-maximal (since $\mathcal{A} \subseteq R_u(P_\alpha)$ and $\mathcal{A} \subseteq R_u(P_\beta)$). For an example where $P_{\mathrm{opt}}(\mathcal{A})$ is not minimal, take $H = U_\alpha U_{\alpha+\beta} = R_u(P_\beta)$: then $P_{\mathrm{opt}}(\mathcal{A}) = P_\beta$ by Exercise 8.14, but $\mathcal{A} \subseteq R_u(B)$.

By considering the various possibilities for $\mathrm{supp}_T(\mathcal{A})$, we see that in all cases there is a unique minimal support set, and it is one of $\{\alpha\}$, $\{\beta\}$, $\{\alpha+\beta\}$, $\{\alpha, \beta\}$, $\{\alpha, \alpha+\beta\}$, $\{\beta, \alpha+\beta\}$ and $\{\alpha, \beta, \alpha+\beta\}$. By the recipe described before Example 8.6.22, it is enough to calculate the optimal destabilising cocharacter $\lambda_{\mathrm{opt}} \in Y_T$ for each of these possible support sets. We have calculated λ_{opt} in the first four cases in Example 8.6.22. For the last three cases we can show using Exercises 8.13 and 8.14 that P_{opt} is P_β, P_α and B, respectively. To calculate λ_{opt}, we argue as follows. In the sixth case we can ignore the $\langle \lambda, \alpha + \beta \rangle$ term when computing $\min\{\langle \lambda, \alpha \rangle, \langle \lambda, \beta \rangle, \langle \lambda, \alpha + \beta \rangle\}$ (why?), so we get the same λ_{opt} as in Case 4 of Example 8.6.22. For the fourth case there is a unique ray of cocharacters λ such that $P_\lambda = P_\beta$, so $\lambda_{\mathrm{opt}} = (2, -1, -1)$. Likewise $\lambda_{\mathrm{opt}} = (1, 1, -2)$ in the fifth case.

Solution 8.12. We use the method of Example 8.6.22/Exercise 8.11, and we take $\| \cdot \|$, T, B, α, β, etc., as in that example. We define $\psi \colon (\mathrm{SL}_3)^n \to (\mathrm{Mat}_3)^n$ by $\psi(g_1, \ldots, g_n) = (g_1 - 1, \ldots, g_n - 1)$. To calculate $P_{\lambda_{\mathrm{opt}}}$, we consider $\mathrm{supp}_T(u^{-1} \cdot \psi(\mathcal{A}_n))$ for $u \in R_u(B)$. Now it is clear from the definition of H_2 that $\mathrm{supp}_T(\psi(\mathcal{A}_n)) = \{\alpha, \alpha + \beta\}$ is minimal. As we saw in Exercise 8.11, this gives $P_{\lambda_{\mathrm{opt}}} = P_\beta$. Note that this argument works for any $n \in \mathbb{N}$, including $n = 1$: we don't need H_2 to admit a generic n-tuple.

Solution 8.13. First observe that P satisfies the properties of $\mathcal{P}(P)$ from Theorem 8.1.1. For if we choose any Levi subgroup L of P, then $c_L(P) = L$ is G-cr, so P does provide a semisimplification of P. Let $\Delta = \Delta_G$. Then Δ^P is precisely the set of simplices corresponding to parabolics of G containing P, and so the stabiliser Γ of this set in G is also P itself, since P is the unique minimal element of Δ_P.

Here is an example for the second part. Let G be the extension of the direct product $G^0 = \mathrm{SL}_2 \times \mathrm{SL}_2 \times \mathrm{SL}_2$ by the symmetric group S_3 permuting the three factors. Let $\sigma = (1,2) \in S_3$, let B be the upper triangular Borel subgroup of SL_2, and let $P = \langle B \times B \times B, \sigma \rangle$. Then it is easy to realise P as an R-parabolic subgroup of G (just take a cocharacter of the form $(\lambda, \lambda, 2\lambda)$, for example, where $P_\lambda(\mathrm{SL}_2) = B$). Now Δ^P consists of the simplices of Δ fixed by P, and these correspond to the following parabolic subgroups of G^0:

$$B \times B \times B; \quad B \times B \times \mathrm{SL}_2; \quad \mathrm{SL}_2 \times \mathrm{SL}_2 \times B; \quad G \text{ itself.}$$

Note that Δ^P is not a subcomplex, which is due to the fact that the action of the element σ is not type-preserving. The stabiliser Γ in G of Δ^P must have $\Gamma^0 = B \times B \times B$, since Γ has to normalise this parabolic subgroup of G^0. It also contains σ, but it cannot meet any other component of G, so $\Gamma = P$. We see that P is an R-parabolic subgroup of G satisfying the properties of Theorem 8.1.1. But it is easy to see that $Q = N_G(B \times B \times B)$ also satisfies these properties: certainly Q contains $\Gamma = P$, and Q does semisimplify P by Lemma 2.9.20.

Solution 8.14. Note that $N_G(P)$ normalises $R_u(P)$, so $N_G(P)$ is contained in $P_{\mathrm{opt}}(\mathscr{A})$ by Theorem 8.6.15. Conversely, $R_u(P)$ is contained in $R_u(P_{\mathrm{opt}}(\mathscr{A}))$ since $\lambda_{\mathrm{opt}}(\mathscr{A})$ destabilises $R_u(P)$ to 1, so $P_{\mathrm{opt}}(\mathscr{A})^0 \subseteq P$. This forces $P_{\mathrm{opt}}(\mathscr{A})^0 = P$, so $P_{\mathrm{opt}}(\mathscr{A}) \subseteq N_G(P)$. Hence $P_{\mathrm{opt}}(\mathscr{A}) = N_G(P)$.

Solution 8.15. We get $(\alpha^\vee, \alpha^\vee) = 4$, $(\beta^\vee, \beta^\vee) = 2$ and $(\alpha^\vee, \beta^\vee) = (\beta^\vee, \alpha^\vee) = -2$. The pairings of the coroots with the roots are given by $\langle \alpha^\vee, \alpha \rangle = \langle \beta^\vee, \beta \rangle = 2$, $\langle \alpha^\vee, \beta \rangle = -2$ and $\langle \beta^\vee, \alpha \rangle = -1$. Following the method of Example 8.6.22/Exercise 8.11/Exercise 8.12, we need to maximise the function

$$f(r,s) = \frac{\min\{\langle r\alpha^\vee + s\beta^\vee, \alpha + \beta \rangle, \langle r\alpha^\vee + s\beta^\vee, 2\alpha + \beta \rangle\}}{\| r\alpha^\vee + s\beta^\vee \|} = \frac{\min\{s, 2r\}}{\sqrt{4r^2 - 4rs + 2s^2}}.$$

We can rewrite the expression in the denominator as

$$4r^2 - 4rs + 2s^2 = (2r - s)^2 + s^2,$$

and now it is clear that for any fixed value of s the denominator is minimised by taking $2r = s$. Applying this condition maximises $f(r,s)$ globally, since when $2r = s$ we have $f(r,s) = 1$ independent of the actual value of s. With the additional indivisibility condition, we get $\lambda_{\mathrm{opt}} = \alpha^\vee + 2\beta^\vee$ and so $\langle \lambda_{\mathrm{opt}}, \alpha \rangle = 0$, so $P_{\lambda_{\mathrm{opt}}} = P_\beta$. Note that this argument works for all $n \in \mathbb{N}$, as in Exercise 8.12.

Solution 9.1. (i). If H is G-cr, then so is $M = N_G(H)$, by Corollary 9.2.1. Hence, in particular, M is reductive, and so $N_G(M)^0 = M^0 C_G(M)^0$, by Proposition 2.4.8. But any element of $C_G(M)$ centralises $H \subseteq M$, so lies in M, so $C_G(M)^0 \subseteq M^0$, so $N_G(M)^0 = M^0$ and we see that $W(M)$ is finite.

(ii). The argument of part (i) actually goes through under the weaker hypothesis that H and $M = N_G(H)$ are reductive, so we need to find an example where H is reductive but $N_G(H)$ is not. The easiest thing to do is take an example with finite H: say $p = 2$, $G = \mathrm{SL}_2$ and H is the finite group of order 2 generated by $\begin{pmatrix} 1 & 1 \\ 0 & 1 \end{pmatrix}$. Then it is easy to see that $M = N_G(H)$ is just the full upper right root group in SL_2, and the normaliser of that is the upper triangular Borel subgroup, so $N_G(M)/M$ is not finite.

Solution 9.2. Since H is normal in HS, one direction follows from Theorem 9.1.1. For the other direction, if H is G-cr then $C_G(H)$ is reductive by Theorem 5.4.1, and S is $C_G(H)$-cr by Lemma 4.2.1, since S is linearly reductive. Hence HS is G-cr by Proposition 9.3.3.

Solution 9.3. Following the hint, suppose H is not G-cr and let $P = \mathcal{P}(H)$ be the corresponding optimal destabilising R-parabolic subgroup witnessing that fact, so that $C_G(H) \subseteq P$. Since $H \subseteq M = C_G(S)$, $S \subseteq C_G(H) \subseteq P$. Since S is linearly reductive, it is G-cr, and hence there is some R-Levi subgroup L of P containing S. But this means that we can write $P = P_\lambda$ with $\lambda \in Y_M$, and so $P_\lambda(M)$ witnesses the fact that H is not M-cr.

Solution 9.4. Recall Exercise 4.10. We have inclusions $H \leq K \leq M \leq G$, where $p = 2$, $M = \mathrm{Sp}_{2m}$ and $K = \mathrm{Sp}_m \times \mathrm{Sp}_m$; the subgroup H which is the diagonal copy of Sp_m in K is K-cr but not M-cr. We showed that $K = C_M(z)$, where z is a suitable element of the Lie algebra of M. This is enough to see that $K = C_M(S)$ where S is the *scheme-theoretic* centre of K viewed as a subgroup scheme of M. Since S is contained in a torus it is linearly reductive, but it is not smooth.

Solution 9.5. We saw in Exercise 4.10 that the only H-stable isotropic subspace is $U = \{v + \phi(v) \mid v \in W\}$, where $\phi : W \to W'$ is the chosen form-preserving isomorphism. So a semisimplification H' of H in M will be found by projecting to a Levi of the corresponding parabolic $P = N_M(U)$. Finding a Levi means finding an isotropic complement to U. We can take U' as follows: if we take a decomposition of W into two maximal isotropic subspaces, say $W = X \oplus Y$, then we can take $U' = \{x + \phi(y) \mid x \in X, y \in Y\}$. It is not hard to check that U' is isotropic (since X and Y are), and that U' is a complement to U in V. The decomposition $V = U \oplus U'$ gives a Levi subgroup L of P. The projection of H to L will give a different diagonal embedding H' of $\mathrm{Sp}(W)$, this time according to the decomposition $V = U \oplus U'$ rather than $V = W \oplus W'$. Now H' is G-conjugate to H (since we can easily conjugate W to U and W' to U' with an element of $\mathrm{GL}(V)$), but not M-conjugate to H (since the spaces W and W' are non-degenerate and the spaces U

and U' are isotropic, so we can't move them to each other with elements of the symplectic group). This illustrates the conclusion of Lemma 9.4.3.

Solution 10.1. (See the proof of [143, Lem. 5.1] for this argument.) Let V be the natural module for GL_{2m} or GL_m. If $(\cdot\,,\cdot)$ is the bilinear form defining $H = \mathrm{Sp}_{2m}$ or SO_m, then for any endomorphism $\phi : V \to V$ we can define an endomorphism $\phi^* : V \to V$ uniquely by the condition

$$(\phi(v), w) = (v, \phi^*(w))$$

for all $v, w \in V$. Then $\mathrm{Lie}(H) = \{\phi \mid \phi + \phi^* = 0\}$. Since $\mathrm{char}(k) \neq 2$, we can form a complementary subspace W of $\mathrm{End}(V)$ given by

$$W = \{\phi \in \mathrm{End}(V) \mid \phi - \phi^* = 0\}.$$

It is easy to check that W is H-stable, and this finishes the proof.

Solution 10.2. (i). Standard properties of the trace imply that the trace form is bilinear and GL_m-invariant. To see that it is non-degenerate, given any nonzero matrix X there is at least one entry $x_{ij} \neq 0$. Let Y be the matrix with a 1 in position (j, i) and 0 elsewhere. Then XY has as its ith column the jth column of X, and every other entry is 0. In particular, $\mathrm{tr}(XY) = x_{ij} \neq 0$.

(ii). Since the differential of Ad is ad, this follows from definition of the Killing form on a Lie algebra G: for $X, Y \in \mathfrak{g}$, their images in Mat_m are $\mathrm{ad}(X), \mathrm{ad}(Y)$, so the trace form restricted to $\mathfrak{g}$ sends the pair (X, Y) to $\mathrm{tr}(\mathrm{ad}(X)\mathrm{ad}(Y))$, and this is the Killing form.

(iii). If the Killing form on $\mathfrak{g}$ is non-degenerate, then the orthogonal complement to $\mathfrak{g}$ inside Mat_m with respect to the trace form is a vector space complement, that is $\mathrm{Mat}_m = \mathfrak{g} \oplus \mathfrak{g}^\perp$. Since the trace form is GL_m-invariant and $\mathfrak{g}$ is G-stable, so is $\mathfrak{g}^\perp$, and we're done.

Solution 10.3. Following the hint, there is only one possible candidate for an M-stable complement to $\mathfrak{m}$ in $\mathfrak{g}$, namely the sum of the root spaces for the long roots. But this subspace is not M-stable: for example, if β is the long simple root for some base of the root system, then acting on $\mathfrak{g}_\beta$ with the short root group $U_\alpha \subseteq M$ can produce elements with nonzero contribution from the root subalgebra $\mathfrak{g}_{\alpha+\beta}$, and $\alpha + \beta$ is not long.

Solution 10.4. We claim that $C_M(H)$ is trivial, and then the result follows from the fact that $C_{\mathfrak{m}}(H)$ contains all the elements of the form z from Exercise 4.10(iii). To see that $C_M(H)$ is trivial, note that (with notation as in Exercise 4.10) the decomposition $V = W \oplus W'$ is a decomposition of V into simple H-modules, so any element of M centralising

H has to act by scalars on W and W'. But any such element lies in the centre of K, and that is trivial.

Solution 10.5. We know G-cr implies reductive by Lemma 4.1.6. On the other hand, if we let $\rho : G \to \mathrm{GL}(V)$ denote the non-degenerate representation we are given, then Lemma 4.2.7 implies that H is G-cr if and only if $\rho(H)$ is $\rho(G)$-cr. Since $p \geq (\dim(V))^2$, Jantzen's result implies that $V \otimes V^* \cong \mathrm{End}(V) \cong \mathrm{Lie}(\mathrm{GL}(V))$ is semisimple as a G-module, hence $(\mathrm{GL}(V), \rho(G))$ is a reductive pair. Now $p \geq \dim(V)$ also, so if H is reductive then V is semisimple for H by Jantzen's result again. Now Corollary 10.5.24 gives the result. For more on this theme, see [14].

Solution 10.6. (i). In this case $\varepsilon = \begin{pmatrix} 0 & 1 \\ 0 & 0 \end{pmatrix}$, and so $u^t = \begin{pmatrix} 1 & t \\ 0 & 1 \end{pmatrix}$, as expected.

(ii). In this case $\varepsilon = \begin{pmatrix} 1 & 1 \\ 1 & 1 \end{pmatrix}$, and so $w^t = \begin{pmatrix} 1+t & t \\ t & 1+t \end{pmatrix}$.

(iii). Whatever the saturation of $N_G(T)$ is, it contains w, hence the w^t from part (ii), and it contains T. Since $U := \{w^t \mid t \in k\}$ is connected, $(N_G(T)^{\mathrm{sat}})^0$ has dimension at least 2. But clearly U is not contained in either of the Borel subgroups that contain T, so $N_G(T)^{\mathrm{sat}} = (N_G(T)^{\mathrm{sat}})^0 = \mathrm{SL}_2$.

Solution 10.7. (i). Since $\exp$ is the inverse to $\log$, it follows from Theorem 10.7.1(i) that $\sigma \circ \exp = \exp \circ (d\sigma)_1$, and hence we have

$$\sigma(u^t) = \sigma(\exp(t \log u)) = \exp((d\sigma)_1(t \log u)) = \exp(t \log \sigma(u)) = \sigma(u)^t,$$

as required (see [10, Cor. 5.5(i)]).

(ii). Since conjugation by $\lambda(a)$ is an automorphism of G, Eq. (10.9.1) follows from part (i). For fixed t, we have a map $\mathcal{U} \to \mathcal{U}$ given by $u \mapsto u^t$, and it follows from Lemma 2.8.2(i) and (10.9.1) that $\lim_{a \to 0} \lambda(a) u^t \lambda(a)^{-1}$ exists if $\lim_{a \to 0} \lambda(a) u \lambda(a)^{-1}$ exists. This shows that if u belongs to P_λ then u^t belongs to P_λ, which means that P_λ is saturated (see [10, Prop. 5.6]).

Solution 11.1. Tensoring with $\mathbb{C}$, we can write (with a bit of abuse of notation) $X^2 + Y^2 = 1$ as $(X + iY)(X - iY) = 1$, and hence the map given by $Z \mapsto X + iY$, $Z^{-1} \mapsto X - iY$ does work.

Solution 11.2. We have the obvious $\mathbb{R}$-structure given by

$$\frac{\mathbb{R}[X_{11}, X_{12}, X_{21}, X_{22}]}{X_{11}X_{22} - X_{12}X_{21} - 1}.$$

It is easy to see that the corresponding group of $\mathbb{R}$-points is $\mathrm{SL}_2(\mathbb{R})$.

For the other case, recall that a typical matrix in SU(2) looks like $\begin{pmatrix} w & z \\ -\bar{z} & \bar{w} \end{pmatrix}$, with $|w|^2 + |z|^2 = 1$. Writing $w = a + bi$ and $z = c + di$, we see that we just need to have $a^2 + b^2 + c^2 + d^2 = 1$ (this is the familiar realisation of SU(2) as the unit sphere in real 4-space). So this is how the candidate $\mathbb{R}$-structure

$$\frac{\mathbb{R}[A, B, C, D]}{A^2 + B^2 + C^2 + D^2 - 1}$$

arises. To see that this is an $\mathbb{R}$-structure on $\mathbb{C}[\mathrm{SL}_2]$, note that after tensoring with $\mathbb{C}$, we can rewrite $A^2 + B^2 = (A + iB)(A - iB)$ and $C^2 + D^2 = -(C + iD)(-C + iD)$. Then we can map $A + iB \mapsto X_{11}, A - iB \mapsto X_{22}, C + iD \mapsto X_{12}, -C + iD \mapsto X_{21}$ and all should fall out.

Solution 11.3. (i). Since the map $t \mapsto t^2$ is surjective on $\mathbb{C}$, G has two orbits in X: the orbit $\{0\}$ and the orbit of all nonzero complex numbers. On the other hand, $G(\mathbb{R}) = \mathbb{R}^*$ is the multiplicative group of $\mathbb{R}$, and $X(\mathbb{R}) = \mathbb{R}$. Since G is acting by squares, we get three orbits: $\{0\}$, $\mathbb{R}_+$ and $\mathbb{R}_- = \{a \in \mathbb{R} \mid a < 0\}$.

(ii). The orbit $\{0\}$ is already cocharacter-closed and Zariski closed, so nothing happens. For the orbit $\mathbb{R}_+$, the only extra point we can get as the limit along an $\mathbb{R}$-defined cocharacter of G is the point 0, so $\overline{\mathbb{R}_+}^c = \mathbb{R}_+ \cup \{0\}$. However, since $\mathbb{R}_+$ is infinite, its Zariski closure in $X = \mathbb{A}^1$ is all of X, and so $\overline{\mathbb{R}_+} \cap X(\mathbb{R}) = \mathbb{R}$ is the whole real line. The analogous thing happens for the other orbit $\mathbb{R}_-$.

Solution 11.4. Most of the details are in the example: we can take a purely inseparable extension $k_1 = k(a)/k$ of degree p generated by the element a, and let H be the abelian algebraic group whose $\bar{k}$-points are the units in the ring $A = \bar{k} \otimes_k k_1$. Then A is p-dimensional as a $\bar{k}$-vector space, so H naturally sits as a subgroup of $\mathrm{GL}_p \cong \mathrm{GL}(A)$.

Now the k-structure on H (and the compatible k-structure on GL_p) comes from the fact that $A = \bar{k} \otimes_k k_1$, so the action of $H(\bar{k}) = A^\times$ on A is deduced from the action of $H(k) = k_1^\times$ on k_1. This latter shows that k_1 is obviously a simple module for H over k, so H is GL_p-cr over k. However, we claim that H is not GL_p-cr (over $\bar{k}$).

First note that we have a surjective ring homomorphism $A \to \bar{k}$ given by the map $x \otimes y \mapsto xy$. The kernel M is a maximal ideal generated by elements $1 \otimes x - x \otimes 1$ for $x \in k_1$, and it is $(p - 1)$-dimensional and consists of nilpotent elements (since $m^p = 0$ for all $m \in M$). Now let $U(\bar{k}) = \{1 + m \mid m \in M\} \subseteq A^\times$. Then U is a connected subgroup of H consisting of unipotent elements, and it is relatively straightforward to see that $H \cong \mathbb{G}_m \times U$ is conjugate to a group of upper triangular matrices in GL_p. In particular, since H is not contained in a torus but is contained in a Borel subgroup of GL_p, it is not GL_p-cr.

For a more systematic approach to such examples, the subgroup H we have just described is the *Weil restriction* $\mathrm{R}_{k_1/k}(\mathbb{G}_m)$ of the multiplicative group $\mathbb{G}_m$ over k_1 across the extension k_1/k. Similar examples can be constructed for other reductive k_1-groups in place of $\mathbb{G}_m$. These are examples of so-called *pseudo-reductive groups* (see [47]).

Solution 11.5. Look at Example 11.3.7.

Solution 12.1. Note that M_q surjects onto each factor of G, and hence is not contained in any proper parabolic subgroup of G. So M_q is G-ir, and therefore any overgroup M of M_q is G-ir, hence reductive. But then the only possibilities for a connected reductive overgroup M are $M = M_q$ or $M = G$ (e.g., for reasons of dimension and rank), so M_q is maximal. Clearly $N_G(M_q) = M_q$. Now the two-dimensional simple representations of SL_2 given by different Frobenius twists are pairwise non-conjugate (since, for example, they have different highest weights), so the M_q are pairwise non-conjugate.

Solution 12.2. Since U is maximal in G it is maximal in H, so there is a Borel subgroup B of H containing U as its unipotent radical. Let B' be an opposite Borel to B in H and let U' be the unipotent radical of B', so that U' is another maximal unipotent subgroup of H. Then U' is maximal in G also, by dimension. Choose $\lambda \in Y_H$ such that $B = P_\lambda(H)$ and $B' = P_{-\lambda}(H)$. Since λ has positive pairing with all the roots for the root groups contained in U, and U is maximal, we see that P_λ is a Borel subgroup of G, and hence $P_{-\lambda}$ is an opposite Borel subgroup of G. But then U and U' together contain all the root subgroups for G, and so $H = G$ since the semisimple group G is generated by its root groups.

Solution 12.3. Straightforward. If you need to, refer back to Exercise 4.1.

Solution 12.4. Suppose $H \cdot \mathbf{k}$ is not closed in G^n. By the Hilbert-Mumford Theorem there exists $\lambda \in Y_H$ such that $\mathbf{k}' := \lim_{a \to 0} \lambda(a) \cdot \mathbf{k}$ exists and $\mathbf{k}' \notin H \cdot \mathbf{k}$. Then $K \subseteq P_\lambda$. If we had $K \subseteq L_{u \cdot \lambda}$ for some $u \in R_u(P_\lambda(H))$, then $u \cdot \lambda$ would centralise K and hence fix $\mathbf{k}$; this would imply $\mathbf{k}' = u \cdot \mathbf{k}$ by Lemma 3.2.7, contradicting the choice of λ. Hence K is not relatively G-cr with respect to H, by Lemma 12.4.3(ii).

Conversely, if $H \cdot \mathbf{k}$ is closed in G^n, then for every $\lambda \in Y_H$ with $K \subseteq P_\lambda$, the limit $\lim_{a \to 0} \lambda(a) \cdot \mathbf{k}$ exists (because $K \subseteq P_\lambda$) and is H-conjugate to $\mathbf{k}$ (by the Hilbert-Mumford Theorem 3.3.2) because $H \cdot \mathbf{k}$ is closed. But then Theorem 3.3.6 implies that $\lim_{a \to 0} \lambda(a) \cdot \mathbf{k}$ is $R_u(P_\lambda(H))$-conjugate to $\mathbf{k}$, which puts K inside some $L_{u \cdot \lambda}$, showing that K is relatively G-cr with respect to H, by Lemma 12.4.3(ii) again.

Solution 12.5. This is [27, Prop. 5.1], where the proof is left as an exercise. We give a bit more detail here.

First suppose K is relatively G-cr with respect to H. For condition (i), let $W \subseteq U$ be a K-submodule of V contained in U, and suppose that the dimension of W is r. Choose a basis $\{e_1, \ldots, e_r\}$ for W and extend it to a basis $\{e_1, \ldots, e_m\}$ for U; together with our fixed basis for $\tilde{U}$, this gives us a basis $\{e_1, \ldots, e_n\}$ for all of V. Let $\lambda \in Y_G$ be the cocharacter whose matrix has block form

$$\lambda(a) = \begin{pmatrix} aI_r & 0 \\ 0 & I_{n-r} \end{pmatrix}, \qquad \text{for } a \in k^*,$$

with respect to this basis. Then we can view λ as a cocharacter of H and of G, and P_λ is precisely the stabiliser of W in G, so $K \subseteq P_\lambda$. Since K is relatively G-cr with respect to H, there exists $u \in R_u(P_\lambda(H))$ such that $K \subseteq L_{u \cdot \lambda}$, by Lemma 12.4.3. Now L_λ corresponds to a decomposition $V = W \oplus W'$, for some subspace W' containing $\tilde{U}$, and thus $L_{u \cdot \lambda} = u L_\lambda u^{-1}$ corresponds to a decomposition $V = W \oplus \tilde{W}$, where $\tilde{W} := u \cdot W'$ is now a K-submodule. Also, since $\tilde{U} \subseteq W'$, $u \in H$, and H centralises $\tilde{U}$, we have $\tilde{U} \subseteq \tilde{W}$, as required.

Condition (ii) follows by a similar argument.

Conversely, suppose conditions (i) and (ii) hold, and let $\lambda \in Y_H$ be a cocharacter with $K \subseteq P_\lambda$. We can choose a basis of U so that λ has the following block form

$$\lambda(a) = \begin{pmatrix} a^{n_1} I_{r_1} & 0 & \cdots & 0 & 0 \\ 0 & a^{n_2} I_{r_2} & \cdots & 0 & 0 \\ \vdots & \vdots & \ddots & \vdots & \vdots \\ 0 & 0 & \cdots & a^{n_s} I_{r_s} & 0 \\ 0 & 0 & \cdots & 0 & I_{n-m} \end{pmatrix} \qquad \text{for } a \in k^*,$$

where each $n_i \in \mathbb{Z}$ with $n_1 > n_2 > \ldots > n_s$, and $\sum_{i=1}^{s} r_i = m$. Then $P_\lambda(H)$ is the stabiliser of a flag

$$(U_1 \subset U_2 \subset \cdots \subset U_{s-1})$$

of subspaces in U. Set also $U_s = U$. To see how this flag gives rise to a flag in V corresponding to P_λ, there are two cases to consider: first, when $n_i = 0$ for some i, and second, when we have a sequence $n_1 > \ldots > n_{i-1} > 0 > n_i > \ldots > n_s$ (here we include the extreme cases $0 > n_i$ and $0 < n_i$ for all i).

In the first case, P_λ is the stabiliser of a flag

$$(U_1 \subset \cdots \subset U_{i-1} \subset U_i \oplus \tilde{U} \subset U_{i+1} \oplus \tilde{U} \subset \cdots \subset U_{s-1} \oplus \tilde{U})$$

of subspaces of V. Since $K \subseteq P_\lambda$, each entry in this flag is a K-submodule (here we mean, for example, that $U_{i+1} \oplus \tilde{U}$ is K-stable, not that U_{i+1} and $\tilde{U}$ are K-stable individually); note also that each entry either is contained in U or contains $\tilde{U}$. Then apply (i) and (ii) to successively find an opposite flag. With respect to a basis compatible with the original flag and its opposite, we can find a cocharacter $\mu \in Y_H$ with $P_\lambda = P_\mu$ and $K \subseteq L_\mu$. The argument in the second case is similar.

Solution 12.6. Certainly $H \subseteq P$, since H by definition stabilises the sequence of subspaces (12.11.1), so stabilises its base change to $\bar{k}$. Now choosing a k-defined Levi subgroup L of P is equivalent to choosing a direct sum decomposition of U compatible with the sequence: that is, we choose a vector space decomposition

$$U = W_1 \oplus \cdots \oplus W_r$$

such that $U_i = W_1 \oplus \cdots \oplus W_i$ for each i (again, as vector spaces). Now $c_L(H)$ is still a subgroup of $\mathrm{GL}(U)$, but now it acts semisimply on U because it preserves the direct sum decomposition, and it acts on each W_i as H acts on the quotient U_i/U_{i-1}, which is to say W_i is a simple $c_L(H)$-module. This completes the argument.

Solution 12.7. The subgroup T is a maximal torus of $C_G(S_1)$ and $C_G(S_2)$. Hence, if $g \in G$ with $gS_1g^{-1} = S_2$, then gTg^{-1} is another maximal torus of $C_G(S_2)$, and so there exists $x \in C_G(S_2)$ with $xgTg^{-1}x^{-1} = T$, i.e., $xg \in N_G(T)$. But $xgS_1g^{-1}x^{-1} = xS_2x^{-1} = S_2$, so we see that S_1 and S_2 are $N_G(T)$-conjugate. To finish, note that since $\dim(T) > 1$, there are infinitely many one-dimensional subtori of T, but the $N_G(T)$-orbits on such subtori are all finite, since T fixes them all and $N_G(T)/T$ is finite, so there are infinitely many pairwise non-conjugate one-dimensional subtori as claimed.

Solution 12.8. Finite groups are linearly reductive in characteristic 0, so every $\rho \in \mathrm{Hom}(H, G)$ has G-cr image by Lemma 4.2.1. The result follows because $C(H, G)$ is finite (Theorem 6.2.2).

Solution 12.9. An easy calculation shows that $C_G(H)$ is finite in this case, so $N_G(H)/H$ is finite. Since X^H is one-dimensional (in terms of the adjoint module for SL_2 in characteristic 2, X^H is the centre of the Lie algebra), we see that $X^H /\!/ N_G(H)$ is infinite, being the quotient of a one-dimensional vector space by a finite group. On the other hand, G has a single closed orbit in X, the orbit of the zero vector, so $X /\!/ G$ is a point.

This is [13, Ex. 8.1]. For more such examples, see also [13, Ex. 8.2, Ex. 8.3].

Solution 12.10. Constructible subsets of k^* are either finite or cofinite. However, the subset $G_{\mathrm{gen}} = \{x \in k^* \mid x \text{ is not a root of } 1\}$ is neither. So G_{gen} is not definable, by Theorem 12.9.1.

Solution 12.11. We can write $X = \bigcup_{1 \le j \le r} X_j$ for some $r \in \mathbb{N}$ and some locally closed subsets $X_1, \ldots, X_r$ of k^t. It is enough to show that for each fixed j, X_j admits a finite covering by subsets of the form $X_j \cap Y_i$.

So fix $j \in \{1, \ldots, r\}$ and set $Y_i(j) = X_j \cap Y_i$ for $i \in \Lambda$. For the rest of the proof we take the closure of a subset of X_j to mean the closure in X_j rather than the closure in k^t. Each $Y_i(j)$ is a finite union of locally closed subsets of X_j: say, $Y_i(j) = \bigcup_{1 \le m \le s_{i,j}} Y_i(j)^{(m)}$. It is enough to show that X_j is covered by finitely many of the subsets $Y_i(j)^{(m)}$. Hence we can assume without loss that each $Y_i(j)$ is locally closed in X_j. By Proposition 12.10.3 there is a finite subset F_j of Λ such that $\bigcup_{i \in F_j} \overline{Y_i(j)} = X_j$. It follows that $Y_{i_j}(j)$ contains a non-empty open subset U_j of X_j for some $i_j \in F_j$. Let $X'_j = X_j \setminus U_j$. By noetherian induction on X_j, we can assume that $\bigcup_{i \in F'}(Y_i(j) \cap X'_j) = X'_j$ for some finite $F' \subseteq \Lambda$. Part (a) now follows. Part (b) also follows, since at least one of the subsets U_j has non-empty interior in X.

Solution 12.12. The irreducible components are

$$C_1 := \{(x, h) \in X \times H \mid hxh^{-1} = x \text{ and } h \text{ is unipotent}\}$$

and

$$C_2 := \{1\} \times H.$$

We see this as follows. Clearly C_1, C_2 are proper closed subsets of C, $C = C_1 \cup C_2$ and C_2 is irreducible. C_1 is irreducible because it is the image of the irreducible set $H \times U \times U$ under the morphism $H \times U \times U \to C$, $(m, u, v) \mapsto (mum^{-1}, mvm^{-1})$, where U is the subgroup of H of unipotent upper triangular matrices.

References

1. P. Abramenko, K. Brown, *Buildings. Theory and Applications*. Graduate Texts in Mathematics, vol. 248 (Springer, New York, 2008), xxii+747 pp.
2. P.N. Achar, W.D. Hardesty, S. Riche, Representation theory of disconnected reductive groups. Doc. Math. **25**, 2149–2177 (2020)
3. S. Akbulut, J.D. McCarthy, *Casson's Invariant for Oriented Homology 3-spheres. An Exposition*. Mathematical Notes, vol. 36 (Princeton University Press, Princeton, 1990), xviii+182 pp.
4. C. Attenborough, M. Bate, M. Gruchot, A. Litterick, G. Röhrle, On relative complete reducibility. Q. J. Math. **71**, 321–334 (2020)
5. V. Balaji, P. Deligne, A.J. Parameswaran, On complete reducibility in characteristic p. Épijournal Géom. Algébrique **1**, Art. 3, 27 pp. (2017)
6. A. Balser, A. Lytchak, Centers of convex subsets of buildings. Ann. Global Anal. Geom. **28**(2), 201–209 (2005)
7. F. Bannuscher, A. Litterick, T. Uchiyama, Complete reducibility of subgroups of reductive algebraic groups over non-perfect fields IV: an F_4 example. J. Group Theory **25**(3), 527–541 (2022)
8. P. Bardsley, R.W. Richardson, Étale slices for algebraic transformation groups in characteristic p. Proc. Lond. Math. Soc. (3) **51**(2), 295–317 (1985)
9. M. Bate, Optimal subgroups and applications to nilpotent elements. Transform. Groups **14**(1), 29–40 (2009)
10. M. Bate, S. Böhm, A. Litterick, B. Martin, G. Röhrle, G-complete reducibility and saturation. Pac. J. Math. **337**(1), 1–24 (2025)
11. M. Bate, S. Böhm, B. Martin, G. Röhrle, On good A_1 subgroups, Springer maps, and overgroups of distinguished unipotent elements in reductive groups. Pac. J. Math. **336**(1–2), 29–61 (2025)
12. M. Bate, S. Böhm, B. Martin, G. Röhrle, L. Voggesberger, Complete reducibility for Lie subalgebras and semisimplification. Eur. J. Math. **9**(4), Paper No. 116, 27 pp. (2023)
13. M. Bate, H. Geranios, B. Martin, Orbit closures and invariants. Math. Z. **293**(3–4), 1121–1159 (2019)
14. M. Bate, S. Herpel, B. Martin, G. Röhrle, G-complete reducibility and semisimple modules. Bull. Lond. Math. Soc. **43**(6), 1069–1078 (2011)
15. M. Bate, S. Herpel, B. Martin, G. Röhrle, G-complete reducibility in non-connected groups. Proc. Am. Math. Soc. **143**(3), 1085–1100 (2015)
16. M. Bate, S. Herpel, B. Martin, G. Röhrle, Cocharacter-closure and spherical buildings. Pac. J. Math. **279**(1–2), 65–85 (2015)

© The Author(s), under exclusive license to Springer Nature Switzerland AG 2026
M. Bate et al., *G-Complete Reducibility, Geometric Invariant Theory and Spherical Buildings*, Oberwolfach Seminars 57, https://doi.org/10.1007/978-3-032-08866-6

17. M. Bate, S. Herpel, B. Martin, G. Röhrle, Cocharacter-closure and the rational Hilbert-Mumford theorem. Math. Z. **287**(1–2), 39–72 (2017)
18. M. Bate, B. Martin, G. Röhrle, A geometric approach to complete reducibility. Invent. Math. **161**(1), 177–218 (2005)
19. M. Bate, B. Martin, G. Röhrle, Complete reducibility and commuting subgroups. J. Reine Angew. Math. **621**, 213–235 (2008)
20. M. Bate, B. Martin, G. Röhrle, Complete reducibility and separable field extensions. C. R. Acad. Sci. Paris Ser. I Math. **348**, 495–497 (2010)
21. M. Bate, B. Martin, G. Röhrle, The strong Centre Conjecture: an invariant theory approach. J. Algebra **372**, 505–530 (2012)
22. M. Bate, B. Martin, G. Röhrle, On a question of Külshammer for representations of finite groups in reductive groups. Isr. J. Math. **214**, 463–470 (2016)
23. M. Bate, B. Martin, G. Röhrle, Semisimplification for subgroups of reductive algebraic groups. Forum Math. Sigma **8**, Paper No. e43, 10 pp. (2020)
24. M. Bate, B. Martin, G. Röhrle, Overgroups of regular unipotent elements in reductive groups. Forum Math. Sigma **10**, Paper No. e13, 13 pp. (2022)
25. M. Bate, B. Martin, G. Röhrle, Edifices: building-like spaces associated to linear algebraic groups. Innov. Incidence Geom. **20**(2–3), 79–134 (2023)
26. M. Bate, B. Martin, G. Röhrle, R. Tange, Complete reducibility and separability. Trans. Am. Math. Soc. **362**(8), 4283–4311 (2010)
27. M. Bate, B. Martin, G. Röhrle, R. Tange, Complete reducibility and conjugacy classes of tuples in algebraic groups and Lie algebras. Math. Z. **269**(3–4), 809–832 (2011)
28. M. Bate, B. Martin, G. Röhrle, R. Tange, Closed orbits and uniform S-instability in geometric invariant theory. Trans. Am. Math. Soc. **365**(7), 3643–3673 (2013)
29. M. Bate, G. Röhrle, D. Sercombe, D. Stewart, A construction of pseudo-reductive groups with non-reduced root systems. Transf. Groups **30**, 455–480 (2025)
30. D. Birkes, Orbits of linear algebraic groups. Ann. Math. (2) **93**, 459–475 (1971)
31. A. Borel, Properties and linear representations of Chevalley groups, in *Seminar on Algebraic Groups and Related Finite Groups*. Lecture Notes in Mathematics, vol. 131 (Springer, Berlin, 1970), pp. 1–55
32. A. Borel, *Linear Algebraic Groups*. Graduate Texts in Mathematics, vol. 126 (Springer-Verlag, Berlin, 1991)
33. A. Borel, J-P. Serre, Théorèmes de finitude en cohomologie galoisienne. Comment. Math. Helv. **39**, 111–164 (1964)
34. A. Borel, J. de Siebenthal, Les sous-groupes fermés de rang maximum des groupes de Lie clos. Comment. Math. Helvet. **23**, 200–221 (1949)
35. A. Borel, J. Tits. Groupes réductifs. Inst. Hautes Études Sci. Publ. Math. **27**, 55–150 (1965)
36. A. Borel, J. Tits, Éléments unipotents et sous-groupes paraboliques des groupes réductifs, I. Invent. Math. **12**, 95–104 (1971)
37. N. Bourbaki, *Groupes et algèbres de Lie*, Chapitres 4, 5 et 6 (Hermann, Paris, 1975)
38. N. Bourbaki, *Groupes et algèbres de Lie*, Chapitres 7 et 8 (Hermann, Paris, 1975)
39. M.R. Bridson, A. Haefliger, *Metric Spaces of Non-positive Curvature*. Grundlehren der Mathematischen Wissenschaften, vol. 319 (Springer, Berlin, 1999)
40. M. Brion, On extensions of algebraic groups with finite quotient. Pac. J. Math. **279**(1–2), 135–153 (2015)
41. K.S. Brown, *Cohomology of Groups*. Graduate Texts in Mathematics, vol.87 (Springer, New York, 1982)
42. T. Brüstle, L. Hille, G. Röhrle, G. Zwara, The Bruhat-Chevalley order of parabolic group actions in general linear groups and degeneration for Δ-filtered modules. Adv. Math. **148**(2), 203–242 (1999)

43. T.C. Burness, S. Gerhardt, R.M. Guralnick, Topological generation of exceptional algebraic groups. Adv. Math. **369**, 107177, 50 pp. (2020)
44. R.W. Carter, Centralizers of semisimple elements in finite groups of Lie type. Proc. Lond. Math. Soc. (3) **37**(3), 491–507 (1978)
45. R.W. Carter, *Finite Groups of Lie type. Conjugacy Classes and Complex Characters*. Pure and Applied Mathematics (New York). A Wiley-Interscience Publication (John Wiley & Sons, Inc., New York, 1985)
46. A. Charnes, W.W. Cooper, The strong Minkowski-Farkas-Weyl Theorem for vector spaces over ordered fields. Proc. Nat. Acad. Sci. U.S.A. **44**, 914–916 (1958)
47. B. Conrad, O. Gabber, G. Prasad, *Pseudo-reductive Groups*. New Mathematical Monographs, vol. 26, 2nd edn. (Cambridge University Press, Cambridge, 2015)
48. S. Cotner, Morphisms of character varieties. Int. Math. Res. Not. IMRN **16** 11540–11548 (2024)
49. S. Cotner, Hom schemes for algebraic groups. Algebra Number Theory (to appear). https://doi.org/10.48550/arXiv.2309.16458
50. G.-M. Cram, On a Question of Külshammer About Algebraic Group Actions: An Example. Australian Mathematical Society Lecture Series, vol. 9. Algebraic Groups and Lie Groups (Cambridge University Press, Cambridge, 1997), pp. 346–348. Appendix to [164]
51. D.A. Craven, Groups with a p-element acting with a single non-trivial Jordan block on a simple module in characteristic p. J. Group Theory **21**(5), 719–787 (2018)
52. D. Craven, D.I. Stewart, A.R. Thomas, A new maximal subgroup of E_8 in characteristic 3. Proc. Am. Math. Soc. **150**(4), 1435–1448 (2022)
53. M. Culler, P.B. Shalen, Varieties of group representations and splittings of 3-manifolds. Ann. Math. (2) **117**(1), 109–146 (1983)
54. C.W. Curtis, G.I. Lehrer, A new proof of a theorem of Solomon-Tits. Proc. Am. Math. Soc. **85**(2), 154–156 (1982)
55. C.W. Curtis, G.I. Lehrer, J. Tits, Spherical buildings and the character of the Steinberg representation. Invent. Math. **58**(3), 201–210 (1980)
56. C. De Concini, C. Procesi, *The Invariant Theory of Matrices*. University Lecture Series, vol. 69 (American Mathematical Society, Providence, 2017), v+151 pp.
57. M. Demazure, P. Gabriel, *Groupes algébriques. Tome I: Géométrie algébrique, généralités, groupes commutatifs* (Masson & Cie, Éditeur, Paris, 1970)
58. H. Derksen, G. Kemper, Computational invariant theory. in *Invariant Theory Algebra Transformation Groups, I*. Encyclopaedia of Mathematical Sciences, vol. 130 (Springer-Verlag, Berlin, 2002), x+268 pp.
59. S. Donkin, Invariants of several matrices. Invent. Math. **110**(2), 389–401 (1992)
60. S. Donkin, On tilting modules for algebraic groups. Math. Z. **212**, 39–60 (1993)
61. E.B. Dynkin, Semisimple subalgebras of semisimple Lie algebras. (Russian) Mat. Sbornik N.S. **30**(72), 349–462 (1952); Trans. AMS **6**, Series 2, 111–244 (1957)
62. E.B. Dynkin, *Selected Papers of E. B. Dynkin with Commentary*, ed. by A.A. Yushkevich, G.M. Seitz, A.L. Onishchik (American Mathematical Society/International Press, Providence/Cambridge, 2000)
63. S. Friedl, M. Gill, S. Tillmann, Linear representations of 3-manifold groups over rings. Proc. Am. Math. Soc. **146**(11), 4951–4966 (2018)
64. G.S. Garden, S.Tillmann, An invitation to Culler-Shalen theory in arbitrary characteristic. Preprint (2024). https://arxiv.org/abs/2411.06859/
65. W.M. Goldman, The symplectic nature of fundamental groups of surfaces. Adv. Math. **54**(2), 200–225 (1984)

66. D. Gorenstein, R. Lyons, R. Solomon. *The classification of the finite simple groups*. Part I. Chapter A: Almost simple $\mathcal{K}$-groups. Mathematical Surveys and Monographs, vol. 40, No. 3 (American Mathematical Society, Providence, 1998)

67. A. Grothendieck, J.A. Dieudonné, Eléments de Géométrie Algébrique IV. Publ. Math. IHÉS **20**, 5–259 (1964); **24**, 5–231 (1965); **28**, 5–255 (1966); **32**, 5–361 (1967)

68. M. Gruchot, A. Litterick, G. Röhrle, Complete reducibility: variations on a theme of Serre. Manuscripta Math. **168**(3–4), 439–451 (2022)

69. R.M. Guralnick, Intersections of conjugacy classes and subgroups of algebraic groups. Proc. Am. Math. Soc. **135**(3), 689–693 (2007)

70. R. Guralnick, G. Malle, Rational rigidity for $E_8(p)$. Compos. Math. **150**(10), 1679–1702 (2014)

71. R. Guralnick, G. Malle, Products and commutators of classes in algebraic groups. Math. Ann. **362**(3–4), 743–771 (2015)

72. W.J. Haboush, Reductive groups are geometrically reductive. Ann. Math. (2) **102**(1), 67–83 (1975)

73. S. Herpel, On the smoothness of centralizers in reductive groups. Trans. Am. Math. Soc. **365**(7), 3753–3774 (2013)

74. S. Herpel, G. Röhrle, D. Gold, Complete reducibility and Steinberg endomorphisms. C. R. Math. Acad. Sci. Paris **349**(5–6), 243–246 (2011)

75. W.H. Hesselink, Uniform instability in reductive groups. J. Reine Angew. Math. **303/304**, 74–96 (1978)

76. D. Hilbert, Über die vollen Invariantensysteme. Math. Ann. **42**, 313–373 (1893)

77. J.E. Humphreys, *Linear Algebraic Groups* (Springer-Verlag, New York, 1975)

78. J.E. Humphreys, *Introduction to Lie Algebras and Representation Theory* (Springer-Verlag, New York-Berlin, 1978), xii+171 pp.

79. J.E. Humphreys, *Conjugacy Classes in Semisimple Algebraic Groups*. Mathematical Surveys and Monographs, vol. 43 (American Mathematical Society, Providence, 1995)

80. I.M. Isaacs, *Finite Group Theory*. Graduate Students in Mathematics, vol. 92 (American Mathematical Society, Providence, 2008)

81. J.C. Jantzen, Low-dimensional representations of reductive groups are semisimple. *Algebraic Groups and Lie Groups*. Australian Mathematical Society Lecture Series, vol. 9 (Cambridge University Press, Cambridge, 1997), pp. 255–266

82. J.C. Jantzen, *Representations of Algebraic Groups*. Mathematical Surveys and Monographs, vol. 107, 2nd edn. (American Mathematical Society, Providence, 2003)

83. J.C. Jantzen, Nilpotent orbits in representation theory, in *Lie Theory. Lie Algebras and Representations*. Progress in Mathematics, vol. 228, ed. by J.-P. Anker, B. Orsted (Birkhäuser, Boston, 2004)

84. T. Kaletha, G. Prasad, *Bruhat-Tits Theory: A New Approach*. New Mathematical Monographs, vol. 44 (Cambridge University Press, Cambridge, 2023)

85. W. Kaplan, *Maxima and Minima with Applications*. Wiley-Interscience Series Discrete Mathematics and Optimization. Wiley-Interscience Publisher (John Wiley & Sons, Inc., New York, 1999), x+284 pp.

86. M. Kapovich, J.J. Millson, On representation varieties of Artin groups, projective arrangements and the fundamental groups of smooth complex algebraic varieties. Inst. Hautes Études Sci. Publ. Math. **88**, 5–95 (1998)

87. M. Kapovich, J.J. Millson, On representation varieties of 3-manifold groups. Geom. Topol. **21**(4), 1931–1968 (2017)

88. G.R. Kempf, Instability in invariant theory. Preprint (1976). https://arxiv.org/abs/1807.02890

89. G.R. Kempf, Instability in invariant theory. Ann. Math. **108**(2), 299–316 (1978)

90. M. Korhonen, Unipotent elements forcing irreducibility in linear algebraic groups J. Group Theory **21**(3), 365–396 (2018)

91. H. Kraft, Geometric methods in representation theory. *Representations of Algebras (Puebla, 1980)*. Lecture Notes in Mathematics, vol. 944 (Springer, Berlin-New York, 1982), pp. 180–258

92. B. Külshammer, Donovan's conjecture, crossed products and algebraic group actions. Isr. J. Math. **92**(1–3), 295–306 (1995)

93. S. Lang, *Algebra*, 3rd edn. (Addison-Wesley, Reading, 1993)

94. M. Larsen, A. Shalev, Fibers of word maps and some applications. J. Algebra **354**, 36–48 (2012)

95. R. Lawther, D.M. Testerman, A_1 Subgroups of exceptional algebraic groups. Mem. Am. Math. Soc. **141**(674) (1999)

96. S. Lawton, A.S. Sikora, Varieties of characters. Algebr. Represent. Theory **20**(5), 1133–1141 (2017)

97. B. Leeb, C. Ramos-Cuevas, The center conjecture for spherical buildings of types F_4 and E_6. Geom. Funct. Anal. **21**(3), 525–559 (2011)

98. M.W. Liebeck, B.M.S. Martin, A. Shalev, On conjugacy classes of maximal subgroups of finite simple groups, and a related zeta function. Duke Math. J. **128**(3), 541–557 (2005)

99. M.W. Liebeck, G.M. Seitz, Reductive subgroups of exceptional algebraic groups. Mem. Am. Math. Soc. **580** (1996)

100. M.W. Liebeck, G.M. Seitz, Variations on a theme of Steinberg. Special issue celebrating the 80th birthday of Robert Steinberg. J. Algebra **260**(1), 261–297 (2003)

101. M.W. Liebeck, G.M. Seitz, The maximal subgroups of positive dimension in exceptional algebraic groups. Mem. Am. Math. Soc. **169** (2004)

102. M.W. Liebeck, G.M. Seitz, *Unipotent and Nilpotent Classes in Simple Algebraic Groups and Lie Algebras*. Mathematical Surveys and Monographs, vol. 180 (American Mathematical Society, Providence, 2012)

103. M.W. Liebeck, A. Shalev, Fuchsian groups, finite simple groups and representation varieties. Invent. Math. **159**(2), 317–367 (2005)

104. M.W. Liebeck, D.M. Testerman, Irreducible subgroups of algebraic groups. Q. J. Math. **55**, 47–55 (2004)

105. A.J. Litterick, D.I. Stewart, A.R. Thomas, Complete reducibility and subgroups of exceptional algebraic groups, in *Groups St Andrews 2022 in Newcastle*, ed. by C.M. Campbell et al. London Mathematical Society Lecture Note Series, vol. 496 (Cambridge University Press, Cambridge, 2025), pp. 191–245

106. A.J. Litterick, A.R. Thomas, Complete reducibility in good characteristic. Trans. Am. Math. Soc. **370**(8), 5279–5340 (2018)

107. A.J. Litterick, A.R. Thomas, Reducible subgroups of exceptional algebraic groups. J. Pure Appl. Algebra, **223**(6), 2489–529 (2019)

108. A.J. Litterick, A.R. Thomas, Complete reducibility in bad characteristic. Algebra Number Theory (to appear). https://doi.org/10.48550/arXiv.2304.08388

109. D. Lond, *On Reductive Subgroups of Algebraic Groups and a Question of Külshammer*. Ph.D. Thesis, University of Canterbury, 2013

110. D. Lond, B.M.S. Martin, On a question of Külshammer for homomorphisms of algebraic groups. J. Algebra **497**, 164–198 (2018)

111. A. Lubotzky, A.R. Magid, Varieties of representations of finitely generated groups. Mem. Am. Math. Soc. **58**(336), xi+117 pp. (1985)

112. D. Luna, Adhérences d'orbite et invariants. Invent. Math. **29**(3), 231–238 (1975)

113. G. Lusztig, On the finiteness of the number of unipotent classes. Invent. Math. **34**, 201–213 (1976)

114. G. Malle, D.M. Testerman, Overgroups of regular unipotent elements in simple algebraic groups. Trans. Am. Math. Soc. Ser. B **8**, 788–822 (2021)

115. D. Marker, *Model theory*. Graduate Texts in Mathematics, vol. 217 (Springer-Verlag, New York, 2002), viii+342 pp.

116. B.M.S. Martin, Étale slices for representation varieties in characteristic p. Indag. Math. **10**(1), 555–564 (1999)

117. B.M.S. Martin, Reductive subgroups of reductive groups in nonzero characteristic. J. Algebra **262**(2), 265–286 (2003)

118. B.M.S. Martin, A normal subgroup of a strongly reductive subgroup is strongly reductive. J. Algebra **265**(2), 669–674 (2003)

119. B.M.S. Martin, Generic stabilisers for actions of reductive groups. Pac. J. Math. **279**(1–2), 397–422 (2015)

120. B.M.S. Martin, Algebraic groups and G-complete reducibility: a geometric approach. Unpublished Notes. https://doi.org/10.48550/arXiv.2207.12169

121. B.M.S. Martin, Powers of commutators in linear algebraic groups. Proc. Edinb. Math. Soc. (2) **67**(3), 830–837 (2024)

122. B.M.S. Martin, A. Neeman, The map $V \to V/\!/G$ need not be separable. Math. Res. Lett. **8**(5–6), 813–817 (2001)

123. B.M.S. Martin, D.I. Stewart, L. Topley, A proof of the first Kac-Weisfeiler conjecture in large characteristics. Represent. Theory **23**, 278–293 (2019)

124. G. McNinch, Dimensional criteria for semisimplicity of representations. Proc. Lond. Math. Soc. (3) **76**(1), 95–149 (1998)

125. G. McNinch, Nilpotent orbits over ground fields of good characteristic. Math. Ann. **329**, 49–85 (2004)

126. G. McNinch, Completely reducible Lie subalgebras. Transform. Groups **12**(1), 127–135 (2007)

127. J.S. Milne, *Algebraic Groups: The Theory of Group Schemes of Finite Type over a Field*. Cambridge Studies in Advanced Mathematics, vol. 170 (Cambridge University Press, Cambridge, 2017)

128. B. Mühlherr, J. Tits, The Centre Conjecture for non-exceptional buildings. J. Algebra **300**(2), 687–706 (2006)

129. B. Mühlherr, R.M. Weiss, Receding polar regions of a spherical building and the center conjecture. Ann. Inst. Fourier **63**(2), 479–513 (2013)

130. D. Mumford, J. Fogarty, F. Kirwan, *Geometric Invariant Theory*. Ergebnisse der Mathematik und ihrer Grenzgebiete, vol. 34, 3rd edn. (Springer-Verlag, Berlin, 1994)

131. M. Nagata, Complete reducibility of rational representations of a matric group. J. Math. Kyoto Univ. **1**, 87–99 (1961)

132. P.E. Newstead, *Introduction to Moduli Problems and Orbit Spaces*. Tata Institute of Fundamental Research Lectures on Mathematics and Physics, vol. 51 (Tata Institute of Fundamental Research, Bombay, 1978)

133. B. Nica, *Three Theorems on Linear Groups* (2013). https://www.math.mcgill.ca/bnica/threetheorems.pdf

134. M.V. Nori, On subgroups of $GL_n(\mathbb{F}_p)$. Invent. Math. **88**, 257–275 (1987)

135. L. Paoluzzi, J. Porti, Examples of character varieties in characteristic p and ramification, in *Characters in Low-dimensional Topology*. Contemporary Mathematics, vol. 760. Centre de Recherches Mathématiques Proceedings (American Mathematical Society, Providence, 2020), pp. 229–261

136. C. Parker, K. Tent, Completely reducible subcomplexes of spherical buildings. Arch. Math. **97**(2), 125–128 (2011)

137. V.P. Platonov, A certain problem for finitely generated groups (Russian). Dokl. Akad. Nauk BSSR **12**, 492–494 (1968)
138. C. Procesi, The invariant theory of $n \times n$ matrices. Adv. Math. **19**(3), 306–381 (1976)
139. M.S. Raghunathan, A note on orbits of reductive groups. J. Indian Math. Soc. New Ser. **38**, 65–70 (1975)
140. C. Ramos-Cuevas, The center conjecture for thick spherical buildings. Geom. Dedicata **166**, 349–407 (2013)
141. I. Rapinchuk, On the character varieties of finitely generated groups. Math. Res. Lett. **22**(2), 579–604 (2015)
142. L. Renner, Orbits and invariants of visible group actions. Transform. Groups **17**(4), 1191–1208 (2012)
143. R.W. Richardson, Conjugacy classes in Lie algebras and algebraic groups. Ann. Math. **86**, 1–15 (1967)
144. R.W. Richardson, Principal orbit types for algebraic transformation spaces in characteristic zero. Invent. Math. **16**, 6–14 (1972)
145. R.W. Richardson, Affine coset spaces of reductive algebraic groups. Bull. Lond. Math. Soc. **9**(1), 38–41 (1977)
146. R.W. Richardson, On orbits of algebraic groups and Lie groups. Bull. Aust. Math. Soc. **25**(1), 1–28 (1982)
147. R.W. Richardson, Conjugacy classes of n-tuples in Lie algebras and algebraic groups. Duke Math. J. **57**(1), 1–35 (1988)
148. C. Riedtmann, Degenerations for representations of quivers with relations. Ann. Sci. École Normale Sup. (4) **19**(2), 275–301 (1986)
149. M. Ronan, *Lectures on Buildings*, 2nd updated and revised edn. (University of Chicago Press, Chicago, 2009)
150. G. Rousseau, Immeubles sphériques et théorie des invariants. C.R.A.S. **286**, 247–250 (1978)
151. J. Saxl, G.M. Seitz, Subgroups of algebraic groups containing regular unipotent elements. J. Lond. Math. Soc. (2) **55**, 370–386 (1997)
152. G.M. Seitz, The maximal subgroups of classical algebraic groups. Mem. Am. Math. Soc. **67**(365), iv+286 pp. (1987)
153. G.M. Seitz, Maximal subgroups of exceptional algebraic groups. Mem. Am. Math. Soc. **90**(441), iv+197 pp. (1991)
154. G.M. Seitz, Unipotent elements, tilting modules, and saturation. Invent. Math. **141**(3), 467–502 (2000)
155. A. Selberg, On discontinuous groups in higher-dimensional symmetric spaces. *Contributions to Function Theory* (Tata Institute of Fundamental Research, Bombay, 1960), pp. 147–164
156. J-P. Serre, Semisimplicity and tensor products of group representations: converse theorems. J. Algebra **194**(2), 496–520 (1997). With an appendix by Walter Feit
157. J-P. Serre, *La notion de complète réductibilité dans les immeubles sphériques et les groupes réductifs* (Séminaire au Collége de France, Résumé dans [184], 1997), pp. 93–98.
158. J-P. Serre, *The Notion of Complete Reducibility in Group Theory*. Moursund Lectures, Part II (University of Oregon, Eugene, 1998). https://doi.org/10.48550/arXiv.math/0305257
159. J-P. Serre, *Complète réductibilité*, Séminaire Bourbaki, 56ème année (2003–2004), n° 932
160. K.S. Sibirskiĭ, Unitary and orthogonal invariants of matrices. Dokl. Akad. Nauk SSSR **172**, 40–43 (1967)
161. A. Sikora, Character varieties. Trans. Am. Math. Soc. **364**(10), 5173–5208 (2012)
162. A. Sikora, SO$(2n, \mathbb{C})$-character varieties are not varieties of characters. J. Algebra **478**, 195–214 (2017)

163. P. Slodowy, Theorie der Optimalen Einparameteruntergruppen. *Algebraische Transformations-gruppen und Invariantentheorie*, ed. by H. Kraft, P. Slodowy, T.A. (Springer). DMV Seminar, vol. 13 (Birkhäuser Verlag, Basel, 1989)

164. P. Slodowy, Two notes on a finiteness problem in the representation theory of finite groups. Australian Mathematical Society Lecture Series, vol. 9. Algebraic groups and Lie groups (Cambridge University Press, Cambridge, 1997), pp. 331–348

165. P. Sobaje, Exponentiation of commuting nilpotent varieties. J. Pure Appl. Algebra **219**(6), 2206–2217 (2015)

166. P. Sobaje, Unipotent elements and generalized exponential maps. Adv. Math. **333**, 463–496 (2018)

167. L. Solomon, The Steinberg Character of a Finite Group with BN-pair, in *Theory of Finite Groups*. Symposium (Harvard University/W. A. Benjamin, Inc., Cambridge/New York-Amsterdam, 1968), pp. 213–221

168. T.A. Springer, *Linear Algebraic Groups*. Progress in Mathematics, vol. 9, 2nd edn. (Birkhäuser Boston, Inc., Boston, 1998)

169. T.A. Springer, R. Steinberg, Conjugacy classes. *Seminar on Algebraic Groups and Related Finite Groups*. Lecture Notes in Mathematics, vol. 131 (Springer-Verlag, Heidelberg, 1970), pp. 167–266

170. The Stacks Project, *The Stacks Project Team* (2024). http://stacks.math.columbia.edu

171. R. Steinberg, Regular elements of semisimple algebraic groups. Inst. Hautes Études Sci. Publ. Math. **25**, 49–80 (1965)

172. R. Steinberg, *Lectures on Chevalley Groups*. Notes Prepared by John Faulkner and Robert Wilson, Lecture Notes (Yale University, New Haven, 1968), iii, 277 pp.

173. R. Steinberg, *Endomorphisms of Linear Algebraic Groups*. Memoirs of the American Mathematical Society, vol. 80 (American Mathematical Society, Providence, 1968)

174. R. Steinberg, *Conjugacy Classes in Algebraic Groups*. Notes by Vinay V. Deodhar. Lecture Notes in Mathematics, vol. 366 (Springer-Verlag, Berlin-New York, 1974)

175. D.I. Stewart, The reductive subgroups of G_2. J. Group Theory **13**(1), 117–130 (2010)

176. D.I. Stewart, The reductive subgroups of F_4. Mem. Am. Math. Soc. **223**(1049), vi+88pp. (2013)

177. D.I. Stewart, On unipotent algebraic G-groups and 1-cohomology. Trans. Am. Math. Soc. **365**(12), 6343–6365 (2013)

178. I. Suprunenko, Irreducible representations of simple algebraic groups containing matrices with big Jordan blocks. Proc. Lond. Math. Soc. (3) **71**, 281–332 (1995)

179. D.M. Testerman, Irreducible subgroups of exceptional algebraic groups. Mem. Am. Math. Soc., **75**(390), iv+190 pp. (1988)

180. D.M. Testerman, A_1-type overgroups of elements of order p in semisimple algebraic groups and the associated finite groups. J. Algebra **177**(1), 34–76 (1995)

181. D. M. Testerman, A.E. Zalesski, Irreducibility in algebraic groups and regular unipotent elements. Proc. Am. Math. Soc. **141**(1), 13–28 (2013)

182. A.R. Thomas, The irreducible subgroups of exceptional algebraic groups. Mem. Am. Math. Soc. **268**(1307), v+191 pp. (2020)

183. J. Tits, *Buildings of Spherical Type and Finite BN-pairs*. Lecture Notes in Mathematics, vol. 386 (Springer-Verlag, Berlin, 1974)

184. J. Tits, *Théorie des Groupes*. Résumés des Cours et Travaux. Annuaire du Collège de France, 97e année (1996–1997), pp. 89–102

185. J. Tits, R. Weiss, *Moufang Polygons*. Springer Monographs in Mathematics (Springer, Berlin, 2002), ix, 535 pp.

186. T. Uchiyama, Separability and complete reducibility of subgroups of the Weyl group of a simple algebraic group of type E_7. J. Algebra **422**, 357–372 (2014)

187. T. Uchiyama, Complete reducibility of subgroups of reductive algebraic groups over nonperfect fields I. J. Algebra **463**, 168–187 (2016)
188. T. Uchiyama, Complete reducibility of subgroups of reductive algebraic groups over nonperfect fields II. Comm. Algebra **45**(11), 4833–4845 (2017)
189. T. Uchiyama, Non-separability and complete reducibility: E_n examples with an application to a question of Külshammer. J. Group Theory **20**(5), 925–944 (2017)
190. T. Uchiyama, Complete reducibility, Külshammer's question, conjugacy classes: a D_4 example. Commun. Algebra **46**(3), 1333–1343 (2018)
191. T. Uchiyama, Complete reducibility of subgroups of reductive algebraic groups over nonperfect fields III. Commun. Algebra **47**(12), 4928–4944 (2019)
192. E.B. Vinberg, On invariants of a set of matrices. J. Lie Theory **6**, 249–269 (1996)
193. A. Weil, Remarks on the cohomology of groups. Ann. Math. (2) **80**, 149–157 (1964)
194. G. Zwara, Degenerations for modules over representation-finite algebras. Am. Math. Soc. **127**, 1313–1322 (1999)
195. G. Zwara, Degenerations of finite-dimensional modules are given by extensions. Compos. Math. **121**(2), 205–218 (2000)

Index

© The Editor(s) (if applicable) and The Author(s), under exclusive license to Springer Nature Switzerland AG 2026

M. Bate et al., *G-Complete Reducibility, Geometric Invariant Theory and Spherical Buildings*, Oberwolfach Seminars 57, https://doi.org/10.1007/978-3-032-08866-6

MIX
Papier aus verantwortungsvollen Quellen
Paper from responsible sources
FSC® C105338

If you have any concerns about our products,
you can contact us on
ProductSafety@springernature.com

In case Publisher is established outside the EU,
the EU authorized representative is:
**Springer Nature Customer Service Center GmbH
Europaplatz 3, 69115 Heidelberg, Germany**

Printed by Libri Plureos GmbH
in Hamburg, Germany